Biowaste to Biofuel

This book encompasses a wide range of topics related to biowaste to biofuels, covering agro-wastes, food waste, wastewater, CO_2, and so forth. Each chapter provides an in-depth examination of a specific biowaste to biofuel conversion technology discussing the underlying science, production processes, challenges, and potential applications. It presents practical insights into microbial biofuel production, including detailed explanations of fermentation processes, microbial pathways, conversion techniques, and technological advancements backed by case studies.

- Includes theory and delves into practical applications of microbial biofuels
- Reviews sustainability and environmental benefits with a focus on renewable energy
- Explores how these fuels can help reduce greenhouse gas emissions and combat climate change
- Discusses bioalcohol, biodiesel, biohydrogen, biomethane, and so forth
- Provides real-world examples and case studies showcasing biofuel potential for a greener future

This book is aimed at graduate/master's students and researchers in bioenergy, fermentation, and chemical engineering.

Biowaste to Biofuel

A Microbial Approach

Edited by
Shashi Kant Bhatia
Parmjit Singh Panesar
Ranjit Gurav

CRC Press
Taylor & Francis Group
Boca Raton London New York

CRC Press is an imprint of the
Taylor & Francis Group, an **informa** business

Designed cover image: Shutterstock

First edition published 2026
by CRC Press
2385 NW Executive Center Drive, Suite 320, Boca Raton FL 33431

and by CRC Press
4 Park Square, Milton Park, Abingdon, Oxon, OX14 4RN

CRC Press is an imprint of Taylor & Francis Group, LLC

ISBN: 978-1-032-98155-0 (hbk)
ISBN: 978-1-041-06091-8 (pbk)
ISBN: 978-1-003-63370-9 (ebk)

DOI: 10.1201/9781003633709

Typeset in Times
by SPi Technologies India Pvt Ltd (Straive)

Contents

About the Editors

Shashi Kant Bhatia is an Associate Professor in the Department of Biological Engineering, Konkuk University, Seoul, South Korea, and has more than ten years' experience in biowastes valorization into bioenergy, biochemicals, and biomaterials. He holds an MSc and a PhD in biotechnology from Himachal Pradesh University (India). Dr. Bhatia has worked as a Brain Pool Post-Doc Fellow at Konkuk University (2014–2016) and has contributed extensively to the industrial press and served as an editorial board member of the *Sustainability and Energies Journal* and associate editor of *Frontier of Microbiology*, *Microbial Cell Factories*, *Bioprocess and Biosystem Engineering*, *Carbohydrate Technology and Applications*, *3 Biotech*, *PLOS One*, *International Journal of Biomaterials*, *International Journal of Energy Research*, and *Marine Life Science and Technology*. He is editor in chief of *Biotechnology for Sustainable Materials Journal* and serves as a reviewer for more than 100 peer-reviewed international journals. He has published more than 200 research and review articles on industrial biotechnology, bioenergy production, biomaterial, biotransformation, microbial fermentation, and enzyme technology in international scientific peer-reviewed journals and holds 15 international patents. He has also edited eight books on microbial biotechnology. With more than 15,000 citations with h-index 66 and i10 index 215 (as per Google Scholar), he is included in the top 2% scientist list of Stanford University. Dr. Bhatia has successfully supervised and completed three projects funded by the National Research Foundation of Korea ($600,000).

Parmjit Singh Panesar is working as Professor, Sant Longowal Institute of Engineering and Technology Longowal, Punjab, India. Prof. Panesar has also worked as Dean (Planning and Development), Dean (Research and Consultancy), and Head, Department of Food Engineering and Technology, SLIET Longowal, India. His research is focused in the area of Food Biotechnology especially bioprocessing of food industry by-products, food enzymes, prebiotics (lactulose, GOS), biopigments, etc. Prof. Panesar has more than 10,000 citations and h-index of 54 (as per Google scholar). In 2005, he has been awarded BOYSCAST (Better Opportunities for Young Scientists in Chosen Areas of Science and Technology) fellowship by Department of Science & Technology (DST), Government of India, to carry out advance research at Chembiotech labs, University of Birmingham Research Park, United Kingdom. Prof. Panesar has successfully completed nine research projects funded by Department of Biotechnology (DBT,) Council of Scientific & Industrial Research (CSIR), Ministry of Human Resource Development (MHRD), All India Council for Technical Education (AICTE), New Delhi, and two projects are in progress. He has published more than 210 international/national scientific papers, 50 book reviews in peer-reviewed journals, 50 chapters, and has authored/edited 18 books. He has guided 19 Ph.D. students, and 5 students are under progress. He is a member of the editorial advisory boards of national/international journals, including the *International Journal of Biological Macromolecules*, *Journal of Food Science and Technology*, and *Carbohydrate Polymer Technologies and Applications*. In recognition of his work, Prof. Panesar was elected as "Fellow 2018" by the Biotech Research Society of India (BRSI); "Fellow 2019" by the "National Academy of Dairy Science India (NADSI); "Fellow 2021" by the International Society for Energy, Environment, and Sustainability (ISEES); "Fellow 2021" by the Academy of Microbiological Sciences; and "Fellow 2022" by the Association of Food Scientists and Technologists, India. He was selected for the award of prestigious "INSA Teachers Award (2020)" by the Indian National Science Academy (INSA). Prof. Panesar has also been listed five times, consequently, in the most coveted list of "World Ranking of Top 2% Scientists" published by Stanford University, USA (2020, 2021, 2022, 2023, 2024). He has also visited several countries like the United Kingdom, United States, Canada, Switzerland, New Zealand, Australia, Germany, France, Singapore, Malaysia, China, Iran, Thailand, etc.

Ranjit Gurav is an Associate Professor at UPES Dehradun, India. He received his Ph.D. in biotechnology in 2013 and has since conducted extensive research for a decade in various countries, including the USA, South Korea, India, and China. Throughout his career, he has held roles such as Associate Professor, Research Professor, Assistant Professor, Postdoctoral Research Associate, and Young Scientist, contributing significantly to advancements in his field. His research primarily focuses on harnessing agriculture, forestry, food industry waste biomass, and by-products to produce bioelectricity, polyhydroxyalkanoates, biodiesel, exopolysaccharides, and biohydrogen. Dr. Gurav is highly skilled in implementing microbial, bioelectrochemical, hydrogel, biochar, and self-activated carbon-based technologies for the remediation of organic contaminants and heavy metals in water and soil. Using a microbial system, he has established a unique technology for converting poultry feather biomass into a growth stimulator, and chitin biomass into an anti-fungal agent. In addition, he developed green infrastructure to manage urban storm-water runoff through biologically active previous structures. Dr. Gurav has secured research funding totaling US$ 130,000 from organizations such as the NSF I-Corps of the United State America (USA), Faculty Innovation Accelerator Program (FIAP) of Texas State University USA, Department of Science and Technology-Science and Engineering Research Board (DST-SERB) India, and University of Petroleum and Energy Studies (UPES) Dehradun, India. He has 125+ Scopus-indexed international publications and holds 4 international patents. He has been listed among the top 2% of global scientists listed by Stanford University, USA, for the years 2022, 2023, and 2024. Dr. Gurav is deeply involved in the academic community, serving as an editor, associate editor, academic editor, topical editor, and editor for ten international journals, and reviewer for over 70 international journals. He holds life memberships with the BRSI and the Association of Microbiologists of India.

Contributors

Sahar Akhavan-Mahdavi
Gorgan University of Agricultural Sciences and Natural Resources
Gorgan, Iran

Mariana Albarell
Laboratory of Bioprocess
Federal University of Santa Maria
Santa Maria, RS, Brazil

Tarun Kanti Bandyopadhyay
Department of Chemical Engineering
National Institute of Technology Agartala
Jirania, India

Ajaya Kumar Behera
Department of Chemistry
Utkal University
Bhubaneswar, India

Shashi Kant Bhatia
Department of Biological Engineering
Konkuk University
Seoul, South Korea

Biswanath Bhunia
Department of Bio Engineering
National Institute of Technology Agartala
Jirania, India

Shubhankar Debnath
Department of Bio Engineering
National Institute of Technology Agartala
Jirania, India

Crisleine Perinazzo Draszewski
Laboratory of Biofuels (LabBioc)
Federal University of Santa Maria
Santa Maria, RS, Brazil

Steven Ferguson
School of Chemical and Bioprocess Engineering
University College Dublin
Dublin, Ireland

Lalit M. Gantayet
Remed Care Technologies
Navi Mumbai, India

C. Gayathri
Microbial Processes and Technology Division
CSIR – National Institute for Interdisciplinary Science and Technology (NIIST)
Thiruvananthapuram, India

Diksha Garg
Department of Microbiology
DAV University
Jalandhar, India

Xinchen Gong
Tianjin Key Laboratory of Clean Energy and Pollution Control
School of Energy and Environmental Engineering
Hebei University of Technology
Tianjin, China

Richa Guleria
Indian Institute of Technology Delhi
New Delhi, India

Ranjit Gurav
Sustainability Cluster
School of advanced Engineering
UPES
Dehradun, India

Navneet Kaur
Department of Biotechnology and Food Technology
Punjabi University
Patiala, India

Simran Jot Kaur
Department of Biotechnology
Sri Guru Granth Sahib World University
Fatehgarh Sahib, India

Sunil H. Koli
Department of Microbiology
Yashwantrao Chavan College of Science
Karad, India

Mayur R. Ladole
Department of Biological Sciences and Biotechnology
Institute of Chemical Technology
Matunga (E), India

Honghong Lyu
Tianjin Key Laboratory of Clean Energy and Pollution Control
School of Energy and Environmental Engineering
Hebei University of Technology
Tianjin, China

Suvendu Manna
Sustainability Cluster
UPES University
Dehradun, India

Pinku Chandra Nath
Food Science and Technology Division
Department of Applied Biology
University of Science and Technology Meghalaya
Baridua, India

Amiya Ojha
Department of Bio Engineering
National Institute of Technology Agartala
Jirania, India

Sachin V. Otari
Department of Microbiology
Yashwantrao Chavan College of Science
Karad, India

Parmjit Singh Panesar
Food Biotechnology Research Laboratory
Department of Food Engineering and Technology
Sant Longowal Institute of Engineering and Technology
Longowal, India

Ruchi Pathania
Department of Medicinal Chemistry
Center for Natural Products, Drug Discovery and Development
University of Florida
Gainesville, Florida, USA

Sujata S. Patil
Aachen Institute of Biobased Materials (AMIBM)
Maastricht University
Maastricht, Netherlands

Pravin B. Pokle
Department of Electronics and Telecommunications
Priyadarshini J. L. Chaturvedi College of Engineering and Technology
Nagpur, India

Prem Pritam
Department of Chemical Engineering
Indian Institute of Technology Bombay
Mumbai, India

Maicon Sérgio Nascimento dos Santos
Laboratory of Agroindustrial Processes Engineering (LAPE)
Federal University of Santa Maria (UFSM)
Cachoeira do Sul, RS, Brazil

Dinalva Schein
Laboratory of Bioprocess
Federal University of Santa Maria
Santa Maria, RS, Brazil

Ramesh Sharma
Department of Food Technology
Sri Shakthi Institute of Engineering and Technology
Coimbatore, India

Shubhnam
Department of Microbiology
DAV University
Jalandhar, India

S. Sree Priya
Microbial Processes and Technology Division
CSIR – NIIST
Thiruvananthapuram, India
India and Academy of Scientific and Innovative Research (AcSIR)
Ghaziabad, India

N. Sridhar
Department of Food Technology
Sri Shakthi Institute of Engineering and Technology
Coimbatore, Tamil Nadu, India

Francisco Dalcin Vezaro
Laboratory of Biomass and Biofuels (L2B)
Federal University of Santa Maria
Santa Maria, RS, Brazil

Guilherme Cabral Wancura
Laboratory of Microbiology and Bioprocess (LAMIBI)
Federal University of Fronteira Sul
Erechim, RS, Brazil

João Henrique Cabral Wancura
Laboratory of Biomass and Biofuels (L2B)
Federal University of Santa Maria
Santa Maria, RS, Brazil

Rakesh K. Yasarla
Microbial Processes and Technology Division
CSIR – NIIST
Thiruvananthapuram, India
AcSIR
Ghaziabad, India

Zehua Zhao
Tianjin Key Laboratory of Clean Energy and Pollution Control
School of Energy and Environmental Engineering
Hebei University of Technology
Tianjin, China

Zixin Zhou
Tianjin Key Laboratory of Clean Energy and Pollution Control
School of Energy and Environmental Engineering
Hebei University of Technology
Tianjin, China

Preface

The rapid depletion of fossil fuel reserves, coupled with increasing environmental concerns, has led to an urgent search for sustainable and renewable energy alternatives. Among the promising solutions, biofuels have emerged as a viable and ecofriendly option. In particular, microbial biofuel production has garnered significant attention due to its potential to convert biowaste into clean energy through innovative and sustainable processes. This book, *Biowaste to Biofuel: A Microbial Approach*, explores the transformative potential of microbial systems in addressing the dual challenges of waste management and renewable energy production. The journey of this book begins with **Chapter 1**, an "Introduction to Biowaste to Biofuel: A Microbial Approach," where the fundamental concepts, importance, and scope of biofuel production from microbial systems are thoroughly examined. Following this foundation, **Chapter 2**, "Biomass to Biofuel Conversion: Biomass Pretreatment Strategies for Microbial Feedstock," explores the crucial steps needed to make biomass suitable for microbial processing. This chapter emphasizes the significance of efficient pretreatment techniques in enhancing biofuel yields. Food waste is a global issue, and **Chapter 3**, "Food Waste Upcycling to Biofuel by Microbial System," discusses how microbial technologies can transform food residues into biofuels, offering a practical approach to waste valorization. Similarly, **Chapter 4**, "Resource Recovery from Wastewater and Microbial Biofuel Production," highlights the integration of wastewater treatment with biofuel production, presenting a holistic approach to resource management. One of the most pressing environmental challenges today is carbon capture, and **Chapter 5**, "Carbon Capture and Biofuel: A Sustainable Synergy for a Greener World," examines how microbial biofuel production can contribute to reducing atmospheric carbon dioxide levels. Following this, **Chapter 6**, "Microbial Enzymes Involved in Biofuel Production," explores the enzymatic processes that highlight the microbial transformation of biowaste into fuel. The technological advancements in microbial biofuel production are presented in **Chapter 7**, "Advances in Immobilization Technologies for Biofuel Production," and **Chapter 8**, "Bioreactors Designs for Microbial Biofuel Production," where the focus is on enhancing efficiency and scalability through innovative engineering solutions. Genetic modification plays a crucial role in optimizing microbial performance, and **Chapter 9**, "Renewable energy Policies and Their Impact on Biofuel Development," addresses the latest breakthroughs in tailoring microbial genomes for increased biofuel yields. Moreover, the path from lab to market is explored in **Chapter 10**, "Commercialization of Microbial Biofuels Around the World," highlighting success stories and challenges. An essential aspect of any biofuel technology is its environmental and economic sustainability, covered in **Chapter 11**, "Life Cycle Assessment of Microbial Biofuel: Environmental and Economic Impacts." This chapter provides a comprehensive analysis of biofuel production's real-world impacts. Lastly, **Chapter 12**, "Integration of Biofuel Production with Other Biorefinery Processes," discusses how integrating biofuel production with other biotechnological processes can enhance overall sustainability.

This book serves as a comprehensive guide for researchers, academics, and professionals who are keen to explore microbial approaches to biofuel production. We hope it not only enhances understanding but also inspires innovative solutions to global energy challenges.

1 Introduction to Biowaste to Biofuel

A Microbial Approach

Shashi Kant Bhatia, Parmjit Singh Panesar, and Ranjit Gurav

Biowaste has emerged as an important renewable feedstock for biofuel production employing various physical as well as biological processes. The transformation of organic waste into biofuels through microbial processes offers an eco-friendly alternative to fossil fuels while addressing waste management challenges (Bhatia et al., 2018). The shift from fossil fuels to biofuels is essential to mitigate climate change, reduce greenhouse gas emissions, and establish a circular bioeconomy. Lignocellulosic biomass, abundant and renewable, serves as a prime feedstock for microbial biofuel production (Li et al., 2024; Vedernikovs et al., 2023). Yet, the complex structure of lignocellulose necessitates pretreatment to break down its components into fermentable sugars. Various methods like physical (grinding, milling), chemical (acid and alkali), and biological (enzymatic) methods are used to pretreat biomass (Bhatia et al., 2020; Chen et al., 2024). Biological methods are more eco-friendly and involve the use of various enzymes, including cellulases, hemicellulases, and ligninases (Bhatia et al., 2021b; Guo et al., 2023). Cellulases degrade cellulose into glucose, while hemicellulases target hemicellulose, yielding xylose and other sugars. Ligninases, on the other hand, break down lignin, facilitating access to cellulose and hemicellulose (Benatti & Polizeli, 2023). These enzymes, often derived from fungi or bacteria, work synergistically to enhance the efficiency of lignocellulosic biomass conversion. Hydrolyzed biomass is rich in fermentable sugar and can be used as a feedstock for microbes to produce biofuel as well as various products (Bhatia et al., 2023; Gurav et al., 2025).

Food waste, another abundant biowaste, presents unique opportunities for microbial valorization through anaerobic digestion and fermentation, enabling the production of bioethanol, biogas, and other biofuels (Narisetty et al., 2022; Zeng et al., 2022). The heterogeneous composition of food waste, consisting of carbohydrates, proteins, and lipids, makes it an excellent substrate for microbial consortia that efficiently convert organic matter into valuable energy carriers. Anaerobic digestion, a widely implemented microbial technology, utilizes hydrolytic, acidogenic, acetogenic, and methanogenic microbial communities to transform food waste into methane-rich biogas (Bhatia et al., 2025). Additionally, fermentation by yeast and bacteria can convert

DOI: 10.1201/9781003633709-1

sugars and starches from food waste into ethanol and butanol, contributing to biofuel production while simultaneously reducing landfill waste (Passadis et al., 2022; Wang et al., 2023). Microbial electrochemical systems, such as microbial fuel cells (MFCs) and microbial electrolysis cells (MECs), provide innovative approaches to harness energy from food waste, directly converting organic matter into electricity or hydrogen gas (Gurav et al., 2021).

Similarly, wastewater treatment systems can serve dual purposes by integrating microbial biofuel production with nutrient recovery, as microalgae and electrogenic bacteria facilitate both pollutant removal and energy generation (Deng et al., 2022). MFCs in wastewater treatment utilize electroactive bacteria to oxidize organic matter, generating bioelectricity while simultaneously degrading pollutants (Bazina et al., 2023). Additionally, microalgae cultivation in wastewater provides a sustainable feedstock for biodiesel production, as algae can assimilate nitrogen and phosphorus while capturing CO_2, thus contributing to wastewater bioremediation and biofuel synthesis (Lee et al., 2023). High-rate algal ponds and photobioreactors have been developed to optimize algal growth for biofuel applications (Penloglou et al., 2024). Furthermore, anaerobic membrane bioreactors (AnMBRs) have been explored to enhance microbial biogas recovery from wastewater, improving energy efficiency and sustainability in water treatment processes (Zhang et al., 2021).

The integration of microbial biofuel production with carbon capture technologies is another promising avenue for achieving sustainability. Microbial carbon fixation offers an innovative strategy to convert CO_2 into biofuels, effectively transforming a greenhouse gas into a valuable energy source (Bhatia et al., 2019). Cyanobacteria and microalgae utilize photosynthesis to assimilate CO_2, producing biomass that can be processed into biofuels such as biodiesel, hydrogen, and bioethanol (Condor et al., 2022; Kumar et al., 2022). Researchers are pioneering approaches to close the carbon cycle, capturing CO_2 emissions from industrial processes or directly from the atmosphere and converting them into valuable biofuels and bioproducts. Various microbes such as *Acetobacterium woodii*, *Clostridium aceticum*, *Chlorella vulgaris*, *Rhodococcus erythropolis*, and *Ralstonia eutropha* have shown potential to capture CO_2 effectively (Bhatia et al., 2019; Wan et al., 2024). Mechanisms involved in CO_2 capture encompass processes such as carbonic anhydrase-mediated hydration of CO_2 to bicarbonate, CO_2 reduction into methane, and the conversion of CO_2 into formate and methanol (Bierbaumer et al., 2023). Engineered microbial strains have been developed to enhance carbon fixation and direct metabolic flux toward biofuel synthesis (Lauer et al., 2022). Bioenergy with carbon capture and storage systems integrate microbial biofuel production with CO_2 sequestration, offering a strategy for net-negative emissions and climate change mitigation.

Similarly, in the conversion of oils into biodiesel, enzymes play a central role. Lipases, enzymes that catalyze the hydrolysis of triglycerides into glycerol and fatty acids, are crucial in this process (Bhatia et al., 2021b). Moreover, during transesterification, where fatty acids are esterified with alcohol to form biodiesel, lipases can also act as catalysts (Bhatia et al., 2021a). These enzymes, sourced from microbial or plant sources, enable the production of biodiesel with high purity and yield. To optimize enzyme activity and stability, various strategies are employed, with enzyme immobilization being a prominent approach. Immobilization involves fixing enzymes onto solid supports or matrices, thereby enhancing their recyclability, stability, and activity Mirsalami et al., (2024). Immobilized enzymes can be utilized in batch, continuous, or fed-batch reactor systems, allowing for prolonged enzyme usage and reduced operational costs. Moreover, immobilization techniques such as covalent binding, adsorption, entrapment, or encapsulation offer flexibility in enzyme immobilization, catering to diverse industrial applications. Overall, the integration of innovative reactor designs and enzymatic advancements holds immense potential in advancing microbial biofuel production. Through continued research and development efforts, the optimization of enzymatic processes and reactor technologies will drive progress toward a sustainable and environmentally friendly energy landscape.

In the area of microbial biofuel production, various reactor designs are employed to optimize fermentation processes and enhance biofuel yields. One common type is the stirred-tank reactor (STR), characterized by its agitator that maintains uniform mixing of nutrients and microorganisms throughout the fermentation broth (Limberg et al., 2016). These reactors allow for precise control of temperature, pH, and oxygen levels, essential for maximizing microbial growth and biofuel production. Additionally, airlift reactors (ALRs) utilize gas sparging to circulate the fermentation broth, promoting efficient mass transfer and oxygenation while minimizing shear stress on fragile microbial cells. Another innovative design is the packed-bed reactor (PBR), where immobilized microbial cells or enzymes are housed within a packed bed of solid support material (Bhatia & Yang, 2017). This configuration offers high surface area-to-volume ratios, facilitating efficient substrate utilization and product formation. Furthermore, membrane bioreactors (MBRs) integrate membranes to separate microorganisms from the fermentation broth, enabling continuous operation and preventing cell washout (Iorhemen et al., 2016). Lastly, photobioreactors harness light energy to drive photosynthetic microorganisms, such as algae, for biofuel production (Bhatia et al., 2021a). These reactors provide precise control over light intensity, temperature, and nutrient availability, optimizing algal growth and lipid accumulation. By leveraging diverse reactor designs, microbial biofuel production can be tailored to specific feedstocks, microorganisms, and process requirements, advancing the quest for sustainable energy solutions.

Genetic engineering has revolutionized microbial biofuel production by enabling the modification of metabolic pathways to enhance fuel yields, tolerance to inhibitors, and substrate flexibility (Adegboye et al., 2021; Keasling et al., 2021). Engineered yeast and bacteria have been developed to ferment pentose and hexose sugars from lignocellulose, tolerate high ethanol concentrations, and produce advanced biofuels such as butanol, fatty acid-derived hydrocarbons, and isoprenoids. CRISPR/Cas9 genome editing has facilitated precise genetic modifications, allowing researchers to optimize microbial metabolism for biofuel production (Garg et al., 2023; Parsaeimehr et al., 2022). Synthetic biology has enabled the construction of entirely new metabolic pathways, expanding the range of biofuels that can be synthesized from microbial hosts.

Despite these technological advancements, commercialization remains a major challenge, as scaling up microbial biofuel production requires overcoming economic, logistical, and policy barriers. While first-generation biofuels such as ethanol from sugarcane have seen widespread adoption, second-generation biofuels derived from lignocellulose and third-generation algal biofuels are still striving for market competitiveness. Challenges include feedstock availability, production costs, and competition with fossil fuels. Large-scale demonstration projects and government incentives have been crucial in advancing microbial biofuel commercialization. Commercialization of biofuels requires collaboration between industry stakeholders, policymakers, and research institutions to address technical, regulatory, and market challenges. Government incentives, such as tax credits and renewable fuel standards, play a crucial role in incentivizing investment in biofuel production infrastructure and stimulating market demand. Moreover, partnerships between academia and industry facilitate technology transfer and scale-up of innovative biofuel production processes, accelerating their commercialization and market penetration.

Even if a microbial biofuel process is technically feasible, its sustainability must be critically evaluated. Life Cycle Assessment (LCA) is the tool that provides a holistic evaluation of environmental impacts (Reijnders, 2021). A key goal is to assess whether microbial biofuels indeed offer net positive environmental outcomes compared to fossil fuels or other biofuel routes. Through LCA studies, it can be concluded that microbial biofuels can greatly reduce net CO_2 emissions relative to petroleum fuels, especially when waste feedstocks are used (avoiding emissions from decomposition or incineration) and when renewable energy powers the process. However, the benefits can be offset by factors like indirect land-use change for feedstock cultivation, high energy usage in processing (e.g., distillation of ethanol), or the production of chemical inputs (like enzymes or nutrients for microbes). Thus, LCA provides a quantified balance sheet of these trade-offs (Love, 2022). It also includes economic considerations via techno-economic analysis (TEA) since sustainability encompasses economic viability, as well as environmental impact. For microbial processes, LCA/TEA may reveal, for instance, that a certain pretreatment yields better energy return on investment despite higher upfront emissions or that a biogas system has a shorter payback period than a fermentation-based system.

Microbial biofuel production does not occur in isolation; increasingly, it is integrated into broader biorefinery systems that maximize the value derived from biomass. In an integrated biorefinery, a single biomass or waste feedstock is processed via multiple pathways (some microbial, some thermochemical or chemical) to generate a portfolio of products – for example, ethanol, biogas, bioplastics, and fertilizers from the same feedstock (Nath et al., 2025; Tsachidou et al., 2019). In a lignocellulosic biorefinery where cellulose is fermented to ethanol by engineered microbes, hemicellulose sugars are converted to biochemicals or biogas, and lignin residues are burned for process heat or used to create value-added materials. Another example is the integration of algal biofuel production with wastewater treatment and CO_2 utilization (the algae clean the water and capture carbon and then are processed into fuels and fertilizers) – a multi-output system improving overall sustainability (Bhatia et al., 2021a). By integrating processes, the biorefinery achieves higher resource efficiency and can tap multiple revenue streams, which is often necessary for profitability in biofuel ventures.

Policy frameworks play a decisive role in shaping the biofuel industry, with renewable energy mandates, carbon pricing mechanisms, and government subsidies influencing the feasibility and expansion of microbial biofuels. Countries with strong biofuel policies, such as the United States, Brazil, and the European Union, have driven innovation and investment in biofuel technologies. However, inconsistent policies, fluctuating oil prices, and regulatory challenges have hindered widespread adoption. Future policy directions should focus on long-term support for advanced biofuels, sustainability certifications, and incentives for integrated biorefinery development.

Government policies have historically been a major driving force behind biofuel expansion. For example, blending mandates (which require a certain percentage of biofuel in gasoline or diesel) in the United States, European Union, Brazil, and many other countries have created guaranteed markets for bioethanol and biodiesel, fostering greater utilization of these fuels (Araújo et al., 2017). Subsidies, tax incentives, and research grants have underwritten the growth of biofuel companies and the advancement of technologies from lab to pilot scale. Supportive policies can accelerate deployment (and even insulate biofuel industries from oil price volatility, poorly designed or inconsistent policies can hinder progress or create uncertainty. Case in point: sudden policy shifts or the lapse of tax credits have led to boom-and-bust cycles in investment. Additionally, international collaborations and agreements (like those under the Paris Climate Accord) are mechanisms pushing countries to adopt cleaner fuels.

In conclusion, microbial biofuel production represents a promising pathway toward a circular bioeconomy, leveraging biological systems for waste valorization and renewable energy generation. Future advancements in metabolic engineering, process intensification, and policy support will be critical in realizing the full potential of microbial biofuels as a sustainable energy solution.

REFERENCES

Adegboye, M.F., Ojuederie, O.B., Talia, P.M., Babalola, O.O. 2021. Bioprospecting of microbial strains for biofuel production: metabolic engineering, applications, and challenges. *Biotechnology for Biofuels*, 14(1), 5.

Araújo, K., Mahajan, D., Kerr, R., Silva, M.D. 2017. Global biofuels at the crossroads: An overview of technical, policy, and investment complexities in the sustainability of biofuel development. *Agriculture*, 7(4), 32.

Bazina, N., Ahmed, T.G., Almdaaf, M., Jibia, S., Sarker, M. 2023. Power generation from wastewater using microbial fuel cells: A review. *Journal of Biotechnology*, 374, 17–30.

Benatti, A.L.T., Polizeli, M. 2023. Lignocellulolytic biocatalysts: The main players involved in multiple biotechnological processes for biomass valorization. *Microorganisms*, 11(1) 162.

Bhatia, S.K., Bhatia, R.K., Jeon, J.-M., Kumar, G., Yang, Y.-H. 2019. Carbon dioxide capture and bioenergy production using biological system – A review. *Renewable and Sustainable Energy Reviews*, 110, 143–158.

Bhatia, S.K., Gurav, R., Cho, D.-H., Kim, B., Jung, H.J., Kim, S.H., Choi, T.-R., Kim, H.-J., Yang, Y.-H. 2023. Algal biochar mediated detoxification of plant biomass hydrolysate: Mechanism study and valorization into polyhydroxyalkanoates. *Bioresource Technology*, 370, 128571.

Bhatia, S.K., Jagtap, S.S., Bedekar, A.A., Bhatia, R.K., Patel, A.K., Pant, D., Rajesh Banu, J., Rao, C.V., Kim, Y.-G., Yang, Y.-H. 2020. Recent developments in pretreatment technologies on lignocellulosic biomass: Effect of key parameters, technological improvements, and challenges. *Bioresource Technology*, 300, 122724.

Bhatia, S.K., Joo, H.-S., Yang, Y.-H. 2018. Biowaste-to-bioenergy using biological methods—A mini-review. *Energy Conversion and Management*, 177, 640–660.

Bhatia, S.K., Mehariya, S., Bhatia, R.K., Kumar, M., Pugazhendhi, A., Awasthi, M.K., Atabani, A.E., Kumar, G., Kim, W., Seo, S.-O., Yang, Y.-H. 2021a. Wastewater based microalgal biorefinery for bioenergy production: Progress and challenges. *Science of the Total Environment*, 751, 141599.

Bhatia, S.K., Patel, A.K., Ravindran, B., Yang, Y.-H. 2025. Renewable bioenergy from palm oil mill effluent (POME): A sustainable solution for net-zero emissions in palm oil production. *Journal of Water Process Engineering*, 70, 107136.

Bhatia, S.K., Vivek, N., Kumar, V., Chandel, N., Thakur, M., Kumar, D., Yang, Y.-H., Pugazendhi, A., Kumar, G. 2021b. Molecular biology interventions for activity improvement and production of industrial enzymes. *Bioresource Technology*, 324, 124596.

Bhatia, S.K., Yang, Y.-H. 2017. Microbial production of volatile fatty acids: Current status and future perspectives. *Reviews in Environmental Science and Bio/Technology*, 16(2), 327–345.

Bierbaumer, S., Nattermann, M., Schulz, L., Zschoche, R., Erb, T.J., Winkler, C.K., Tinzl, M., Glueck, S.M. 2023. Enzymatic conversion of CO_2: From natural to artificial utilization. *Chemical Reviews*, 123(9), 5702–5754.

Chen, J., Ma, X., Liang, M., Guo, Z., Cai, Y., Zhu, C., Wang, Z., Wang, S., Xu, J., Ying, H. 2024. Physical–chemical–biological pretreatment for biomass degradation and industrial applications: A review. *Waste*, 2(4), 451–473.

Condor, B.E., de Luna, M.D.G., Chang, Y.-H., Chen, J.-H., Leong, Y.K., Chen, P.-T., Chen, C.-Y., Lee, D.-J., Chang, J.-S. 2022. Bioethanol production from microalgae biomass at high-solids loadings. *Bioresource Technology*, 363, 128002.

Deng, Z., Khatiwada, J.R., Fan, L., Qin, W. 2022. Algae for wastewater treatment and production of biofuels and bioproducts. *International Journal of Environmental Research*, 16(3), 34.

Garg, D., Samota, M.K., Kontis, N., Patel, N., Bala, S., Rosado, A.S. 2023. Revolutionizing biofuel generation: Unleashing the power of CRISPR-Cas mediated gene editing of extremophiles. *Microbiological Research*, 274, 127443.

Guo, H., Zhao, Y., Chang, J.-S., Lee, D.-J. 2023. Enzymes and enzymatic mechanisms in enzymatic degradation of lignocellulosic biomass: A mini-review. *Bioresource Technology*, 367, 128252.

Gurav, R., Bhatia, S.K., Choi, T.-R., Kim, H.-J., Lee, H.-J., Cho, J.-Y., Ham, S., Suh, M.-J., Kim, S.-H., Kim, S.-K., Yoo, D.-W., Yang, Y.-H. 2021. Seafood processing chitin waste for electricity generation in a microbial fuel cell using halotolerant catalyst *Oceanisphaera arctica* YHY1. *Sustainability*, 13(15), 8508.

Gurav, R., Hwang, S., Bhatia, S.K., Kshirsagar, A.D., Mandal, S., Yang, Y.-H. 2025. Exopolysaccharide production using pinewood hydrolysate as a substrate for psychrotrophic bacterium isolated from Svalbard glacier soil. *Biomass Conversion and Biorefinery*, 15(2), 2263–2273.

Iorhemen, O.T., Hamza, R.A., Tay, J.H. 2016. Membrane bioreactor (MBR) technology for wastewater treatment and reclamation: Membrane fouling. *Membranes*, 6(2), 33.

Keasling, J., Garcia Martin, H., Lee, T.S., Mukhopadhyay, A., Singer, S.W., Sundstrom, E. 2021. Microbial production of advanced biofuels. *Nature Reviews Microbiology*, 19(11), 701–715.

Kumar, M.D., Kavitha, S., Tyagi, V.K., Rajkumar, M., Bhatia, S.K., Kumar, G., Banu, J.R. 2022. Macroalgae-derived biohydrogen production: Biorefinery and circular bioeconomy. *Biomass Conversion and Biorefinery*, 12(3), 769–791.

Lauer, I., Philipps, G., Jennewein, S. 2022. Metabolic engineering of Clostridium ljungdahlii for the production of hexanol and butanol from CO_2 and H_2. *Microbial Cell Factories*, 21(1), 85.

Lee, J.-C., Moon, K., Lee, N., Ryu, S., Song, S.H., Kim, Y.J., Lee, S.M., Kim, H.-W., Joo, J.-H. 2023. Biodiesel production and simultaneous treatment of domestic and livestock wastewater using indigenous microalgae, Chlorella sorokiniana JD1-1. *Scientific Reports*, 13(1), 15190.

Li, W., Tan, X., Li, L., Cheng, X., Song, L., Xie, G., Zhang, Y., Zhang, Q., Miao, C., Zhuang, X. 2024. Biofuels production from organosolv treated lignocellulose: Nondetoxified fermentation and microbial community analysis. *ACS Sustainable Chemistry & Engineering*, 12(41), 14988–14998.

Limberg, M.H., Pooth, V., Wiechert, W., Oldiges, M. 2016. Plug flow versus stirred tank reactor flow characteristics in two-compartment scale-down bioreactor: Setup-specific influence on the metabolic phenotype and bioprocess performance of Corynebacterium glutamicum. *Engineering in Life Sciences*, 16(7), 610–619.

Love, J. 2022. Microbial pathways for advanced biofuel production. *Biochemical Society Transactions*, 50(2), 987–1001.

Mirsalami, S.M., Mirsalami, M., Ghodousian, A. 2024. Techniques for immobilizing enzymes to create durable and effective biocatalysts. *Results in Chemistry*, 7, 101486.

Narisetty, V., Zhang, L., Zhang, J., Sze Ki Lin, C., Wah Tong, Y., Loke Show, P., Kant Bhatia, S., Misra, A., Kumar, V. 2022. Fermentative production of 2,3-Butanediol using bread waste – A green approach for sustainable management of food waste. *Bioresource Technology*, 358, 127381.

Nath, P., Borah, D., Paul, P., Rout, J. 2025. Integrated biorefinery approach for sustainable production of biodiesel, bioplastics and high value bioproducts from a bloom forming alga, Botryococcus braunii. *Science of the Total Environment*, 964, 178599.

Parsaeimehr, A., Ebirim, R.I., Ozbay, G. 2022. CRISPR-Cas technology a new era in genomic engineering. *Biotechnology Reports*, 34, e00731.

Passadis, K., Christianides, D., Malamis, D., Barampouti, E.M., Mai, S. 2022. Valorisation of source-separated food waste to bioethanol: Pilot-scale demonstration. *Biomass Conversion and Biorefinery*, 12(10), 4599–4609.

Penloglou, G., Pavlou, A., Kiparissides, C. 2024. Recent advancements in photo-bioreactors for microalgae cultivation: A brief overview. *Processes*, 12(6), 1104.

Reijnders, L. 2021. Life cycle assessment of biofuels. *Methods in Molecular Biology*, 2290, 53–67.

Tsachidou, B., Scheuren, M., Gennen, J., Debbaut, V., Toussaint, B., Hissler, C., George, I., Delfosse, P. 2019. Biogas residues in substitution for chemical fertilizers: A comparative study on a grassland in the Walloon Region. *Science of The Total Environment*, 666, 212–225.

Vedernikovs, N., Khroustalyova, G., Muiznieks, I., Rapoport, A. 2023. New concept for conversion of lignocellulose to ethanol and furfural. *Applied Microbiology and Biotechnology*, 107(2), 535–542.

Wan, S., Lai, M., Gao, X., Zhou, M., Yang, S., Li, Q., Li, F., Xia, L., Tan, Y. 2024. Recent progress in engineering Clostridium autoethanogenum to synthesize the biochemicals and biocommodities. *Synthetic and Systems Biotechnology*, 9(1), 19–25.

Wang, X., Sun, H., Wang, Y., Wang, F., Zhu, W., Wu, C., Wang, Q., Gao, M. 2023. Feasibility of efficient, direct, butanol production from food waste without nutrient supplement by *Clostridium saccharoperbutylacetonicum* N1-4. *Sustainability*, 15(7), 6061.

Zeng, J., Zeng, H., Wang, Z. 2022. Review on technology of making biofuel from food waste. *International Journal of Energy Research*, 46(8), 10301–10319.

Zhang, J., Li, Y., Wu, B., Huang, X., Hou, Z., Chen, R. 2021. Performance and mechanism of in-situ biogas upgrading using anaerobic membrane bioreactor effluent. *Journal of Water Process Engineering*, 44, 102323.

2 Biomass to Biofuel Conversion

Biomass Pretreatment Strategies for Microbial Feedstock

João Henrique Cabral Wancura, Francisco Dalcin Vezaro, Crisleine Perinazzo Draszewski, Mariana Albarello, Dinalva Schein, Maicon Sérgio Nascimento dos Santos, and Guilherme Cabral Wancura

2.1 INTRODUCTION

Recently, scientific projections indicate that global biofuel production has grown from 9.2 million metric tons of oil equivalent (Mtoe) in 2000 to 95.38 Mtoe in 2018, indicating advances that could reach US$132.7 billion by 2023 (Awogbemi et al., 2021). Based on a strongly promising scenario, the exploration of biofuels plays a key factor in minimizing dependence on finite fossil fuels, reducing greenhouse gas emissions, and promoting an environmentally friendly energy matrix (Gonçalves et al., 2023; Wancura et al., 2023a). The current performance of biofuel production and consumption is an indication of constant innovations in the processes for obtaining alternative fuels, highlighting a revolutionary change in research related to sustainable energy sources. Furthermore, its innovation lies in the use of renewable plant materials, crops, and even microbial species, which synthesize clean-burning fuels while promoting a significant reduction in carbon emissions and pollutants in the atmosphere (Boly and Sanou, 2022).

The production of microbial biofuels from lignocellulosic biomass involves a multistage process that leverages microbial metabolic capacity readily to convert lignocellulosic biomass into usable fuels, such as pretreatment, enzymatic hydrolysis, and microbial fermentation processes. Pretreating lignocellulosic biomass is a fundamental step in the production of microbial biofuels, involving the disruption of the complex structure of lignocellulose and promoting increased contact between enzymes and microorganisms (Sharma et al., 2023a). Lignocellulosic biomass

DOI: 10.1201/9781003633709-2

have a complex structure encompassing cellulose, hemicellulose, and lignin, where the pretreatment stage acts to break down this structure, increasing the accessibility of cellulose and hemicellulose for subsequent microbial or enzymatic conversion (Singhania et al., 2022). Furthermore, lignin, a rigid and complex polymer, contributes substantially to biomass recalcitrance, hindering enzymatic access to cellulose, where the pretreatment helps to break down or remove this polymer, reducing recalcitrance and facilitating the bioconversion synthesis (Bhatia et al., 2020). With the disruption of the lignocellulosic matrix, the cellulose and hemicellulose become more accessible to enzymes, enhancing the efficiency of enzymatic digestibility and leading to higher yields of fermentable sugars (Scapini et al., 2021).

The operational approaches from lignocellulosic biomass to the final biofuel product exemplify the intricate interplay between enzymatic actions, microbial metabolism, and bioconversion procedures, offering a sustainable route to renewable energy sources. The success of all these steps is closely linked to the necessary pretreatment process of biomass considered for the production of microbial biofuels. Accordingly, this chapter presents a current description of modern approaches aimed at the distinct pretreatment techniques considered for microbial biofuel synthesis under the biorefinery concepts. Its key purposes encompass exploring the current panorama of the pretreatment processes, including an objective description of the lignocellulosic composition of biomass, serving as an explanation of the reasons why these materials need to go through the aforementioned process, as well as the different categories selected for this purpose. Finally, the text also presents data on the current technological maturity scenario of biomass pretreatment processes through a search accomplished on a patent basis. This chapter aims to evaluate the effectiveness and versatility of biomass pretreatment strategies in the field of microbial biofuels, highlighting their potential as sustainable alternatives regarding the "biorcfincry approach."

2.2 WHY PRETREAT BIOMASS?

Biomass are extremely abundant owing to the countless options available *in nature* and as agro-industrial wastes. Furthermore, due to the lignocellulosic compositions, each biomass has a different potential for conversion into microbial biofuels. These differences end up attracting the attention of researchers who aim to find the best alternatives for biomass selection, raw material processing methods, and pretreatments to optimize the process yield (Abolore et al., 2023).

Cellulose is the main component of interest for obtaining microbial biofuels, as it is a polysaccharide formed by β-1,4 bonds of glucose monosaccharides, giving the cellulose a crystalline characteristic. Contextually, hemicellulose can be hydrolyzed into pentose and hexose sugars, but it can also generate organic acids. Furthermore, its structure is amorphous, with a lower molecular weight than cellulose (Afraz et al., 2024). Lignin is formed by coniferyl alcohol, sinapyl alcohol, and coumaryl alcohol in different proportions for each type of biomass, which are responsible for giving rigidity to the biomass (Li et al., 2024b). Lignin is recalcitrant to hydrolysis and tends to prevent the access of enzymes to cellulose and hemicellulose, in addition to hindering the cleavage of hydrogen ions in acid and subcritical hydrolysis. Furthermore, although hemicellulose generates some fermentable sugars, it may also act as a physical barrier to cellulose. Therefore, the natural lignocellulosic structure usually does not present good conversions of polysaccharides into fermentable sugars, and it may be more advantageous to pretreat the biomass to optimize hydrolysis. Thus, chemical, physical, thermochemical, or biological pretreatments usually must be performed to overcome or break the resistance of the lignocellulosic structure through fragmentation or removal of lignin and hemicellulose, as well as to reduce the biomass particle size, polymerization, and crystallinity of the cellulose to increase the pores and surface area of the biomass (Bhatia et al., 2020).

Table 2.1 presents the lignocellulosic composition of several biomass. It is possible to verify in these data that the lignocellulosic composition of these materials can vary substantially in terms of cellulose (11.7–56.5 wt.% for *Butia odorata* seed and poplar wood, respectively), hemicellulose (4.6–37.0 wt.% for castor seed

TABLE 2.1
Lignocellulosic Composition (wt.%) of Different Biomass

Biomass	Cellulose	Hemicellulose	Lignin	Reference
Rice bran	27.0–42.0	11.0–26.0	8.0–22.0	Moreira et al. (2022); Wancura et al. (2023b)
Butia odorata seed	11.7	17.7	8.4	Gallon et al. (2023)
Rice husk	32.0–45.0	19.0–30.0	22.0–26.0	Draszewski et al. (2021); Kartal and Özveren (2021)
Rice straw	22.0–40.0	24.0–30.0	9.0–23.0	Abaide et al. (2019), Kartal and Özveren (2021)
Corn stover	28.0–35.0	16.0–24.0	8.0–21.0	Li et al. (2024a), Zabed et al. (2023)
Corn cob	35.0–39.0	35.0–39.0	14.0–15.0	Malik et al. (2022)
Cornstalk	29.0–36.0	19.0–30.0	15.0–33.0	Jia et al. (2023), Zhang et al. (2016)
Bamboo	38.0–42.0	18.0–28.0	15.0–30.0	Cai et al. (2017), Wang et al. (2023)
Pinewood	40.7	25.8	26.8	Wang et al. (2023)
Soybean hulls	34.0	11.0	7.0	Martinez-Burgos et al. (2023)
Soybean straw	31.0	20.6	17.6	Zhang et al. (2020)
Wheat straw	35.0–40.0	21.0–36.0	18.0–29.0	Kartal and Özveren (2021), Zhan et al. (2023)
Sugarcane bagasse	36.0–43.0	18.0–26.0	12.0–19.0	Kartal and Özveren (2021), Mishra and Mohanty (2019)
Banana trunk	25.0	25.0	15.0	Kartal and Özveren (2021)
Pine sawdust	36.5	14.2	21.9	Puhl et al. (2024)
Mustard stalk	39.5	18.7	22.5	Mishra and Mohanty (2019)
Castor seed	46.9	4.6	32.3	Mishra and Mohanty (2019)
Cotton stalk	39.4	19.2	23.2	Mishra and Mohanty (2019)
Walnut shell	32.2	26.2	36.9	Özsin and Pütün (2017a)
Peach stone	29.5	25.1	39.3	Özsin and Pütün (2017a)
Chestnut shell	31.6	22.6	42.7	Özsin and Pütün (2017b)
Cherry stone	26.9	26.9	42.2	Özsin and Pütün (2017b)

(Continued)

TABLE 2.1 (Continued)
Lignocellulosic Composition (wt.%) of Different Biomasses

Biomass	Cellulose	Hemicellulose	Lignin	Reference
Grape seed	13.8	18.7	49.2	Özsin and Pütün (2017b)
Peanut shell	37.5	21.2	29.7	Yousaf et al. (2017)
Elephant grass	31.3	29.0	15.8	Scholl et al. (2015)
Eucalyptus wastes	36.0–42.0	12.0–14.0	27.0–30.0	Cedeno et al. (2022)
Napier grass	34.9	20.2	25.6	Gundupalli et al. (2023)
Yellow *passion* fruit	52.0	18.9	13.6	Teles et al. (2023)
Limonia acidissima shell	41.3	27.0	28.3	Nagarajasetty et al. (2022)
Spruce wood	42.9	20.9	26.7	Sejati et al. (2024)
Poplar wood	56.5	20.9	14.4	Sun et al. (2023)
Oat straw	26.0–35.0	25.0–28.0	12.0–16.0	Rencoret et al. (2023)
Barley straw	34.3	22.6	20.7	Liu et al. (2023)
Olive stones	20.9	26.0	35.6	Rubio-Valle et al. (2024)
Coffee husk	39.2	12.6	26.2	Van Nguyen et al. (2023)
Bean straw	21.4	19.6	10.2	Okot et al. (2022)
Olive leaves	12.4	37.0	17.0	Mabrouk et al. (2023)
Vegetable crop residues	25.0–33.0	14.0–19.0	8.0–17.0	Li et al. (2021)
Herb residues	36.5	18.5	24.2	Wei et al. (2021)
Nettle herb	18.1	10.3	60.2	Gęca et al. (2023)
Sage herb	19.4	19.1	39.2	Gęca et al. (2023)
Mint herb	16.3	27.4	43.1	Gęca et al. (2023)

and olive leaves, respectively), and lignin (8.0–60.2 wt.% for rice bran and corn stover as well as nettle herbs, respectively). These variations are common and are typically associated with distinct factors, such as cultivation method, climate, soil conditions, plant variety, plant characteristics, and stage of development (Abolore et al., 2023). Differences in lignocellulosic composition may end up requiring different biomass processing. For example, for a biomass with high lignin content, a more aggressive pretreatment may be necessary so that it is sufficient to overcome the recalcitrance of the lignin by removing or depolymerizing part of it. On the other hand, if there is low lignin and higher concentrations of hemicellulose, some milder pretreatments may be sufficient, considering that hemicellulose is more easily dissociated than lignin and cellulose if only the polysaccharide is considered as the target compound (Pendse et al., 2023).

Depending on the methodology used to obtain sugars, cellulose and hemicellulose can be converted into fermentable sugars. In these cases, improving permeability and access to cellulose-hemicellulose from biomass is essential, but energy viability, generation of less inhibitory co-products, and an environmentally friendly practice must be taken into consideration when selecting a pretreatment. Furthermore, it is important that the pretreatment is capable of partially removing or depolymerizing lignin, is economical, can reduce cellulose crystallinity, reduces particle size, increases the surface area of the biomass, and applies to various biomass (Yadav et al., 2023).

2.3 BIOMASS PRETREATMENT STRATEGIES

The pretreatment of lignocellulosic biomass is a crucial step in the production of microbial biofuels, and its importance cannot be overstated. Lignocellulosic biomass, such as agricultural residues, forestry waste, and energy crops, are abundant and sustainable feedstocks for biofuel production. However, their complex and recalcitrance structure presents challenges for efficient conversion by microorganisms. Pretreatment processes serve several vital purposes in this context, disrupting the lignocellulosic matrix of the biomass, and breaking down the lignin, cellulose, and hemicellulose components to make them more accessible to enhance the conversion efficiency in the biofuel-producing microorganisms. In this section, the main categories of pretreatments used in biomass selected for the production of microbial biofuels are presented, according to the illustration in Figure 2.1. For each category, processes commonly considered to perform such important stages of treatment are discussed.

2.3.1 Physical Pretreatments

Physical pretreatments employ shear (mechanical) force to break the rigid structure of lignocellulose, enhancing accessibility for chemicals, microorganisms, or enzymes to interact with cellulose and hemicellulose. Usually, physical pretreatment methods are employed to reduce the particle size of the biomass and considered as a preliminary step to subsequently send the biomass to be pretreated (Zhao et al., 2022). Mechanical implements and methodologies, including mills, grinders, screws, ultrasound, or microwave radiations, are typically utilized during physical pretreatments (Mankar et al., 2021).

Milling is frequently considered a pretreatment technique for reducing the biomass particle size. It can be accomplished through various methods, such as grinding, milling, shredding, and chipping. Nevertheless, the ultimate particle size is contingent

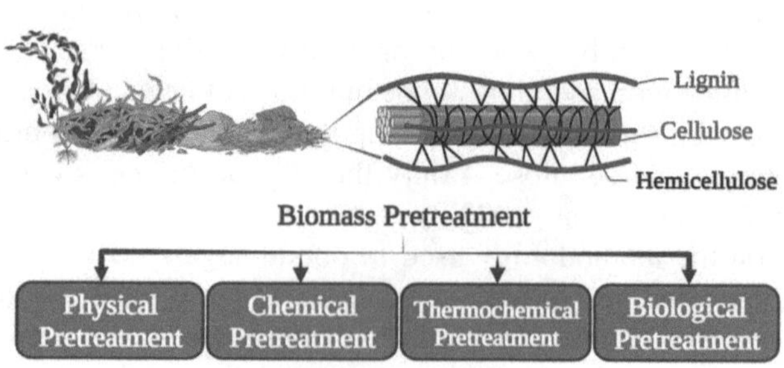

FIGURE 2.1 Representation of the lignocellulosic structure of diverse biomass and the typical pretreatments applied to reduce their characteristic recalcitrance.

upon the methodology utilized, e.g., grinding or milling of biomass results in a typical particle size of 0.2 to 2.0 mm, while 10.0 to 30.0 mm particles are obtained on chipping (Mankar et al., 2021). Milling procedures can be employed to enhance the susceptibility of lignocelluloses to cellulases. It is advisable to conduct milling to size reduction procedures on the lignocellulosic substrate to benefit the heat and mass transfer (Aftab et al., 2019). As an illustration, extrusion, grinding, cutting, and ball milling applied to distinct substrates with low lignin content not only markedly decrease substrate size and polymerization degree but also can lower lignin content up to 54% (Dahunsi, 2019).

Common mechanical size reduction processes encompass dry milling, compression milling, vibratory ball milling, and wet milling. The energy consumption for grinding lignocellulosic biomass is dependent upon machine variables, feed material rate, material properties, moisture content, and initial particle size. Grinders are classified primarily based on the grinding mechanism. Fundamentally, size reduction is achieved through splitting or shearing with sharp knives, wherein particle geometry is disrupted through impact or compression (Kamarludin et al., 2014). The choice of a grinder must be based on a good knowledge of the technology and the raw materials (Mayer-Laigle et al., 2018). This process provides significant advantages, including reduced cellulose crystallinity, enhanced surface area for enzymatic hydrolysis, decreased cellulose polymerization, and improved mass transfer through particle size reduction (Veluchamy et al., 2019).

Extrusion constitutes a thermo-mechanical, continuous, and cost-effective process widely employed in the metal, food processing, chemical, and polymer industries. Its application in the context of lignocellulosic biomass pretreatment can offer several advantages. Notably, extrusion is characterized by its flexibility, allowing for combination with chemicals and/or other processes. The method entails high shear forces, rapid heat transfer, and efficient mixing, collectively enhancing the efficacy of the pretreatment. Additionally, the extrusion pretreatment process accommodates high biomass load, owing to its capacity for intense shear, leading to successful biomass pulverization and fibrillation (Guiao et al., 2022). The primary differentiation among extrusion machines is the distinction between single-screw and twin-screw extruders. Single-screw extruders consist of a singular, solid piece, whereas twin-screw extruders are constructed using smaller components referred to as screw elements, which are assembled onto the shafts. Within twin-screw extruders, three fundamental types of screw elements find application: kneading blocks, reverse screws, and conveying screws. Moreover, twin-screw extruders can be categorized as either counter-rotating or co-rotating (Duque et al., 2017). Extrusion has been demonstrated to be an effective pretreatment method for various types of lignocellulosic biomass. It is frequently implemented with the incorporation of additives, including ethylene glycol, alkali, glycerol, and deep eutectic solvents (DESs). The inclusion of additives is an interesting approach, as it is commonly reported that extrusion without such components can lead to issues such as material buildup, burning, and blockage. These problems arise due to the suboptimal flow characteristics inherent in lignocellulosic feedstocks (Ai et al., 2020). José et al. (2022) investigated the extrusion process (115°C–130°C, 14 rpm) to pretreat corn cob. The authors applied a water-to-glycerol mass ratio of 25:25 (wt.%) as an additive to improve the process efficiency, reaching a cellulose conversion to glucose of 90.4 wt.%, as well as 44.0 wt.% of hemicellulose (xylose and arabinose). Moro et al. (2017) analyzed the pretreatment of sugarcane bagasse and straw by extrusion, investigating three screw profiles on the crystallinity and glucose released from the biomass, using 20–100 rpm, 85°C–150°C, 0.53–0.75 wt.% of glycerol, and three extrusion passes, achieving glucose yields higher than 50 wt.% after 48 h of enzymatic hydrolysis.

Microwaves (MWs), electromagnetic energy waves with wavelengths and frequencies ranging from 1 m to 1 mm and 0.3 GHz to 300 GHz, are a frequently considered method for biomass pretreatment. MWs-assisted pretreatment utilizes microwave energy to heat solvents and plant tissues, enhancing the process kinetics. This technique has several advantages that include ease of handling, increased heating capacity, short processing time (usually between 20 and 60 min), minimal generation of inhibitors, and less energy requirement (typical work temperature between 60 and 90°C) (Aftab et al., 2019; Ceaser et al., 2023). The mechanism underlying MW-assisted pretreatment of lignocellulosic biomass entails the interaction between the microwave electromagnetic field and the polar molecules within the biomass. This interaction leads to rapid heating and the breakdown of the lignocellulosic structure (Bundhoo, 2018). When microwaves are transmitted, they are converted to suitable frequencies so that the material being heated can absorb the energy from the waves. Afterward, the constantly fluctuating electric fields cause continuous alignment and realignment of dipoles in polar liquids. The dynamic alterations in dipoles, coupled with ion migration induced by the electromagnetic field, generate internal friction within the material. This internal energy, subsequently dissipated as heat, results in the elevation of the overall temperature of the material (Abolore et al., 2023; Hoang et al., 2021). Azlan et al. (2022) described that the MW heating process exhibits rapid and efficient rates, achieving the desired temperature within short time intervals, where high selective efficiency is observed, particularly for polar substances; moreover, the system efficiency is significantly contingent on the material properties, and precise control is maintained over the heating process. On the other hand, the selective efficiency is diminished for apolar substances, where the operational efficiency exhibits reduced dependence on material properties and the heating system demonstrates diminished controllability in comparison to alternative methods (Hoang et al., 2021). In an interesting work, Amini et al. (2021) applied MW to pretreat woody biomass (*Eucalyptus regnans* and *Pinus radiata* sawdust), investigating the effects of the dipolar polarization mechanism of polar solvents (ethylene glycol, isopropyl alcohol, formaldehyde, and water), to enhance the process efficiency. At optimized conditions (3.7 wt.% of formaldehyde solution and 180°C for 10 min), the authors observed an increase in biomass saccharification by 56% with the production of stabilized lignin monomers. Contextually, Shangdiar et al. (2022) pretreated sugarcane bagasse *via* the MW technique, reaching 43.62% of saccharification rate when the process was performed at 100°C, 500 W, and 2.5 g of H_2SO_4 per gram of biomass.

Ultrasound (US) is another method usually considered for biomass pretreatment aiming at process intensification. In the context of biomass treatment, the primary effects are associated with the increase of convective forces, acoustic cavitation, and the production of highly reactive radicals (Flores et al., 2021). The propagation of ultrasound waves in any medium leads to the generation of acoustic cavitation, subsequently instigating the spontaneous formation of microbubbles. The collapse of these microbubbles results in the creation of shockwaves and local hotspots, releasing high temperatures and pressures, where the crystalline structure of the solid particles present in the medium is disintegrated upon exposure to the shockwaves and hotspots at these conditions (Mankar et al., 2021). Alongside the disintegration of the complex lignocellulosic structure, ultrasound additionally enhances the process mass transfer, where the selection of an adequate solvent (comprising inorganic acids, alkalis, ionic liquids (IL), or organic solvents) to the process is crucial for achieving desired outcomes. Principal factors influencing the success of ultrasound treatment encompass sonication amplitude, duration of sonication, and reaction time, where the typical US-assisted pretreatment is conducted within the range of 20–80 kHz for 20–150 min (Sidana and Yadav, 2022). This process has been shown to reduce the degree of polymerization and cellulose crystallinity in grasses and bagasses by 1.5%–29.2% (Zhao et al., 2022). US-assisted pretreatment of biomass has yielded interesting results in the disintegration of distinct biomass. Subhedar et al. (2018) used US combined NaOH solution for the delignification of groundnut shells, coconut coir, and pistachio shells, reporting 41.8%, 45.9%, and 38% of reducing lignin content, respectively, representing a lignin removal of 80%–100%. Singh et al. (2020) evaluated a US-assisted-alkali-urea pretreatment of *miscanthus* biomass for improved delignification and cellulose fiber extraction, where at optimized conditions (2.1 wt.% NaOH, 1.7 wt.% urea, and 15.5 min of sonication time), a cellulose and lignin contents of 47.8 wt.% and 27.5 wt.% (delignification of 52.5%), respectively.

Each physical pretreatment modality has distinct advantages and disadvantages. In general, mechanical pretreatment is regarded as the most important initial step in biomass conversion to microbial biofuels, preceding subsequent pretreatment processes. The advantages and disadvantages of physical pretreatment methods are summarized in Table 2.2.

2.3.2 Chemical Pretreatments

Chemical pretreatments use chemical agents to make changes to the lignocellulosic structure of biomass. The choice of reagent used will be responsible for the main effect on the structure, such as removal of lignin, partial or total hydrolysis of hemicelluloses, resulting in a solid rich in cellulose, or dissolution of cellulose, obtaining liquid rich in cellulose (Liao et al., 2020). Among the chemical pretreatment techniques considered for biomass to biofuel conversions, acid and alkaline hydrolysis, organosolv processes, ILs, and DESs stand out.

2.3.2.1 Acid Hydrolysis

Acid hydrolysis can be performed with concentrated or diluted acids. Hydrochloric, phosphoric, and sulfuric acids are compounds typically used. The acid treatment results in a high concentration of hemicellulose sugars in the hydrolyzed solution, with a solid cellulose fraction with a high capacity for enzymatic conversion. However, enzymatic digestibility is commonly impaired due to the presence of inhibitory compounds formed during this process (Gomes et al., 2020). Issues associated

TABLE 2.2
Strengths and Limitations Associated With the Main Physical Biomass Pretreatment Techniques

Technique	Strengths	Limitations
Milling	• High efficiency when combined with other pretreatment process • No release of toxic compounds • Simple operation • Facilitate material handling • Mass transfer improved due to particle size reduction	• Low efficiency • Generation of waste materials • High energy consumption • Not suitable for removing lignin • Need of secondary pretreatment
Extrusion	• Moderate temperature requirements • No inhibitors generation • Adaptability to modifications • No degradation products • Controllable environment	• Partial degradation of hemicellulose • Mostly effective for herbaceous biomass
Ultrasound	• Increasing porosity and surface area • Higher efficiency • Short residence time • Low severity of process • Environmentally friendly process	• High energy demand • High temperature • High pressure • Size reduction of biomass is required
Microwave	• Increasing porosity and surface area • Less energy consumption • Short reaction time • Nonpolluting method • Accurate process control	• Generation of waste materials • High energy demand • Generation of fermentation inhibitors • The working efficiency depends on the material properties

with corrosion and dangers in the handling of these compounds are frequently mentioned, requiring reactors that must be built with corrosion-resistant materials, as concentrated acids are extremely corrosive and harmful (Westensee et al., 2018).

2.3.2.2 Alkaline Hydrolysis

Alkaline hydrolysis can also be applied as a biomass pretreatment where a high rate of delignification can be obtained for materials with low lignin content. Usually, only a small portion of the hemicelluloses is removed during the process, obtaining solids rich in carbohydrates (Sharma et al., 2023b). One of the most considered alkaline agents is lime (CaOH), due to its low cost and easy use, with chemical compounds such as NaOH and KOH also frequently applied (Vaithyanathan et al., 2023).

2.3.2.3 Organosolv Method

In the organosolv technique, different organic solvents are used, such as alcohols, organic acids, ketones, and ethers, with the presence of acid or basic catalyst aiming to aid in the lignin separation from the plant cell structures (Hrůzová et al., 2021).

This pretreatment results in solids rich in cellulose, with a broth rich in hemicelluloses and a precipitate rich in lignin (Liao et al., 2020). The precipitated lignin typically possesses a low molecular weight, with sulfur traces lower than detected in acid pretreatments, and has a carbohydrate content usually inferior to 1 wt.% (Sharma et al., 2023b). Consequently, this lignin can be recovered and utilized independently.

2.3.2.4 Ionic Liquids and Deep Eutectic Solvents

ILs are highly effective in biomass fragmentation, delignification, and, consequently, in reducing the crystallinity of the biomass, which is extremely important in subsequent stages, especially in enzymatic hydrolysis. However, most ILs are expensive and can be environmentally toxic, where a post-processing step is required to remove residual traces of the compounds, as they can inhibit enzymatic activity (Dharmaraja et al., 2023). Conversely, DESs offer a cost-effective alternative to ILs, boasting environmental friendliness and efficient delignification rates in pretreated materials. DESs stand out for their simplified synthesis process and the versatility of hydrogen donors and receptors they can accommodate, contributing to their appeal in various applications within the field. However, as occurs with ILs, post-processing of washing and filtration with water/acetone solutions is also necessary to remove residual DES and soluble lignin (Muley et al., 2019).

Accordingly, Table 2.3 presents recent scientific research that evaluated the effect of different chemical pretreatments aimed at reducing biomass recalcitrance, as well as the usual process conditions adopted. Reported data regarding changes in the lignocellulosic structure of biomass are also presented. It is undeniably observed among published research that the pretreatment with diluted acid is the most used when the goal is the production of bioethanol, even though it is not efficient in removing lignin and removing parts of the cellulose (Kumar et al., 2020). Among the pretreatment methods that can be considered for this proposal, the use of DESs stands out, with a clear removal capacity of lignin, preserving the carbohydrates fraction (Draszewski et al., 2021). Contextually, alkaline hydrolysis and organosolv processes generally present high efficiency in the removal of lignin and hemicelluloses, obtaining solids rich in cellulose (Madadi et al., 2022; Zhang et al., 2023a).

2.3.3 Thermochemical Pretreatments

Thermochemical pretreatments are hybrid processes involving chemical and physical approaches. These systems have attracted the attention of researches mainly due to the elimination of the necessity to work with corrosion-resistant reactors in addition to presenting efficient conversion of hemicelluloses, obtaining solids rich in cellulose (Sarker et al., 2021; Scapini et al., 2021). Furthermore, thermochemical conversions are clearly more commercially attractive due to their high efficiency, better flexibility, and fast conversion rates (Malik et al., 2022). Among the main thermochemical pretreatments considered for biomass management, the processes of liquid hot water (LHW or autohydrolysis), steam explosion (SE), ammonia fiber expansion (AFEX), and ammonia recycle percolation (ARP) can be highlighted.

TABLE 2.3
Examples of Chemical Pretreatments Applied to Different Biomass

Pretreatment	Biomass	Process Conditions	Modifications in the Biomass	Reference
Acid hydrolysis	Poplar	1:10 (mass ratio, biomass:H_2SO_4 solution), 3.5 wt.% H_2SO_4, 180°C, 1 h	Removal of cellulose, hemicellulose, and lignin: 3.5%, 88.9%, and 9.2%, respectively	Tong et al. (2023)
Acid hydrolysis	Corn stover	1:10 (mass ratio, biomass:H_2SO_4 solution), 3.5 wt.% H_2SO_4, 130°C, 30 min	Removal of cellulose, hemicellulose, and lignin: 3.0%, 84.0%, and 4.0%, respectively	Jiang et al. (2023)
Alkaline hydrolysis	Bamboo wastes	1:10 (mass ratio, biomass:NaOH solution), 2 wt.% NaOH, 70°C, 3 h	Removal of hemicellulose and lignin: 80.8% and 55.3%, respectively	Zhang et al. (2023b)
Alkaline hydrolysis	Poplar	1:10 (mass ratio, biomass:NaOH solution), 5 wt.% NaOH, 70°C, 1 h	Removal of hemicellulose and lignin: 68.9% and 22.6%, respectively	Ding et al. (2019)
Organosolv	Poplar sawdust	1:10 (mass ratio, biomass:p-TsOH/ etanol), 85°C, 5 h	Hemicellulose removed: 35.7%, lignin recovery: 83.7%	Fan et al. (2021)
Organosolv	Aspen chips	1:10 (mass ratio, biomass:solution 1 wt.% H_2SO_4/ 60 wt.% pentanol), 160°C, 1 h	Cellulose enrichment: 78%, lignin recovery: 85%	Madadi et al. (2022)
ILs	Pine	1:10 (mass ratio, biomass:[Bmim] [$MeCO_2$]), 120°C, 2 h	Cellulose solubilization: 26.2%	Gschwend et al. (2019)
ILs	Switchgrass	Biomass:[Bmim] [PF_6]: 15 g/135 g, 160°C, 3 h	Cellulose solubilization: 65.0%	Williams et al. (2018)
DESs	Watermelon rind	Choline chloride:lactic acid, 1:20 (mass ratio, biomass/DES), 120°C, 3 h	Cellulose enrichment: 57.4%, lignin removed: 43.6%	Fakayode et al. (2020)
DESs	Rice husks	Choline chloride:glycerol plus 20 wt.% water, 1:10 (mass ratio, biomass/DES), 120°C, 4h	Cellulose enrichment: 60.1%, lignin removed: 43.6%	Draszewski et al. (2021)

The LHW pretreatment is typically performed at temperatures ranging between 150°C and 260°C with pressures high enough to keep the water in the liquid state, generally between 0.5 and 2.4 MPa (Chen et al., 2022). Under these conditions, the water has an acidic character (pH = 5.0), which leads to alterations in its properties, such as an increase in diffusion coefficients and drops in viscosity (Cocero, 2018). According to the increase in the process temperatures, the hydrogen bonds of the water are broken, which results in an increase in the ionization constant of the compound (Chen et al., 2022). These changes help to dissolve the acetyl groups of hemicelluloses into acetic, formic, and levulinic acids, which contribute to carbohydrate depolymerization (Ruiz et al., 2023). A usual disadvantage reported in this process is the elevated water consumption, where generally a solvent-to-solid mass ratio of 3:1 is chosen, which also results in hydrolyzed solutions with low concentrations of polysaccharides (Dharmaraja et al., 2023).

An alternative to LHW pretreatment is the process of SE, which generally requires inferior amounts of water, where usually a solvent to biomass mass ratio ranging between 1:1 and 2:1 is utilized (Hoang et al., 2023). SE technique is a widely used pretreatment involving the submission of the biomass to high temperatures (160°C–280°C) under pressures 0.7–5 MPa) – water in vapor conditions – for a few minutes (1–30 min; Zhang et al., 2023a). The pressurized biomass is rapidly depressurized, causing an explosive release of steam. This sudden release leads to the rupture of cell walls and the formation of cavities within the biomass, creating fissures and cracks in the cellulose and hemicellulose components. The ruptured cell walls expose the cellulose, hemicellulose, and lignin components to the surrounding environment, causing a reduction in cellulose crystallinity and the removal of hemicellulose and lignin layers (Beig et al., 2021). Commonly, the SE process is assisted by (acid) catalysts, where the main considered compounds are H_2SO_4 and H_2S in concentrations no greater than 3 wt.% (Yu et al., 2022).

The AFEX technique is conducted in a similar operational way as the SE. However, for this option, an NH_3 solution in concentrations between 40 and 80 wt.% is used instead of pure water, with a slightly superior process time of around 60 min (Juneja et al., 2023). The main advantage of using the AFEX process in relation to SE is the lower energy consumption involved since NH_3 reaches thermodynamic vapor states in milder conditions than water. Similarly, the ARP technique uses NH_3 in a liquid state with continuous percolation of the organic compound typically in an aqueous solution up to 20 wt.% in temperatures ranging between 90°C and 180°C for pressures of 0.1–0.3 MPa (Latif et al., 2018). During both processes (AFEX and ARP), NH_3 penetrates into the biomass and reacts exothermically with the water present in the substrate to form ammonium hydroxide, which in turn acts as a catalyst in the hemicelluloses conversion and lignin solubilization (Zhao et al., 2020).

Table 2.4 presents interesting information that evaluates the use of the thermochemical processes described to pretreat different biomass and the usually selected process conditions. From these data, it is possible to note that pretreatments involving the use of NH_3 usually present a lower fraction of cellulose in the residual solid when compared to LHW and SE. This occurs because the organic compound not only hydrolyzes hemicellulose but also some parts of the cellulose present in the

TABLE 2.4
Examples of Thermochemical Pretreatment Employed Into Different Biomass for Cellulose Content Enrichment and Lignin Removal

Pretreatment	Biomass	Process Condition	Cellulose Enrichment (%)	Lignin Removed (%)	Reference
LHW	Agave bagasse	190°C, 10 min	83.5	26.9	Pino et al. (2019)
LHW	Sorghum bagasse	190°C, 10 min	5.0	10.2	Cheng et al. (2019)
SE	Corn stover	190°C, 1.2 MPa, 3 min	25.7	38.1	Zhang et al. (2023a)
SE	Cardoon stalk	165°C, 15 min	52.6	not reported	Cavalaglio et al. (2021)
AFEX	Bamboo moss	150°C, 60 wt.% NH_3, 10 min	1.1	13.5	Shao et al. (2013)
AFEX	Corn stover	130°C, 80 wt.% NH_3, 10 min	9.4	25.0	Zhao et al. (2014)
ARP	Poplar	180°C, 15 wt.% NH_3, 3 mL min^{-1} of feed, 90 min	not reported	87.0	Bouxin et al. (2014)
ARP	Blend of corn cob/ stover	170°C, 10 wt.% NH_3, 1 mL min^{-1} of feed, 60 min	−7.4	79.6	Zhao et al. (2020)

biomass (Zhao et al., 2020). The AFEX is also not efficient for biomass with high lignin content (between 25 and 30 wt.%), where for this type of biomass, the most indicated process is the ARP, which is recognized for its ability to dissolve high loads of lignin among all the common pretreatments considered to biomass management (Malik et al., 2022). The LHW and SE methods presented a high cellulose enrichment, in addition to not requiring the use of chemicals that increase the cost of the process, making these pretreatments more suitable when the objective is solids rich in cellulose (Aparicio et al., 2021).

2.3.4 Biological Pretreatments

Biomass biological pretreatments involve the use of living organisms to transform cellulose into fermentable sugars (Roukas and Kotzekidou, 2022; Soni et al., 2023). It has broad applications, requires low energy, generates less pollution, and provides good efficiencies (Aditiya et al., 2016; Sharma et al., 2019). It can be performed through direct biomass fermentation or by applying hydrolytic enzymes previously produced by microorganisms (Bhushan et al., 2023). These enzymes alter or degrade the lignocellulosic structure of the biomass, including ligninolytic enzymes that depolymerize lignin and hydrolytic enzymes that break cellulose chains into smaller saccharides (Sharma et al., 2019).

Ramamoorthy and Sahadevan (2020) employed the microorganism *Paecilomyces inflatus* ATCC 32919 for the conversion of surgical cotton and cardboard residues, achieving a degradation of 25 wt.% of hemicellulose and 76 wt.% of lignin. Accordingly, saccharification reached an efficient rate of 57.09% using 35 FPU/mL of in-house cellulases incubated in a stirred tank reactor operated at 50°C for 48 h. Sugarcane bagasse, previously treated with NaOH, underwent enzymatic hydrolysis with chimeric enzyme and cellobiohydrolase, yielding 230 mg/g of total reducing sugars, with 137 mg/g attributed to glucose (Nath et al., 2021). Basak et al. (2020) successfully converted 54% of cellulose and 63% of hemicellulose from a mixture of fruit and vegetable residues using cellulases and xylanases produced by the fungus *Aspergillus fumigatus*. Additionally, enzymes from *Penicillium echinulatum* S1M29 decreased the lignin content by 31 wt.%, leading to a 10 wt.% increase in glucose yield after hydrolysis (Schneider et al., 2020). Matei et al. (2020) submitted sugarcane bagasse and rice husk to saccharification using laccase (9,467.8 U/L) obtained from *Trametes villosa*, resulting in a remarkable tenfold increase in the concentration of reducing sugars compared to the control. The most noteworthy outcomes reported by the authors were achieved with sugarcane bagasse, reaching 291 mg/g of sugars, while the rice husk achieved 193 mg/g.

Currently, the main scientific focus of research considering pretreatment of biomass by biological means is to evaluate the production of microorganisms from varied sources, mostly waste. In the fermentative process, microorganisms can produce hydrolytic enzymes, such as cellulases and hemicellulases (Singhania, 2011). Cellulases are a multi-enzymatic complex, and within this group, three enzymes stand out for specific functions: endoglucanase, exoglucanase, and β-glucosidases (Chandel et al., 2012; Singhania, 2011; Thapa et al., 2020). Contextually, a series of enzymes is required to hydrolyze lignin and xylan, including xylanase, β-xylosidase, α-L-arabinofuranosidase, α-D-glucuronidases, 1-5-α-L-arabinase, acetyl xylan esterase (Thapa et al., 2020).

Endoglucanase (or endocellulase or CMCase) is the enzyme that initiates the hydrolysis process, playing its role in the middle of the cellulose chain. Its action promotes the generation of smaller fragments composed of a few glucose units, named oligosaccharides. By randomly hydrolyzing the internal regions of the amorphous structure of the cellulose fiber, endocellulase releases oligosaccharides with varying degrees of polymerization. Most cellulases are endocellulases, characterized by having open active sites, allowing the binding and cleavage of cellulose molecules at accessible points along the chain (Singhania, 2011; Wilson, 2011). Exoglucanase (or exocellulase) takes on the responsibility of breaking the cellulose chain at its end, resulting in the release of cellobiose, a smaller compound formed by two glucose units. Exocellulases remain attached to the cellulose chain, cleaving cellobiose residues from the end of the chain until complete dissociation (Chandel et al., 2012; Wilson, 2011). β-Glucosidase also acts on the ends of cellulose chains, specifically on non-reducing ends, where a glucose unit does not have a free aldehyde. These microorganisms play a crucial role in the final stage of cellulose hydrolysis, converting cellobiose units into glucose (Thapa et al., 2020; Wilson, 2011).

Several microorganisms have been investigated for the production of hydrolytic enzymes for biomass pretreatment due to their potential for cellulose degradation.

A wide variety of bacteria, both aerobic and anaerobic, white, and soft rot fungi, and anaerobic fungi, have stood out as cellulase producers. In aerobic bacteria, *actinomycetes*, and filamentous fungi, cellulases are predominantly secreted as free molecules (Singhania, 2011).

Fungi belonging to the genus *Trichoderma*, notably recognized for their ability to thrive on various cellulose substrates, have been identified as remarkable cellulase producers (Ramamoorthy and Sahadevan, 2019). Different studies in the literature highlight various fungi as prodigies in the production of hydrolytic enzymes, including *Trichoderma reesei* QM9414 e *Trichoderma reesei* MCG77 (Latifian et al., 2007), *Trichoderma harzanium* (Ramamoorthy and Sahadevan, 2019), *Aspergillus terreus* MS105 (Menendez et al., 2015), *Aspergillus fumigatus* NITDGPKA3 (Sohail et al., 2016), *Trichoderma viride* NCIM 1051 (Adsul et al., 2004), *Penicillium funiculosum* (Burugu et al., 2021).

Bacteria have been widely explored for cellulase production due to their high growth rate, ability to express multi-enzymatic complexes, stability under extreme temperature and pH conditions, as well as capacity to withstand environmental stress (Bhat, 2000; Deka et al., 2013). Among the bacteria investigated in the literature, *Bacillus subtilis* AS3 (Deka et al., 2011), *Bacillus subtilis* (Deka et al., 2013), *Pseudomonas coleopterorum* sp. (Menendez et al., 2015), *Bacillus aestuarii* UE25 (Ejaz et al., 2019), *Brevibacillus borstelensis* (Ejaz et al., 2020), *Aneurinibacillus thermoaerophilus* (Ejaz et al., 2020), *Penicillium decumbens* JUA10 (Fujian et al., 2002), *Penicillium occitanis* (Belghith et al., 2001), *Penicillium janthinellum* NCIM 1171 (Adsul et al., 2004), *Rhodothermus marinus* (Hreggvidsson et al., 1996), *Bacillus pseudomycoides* (Pramanik et al., 2021), *Bacillus velezensis* (Li et al., 2020), and *Paenibacillus* sp. (Islam, 2019) stand out. Bacteria of the genus *Bacillus* have stood out, being dominant due to their significant ability to produce and secrete considerable amounts of enzymes and extracellular substances (Ariffin et al., 2006; Arsov et al., 2021; Bhagat and Kokitkar, 2021).

In the study reported by Narendra et al. (2023), *Chryseobacterium junjuense* Bp17, isolated from forest soil, was utilized for cellulase production. The authors characterized β-1,4-endoglucanase, β-1,4 exoglucanase, and β-glucosidase, initially obtaining enzymatic activities of 148.55 U/mL, 37.21 U/mL, and 42.21 U/mL, respectively. When using others biomass sources, activities for β-1,4-endoglucanase, β-1,4 exoglucanase, and β-glucosidase were obtained as follows: 223.72 U/mL, 112.88 U/mL, and 133.43 U/mL (rice bran); 196.86 U/mL, 138.90 U/mL, and 168.84 U/mL (wood residues); 389.69 U/mL, 205.92 U/mL, and 187.99 U/mL (*Pongamia* leaves); and 212.88 U/mL, 151.65 U/mL, and 176.63 U/mL (coconut husk), respectively. The hydrolysis zone was tested on CMC agar medium, demonstrating the strain's effectiveness for use for the biomass pretreatment, particularly with *Pongamia* leaves, which exhibited the largest hydrolysis activity zone, reaching a diameter of 34.66 mm.

Regarding the main challenges regarding the production of enzymes for biomass pretreatment, issues involving the efficiency in the cellulase preparation by microorganisms face tasks such as cultivation, humidity, particle size, temperature, time, and agitation. Biological processes poses obstacles such as space requirements, prolonged time, and continuous monitoring of microbial growth (Bhushan et al., 2023;

Pinheiro et al., 2023). Contextually, optimization of the culture medium and the process becomes crucial to mitigate costs and reduce the time associated with the production process. Optimization of *Bacillus subtilis* AS3 using lignocellulosic biomass from *Achnatherum hymenoides* achieved optimal levels of initial pH (7.2), temperature (39°C), and agitation (121 rpm), resulting in a 33% increase in CMCase activity (0.57 U/mL), where the maximum production in a 2 L bioreactor under optimized conditions was 0.75 U/mL (Deka et al., 2013). The optimization of cellulase production was investigated with the fungal strain *Penicillium funiculosum*, assessing optimal concentrations of initial substrate (10 wt.% rice husk), pH 6.0, and supplementary carbon source (5 wt.%), resulting in a cellulase concentration of 134 UI/mL (Burugu et al., 2021). Using carboxymethylcellulose as a carbon source for the optimization of cellulases produced by *Bacillus* sp., it reached 7.82 UI/mL in a basal medium at pH 7.0, with 1% (w/v) peptone, 2 vol.% inoculum, incubation at 37°C for 96 h at 120 rpm (Lugani et al., 2015). Maximum cellulase activities with sugarcane bagasse reached 3.5 U/mg with *Bacillus pseudomycoides*, incubated at 40°C for 72 h at pH 7.0 (Pramanik et al., 2021). A promising strategy to overcome these challenges and increase the production of hydrolytic enzymes is co-cultivation. This approach includes both the association of distinct bacterial cultures and the simultaneous cultivation of fungi and bacteria. The emerging synergy from this co-cultivation has proven promising to maximize cellulase production (Sharma et al., 2019).

2.4 TECHNOLOGICAL DEVELOPMENT OF THE PROCESSES: A PATENT DATABASE RESEARCH

Exploring a targeted technology within a patent database offers intriguing insights into the evolution of a specific technological domain. Contextually, this section presents findings from a comprehensive search conducted using the *Questel Orbit Intelligence®* system, renowned for assessing patents filed across over 100 countries. The search, guided by a strategic selection of keywords, specifically aimed to unveil inventive technologies related to the pretreatment of biomass for biofuel production. This methodological approach not only provides a snapshot of current innovations but also sheds light on the global landscape of advancements in microbial biofuel-related technologies, offering a valuable resource for researchers and industry professionals alike. Accordingly, in the system background, entering the search fields with the string "*((biomass OR microbial biomass OR microbial raw material OR microbial feedstock OR lignocellulosic biomass OR lignocellulosic material OR lignocellulosic feedstock) AND (pretreatment OR pretreatment OR treatment) AND (physical OR chemical OR thermochemical OR thermochemical OR physicochemical OR physico-chemical OR biological) AND (*fuel))*," it verified the existence of 35,767 registers since January 2013 (data of the research: November 8, 2023). Some important insights from these data are discussed next.

From this load of patents, it was found that the legal status of the deposits is 51.83% granted, 22.28% pending, 17.34% lapsed, and 8.13% revoked. Another interesting point is that 14% of patented inventions are owned by the main ten players of the cluster, including companies such as Pionner Hi Bred International Inc., MS Technologies, Monsanto Technologies, China Petroleum & Chemical, IFP

Energies Nouvelles, Novozymes A/K, and the academic sector, represented by the South China University of Technology. This data demonstrates a mild concentration of technologies owned by the main investors, indicating that this technological area is evaluated by a wide variety of players since it is common to check concentrations of patents superior to 30% in a restricted group of investors/inventors.

Figure 2.2 presents a general overview of the location of deposits considering the search performed in terms of technological density (Figure 2.2a) and main inventor nations (Figure 2.2b). From these data, it is possible to ascertain the massive participation of China and the United States in the number of patents registered in the last decade, responsible for more than 62% of the number of deposits (Figure 2.2b).

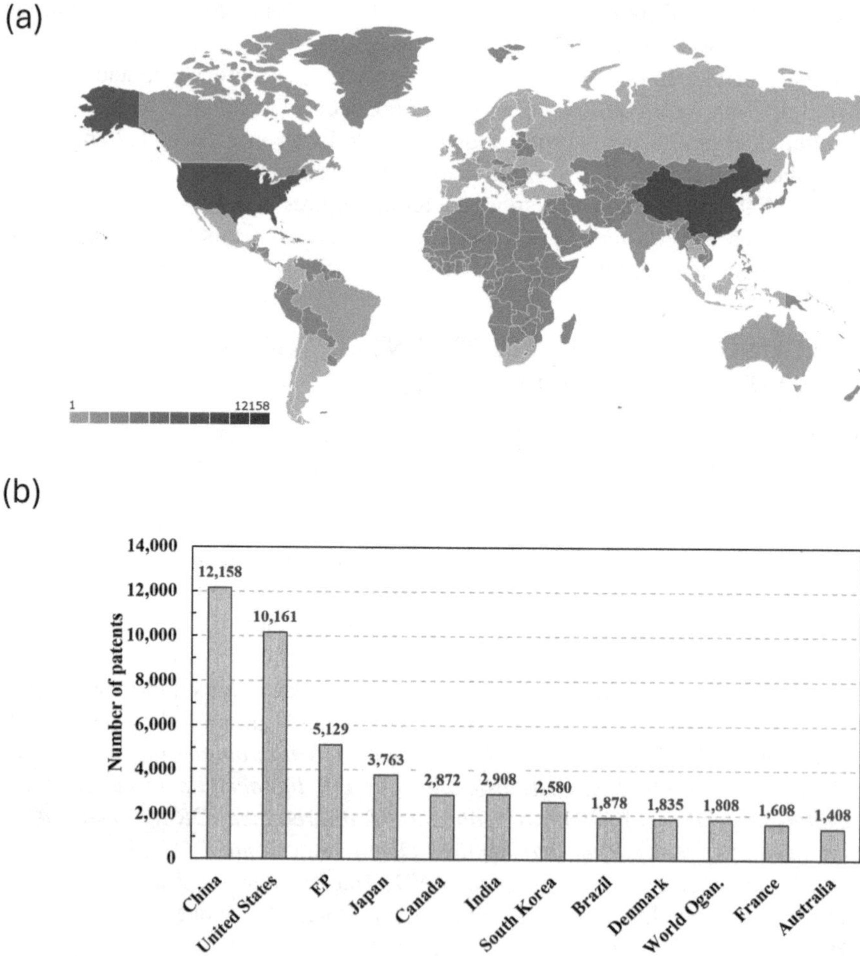

FIGURE 2.2 (a) Technology density of protected inventions at a worldwide panorama and (b) main inventor countries of technologies regarding biomass pretreatment for microbial fuels synthesis.

(Source: ***Questel Orbit Intelligence®***.)

Even though these two countries are globally recognized for their development, not only scientific but also technological, it is undeniable that this concentration of inventions in just two nation-states is significantly elevated. On the other hand, this information indicates that the development of processes aimed at pretreating biomass for energy purposes is an attractive topic not only for environmental but also for economic issues. Another question that must be described, even though it addresses already expected questions, is the concentration of patents in nations that present certain technological development (Figure 2.2.a) and/or that are major agro-industrial powers since such processes of pretreatment are usually designed to deal with biomass, which comes, in the vast majority, from wastes of this sector.

In contrast to scientific papers, usually patent documents provide fewer details concerning the methodologies and results of their inventions. In any case, the research accomplished returned some interesting inventions, such as those deposited by Hackl (2021), Ettore and Ghisolfi (2015), Dadi and Paripati (2013), Sousa and Balan (2020), Duque et al. (2012), and Smit et al. (2013). Aymard et al. (2021) – *IFP Energies Nouvelles* – patented a multi-step process involving the possibility of pretreating biomass by impregnating the material with an acid, alkaline or oxidizing aqueous solution, and posteriorly used a process of SE for subsequent extraction of 2G-sugary liquors. In another deposit, Foody and Dechman (2016) – *Iogen Co.* – pertain to a process for the pretreatment (and subsequent hydrolysis) of lignocellulosic biomass involving processing material with a reduced amount of sulfur dioxide and/or sulfurous acid. According to the document, a designed pretreatment reactor can be charged with sulfur dioxide, and a significant portion remains in the vessel's headspace when the pretreated material is discharged. The retained sulfur dioxide is then available to facilitate the pretreatment of additional lignocellulosic biomass in subsequent batches. Consequently, there is a reduced need to add more sulfur dioxide, leading to improved process economics. Foody et al. (2020) – *Iogen Co.* – also used a similar approach to patented a low-temperature sulfur dioxide pretreatment.

Larsen et al. (2013) – *Inbicon A/S* – registered an invention that pertains to a technique for processing lignocellulosic biomass through a hydrothermal pretreatment. Specifically, it focuses on a method where the pretreatment is regulated through feedback monitoring of a defined "endpoint," assessed in the output stream of the pretreated biomass. The "endpoint" referred to by the inventors is associated with the "lignin number" of the hydrolyzed stream. This measure integrates evaluations of both lignin content in insoluble solids and the unhydrolyzed solid, including products from hemicellulose degradation during the pretreatment, such as condensation and polymerization products. According to the inventors, the "lignin number" can offer a convenient online follow-up of the process, serving as a reliable indicator of digestibility at low severity for hydrothermally pretreated biomass. Furthermore, by employing feedback control of pretreatment, and utilizing the composite "endpoint" measures, reactor conditions can be promptly adapted to accommodate the variability of incoming feedstocks at the production scale, helping to minimize the occurrence of pretreatment inefficiencies.

In an interesting deposit, Laosiripojana et al. (2015) – *PTT Global Chemical Co.* – focused on devising a lignocellulosic biomass fractionation process. The technique involves treating the biomass with a liquid mixture in a closed system, maintaining the temperature below 200°C and the pressure between 0.5 and 2 MPa. The treatment liquid consists of a water-miscible component, containing formic acid, ethanol, water, and a water-immiscible component containing alkyl acetate. The lignocellulosic biomass eligible for this process includes rice straw, bagasse, corncobs, corn stem, palm waste, grass, pineapple shell, bamboo, or a combination thereof. Tests performed by the inventors returned good results to the pretreatment of sugarcane bagasse: the employment of formic acid yielded between 8.2 and 10.7 g of lignin in the solid fraction per 100 g of initial bagasse (around 26.6%–34.9% of lignin in the solid fraction compared to the original bagasse material). Notably, the amount of lignin obtained using formic acid was lower than that obtained with sulfuric acid or acetic acid, indicating formic acid's heightened specificity for lignin removal, resulting in higher cellulose purity. Additionally, elevating the process temperature increased the enzymatic digestibility of biomass. Specifically, when using formic acid, it achieved between 62.0 and 77.0% of enzymatic digestibility. At 160°C, formic acid yielded the highest glucose amount (approximately 68.0 wt.%.

Bhave et al. (2018) patented a process for lignocellulosic biomass pretreatment using inorganic nanoporous membranes. Basically, the technique involves passing water through a pretreatment reactor with a bed of particulate lignocellulosic biomass, generating a pressurized, high-temperature (170°C–230°C) hydrolyzed stream. An inorganic nanoporous membrane element is then utilized to separate solubilized compounds from the hydrolyzate stream. This separation excludes solutes with dimensions in the order of 1.0 nm, encompassing molecular weights less than 200 Da. Typical differential pressure in the nanoporous membrane is between 200 and 550 psi. The permeate, enriched with solvent, is subsequently returned to the pretreatment reactor. The retentate is then subjected to fractionation and fermentation processes to generate ethanol, among other potential products.

Figure 2.3 presents data concerning the evolutionary trend of investments considering the technological platform "*processes of biomass pretreatment for the production of biofuels*" in recent years. Examining Figure 2.3(a) reveals a remarkable surge of over 1,500% in registered inventions between 2011 and 2013, following a comparatively sluggish period in the 2000s. From 2013 to 2021, the data in the same graph demonstrates notable consistency in deposits, averaging over 3,000 annually. The peak occurred in 2019 with 3,573 patents, highlighting sustained interest and activity in the considered technological platform. This trend signifies a robust and enduring focus on innovation within this domain, underscoring its enduring significance and attractiveness for inventors and investors alike. However, a clear downward trend is verified in the number of inventions registered after 2021, where 2,428 patents were granted in 2022, a level similar to that observed 10 years earlier. Furthermore, until the date of this research (November 2023), only 1,065 deposits were found regarding the technology, a value that requires attention, although there are still 22% of registrations pending concession, as previously described. Considering that the number of registered inventions is an important measure of the progress in the development of a given technology, the data presented in Figure 2.3(a) provide a strong indication that biomass pretreatment processes for biofuel production purposes are already reaching its technological maturity. In other words, these data indicate that possibly the

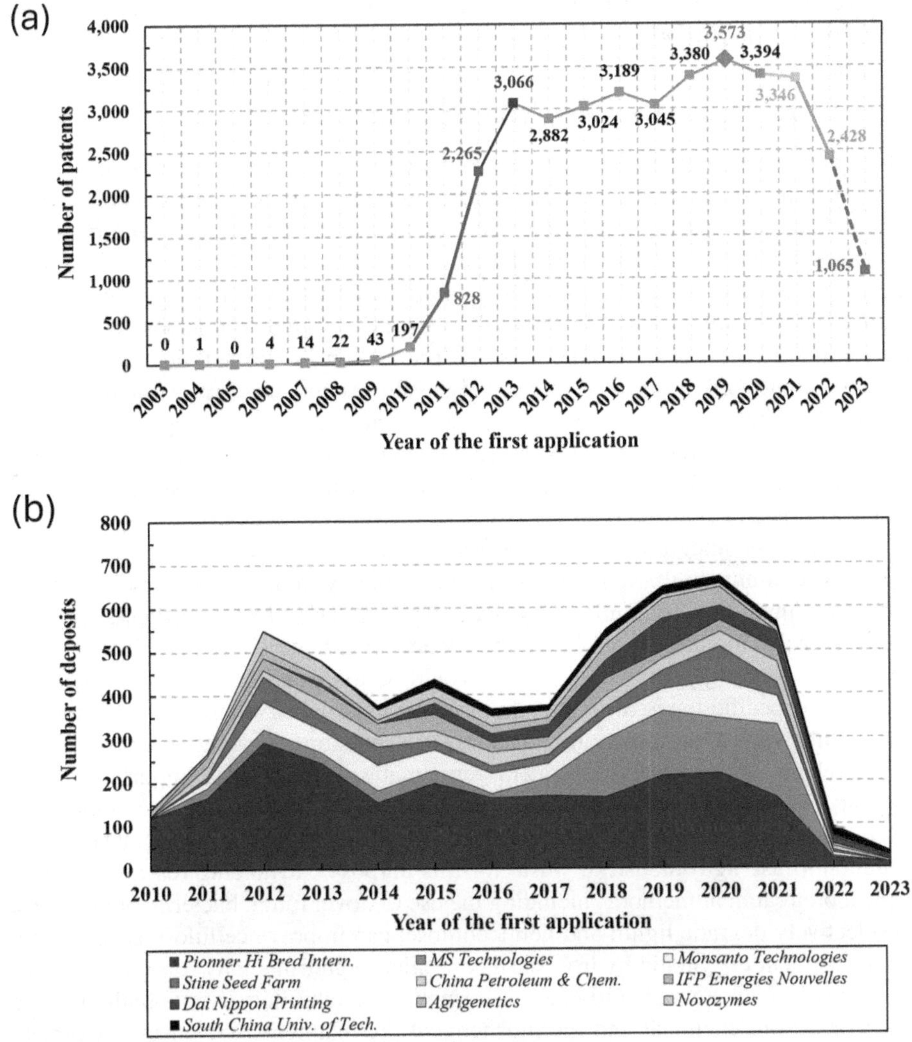

FIGURE 2.3 (a) Trend of technology investment over the last 20 years regarding biomass pretreatment to microbial fuel production and (b) number of deposits related to the considered technology platform – since 2010 – of the main inventor companies.

(Source: *Questel Orbit Intelligence*®.)

techno-scientific gaps involving this domain have already been filled, consolidating the technological advances in this area.

This assumption can be corroborated by the data presented in Figure 2.3(b), which shows the evolution of inventions registered by the ten main players involving this technological platform since 2010. The data presented is in consonance with that observed in Figure 2.3(a), showing peaks in the number of deposits between 2019 and 2021. Contextually, it is also possible to observe a severe drop in deposits since then; with these organizations being important investors in this technological sector, the decline in registered patents supports the hypothesis of reaching technological maturity for this platform.

2.5 CONCLUSIONS AND FUTURE PERSPECTIVES

The pretreatment of lignocellulosic biomass for the production of microbial biofuels is an essential step in the search for interesting yields in the processes considered. Among the techniques usually evaluated to reduce the recalcitrance of these materials and, consequently, allow fermentative microorganisms access to their cellulose layer, it is clear that each process has its applications that lead to advantages and limitations typically reported in scientific works. Contextually, it is possible to state that, although the use of biomass for the production of biofuels is still a widely explored topic, researches that exclusively evaluates just one pretreatment process will tend to become scarcer, given that scientific novelties concerning the theme are consolidated. Such findings can be corroborated by the data presented regarding the technological maturity of the processes discussed, which indicated a decline in the number of patents registered in recent years.

Nevertheless, integrative approaches aiming to develop hybrid pretreatment strategies that combine multiple methods to synergistically improve biomass conversion efficiency, involving combining physical, chemical, and biological approaches for comprehensive biomass deconstruction, still appear to have a margin for exploration. Another interesting point to investigate would be to design pretreatment processes adapted to specific types of lignocellulosic raw materials. As biomass have unique and highly varied compositions, customized pretreatment methods can optimize sugar release to improve subsequent fermentation processes. Similarly, another subject that still appears to be under development is the production of biological microorganisms capable of pretreating lignocellulosic biomass, mainly because it is common to use agro-industrial waste for this purpose. Advancing research on biological pretreatment methods, including the use of novel fungi, bacteria, or enzymes, to selectively degrade lignin and hemicellulose, can improve cellulose accessibility for subsequent enzymatic hydrolysis. These microorganisms may develop effective strategies to manage and mitigate the formation of inhibitory compounds that are released during chemical and thermochemical pretreatments, which can adversely affect downstream fermentation processes.

Addressing these future perspectives has the potential to contribute to the development of more efficient, sustainable, and economically viable pretreatment methods, thereby accelerating the realization of lignocellulosic biomass as a key feedstock for the production of microbial biofuels.

ACKNOWLEDGMENTS

J. H. C. Wancura, C. P. Draszewski, and D. Schein are grateful for the scholarships of the Human Resources Program of the Brazilian Agency of Petroleum, Natural Gas and Biofuels – PRH/ANP through the Human Resources Training Program for Petroleum and Biofuels Processing (PRH 52.1). F. D. Vezaro and M. Albarello thank the Coordination for the Improvement of Higher Education Personnel (CAPES, 001) for the scholarships.

REFERENCES

Abaide, E.R., Mortari, S.R., Ugalde, G., Valério, A., Amorim, S.M., Di Luccio, M., Moreira, R. de F.P.M., Kuhn, R.C., Priamo, W.L., Tres, M. V., Zabot, G.L., Mazutti, M.A., 2019. Subcritical water hydrolysis of rice straw in a semi-continuous mode. *J Clean Prod* 209, 386–397. https://doi.org/10.1016/j.jclepro.2018.10.259

Abolore, R.S., Jaiswal, S., Jaiswal, A.K., 2023. Green and sustainable pretreatment methods for cellulose extraction from lignocellulosic biomass and its applications: A review. *Carbohydr Polym Technol Appl* 100396. https://doi.org/10.1016/j.carpta.2023.100396

Aditiya, H.B., Mahlia, T.M.I., Chong, W.T., Nur, H., Sebayang, A.H., 2016. Second generation bioethanol production: A critical review. *Renew Sustain Energy Rev* 66, 631–653. https://doi.org/10.1016/j.rser.2016.07.015

Adsul, M.G., Ghule, J.E., Singh, R., Shaikh, H., Bastawde, K.B., Gokhale, D.V., Varma, A.J., 2004. Polysaccharides from bagasse: Applications in cellulase and xylanase production. *Carbohydr Polym* 57, 67–72. https://doi.org/10.1016/j.carbpol.2004.04.001

Afraz, M., Muhammad, F., Nisar, J., Shah, A., Munir, S., Ali, G., Ahmad, A., 2024. Production of value added products from biomass waste by pyrolysis: An updated review. *Waste Manag Bull* 1, 30–40. https://doi.org/10.1016/j.wmb.2023.08.004

Aftab, M.N., Iqbal, I., Riaz, F., Karadag, A., Tabatabaei, M., 2019. Different pretreatment methods of lignocellulosic biomass for use in biofuel production, in: *Biomass for Bioenergy - Recent Trends and Future Challenges*. pp. 1–24. https://doi.org/10.5772/intechopen.84995

Ai, B., Li, W., Woomer, J., Li, M., Pu, Y., Sheng, Z., Zheng, L., Adedeji, A., Ragauskas, A.J., Shi, J., 2020. Natural deep eutectic solvent mediated extrusion for continuous high-solid pretreatment of lignocellulosic biomass. *Green Chem* 22, 6372–6383. https://doi.org/10.1039/d0gc01560a

Amini, N., Dwivedi, S., Ahmad, W., Haritos, V.S., Tanksale, A., 2021. Polar solvents enhance the efficiency of microwave pre-treatment of woody biomass. *Biomass Bioenergy* 155, 106281. https://doi.org/10.1016/j.biombioe.2021.106281

Aparicio, E., Rodríguez-Jasso, R.M., Pinales-Márquez, C.D., Loredo-Treviño, A., Robledo-Olivo, A., Aguilar, C.N., Kostas, E.T., Ruiz, H.A., 2021. High-pressure technology for Sargassum spp biomass pretreatment and fractionation in the third generation of bioethanol production. *Bioresour Technol* 329, 124935. https://doi.org/10.1016/j.biortech.2021.124935

Ariffin, H., Abdullah, N., Umi Kalsom, M.S., Shirai, Y., Hassan, M.A., 2006. Production and characterization of cellulase by *Bacillus pumilus* EB3. *Int J Eng Technol* 3, 47–53.

Arsov, A., Petrov, K., Petrova, P., 2021. Enhanced activity by genetic complementarity: Heterologous secretion of clostridial cellulases by bacillus licheniformis and bacillus velezensis. *Molecules* 26. https://doi.org/10.3390/molecules26185625

Awogbemi, O., Von Kallon, D.V., Onuh, E.I., Aigbodion, V.S., 2021. An overview of the classification, production and utilization of biofuels for internal combustion engine applications. *Energies (Basel)* 14, 1–42. https://doi.org/10.3390/en14185687

Aymard, C., Aragones, M., Bouras, G.M., 2021. Process for treating lignocellulosic biomass. US20210381017A1.

Azlan, N.S.M., Yap, C.L., Gan, S., Rahman, M.B.A., 2022. Effectiveness of various solvents in the microwave-assisted extraction of cellulose from oil palm mesocarp fiber. *Mater Today Proc* 59, 583–590. https://doi.org/10.1016/j.matpr.2021.12.086

Basak, B., Saha, S., Chatterjee, P.K., Ganguly, A., Woong Chang, S., Jeon, B.H., 2020. Pretreatment of polysaccharidic wastes with cellulolytic *Aspergillus fumigatus* for enhanced production of biohythane in a dual-stage process. *Bioresour Technol* 299, 122592. https://doi.org/10.1016/j.biortech.2019.122592

Beig, B., Riaz, M., Raza Naqvi, S., Hassan, M., Zheng, Z., Karimi, K., Pugazhendhi, A., Atabani, A.E., Thuy Lan Chi, N., 2021. Current challenges and innovative developments in pretreatment of lignocellulosic residues for biofuel production: A review. *Fuel* 287, 119670. https://doi.org/10.1016/j.fuel.2020.119670

Belghith, H., Ellouz-Chaabouni, S., Gargouri, A., 2001. Biostoning of denims by *Penicillium occitanis* (Pol6) cellulases. *J Biotechnol* 89, 257–262. https://doi.org/10.1016/S0168-1656(01)00309-1

Bhagat, S.A., Kokitkar, S.S., 2021. Isolation and identification of bacteria with cellulose-degrading potential from soil and optimization of cellulase production. *J Appl Biol Biotechnol* 9, 154–161. https://doi.org/10.7324/JABB.2021.96020

Bhat, M.K., 2000. Cellulases and related enzymes in biotechnology. *Biotechnol Adv* 18, 355–383. https://doi.org/10.1016/S0734-9750(00)00041-0

Bhatia, S.K., Jagtap, S.S., Bedekar, A.A., Bhatia, R.K., Patel, A.K., Pant, D., Rajesh Banu, J., Rao, C. V., Kim, Y.-G., Yang, Y.-H., 2020. Recent developments in pretreatment technologies on lignocellulosic biomass: Effect of key parameters, technological improvements, and challenges. *Bioresour Technol* 300, 122724. https://doi.org/10.1016/j.biortech.2019.122724

Bhave, R.R., Lee, L., Shao, X., 2018. Flow-through pretreatment of lignocellulosic biomass with inorganic nanoporous membranes. US9932648B2.

Bhushan, S., Jayakrishnan, U., Shree, B., Bhatt, P., Eshkabilov, S., Simsek, H., 2023. Biological pretreatment for algal biomass feedstock for biofuel production. *J Environ Chem Eng* 11, 109870. https://doi.org/10.1016/j.jece.2023.109870

Boly, M., Sanou, A., 2022. Biofuels and food security: Evidence from Indonesia and Mexico. *Energy Policy* 163. https://doi.org/10.1016/j.enpol.2022.112834

Bouxin, F.P., David Jackson, S., Jarvis, M.C., 2014. Isolation of high quality lignin as a by-product from ammonia percolation pretreatment of poplar wood. *Bioresour Technol* 162, 236–242. https://doi.org/10.1016/j.biortech.2014.03.082

Bundhoo, Z.M.A., 2018. Microwave-assisted conversion of biomass and waste materials to biofuels. *Renew Sustain Energy Rev* 82, 1149–1177. https://doi.org/10.1016/j.rser.2017.09.066

Burugu, A., Suman, D.K., Chanda, C., 2021. Production and purification of extracellular fungal cellulases using agricultural waste. *J Appl Biol Biotechnol* 9, 153–156. https://doi.org/10.7324/JABB.2021.9214

Cai, J., Wang, S., Kuang, C., Tang, X., 2017. Insight into the kinetic analysis of catalytic combustion for biomass after alkaline metals loaded pretreatment. *Fuel* 203, 501–513. https://doi.org/10.1016/j.fuel.2017.04.137

Cavalaglio, G., Gelosia, M., Giannoni, T., Barros Lovate Temporim, R., Nicolini, A., Cotana, F., Bertini, A., 2021. Acid-catalyzed steam explosion for high enzymatic saccharification and low inhibitor release from lignocellulosic cardoon stalks. *Biochem Eng J* 174, 108121. https://doi.org/10.1016/j.bej.2021.108121

Ceaser, R., Rosa, S., Montané, D., Constantí, M., Medina, F., 2023. Optimization of softwood pretreatment by microwave-assisted deep eutectic solvents at high solids loading. *Bioresour Technol* 369, 128470. https://doi.org/10.1016/j.biortech.2022.128470

Cedeno, F.R.P., de Siqueira, B.B., Chavez, E.G.S., Roldán, I.U.M., Ropelato, L.M., Galán, J.P.M., Masarin, F., 2022. Recovery of cellulose and lignin from Eucalyptus by-product and assessment of cellulose enzymatic hydrolysis. *Renew Energy* 193, 807–820. https://doi.org/10.1016/j.renene.2022.05.027

Chandel, A.K., Chandrasekhar, G., Silva, M.B., Silvério Da Silva, S., 2012. The realm of cellulases in biorefinery development. *Crit Rev Biotechnol* 32, 187–202. https://doi.org/10.3109/07388551.2011.595385

Chen, W.-H., Nižetić, S., Sirohi, R., Huang, Z., Luque, R.M., Papadopoulos, A., Sakthivel, R., Phuong Nguyen, X., Tuan Hoang, A., 2022. Liquid hot water as sustainable biomass pretreatment technique for bioenergy production: A review. *Bioresour Technol* 344, 126207. https://doi.org/10.1016/j.biortech.2021.126207

Cheng, M.-H., Dien, B.S., Lee, D.K., Singh, V., 2019. Sugar production from bioenergy sorghum by using pilot scale continuous hydrothermal pretreatment combined with disk refining. *Bioresour Technol* 289, 121663. https://doi.org/10.1016/j.biortech.2019.121663

Cocero, M.J., 2018. Supercritical water processes: Future prospects. *J Supercrit Fluids* 134, 124–132. https://doi.org/10.1016/j.supflu.2017.11.018

Dadi, A.P., Paripati, P., 2013. Alkaline treatment of lignocellulosic biomass. BR112015023022.

Dahunsi, S.O., 2019. Mechanical pretreatment of lignocelluloses for enhanced biogas production: Methane yield prediction from biomass structural components. *Bioresour Technol* 280, 18–26. https://doi.org/10.1016/j.biortech.2019.02.006

Deka, D., Bhargavi, P., Sharma, A., Goyal, D., Jawed, M., Goyal, A., 2011. Enhancement of cellulase activity from a new strain of bacillus subtilis by medium optimization and analysis with various cellulosic substrates. *Enzyme Res* 2011. https://doi.org/10.4061/2011/151656

Deka, D., Das, S.P., Sahoo, N., Das, D., Jawed, M., Goyal, D., Goyal, A., 2013. Enhanced cellulase production from *Bacillus subtilis* by optimizing physical parameters for bioethanol production. *ISRN Biotechnol* 2013, 1–11. https://doi.org/10.5402/2013/965310

Dharmaraja, J., Shobana, S., Arvindnarayan, S., Francis, R.R., Jeyakumar, R.B., Saratale, R.G., Ashokkumar, V., Bhatia, S.K., Kumar, V., Kumar, G., 2023. Lignocellulosic biomass conversion via greener pretreatment methods towards biorefinery applications. *Bioresour Technol* 369, 128328. https://doi.org/10.1016/j.biortech.2022.128328

Ding, D., Li, P., Zhang, X., Ramaswamy, S., Xu, F., 2019. Synergy of hemicelluloses removal and bovine serum albumin blocking of lignin for enhanced enzymatic hydrolysis. *Bioresour Technol* 273, 231–236. https://doi.org/10.1016/j.biortech.2018.11.024

Draszewski, C.P., Bragato, C.A., Lachos-Perez, D., Celante, D., Frizzo, C.P., Castilhos, F., Tres, M. V., Zabot, G.L., Abaide, E.R., Mayer, F.D., 2021. Subcritical water hydrolysis of rice husks pretreated with deep eutectic solvent for enhance fermentable sugars production. *J Supercrit Fluids* 178, 105355. https://doi.org/10.1016/j.supflu.2021.105355

Duque, A., Garcia, E.B., Lima, E.V., Lombard, E., Vilarem, G., Usitalo, J.M., Brault, J., Mata-Segreda, J., Rigal, L., Luna, M.H., Ballesteros, M.S.S.-A.M., Melendez, O.H., Manzanares, P., Guillouet, S., Vandenbossche, V., Cameleyre, X., 2012. Method of enzymatic treatment of a solid lignocellulosic material. US20150299751A1.

Duque, A., Manzanares, P., Ballesteros, M., 2017. Extrusion as a pretreatment for lignocellulosic biomass: Fundamentals and applications. *Renew Energy* 114, 1427–1441. https://doi.org/10.1016/j.renene.2017.06.050

Ejaz, U., Muhammad, S., Ali, F.I., Hashmi, I.A., Sohail, M., 2019. Methyltrioctylammonium chloride mediated removal of lignin from sugarcane bagasse for themostable cellulase production. *Int J Biol Macromol* 140, 1064–1072. https://doi.org/10.1016/j.ijbiomac.2019.08.206

Ejaz, U., Muhammad, S., Hashmi, I.A., Ali, F.I., Sohail, M., 2020. Utilization of methyltrioctylammonium chloride as new ionic liquid in pretreatment of sugarcane bagasse for production of cellulase by novel thermophilic bacteria. *J Biotechnol* 317, 34–38. https://doi.org/10.1016/j.jbiotec.2020.04.013

Ettore, B., Ghisolfi, G., 2015. Pretreatment process of a ligno-cellulosic feedstock. BR112017015990.

Fakayode, O.A., Aboagarib, E.A.A., Yan, D., Li, M., Wahia, H., Mustapha, A.T., Zhou, C., Ma, H., 2020. Novel two-pot approach ultrasonication and deep eutectic solvent pretreatments for watermelon rind delignification: Parametric screening and optimization via response surface methodology. *Energy* 203, 117872. https://doi.org/10.1016/j.energy.2020.117872

Fan, D., Xie, X., Li, C., Liu, X., Zhong, J., Ouyang, X., Liu, Q., Qiu, X., 2021. Extraction of noncondensed lignin from poplar sawdusts with *p* -Toluenesulfonic acid and ethanol. *J Agric Food Chem* 69, 10838–10847. https://doi.org/10.1021/acs.jafc.1c03774

Flores, E.M.M., Cravotto, G., Bizzi, C.A., Santos, D., Iop, G.D., 2021. Ultrasound-assisted biomass valorization to industrial interesting products: state-of-the-art, perspectives and challenges. *Ultrason Sonochem* 72, 105455. https://doi.org/10.1016/j.ultsonch.2020.105455

Foody, B., Dechman, J., 2016. Pretreatment of lignocellulosic biomass with sulfur dioxide and/or sulfurous acid. BR112018016366.

Foody, B., Tolan, J.S., MacDonald, D.G., Martens, K., Leduc, N., Mackenzie, D.A., 2020. Low temperature sulfur dioxide pretreatment. US20200354891A1.

Fujian, X., Hongzhang, C., Zuohu, L., 2002. Effect of periodically dynamic changes of air on cellulase production in solid-state fermentation. *Enzyme Microb Technol* 30, 45–48. https://doi.org/10.1016/S0141-0229(01)00454-9

Gallon, R., Draszewski, C.P., Santos, J.A.A., Wagner, R., Brondani, M., Zabot, G.L., Tres, M. V., Hoffmann, R., Castilhos, F., Abaide, E.R., Mayer, F.D., 2023. Obtaining oil, fermentable sugars, and platform chemicals from Butia odorata seed using supercritical fluid extraction and subcritical water hydrolysis. *J Supercrit Fluids* 203, 106062. https://doi.org/10.1016/j.supflu.2023.106062

Gęca, M., Wiśniewska, M., Nowicki, P., 2023. Modified method of lignocellulose content determination and its use for the analysis of selected herbs – Precursors of biochars and activated carbons. *Measurement* 212, 112672. https://doi.org/10.1016/j.measurement.2023.112672

Gomes, M.G., Gurgel, L.V.A., Baffi, M.A., Pasquini, D., 2020. Pretreatment of sugarcane bagasse using citric acid and its use in enzymatic hydrolysis. *Renew Energy* 157, 332–341. https://doi.org/10.1016/j.renene.2020.05.002

Gonçalves, T. dos S., Oro, C.E.D., Wancura, J.H.C., Dos Santos, M.S.N., Junges, A., Dallago, R.M., Tres, M. V., 2023. Challenges for energy guidelines in crop-based liquid biofuels development in Brazil. *Next Sustainability* 2, 100002. https://doi.org/10.1016/j.nxsust.2023.100002

Gschwend, F.J. V., Chambon, C.L., Biedka, M., Brandt-Talbot, A., Fennell, P.S., Hallett, J.P., 2019. Quantitative glucose release from softwood after pretreatment with low-cost ionic liquids. *Green Chem* 21, 692–703. https://doi.org/10.1039/C8GC02155D

Guiao, K.S., Gupta, A., Tzoganakis, C., Mekonnen, T.H., 2022. Reactive extrusion as a sustainable alternative for the processing and valorization of biomass components. *J Clean Prod* 355, 131840. https://doi.org/10.1016/j.jclepro.2022.131840

Gundupalli, M.P., Cheenkachorn, K., Chuetor, S., Kirdponpattara, S., Gundupalli, S.P., Show, P.-L., Sriariyanun, M., 2023. Assessment of pure, mixed and diluted deep eutectic solvents on Napier grass (Cenchrus purpureus): Compositional and characterization studies of cellulose, hemicellulose and lignin. *Carbohydr Polym* 306, 120599. https://doi.org/10.1016/j.carbpol.2023.120599

Hackl, A., 2021. Method and treatment facility for physical and thermochemical treatment of biomass. BR112023002357.

Hoang, A.T., Nguyen, X.P., Duong, X.Q., Ağbulut, Ü., Len, C., Nguyen, P.Q.P., Kchaou, M., Chen, W.-H., 2023. Steam explosion as sustainable biomass pretreatment technique for biofuel production: Characteristics and challenges. *Bioresour Technol* 385, 129398. https://doi.org/10.1016/j.biortech.2023.129398

Hoang, A.T., Nižetić, S., Ong, H.C., Mofijur, M., Ahmed, S.F., Ashok, B., Bui, V.T.V., Chau, M.Q., 2021. Insight into the recent advances of microwave pretreatment technologies for the conversion of lignocellulosic biomass into sustainable biofuel. *Chemosphere* 281. https://doi.org/10.1016/j.chemosphere.2021.130878

Hreggvidsson, G., Kaiste, E., Holst, O., Eggertsson, G., Palsdottir, A., Kristjansson, J., 1996. An extremely thermostable cellulase from the thermophilic eubacterium *Rhodothermus marinus*. *Appl Environ Microbiol* 62, 3047–3049. https://doi.org/10.1128/aem.62.8.3047-3049.1996

Hrůzová, K., Matsakas, L., Rova, U., Christakopoulos, P., 2021. Organosolv fractionation of spruce bark using ethanol–water mixtures: Towards a novel bio-refinery concept. *Bioresour Technol* 341, 125855. https://doi.org/10.1016/j.biortech.2021.125855

Islam, F., 2019. Isolation and characterization of cellulase-producing bacteria from sugar industry waste. *Am J BioSci* 7, 16. https://doi.org/10.11648/j.ajbio.20190701.13

Jia, Z., Wang, S., Yu, H., Li, W., Ye, J., Hu, Y., Liu, C., Ye, Z., Sun, Y., Xu, X., 2023. Novel supramolecular deep eutectic solvent pretreatment for obtaining fluorescent lignin and promoting biomass pyrolytic saccharification. *Bioresour Technol* 388, 129780. https://doi.org/10.1016/j.biortech.2023.129780

Jiang, X., Zhai, R., Li, H., Li, C., Deng, Q., Jin, M., 2023. Understanding acid hydrolysis of corn stover during densification pretreatment for quantitative predictions of enzymatic hydrolysis efficiency using modified pretreatment severity factor. *Bioresour Technol* 386, 129487. https://doi.org/10.1016/j.biortech.2023.129487

José, Á.H.M., Moura, E.A.B., Rodrigues Jr, D., Kleingesinds, E.K., Rodrigues, R.C.L.B., 2022. A residue-free and effective corncob extrusion pretreatment for the enhancement of high solids loading enzymatic hydrolysis to produce sugars. *Ind Crops Prod* 188, 115655. https://doi.org/10.1016/j.indcrop.2022.115655

Juneja, A., Kumar, D., Mittal, A., 2023. 4 – Pretreatment technologies for lignocellulosic biomass refineries, in: Moodley, P., Ray, R.C., Kana, E.B.G. (Eds.), *Advances in Lignocellulosic Biofuel Production Systems*. Elsevier, pp. 81–106. https://doi.org/10.1016/C2020-0-04155-1

Kamarludin, S.N.C., Jainal, M.S., Azizan, A., Safaai, N.S.M., Daud, A.R.M., 2014. Mechanical pretreatment of lignocellulosic biomass for biofuel production, in: *Applied Mechanics and Materials*. pp. 838–841. https://doi.org/10.4028/www.scientific.net/AMM.625.838

Kartal, F., Özveren, U., 2021. An improved machine learning approach to estimate hemicellulose, cellulose, and lignin in biomass. *Carbohydr Polym Technol Appl* 2, 100148. https://doi.org/10.1016/j.carpta.2021.100148

Kumar, M., Mishra, P.K., Upadhyay, S.N., 2020. Thermal degradation of rice husk: Effect of pre-treatment on kinetic and thermodynamic parameters. *Fuel* 268, 117164. https://doi.org/10.1016/j.fuel.2020.117164

Laosiripojana, N., Suriyachai, N., Champreda, V., 2015. Process for fractionation of lignocellulosic biomass. BR112018009790.

Larsen, J., Mogensen, K.K., Jφrgensen, P., 2013. Method of processing lignocellulosic biomass using feedback control of hydrothermal pretreatment. BR112014020181.

Latif, A., Harun, S., Sajab, M.S., Markom, M., Jahim, J.M., 2018. Ammonia-based pretreatment for ligno-cellulosic biomass conversion – An overview. *J Eng Sci Technol* 13, 1595–1620.

Latifian, M., Hamidi-Esfahani, Z., Barzegar, M., 2007. Evaluation of culture conditions for cellulase production by two Trichoderma reesei mutants under solid-state fermentation conditions. *Bioresour Technol* 98, 3634–3637. https://doi.org/10.1016/j.biortech.2006.11.019

Li, F., Xie, Y., Gao, X., Shan, M., Sun, C., Niu, Y.D., Shan, A., 2020. Screening of cellulose degradation bacteria from Min pigs and optimization of its cellulase production. *Electron J Biotechnol* 48, 29–35. https://doi.org/10.1016/j.ejbt.2020.09.001

Li, P., Pei, Z., Liu, D., Shi, F., Wang, S., Li, W., Sun, Y., Liu, J., Gao, Y., Yu, Q., 2021. Application of anaerobic digestion model no. 1 for modeling anaerobic digestion of vegetable crop residues: Fractionation of crystalline cellulose. *J Clean Prod* 285, 124865. https://doi.org/10.1016/j.jclepro.2020.124865

Li, W., Xu, B., Xu, M., Wu, Y., Liu, Z., Li, B., Zhong, C., Jia, S., 2024a. Mild and recyclable ethylenediamine pretreatment for bacterial cellulose fermentation and lignin valorization. *Ind Crops Prod* 207, 117681. https://doi.org/10.1016/j.indcrop.2023.117681

Li, Y., Chen, M., Shi, Q.-S., Xie, X., Guo, Y., 2024b. Biomass fractionation techniques impact on the structure and antioxidant properties of isolated lignins. *Sep Purif Technol* 330, 125499. https://doi.org/10.1016/j.seppur.2023.125499

Liao, J.J., Latif, N.H.A., Trache, D., Brosse, N., Hussin, M.H., 2020. Current advancement on the isolation, characterization and application of lignin. *Int J Biol Macromol* 162, 985–1024. https://doi.org/10.1016/j.ijbiomac.2020.06.168

Liu, Z., Khurshid, K., Saldaña, M.D.A., 2023. Biorefinery of barley straw using pressurized fluids: Biocompounds and biopolymers production. *J Supercrit Fluids* 203, 106078. https://doi.org/10.1016/j.supflu.2023.106078

Lugani, Y., Singla, R., Singh Sooch, B., 2015. Optimization of cellulase production from newly isolated bacillus Sp. Y3. *J Bioprocess Biotech* 5, 3–8. https://doi.org/10.4172/2155-9821.1000264

Mabrouk, A. Ben, Putaux, J.-L., Boufi, S., 2023. Valorization of olive leaf waste as a new source of fractions containing cellulose nanomaterials. *Ind Crops Prod* 202, 116996. https://doi.org/10.1016/j.indcrop.2023.116996

Madadi, M., Zahoor, S. G., Karimi, K., Zhu, D., Elsayed, M., Sun, F., Abomohra, A., 2022. One-step lignocellulose fractionation using acid/pentanol pretreatment for enhanced fermentable sugar and reactive lignin production with efficient pentanol retrievability. *Bioresour Technol* 359, 127503. https://doi.org/10.1016/j.biortech.2022.127503

Malik, K., Sharma, P., Yang, Y., Zhang, P., Zhang, L., Xing, X., Yue, J., Song, Z., Nan, L., Yujun, S., El-Dalatony, M.M., Salama, E.-S., Li, X., 2022. Lignocellulosic biomass for bioethanol: Insight into the advanced pretreatment and fermentation approaches. *Ind Crops Prod* 188, 115569. https://doi.org/10.1016/j.indcrop.2022.115569

Mankar, A.R., Pandey, A., Modak, A., Pant, K.K., 2021. Pretreatment of lignocellulosic biomass: A review on recent advances. *Bioresour Technol* 334, 125235. https://doi.org/10.1016/j.biortech.2021.125235

Martinez-Burgos, W.J., Porto de Souza Vandenberghe, L., Karp, S.G., Murawski de Mello, A.F., Thomaz Soccol, V., Soccol, C.R., 2023. Microbial lipid production from soybean hulls using Lipomyces starkeyi LPB53 in a circular economy. *Bioresour Technol* 372, 128650. https://doi.org/10.1016/j.biortech.2023.128650

Matei, J.C., Soares, M., Bonato, A.C.H., de Freitas, M.P.A., Helm, C. V., Maroldi, W. V., Magalhães, W.L.E., Haminiuk, C.W.I., Maciel, G.M., 2020. Enzymatic delignification of sugar cane bagasse and rice husks and its effect in saccharification. *Renew Energy* 157, 987–997. https://doi.org/10.1016/j.renene.2020.05.028

Mayer-Laigle, C., Blanc, N., Rajaonarivony, R.K., Rouau, X., 2018. Comminution of dry lignocellulosic biomass, a review: Part I. from fundamental mechanisms to milling behaviour. *Bioengineering* 5. https://doi.org/10.3390/bioengineering5020041

Menendez, E., Garcia-Fraile, P., Rivas, R., 2015. Biotechnological applications of bacterial cellulases. *AIMS Bioeng* 2, 163–182. https://doi.org/10.3934/bioeng.2015.3.163

Mishra, R.K., Mohanty, K., 2019. Thermal and catalytic pyrolysis of pine sawdust (Pinus ponderosa) and Gulmohar seed (Delonix regia) towards production of fuel and chemicals. *Mater Sci Energy Technol* 2, 139–149. https://doi.org/10.1016/j.mset.2018.12.004

Moreira, B.P., Draszewski, C.P., Celante, D., Brondani, L., Lachos-Perez, D., Mayer, F.D., Abaide, E.R., Castilhos, F., 2022. Defatted rice bran pretreated with deep eutectic solvents and sequential use as feedstock for subcritical water hydrolysis. *Bioresour Technol* 351, 127063. https://doi.org/10.1016/j.biortech.2022.127063

Moro, M.K., Teixeira, R.S.S., da Silva, A.S., Fujimoto, M.D., Melo, P.A., Secchi, A.R., Bon, E.P. da S., 2017. Continuous pretreatment of sugarcane biomass using a twin-screw extruder. *Ind Crops Prod* 97, 509–517. https://doi.org/10.1016/j.indcrop.2016.12.051

Muley, P.D., Mobley, J.K., Tong, X., Novak, B., Stevens, J., Moldovan, D., Shi, J., Boldor, D., 2019. Rapid microwave-assisted biomass delignification and lignin depolymerization in deep eutectic solvents. *Energy Convers Manag* 196, 1080–1088. https://doi.org/10.1016/j.enconman.2019.06.070

Nagarajasetty, V.K.S., Goud, G., Mavinkere Rangappa, S., Siengchin, S., 2022. Limonia Acidissima (wood-apple) shell: Micro and nanoparticles preparation and chemical treatment. *Mater Today Proc* 52, 2543–2547. https://doi.org/10.1016/j.matpr.2021.12.049

Narendra, K., Chandra, M., Amrutha, H., Rakshith, D., Harini, B.P., Satish, S., 2023. Biomass conversion through optimization of cellulase from Chryseobacterium junjuense Bp17 and their utility in bioethanol production. *Energy* 283, 129187. https://doi.org/10.1016/j.energy.2023.129187

Nath, P., Maibam, P.D., Singh, S., Rajulapati, V., Goyal, A., 2021. Sequential pretreatment of sugarcane bagasse by alkali and organosolv for improved delignification and cellulose saccharification by chimera and cellobiohydrolase for bioethanol production. *3 Biotech* 11, 1–16. https://doi.org/10.1007/s13205-020-02600-y

Van Nguyen, D., Duong, C.T.T., Vu, C.N.M., Nguyen, H.M., Pham, T.T., Tran-Thuy, T.-M., Nguyen, L.Q., 2023. Data on chemical composition of coffee husks and lignin microparticles as their extracted product. *Data Brief* 51, 109781. https://doi.org/10.1016/j.dib.2023.109781

Okot, D.K., Bilsborrow, P.E., Phan, A.N., 2022. Thermo-chemical behaviour of maize cob and bean straw briquettes. *Energy Convers Manag: X* 16, 100313. https://doi.org/10.1016/j.ecmx.2022.100313

Özsin, G., Pütün, A.E., 2017a. Insights into pyrolysis and co-pyrolysis of biomass and polystyrene: Thermochemical behaviors, kinetics and evolved gas analysis. *Energy Convers Manag* 149, 675–685. https://doi.org/10.1016/j.enconman.2017.07.059

Özsin, G., Pütün, A.E., 2017b. Kinetics and evolved gas analysis for pyrolysis of food processing wastes using TGA/MS/FT-IR. *Waste Manag* 64, 315–326. https://doi.org/10.1016/j.wasman.2017.03.020

Pendse, D.S., Deshmukh, M., Pande, A., 2023. Different pre-treatments and kinetic models for bioethanol production from lignocellulosic biomass: A review. *Heliyon* 9, e16604. https://doi.org/10.1016/j.heliyon.2023.e16604

Pinheiro, U.V., Wancura, J.H.C., Brondani, M., da Silva, C.M., Mainardi, M.A., Gai, R.M., Jahn, S.L., 2023. Production of gibberellic acid by solid-state fermentation using wastes from rice processing and brewing industry. *Appl Biochem Biotechnol.* https://doi.org/10.1007/s12010-023-04637-0

Pino, M.S., Rodríguez-Jasso, R.M., Michelin, M., Ruiz, H.A., 2019. Enhancement and modeling of enzymatic hydrolysis on cellulose from agave bagasse hydrothermally pretreated in a horizontal bioreactor. *Carbohydr Polym* 211, 349–359. https://doi.org/10.1016/j.carbpol.2019.01.111

Pramanik, S.K., Mahmud, S., Paul, G.K., Jabin, T., Naher, K., Uddin, M.S., Zaman, S., Saleh, M.A., 2021. Fermentation optimization of cellulase production from sugarcane bagasse by *Bacillus pseudomycoides* and molecular modeling study of cellulase. *Curr Res Microb Sci* 2, 100013. https://doi.org/10.1016/j.crmicr.2020.100013

Puhl, B.A., Draszewski, C.P., Vezaro, F.D., Ten Caten, L.R., Wancura, J.H.C., de Castilhos, F., Mayer, F.D., Abaide, E.R., 2024. Semi-continuous hydrothermal processing of pine sawdust for integrated production of fuels precursors and platform chemicals. *Sci Total Environ* 912, 169168. https://doi.org/10.1016/j.scitotenv.2023.169168

Ramamoorthy, N.K., Sahadevan, R., 2019. Assessment of fed-batch strategies for enhanced cellulase production from a waste lignocellulosic mixture. *Biochem Eng J* 152. https://doi.org/10.1016/j.bej.2019.107387

Ramamoorthy, N.K., Sahadevan, R., 2020. Production of bioethanol by an innovative biological pre-treatment of a novel mixture of surgical waste cotton and waste card board. *Energy Sources, Part A: Recovery Util. Environ. Eff* 42, 942–953. https://doi.org/10.1080/15567036.2019.1602208

Rencoret, J., Marques, G., Rosado, M.J., Benito, J., Barro, F., Gutiérrez, A., del Río, J.C., 2023. Variations in the composition and structure of the lignins of oat (Avena sativa L.) straws according to variety and planting season. *Int J Biol Macromol* 242, 124811. https://doi.org/10.1016/j.ijbiomac.2023.124811

Roukas, T., Kotzekidou, P., 2022. From food industry wastes to second generation bioethanol: A review. *Reviews in Environmental Science and Biotechnology*. Springer Netherlands. https://doi.org/10.1007/s11157-021-09606-9

Rubio-Valle, J.F., Martín-Alfonso, J.E., Eugenio, M.E., Ibarra, D., Oliva, J.M., Manzanares, P., Valencia, C., 2024. Bioethanol lignin-rich residue from olive stones for electrospun nanostructures development and castor oil structuring. *Int J Biol Macromol* 255, 128042. https://doi.org/10.1016/j.ijbiomac.2023.128042

Ruiz, H.A., Sganzerla, W.G., Larnaudie, V., Veersma, R.J., van Erven, G., Shiva, Ríos-González, L.J., Rodríguez-Jasso, R.M., Rosero-Chasoy, G., Ferrari, M.D., Kabel, M.A., Forster-Carneiro, T., Lareo, C., 2023. Advances in process design, techno-economic assessment and environmental aspects for hydrothermal pretreatment in the fractionation of biomass under biorefinery concept. *Bioresour Technol* 369, 128469. https://doi.org/10.1016/j.biortech.2022.128469

Sarker, T.R., Pattnaik, F., Nanda, S., Dalai, A.K., Meda, V., Naik, S., 2021. Hydrothermal pretreatment technologies for lignocellulosic biomass: A review of steam explosion and subcritical water hydrolysis. *Chemosphere* 284, 131372. https://doi.org/10.1016/j.chemosphere.2021.131372

Scapini, T., dos Santos, M.S.N., Bonatto, C., Wancura, J.H.C., Mulinari, J., Camargo, A.F., Klanovicz, N., Zabot, G.L., Tres, M. V., Fongaro, G., Treichel, H., 2021. Hydrothermal pretreatment of lignocellulosic biomass for hemicellulose recovery. *Bioresour Technol* 342, 126033. https://doi.org/10.1016/j.biortech.2021.126033

Schneider, W.D.H., Fontana, R.C., Baudel, H.M., de Siqueira, F.G., Rencoret, J., Gutiérrez, A., de Eugenio, L.I., Prieto, A., Martínez, M.J., Martínez, Á.T., Dillon, A.J.P., Camassola, M., 2020. Lignin degradation and detoxification of eucalyptus wastes by on-site manufacturing fungal enzymes to enhance second-generation ethanol yield. *Appl Energy* 262, 114493. https://doi.org/10.1016/j.apenergy.2020.114493

Scholl, A.L., Menegol, D., Pitarelo, A.P., Fontana, R.C., Filho, A.Z., Ramos, L.P., Dillon, A.J.P., Camassola, M., 2015. Elephant grass pretreated by steam explosion for inducing secretion of cellulases and xylanases by Penicillium echinulatum S1M29 solid-state cultivation. *Ind Crops Prod* 77, 97–107. https://doi.org/10.1016/j.indcrop.2015.08.051

Sejati, P.S., Obounou Akong, F., Fradet, F., Gérardin, P., 2024. Understanding the thermoplasticization mechanism of wood via esterification with fatty acids: A comparative study of the reactivity of cellulose, hemicelluloses and lignin. *Carbohydr Polym* 324, 121542. https://doi.org/10.1016/j.carbpol.2023.121542

Shangdiar, S., Lin, Y.-C., Ponnusamy, V.K., Wu, T.-Y., 2022. Pretreatment of lignocellulosic biomass from sugar bagasse under microwave assisted dilute acid hydrolysis for biobutanol production. *Bioresour Technol* 361, 127724. https://doi.org/10.1016/j.biortech.2022.127724

Shao, Q., Cheng, C., Ong, R.G., Zhu, L., Zhao, C., 2013. Hydrogen peroxide presoaking of bamboo prior to AFEX pretreatment and impact on enzymatic conversion to fermentable sugars. *Bioresour Technol* 142, 26–31. https://doi.org/10.1016/j.biortech.2013.05.011

Sharma, H.K., Xu, C., Qin, W., 2019. Biological pretreatment of lignocellulosic biomass for biofuels and bioproducts: An overview. *Waste Biomass Valoriz* 10, 235–251. https://doi.org/10.1007/s12649-017-0059-y

Sharma, S., Tsai, M.L., Sharma, V., Sun, P.P., Nargotra, P., Bajaj, B.K., Chen, C.W., Dong, C. Di, 2023a. Environment friendly pretreatment approaches for the bioconversion of lignocellulosic biomass into biofuels and value-added products. *Environments*. https://doi.org/10.3390/environments10010006

Sharma, V., Tsai, M.-L., Nargotra, P., Chen, C.-W., Sun, P.-P., Singhania, R.R., Patel, A.K., Dong, C.-D., 2023b. Journey of lignin from a roadblock to bridge for lignocellulose biorefineries: A comprehensive review. *Sci Total Environ* 861, 160560. https://doi.org/10.1016/j.scitotenv.2022.160560

Sidana, A., Yadav, S.K., 2022. Recent developments in lignocellulosic biomass pretreatment with a focus on eco-friendly, non-conventional methods. *J Clean Prod* 335, 130286. https://doi.org/10.1016/j.jclepro.2021.130286

Singh, S.S., Lim, L.-T., Manickavasagan, A., 2020. Ultrasound-assisted alkali-urea pre-treatment of Miscanthus × giganteus for enhanced extraction of cellulose fiber. *Carbohydr Polym* 247, 116758. https://doi.org/10.1016/j.carbpol.2020.116758

Singhania, R.R., 2011. Production of celluloytic enzymes for the hydrolysis of lignocellulosic biomass, 1st ed, *Biofuels: Alternative feedstocks and conversion processes*. Elsevier Inc. https://doi.org/10.1016/B978-0-12-385099-7.00008-5

Singhania, R.R., Patel, A.K., Singh, A., Haldar, D., Soam, S., Chen, C.-W., Tsai, M.-L., Dong, C.-D., 2022. Consolidated bioprocessing of lignocellulosic biomass: Technological advances and challenges. *Bioresour Technol* 354, 127153. https://doi.org/10.1016/j.biortech.2022.127153

Smit, A.T., Huijgen, W.J.J., Grisel, R.J.H., 2013. Improved process for the organosolv treatment of lignocellulosic biomass. US20160152736A1.

Sohail, M., Ahmad, A., Khan, S.A., 2016. Production of cellulase from Aspergillus terreus MS105 on crude and commercially purified substrates. *3 Biotech* 6, 103. https://doi.org/10.1007/s13205-016-0420-z

Soni, S.K., Sharma, A., Soni, R., 2023. Microbial enzyme systems in the production of second generation bioethanol. *Sustainability (Switzerland)* 15, 1–26. https://doi.org/10.3390/su15043590

Sousa, L. da C., Balan, V., 2020. Biomass pretreatment method. US10730958B2.

Subhedar, P.B., Ray, P., Gogate, P.R., 2018. Intensification of delignification and subsequent hydrolysis for the fermentable sugar production from lignocellulosic biomass using ultrasonic irradiation. *Ultrason Sonochem* 40, 140–150. https://doi.org/10.1016/j.ultsonch.2017.01.030

Sun, C., Tan, H., Zhang, Y., 2023. Simulating the pyrolysis interactions among hemicellulose, cellulose and lignin in wood waste under real conditions to find the proper way to prepare bio-oil. *Renew Energy* 205, 851–863. https://doi.org/10.1016/j.renene.2023.02.015

Teles, G.H., dos Santos, E.C., da Silva, G.B., Gabriel Lopes Da Silva, M., Maria da Silva, J., de Moraes Rocha, G.J., de Barros Pita, W., Ribeiro, E., 2023. Full utilization of the yellow passion fruit peel: Chemical characterization and valorization to reduce biomass waste. *Ind Crops Prod* 206, 117593. https://doi.org/10.1016/j.indcrop.2023.117593

Thapa, S., Mishra, J., Arora, N., Mishra, P., Li, H., O'Hair, J., Bhatti, S., Zhou, S., 2020. Microbial cellulolytic enzymes: diversity and biotechnology with reference to lignocellulosic biomass degradation. *Rev Environ Sci Biotechnol* 19, 621–648. https://doi.org/10.1007/s11157-020-09536-y

Tong, W., Fang, H., Song, K., Xie, X., Wang, J., Jin, Y., Wu, S., Hu, J., Chu, Q., 2023. Modified acid pretreatment to alter physicochemical properties of biomass for full cellulose/hemicellulose utilization. *Carbohydr Polym* 299, 120182. https://doi.org/10.1016/j.carbpol.2022.120182

Vaithyanathan, V.K., Goyette, B., Rajagopal, R., 2023. A critical review of the transformation of biomass into commodity chemicals: Prominence of pretreatments. *Environ Chall* 11, 100700. https://doi.org/10.1016/j.envc.2023.100700

Veluchamy, C., Kalamdhad, A.S., Gilroyed, B.H., 2019. Advanced pretreatment strategies for bioenergy production from biomass and biowaste, in: *Handbook of Environmental Materials Management*. Springer International Publishing, Cham, pp. 1507–1524. https://doi.org/10.1007/978-3-319-73645-7_45

Wancura, J.H.C., Brondani, M., dos Santos, M.S.N., Oro, C.E.D., Wancura, G.C., Tres, M. V., Oliveira, J.V., 2023a. Demystifying the enzymatic biodiesel: How lipases are contributing to its technological advances. *Renew Energy* 216, 119085. https://doi.org/10.1016/j.renene.2023.119085

Wancura, J.H.C., Brondani, M., Vezaro, F.D., Martins-Vieira, J.C., Moreira, B.P., dos Santos, M.S.N., Abaide, E.R., de Castilhos, F., Mayer, F.D., 2023b. Motivations to produce biofuels from rice bran: An overview involving a recent panorama. *Ind Crops Prod* 203, 117170. https://doi.org/10.1016/j.indcrop.2023.117170

Wang, H., Han, X., Zeng, Y., Xu, C.C., 2023. Development of a global kinetic model based on chemical compositions of lignocellulosic biomass for predicting product yields from hydrothermal liquefaction. *Renew Energy* 215, 118956. https://doi.org/10.1016/j.renene.2023.118956

Wei, H.-L., Bu, J., Zhou, S.-S., Deng, M.-C., Zhu, M.-J., 2021. A facile ionic liquid and p-toluenesulfonic acid pretreatment of herb residues: Enzymatic hydrolysis and lignin valorization. *Chem Eng J* 419, 129616. https://doi.org/10.1016/j.cej.2021.129616

Westensee, D.K., Rumbold, K., Harding, K.G., Sheridan, C.M., van Dyk, L.D., Simate, G.S., Postma, F., 2018. The availability of second generation feedstocks for the treatment of acid mine drainage and to improve South Africa's bio-based economy. *Sci Total Environ* 637–638, 132–136. https://doi.org/10.1016/j.scitotenv.2018.04.410

Williams, C.L., Li, C., Hu, H., Allen, J.C., Thomas, B.J., 2018. Three way comparison of hydrophilic ionic liquid, hydrophobic ionic liquid, and dilute acid for the pretreatment of herbaceous and woody biomass. *Front Energy Res* 6. https://doi.org/10.3389/fenrg.2018.00067

Wilson, D.B., 2011. Microbial diversity of cellulose hydrolysis. *Curr Opin Microbiol* 14, 259–263. https://doi.org/10.1016/j.mib.2011.04.004

Yadav, A., Sharma, V., Tsai, M.-L., Chen, C.-W., Sun, P.-P., Nargotra, P., Wang, J.-X., Dong, C.-D., 2023. Development of lignocellulosic biorefineries for the sustainable production of biofuels: Towards circular bioeconomy. *Bioresour Technol* 381, 129145. https://doi.org/10.1016/j.biortech.2023.129145

Yousaf, B., Liu, G., Abbas, Q., Wang, R., Ubaid Ali, M., Ullah, H., Liu, R., Zhou, C., 2017. Systematic investigation on combustion characteristics and emission-reduction mechanism of potentially toxic elements in biomass- and biochar-coal co-combustion systems. *Appl Energy* 208, 142–157. https://doi.org/10.1016/j.apenergy.2017.10.059

Yu, Y., Wu, J., Ren, X., Lau, A., Rezaei, H., Takada, M., Bi, X., Sokhansanj, S., 2022. Steam explosion of lignocellulosic biomass for multiple advanced bioenergy processes: A review. *Renew Sustain Energy Rev* 154, 111871. https://doi.org/10.1016/j.rser.2021.111871

Zabed, H.M., Akter, S., Dar, M.A., Tuly, J.A., Kumar Aswathi, M., Yun, J., Li, J., Qi, X., 2023. Enhanced fermentable sugar production in lignocellulosic biorefinery by exploring a novel corn stover and configuring high-solid pretreatment conditions. *Bioresour Technol* 386, 129498. https://doi.org/10.1016/j.biortech.2023.129498

Zhan, Q., Lin, Q., Wu, Y., Liu, Y., Wang, X., Ren, J., 2023. A fractionation strategy of cellulose, hemicellulose, and lignin from wheat straw via the biphasic pretreatment for biomass valorization. *Bioresour Technol* 376, 128887. https://doi.org/10.1016/j.biortech.2023.128887

Zhang, H., Zhang, R., Song, Y., Miu, X., Zhang, Q., Qu, J., Sun, Y., 2023a. Enhanced enzymatic saccharification and ethanol production of corn stover via pretreatment with urea and steam explosion. *Bioresour Technol* 376, 128856. https://doi.org/10.1016/j.biortech.2023.128856

Zhang, J., Sun, H., Pan, C., Fan, Y., Hou, H., 2016. Optimization of process parameters for directly converting raw corn stalk to biohydrogen by Clostridium sp. FZ11 without substrate pretreatment. *Energy & Fuels* 30, 311–317. https://doi.org/10.1021/acs.energyfuels.5b01766

Zhang, M., Hou, Q., Shi, Z., Yang, H., Wang, D., Yang, J., 2023b. Enhanced bio-ethanol production from bamboo residues by alkali-aided Fenton reaction pretreatment and enzymatic saccharification. *Ind Crops Prod* 194, 116339. https://doi.org/10.1016/j.indcrop.2023.116339

Zhang, W., Pan, K., Liu, C., Qu, M., OuYang, K., Song, X., Zhao, X., 2020. Recombinant Lentinula edodes xylanase improved the hydrolysis and in vitro ruminal fermentation of soybean straw by changing its fiber structure. *Int J Biol Macromol* 151, 286–292. https://doi.org/10.1016/j.ijbiomac.2020.02.187

Zhao, C., Ding, W., Chen, F., Cheng, C., Shao, Q., 2014. Effects of compositional changes of AFEX-treated and H-AFEX-treated corn stover on enzymatic digestibility. *Bioresour Technol* 155, 34–40. https://doi.org/10.1016/j.biortech.2013.12.091

Zhao, C., Shao, Q., Chundawat, S.P.S., 2020. Recent advances on ammonia-based pretreatments of lignocellulosic biomass. *Bioresour Technol* 298, 122446. https://doi.org/10.1016/j.biortech.2019.122446

Zhao, L., Sun, Z.-F., Zhang, C.-C., Nan, J., Ren, N.-Q., Lee, D.-J., Chen, C., 2022. Advances in pretreatment of lignocellulosic biomass for bioenergy production: Challenges and perspectives. *Bioresour Technol* 343, 126123. https://doi.org/10.1016/j.biortech.2021.126123

3 Food Waste Upscaling to Biofuel by Microbial System

Sachin V. Otari and Sunil H. Koli

3.1 INTRODUCTION

Global food waste has been the most critical issue that starts with agricultural production, extends with a continual food supply chain, and finally ends in landfills. Annually, more than 30% of food contributes to the food waste. Considering the large population of the poor and starving people, this number is astounding. The unavailability of food contributes to the social injustice that could be utilized to encourage the food recovery hierarchy and divert the loss of good food. With social impact, food waste has a tremendous negative impact on the environment as its organic nature has been the major source of potent greenhouse gas (GHG) emissions and undervalued natural resources. The researchers have estimated the food loss is also contributing to approximately 24% of the total food calories available. In this regard, the United Nations (UN) assessed that 320 million more people than the previous year, or roughly a billion people, lived in low-income and developing nations in 2020 and lacked adequate access to food.

By 2050, the expanding global population will be around ten billion, which will in turn cripple the food supply structure due to raised global demand for food (Haldar et al., 2022). India with over 1.3 billion populations is contributing half a kilogram of organic waste per individual per day. According to the United Nations Environment Programme's (UNEP) Food Waste Index Report 2021, the food waste per capita per year in various income groups – *viz*, high, middle, and low – is 90, 68, and 63 Kg, respectively (Chaudhary et al., 2021). Sources of food waste include unsold items from local markets or retail outlets, such as produce; plate waste from restaurants; uneaten, yet perfectly edible, prepared food; food preparation scraps from homes; cafeterias, or restaurants; and by-products generated during food and beverage processing. Ordinary or no food waste management is responsible for the piling up of food and kitchen waste annually. Sustainable Development Goal (SDG) 12: Ensure sustainable consumption and production. The UN is also trying to ensure food waste management: "By 2030, halve per capita global food waste at the retail and consumer levels and reduce food losses along production and supply chains, including post-harvest losses" (UN, 2015). There is a loss of invested resources like

DOI: 10.1201/9781003633709-3

energy, water, labor, and money with the loss of food waste and food waste disposal leads to the emission of GHGs aggravating climate change. There is an urge in the development of the system by leveraging technological advancement for the sustainable management and solution of food waste, converting it to valuable by-products.

The use of engineered microorganisms for the conversion of organic matter from renewable carbon sources into biofuel has been practiced for decades converting waste into bioethanol and biogas, which is considered an alternative to petroleum fuels (El-Araby, 2024). Food waste, a ubiquitous and abundant bioresource, has been proved to be an effective feedstock for biofuel production, which provides important benefits, food waste management, and renewable energy generation (Zeng et al., 2022). The inherent complexity of food waste, stemming from its heterogeneous composition and variability, necessitates the development of robust and adaptable bioconversion technologies capable of efficiently processing this complex substrate (Zeng et al., 2022). Primarily food waste consists of carbohydrates (35.5%–69%), proteins (3.9%–21.9%), lipids, and small amounts of inorganic compounds (Kiran et al., 2014; Lin et al., 2013). Its composition depends on the type of food and its ingredients. For example, food waste from rice and vegetables is rich in carbohydrates, whereas waste from meat and eggs contains higher levels of proteins and lipids.

Anaerobic digestion and fermentation have been proved to be effective approaches for applying a consortium of microorganisms to convert organic matter into simple molecules in the absence or presence of oxygen, producing value-added by-products, like methane (Bux et al., 2023), hydrogen (Han, 2004), ethanol (Koike et al., 2009), enzymes (Khaswal et al., 2024), organic acid (Ohkouchi and Inoue, 2007), biopolymers (Ranganathan et al., 2020), and bioplastics (De Castro et al., 2022). Across countries, the demand for biofuel is much higher than other by-products to curb the need for conventional fuel. Food waste valorization techniques have been developed to produce various types of biofuels, including ethanol, hydrogen, methane, and biodiesel. The bioconversion of food waste to liquid biofuel has been studied and practiced by several investigators aiming to commercialize with the development of adaptive fuel engines. Most importantly, with advancements in technology and the development of engineered microbial species or consortia, the production of biofuel from food waste has become an integral part of food waste management. This not only reduces dependency on conventional fossil fuels but also helps stabilize food prices globally. The presented chapter provide the information regarding conversion of food waste to biofuels like biodiesel, ethanol, hydrogen, and methane with the contribution of food waste conversion to circular economy.

3.2 TYPES, SOURCES, AND CHARACTERISTICS OF FOOD WASTE

Though food waste is organic, it varies in chemical composition depending on its source. Table 3.1 provides a breakdown of different food waste sources in biofuel, along with their chemical characteristics.

TABLE 3.1
Food Waste Types, Sources, and Characteristics and Biofuel Produced

Food Waste Type	Source	Characteristics	Biofuel Type	References
Kitchen Waste	Household, restaurants	High moisture, rich in carbohydrates and lipids	Biogas, Bioethanol	Kazemi Shariat Panahi et al. (2022)
Fruit Peels	Juice industries, markets	Rich in sugars, fibers, and antioxidants	Bioethanol, Biogas	Mgeni et al. (2024)
Vegetable Residues	Farms, food processing	Contains cellulose, hemicellulose, and lignin	Biogas, Bioethanol	Mujtaba et al. (2023)
Meat and Bone Meal	Slaughterhouses	High protein and fat content	Biodiesel, Biogas	Sarrion et al. (2023)
Dairy Waste	Dairy processing units	Contains lactose, fats, and proteins	Biogas, Bioethanol	Sar et al. (2022)
Used Cooking Oil	Restaurants, households	High in free fatty acids, triglycerides	Biodiesel	Monika and Pathak (2023)
Bakery Waste	Bakeries, food industry	Rich in starch and simple sugars	Bioethanol, biogas	Narisetty et al. (2022)
Cereal by-products	Grain mills, breweries	High lignocellulosic content	Bioethanol, biogas	ElMekawy et al. (2013)
Algal sludge	Algae cultivation sites	High lipid and carbohydrate content	Biodiesel, biohydrogen	Xia et al. (2015)
Brewery waste	Breweries	High in fiber, proteins, and residual sugars	Bioethanol, biogas	Emmanuel et al. (2022)
Coffee grounds	Cafés, beverage industry	Contains lignin, cellulose, and oils	Biodiesel, biogas	Sugebo (2022)
Aquaculture waste	Fish farms, seafood industry	Rich in proteins and oils	Biogas, biodiesel	Alvarado-Ramírez et al. (2023)
Food packaging waste	Processed food industry	Often mixed with plastic and biodegradable materials	Biogas, pyrolysis oil	Singh et al. (2023)

3.3 BIOFUEL GENERATION FROM FOOD WASTE USING THE MICROBIAL SYSTEM

Food waste is a rich source of organic matter that can be effectively converted into biofuels using microbial systems. Various microorganisms, including bacteria, fungi, and yeast, play a crucial role in breaking down food waste and transforming it into biofuels, such as biomethane, bioethanol, biodiesel, and biohydrogen.

3.3.1 Bioethanol Production

The net CO_2 emission during ethanol combustion is approximately zero, demonstrating its importance as a cleaner energy source. Ethanol has been favored as a transport fuel not only due to its liquidity and superior production rate but also due to substantial reduction in GHG emission. So the mixture of gasoline and ethanol – i.e., gasohol – has been widely used for vehicles as an alternative to fossil fuel in several countries contributing to pollution reduction. The production of ethanol using waste feedstock has gained important considering both economic and environmental aspects. Traditionally more than 90% of ethanol has been produced through the fermentation process. Like anaerobic digestion, fermentation is also an anaerobic process where microorganisms like yeast and bacteria breakdown carbohydrate rich food waste in ethanol and acids. Bioethanol can be further blended with petroleum fuel to be utilized as an alternative fuel in ethanol-compatible vehicles (Nigam and Singh, 2011). With the advancement of knowledge and technology, cellulosic materials have shown to be a significant source of carbohydrates and sugars which can be further utilized in ethanol production. Consequently, the promotion of necessary crop cultivation for bioethanol production has stimulated the rural economy in the biofuel sector (Kim and Dale, 2004).

Like lignocellulose biomass, food waste is also a source of complex organic matter composed of starchy, fatty, and cellulose materials which is piling up with increasing global population and food waste mismanagement. The chemical characteristics of food waste described in Table 3.1 are favorable for the microbial growth system with or without the pretreatment process. The fermentation process is an important step in bioethanol production from food waste, where the microorganisms utilize the sugar produced during the chemical or enzymatic hydrolysis of food waste, demonstrating a significant method for food waste management. Ethanol producing fermentation process is simple, cost-effective, low energy consuming, and low waste generating which has been explored for the production of several value-added products (Tlais et al., 2020). Fungi and yeast have been studied extensively for the fermentation process *via* solid state fermentation (SSF) and submerged fermentation (SmF), where a moisture content is between 40% and 80% for SSF and 80% and 95% for SmF. Figure 3.1 gives the outline of the process flow involved in bioethanol production using microorganisms.

Food waste is composed of heterogeneous, complex organic molecules as sources of food waste are varied. During the fermentation process, microorganisms convert large molecules to simple molecules, which might take some time. To speed up and simplify the fermentation process, complex food waste molecules are pretreated using an appropriate pretreatment procedure, depending on the nature of the food waste. Various pretreatment processes are utilized for the hydrolysis process like acid (concentrated or mild), alkali, enzyme, thermal hydrolysis, and wet oxidation. The effectiveness of the treatment process is based on the effective release of the sugar molecules to form the waste and recovery of the bioethanol in the downstream process, including the cost of the pretreatment process. Acid and alkali hydrolysis processes are the most common method for food waste of agriculture origin, rich in

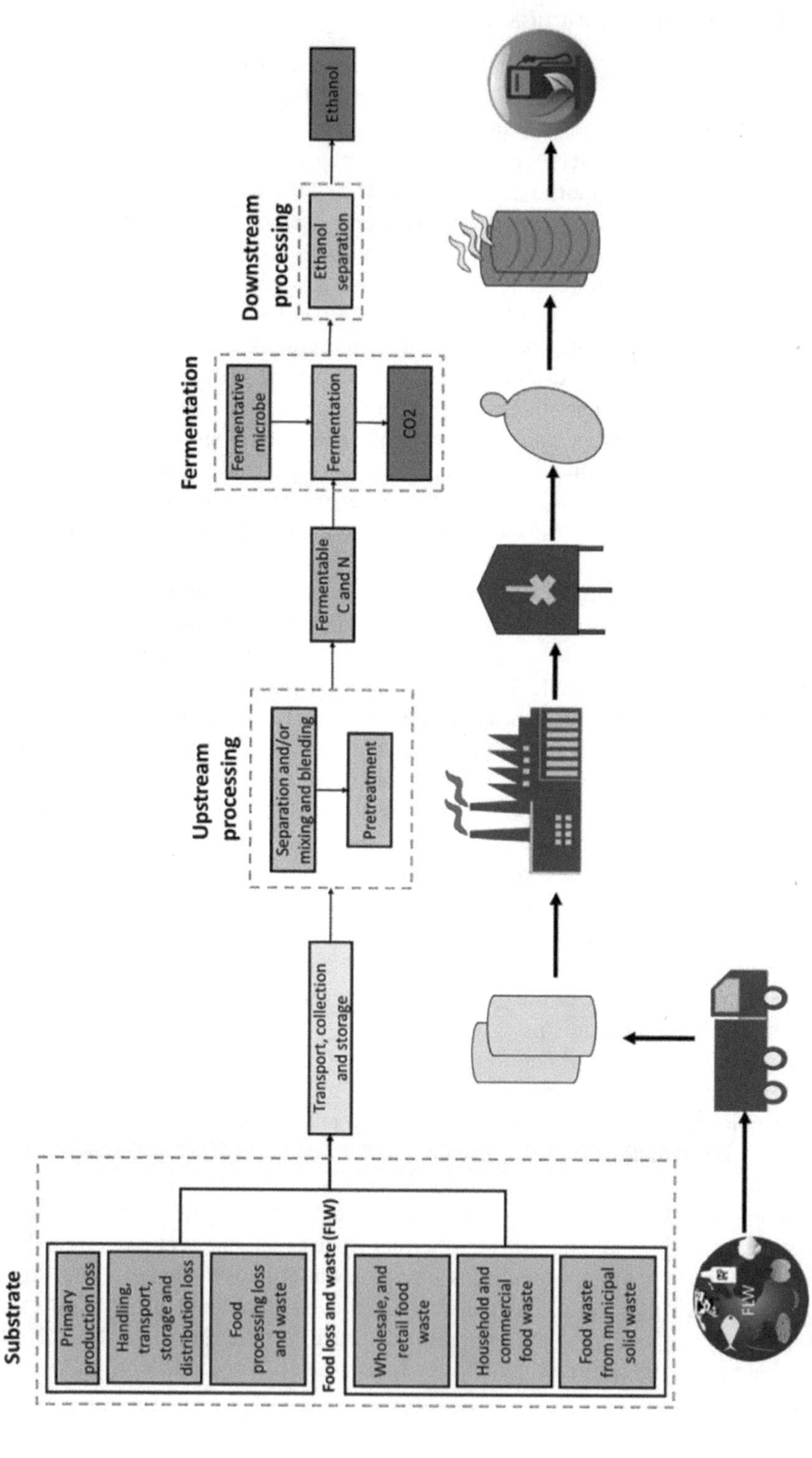

FIGURE 3.1 Operational flow for the bioethanol production using food waste with microbial system.

polysaccharides-cellulose and hemicellulose converting into simple sugar molecules (Favaro et al., 2017; Khedkar et al., 2018). Enzymatic hydrolysis offers an advantage over acid hydrolysis for producing sugars in alcohol fermentations. Different cellulase enzymes like cellobiohydrolase, glucosidases, and exoglucanases are employed for enzymatic hydrolysis (Ingram and Doran, 1995). Different advanced physical technologies for the pretreatment involving milling, extrusion, ultrasound, and microwave pretreatment reduce the complexity of the chemical and enzymatic pretreatment process involved in the bioethanol downstream process (Gallego-García et al., 2023). Biological pretreatment has also been a choice for some food waste where microorganisms like *Bacillus* sp., *Aspergillus* sp., etc., are applied as SSF before actual bioethanol fermentation (Mu et al., 2023).

The choice of the microorganism is critical for bioethanol, as several microorganisms have been demonstrated for ethanol fermentation. Yeast has been the choice among the several microorganisms for the food waste fermentation process. *Saccharomyces cerevisiae* has the potential to convert glucose to ethanol from household waste where the distinct simultaneous liquefication/saccharification and fermentation processes were employed, increasing ethanol production fivefold (Matsakas et al., 2014). The enzyme hydrolyzed starch-based food waste was also utilized for the fermentation process with *Saccharomyces cerevisiae* producing 19.27 g/L of bioethanol after 48 h (Onyeaka et al., 2022). Uncu and Cekmecelioglu achieved 32.2 g/L ethanol production after 59 h of fermentation using food wastes treated with amylases (Uncu and Cekmecelioglu, 2011). *Issatchenkia orientalis*, also known as *Pichia kudriavzevii* or *Candida krusei* yeast, was also employed for ethanol production under stress conditions (Isono et al., 2012). Zwirzitz and coworkers isolated *Issatchenkia orientalis* strain from waste cooking oil (WCO) capable of producing ethanol (10.3 g/ L ethanol within 24 h) from wheat straw hydrolysate (WSHL; Zwirzitz et al., 2021). Hossain and coworkers isolated five yeast species – namely, *Candida wangnamkhiaoensis*, *Hyphopichia pseudoburtonii* (two isolates), *Wickerhamia* sp., and *Streptomyces drozdowiczii* and studied for ethanol production from the food waste where *Wickerhamia* sp. showed the highest ethanol production (30.4 g/L) optimizing the bioethanol fermentation process statically for industrial application (Hossain et al., 2018).

The growth of exogenous microorganisms during yeast mediated food waste fermentation processing has a negative impact on ethanol production (Brexó and Sant'Ana, 2017). Bacterial species are also exploited for possible bioethanol production using several food waste hydrolysates, along with other valuable by-products. Lactic acid bacteria – i.e., *Lactococcus lactis* – was utilized for ethanol production using dairy waste producing 41 g/L of ethanol by fed batch production strategy (Liu et al., 2016). Soleimani and coworkers isolated facultative aerobic bacteria – namely, *Lactobacillus plantarum*, *Lactobacillus sakei*, *Weissella viridescens*, and *Pediococcus acidilactici* – from food waste materials like corn, corncob, paper, and pine cones. These bacteria were evaluated for ethanol production, with *L. plantarum* showing the highest ethanol yield among the isolates (Soleimani et al., 2017). Consolidated bioprocessing (CBP) is a combination of saccharification and fermentation in a single step; it is considered a cost-effective process for ethanol production using microorganisms. Maleki and coworkers developed recombinant species of *Bacillus subtilis* carrying the genes encoding alcohol dehydrogenase from *Z. mobilis* (*adhZ*) and *S. cerevisiae* (*adhS*) named as *B. subtilis* NZ and NZS strains, respectively, where the ethanol production using potato peel waste was 16.3 g/L and 21.5 g/L by NZS and NS:Z during 96 h fermentation, respectively (Maleki et al., 2021).

Co-culture of microorganisms, also known as mixed culture, involves cultivating two or more microorganisms together, allowing them to coexist in the same medium simultaneously. It has been extensively studied for food waste digestion. Some studies involving the hydrolase enzyme producing *Bacillus paralicheniformis* and *Bacillus velezensis* were shown to have enhanced biodegradation of food waste in food waste biodigester (Roslan et al., 2021). Considering the capabilities of the co-culture, the fermentation process is also studied for its utilization of sugars in ethanol production. *Kluyveromyces marxianus* and *Bacillus coagulans* were grown on pomegranate peel waste where simultaneous production of lactic acid (LA) from C5 sugars (i.e., xylose) and bioethanol from C6 (i.e., glucose/fructose) was observed (Demiray et al., 2024). Recent research has demonstrated the application of co-culture systems for ethanol production, including combinations such as immobilized *Z. mobilis* with free cells of *P. stipitis*, ethanologenic *E. coli* strain KO11 with *S. cerevisiae*, *S. cerevisiae* with *Pachysolen tannophilis*, and *Z. mobilis* with *Candida tropicalis*, among others (Chen, 2011). However, the most effective approach for fermenting glucose and xylose is the co-culture of immobilized *Z. mobilis* and *P. stipitis*, enabling ethanol production from various biomass materials, including food waste. The success of a co-culture system primarily depends on the growth rate of different microorganisms in various hydrolysates and their adaptation to fermentation conditions, such as temperature and pH.

Although microbial fermentation using food waste hydrolysate is a promising process for bioethanol production, it faces several challenges that hinder its practical and industrial feasibility (Bibra et al., 2022). The heterogeneous composition of food waste, pretreatment complexity with recovery in the downstream process, inhibitor formation, microbial adaptation, co-fermentations of sugars, sterility, contamination control, process optimization, economic viability, scale-up challenges, and regulatory and environmental concerns are the challenges for the microbial bioethanol production using food waste (Mussatto et al., 2010). Addressing these challenges through advanced microbial engineering, process optimization, and efficient waste management strategies can enhance the feasibility of bioethanol production from food waste. Table 3.2 presents additional examples of microbial ethanol production using different types of food waste with production yield and methods employed for production.

3.3.2 Biomethane Production

The conventional methods of food waste disposal, such as landfilling, not only occupy valuable land but also lead to the release of GHGs like CH_4 and CO_2. Improper disposal of food waste contributes to the emission of GHGs, with CH_4 having a global warming effect 21 times higher than CO_2 (Ferdeş et al., 2022). This has necessitated the exploration of more sustainable waste management practices, with anaerobic digestion (AD) emerging as a promising solution. AD is a biological mechanism

TABLE 3.2
Bioethanol Production Using Food Waste by Microorganisms

Type of Food Waste	Microorganism	Method	Pretreatment	Duration (h)	Y (g/g FW)	Y (g/L·h)	References
Household food waste	*Saccharomyces cerevisiae*	Simultaneous saccharification and fermentation	None	72	0.14	0.08	Matsakas et al. (2014)
Starch-based food waste	*Saccharomyces cerevisiae*	Separate hydrolysis and fermentation (SHF)	Enzymatic hydrolysis	48	0.43	0.18	Onyeaka et al. (2022)
Mixed food waste	*Geobacillus thermoglucosidasius*	Single-pot bioprocessing	None	72	0.31	0.13	Bibra et al. (2020)
Pineapple peel waste	*Clostridium acetobutylicum*	Fermentation	Acid hydrolysis	96	0.13	0.05	Khedkar et al. (2018)
Mixed food waste	Genetically modified *Bacillus subtilis*	Consolidated bioprocessing	None	72	0.31	0.13	Maleki et al. (2021)
Mixed food waste	*Saccharomyces cerevisiae*	Fermentation	None	72	0.14	0.08	Ma et al. (2024)
Potato peel waste	*Wickerhamia* sp. strain SD1 (wild)	Batch, submerged fermentation	None	72	0.54	0.23	Hossain et al. (2018)
Household food waste	Baker's Yeast	Batch fermentation	Milling, liquefaction	16	0.107	2.85	Huang et al. (2015)
Dairy waste	*Lactococcus lactis* subsp. *cremoris* strain MG1363 (Recombinant)	Fed batch, fermentation	None	96	0.38	0.77	Liu et al. (2016)
Food waste	*Issatchenkia orientalis*	Separate hydrolysis and fermentation	Screw pressed and dried	8	0.04	1.45	Kim et al. (2018)
Organic fraction of municipal solid waste	*Mucor indicus* CCUG 22,424	Separate hydrolysis, saccharification and fermentation	Diluted acid and enzymatic hydrolysis	72	0.12	0.38	Mahmoodi et al. (2018)

(*Continued*)

TABLE 3.2 (Continued)
Bioethanol Production Using Food Waste by Microorganisms

Type of Food Waste	Microorganism	Method	Pretreatment	Duration (h)	Y (g/g FW)	Y (g/L·h)	References
Damage rice grains	*Paenibacillus chitinolyticus* strain CKS1	Separate hydrolysis, saccharification and fermentation	Enzymatic hydrolysis	72	0.15	0.62	Mihajlovski et al. (2018)
Apple pomace	*Saccharomyces cerevisiae montrachet*	Fermentation	None	24	0.044	1.48	Hang et al. (1981)
Palm kernel cake hydrolysate	*Geobacillus thermoglucosida-sius* (recombinant)	Saccharification and fermentation	Steam explosion	48	1.15	0.21	Raita et al. (2016)
Mixture of rice milling by-products	*Sacchaaromyces cerevisiae* strain M2	Hydrolysis, saccharification and fermentation	Alkaline peroxide and enzymatic hydrolysis	96	0.24	0.98	Favaro et al. (2017)
Food waste hydrolysate	*Zymomonas mobilis* (Mutant)	Fermentation	None	44	0.49	0.50	Ma et al. (2016)
Kitchen garbage	*Zymomonas mobilis*	Fermentation	None	40		1.35	Ma et al. (2008)
Tea waste	*Saccharomyces cerevisiae* and *Candida boidinii*	Co-fermentation	None	-			Demiray et al. (2022)
Food waste	*Fusarium oxysporum* F3 and *Saccharomyces cerevisiae*	Co-fermentation	Dilute acid and enzymatic hydrolysis	96	-	1.4	Prasoulas et al. (2020)

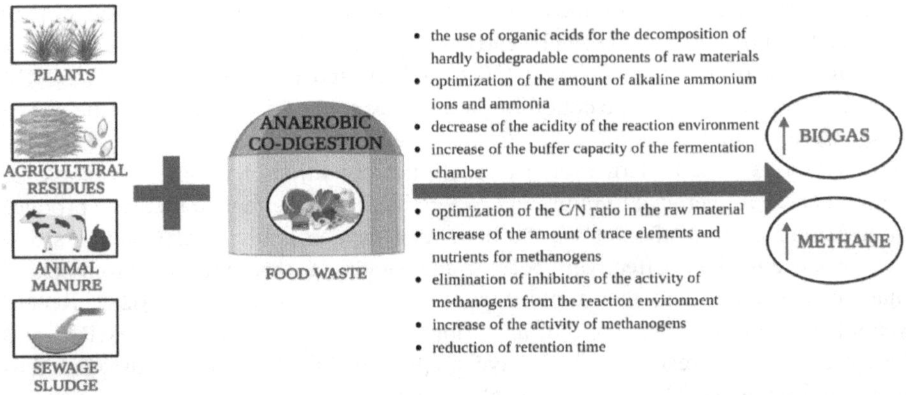

FIGURE 3.2 Effects of anaerobic co-digestion of food waste with distinct feedstock.

where natural matter is degraded by microorganisms devoid of O_2, producing biogas (mainly made of CH_4 and CO_2) and a nutrient-rich material (Figure 3.2) where the produced gases can be used directly utilized as cooking fuel or used in generating power gas engines or upgraded into natural gas-quality biomethane or injection into natural gas grids (Zhang et al., 2014).

The biogas produced can be utilized for various purposes, including electricity generation, heating, and as a transportation fuel in the form of compressed natural gas (CNG). Moreover, the digestate, rich in nutrients, can be used as a bio-fertilizer, thus completing the circular economy loop. Biomethane is a renewable gas predominantly composed of methane (CH_4) that is produced through the biological processing of organic matter. Unlike natural gas, which is a fossil fuel, biomethane is considered carbon neutral as it's derived from recently grown organic materials. According to recent literature, biomethane contains approximately 95%–98% methane content after purification, with the remaining components being primarily carbon dioxide and trace gases (Bux et al., 2023). This high-quality gas can be used interchangeably with natural gas in existing infrastructure, making it a valuable renewable energy source for heating, electricity production, and transportation fuel (Tiago et al., 2025). Food waste, due to its rich composition of carbohydrates, lipids, and proteins, is considered an excellent substrate for AD. During AD, the methane yield largely depends on the biochemical composition of food waste. The generation of biomethane using food waste from China and Southeast Asia is in the range from 116 to 684 LCH_4/kg VS whereas it is in the range of 425–529 LCH_4/kg VS for the food waste from Europe, which is due to cultural and composition differences in two regions. It was found that significant high yield of methane was observed for the food waste (480 ± 88 LCH_4/kg VS) compared to that of energy crops (237 ± 58 LCH_4/kg VS) demonstrating significance of the food waste in bioenergy production (Negri et al., 2020). This is specially very important in tackling food energy conflict and reducing the cost associated with renewable energy production using energy crops.

The AD process is intricate and impacted by several parameters, including temperature, pH, volatile fatty acids (VFAs) concentration, and substrate characteristics.

Temperature is a critical variable influencing the activity of the microbes and, consequently, the biogas yield. AD can be carried out under different temperature regimes: psychrophilic (<20°C), mesophilic (30°C–40°C), and thermophilic (50°C–60°C). Mesophilic and thermophilic conditions are more commonly employed for the AD of food waste, with thermophilic conditions often resulting in higher biogas yields (Pilarska et al., 2023). However, psychrophilic conditions have the advantage of higher bacterial tolerance to inhibitors, which can lead to effective methane production. According to Muñoz et al., the benefit of psychrophilic environments is that they produce methane effectively because the bacteria are more resistant to blockers than in mesophilic and thermophilic environments (Muñoz et al., 2020). Recent years have seen the emergence of anaerobic membrane bioreactors (AnMBR) and bioelectrochemical reactors, which have strengthened the technology customized to these environments (Martí-Herrero et al., 2019).

Another vital parameter that affects the steadiness of the AD procedure is pH. The optimal pH scale for methane-producing bacteria is between 6.5 and 7.5 (Chen et al., 2008). Deviations from this range can inhibit the methanogenic activity and lead to process failure. The concentration of VFAs is quietly connected to pH and serves as an indicator of process stability. Elevated levels of VFAs can lead to a drop in pH, inhibiting methanogenic bacteria. The optimal VFA concentration is considered to be on the scale of 2–3/gL (Yuan et al., 2006). Along with pH, VFAs are commonly used as indicators of AD process stability and consistency. Na concentration over 8/gL, which is increasingly common in food waste, specifically from fast food, can inhibit methanogen activity, though Na in a mixture with K or Mg can have valuable consequences on biogas production (Ferdeş et al. 2022).

Recent years have witnessed significant advancements in improving biogas generation from food waste. One of the approaches is the co-digestion of food waste with other substrates to improve the nutrient balance, increase biogas yield, and mitigate any inhibitory effects. Co-digestion of food waste with fruit and vegetable waste at an optimal ratio of 1:1 led to improved stability and higher biogas generation (Lin et al., 2011). Food waste co-digestion with distinct feedstock materials enhanced total bio-CH_4 output on a mean from 268 ± 199 mL/g VS to 406 ± 137 mL/g VS (Negri et al., 2020). Another strategy is the pretreatment of food waste to enhance its biodegradability and subsequent biogas yield. Pretreatment methods can be mechanical, thermal, chemical, or biological, each with its specific advantages and drawbacks.

An innovative approach in the field is the use of nanocatalysts in the AD process. The use of iron oxide nanoparticles derived from red mud lowered the reaction time for AD by 5.38% conflicted with the noncatalytic mechanism. The methane production was found to be elevated in the presence of a nanocatalyst yielding 50% more CH_4 and 23.5% more total biogas compared to that of in the absence of a nanocatalyst. This indicates the importance of nanotechnology in the improvement of biogas generation efficacy from food waste (Bharathi et al., 2022).

For the boarder implementation of biogas generation from food waste, economic and technical viability in biogas production is a very critical aspect. Techno-economic research was implemented to make feasible the production of biogas from various forms of locally produced food waste. The investigation revealed that

1,550 mL of accumulative gas was produced from every 1 g of dry matter form blended food waste samples at day 21. After economic analysis, it was found that the biogas production project at $0.2944 per cubic meter of biogas, with any price above this rate yielding a positive net present value (NPV; Al-Wahaibi et al., 2020). With a six-year discounted payback period, a positive NPV of $3,108 was evaluated at a biogas price of $0.39/m^3, suggesting the economic viability of the biogas production from food waste.

In view of sustainable development, AD of food waste supports multiple goals of the UN SDGs, which include SDG 7 (Affordable and Clean Energy), SDG 12 (Responsible Consumption and Production), and SDG 13 (Climate Action). AD encourages sustainable waste management through lowering GHG emissions and treating the food waste out of landfills, turning it into renewable energy (Ferdeş et al., 2022).

Several real-world applications and case studies have demonstrated the viability of biogas production from food waste. The Inland Bio-Energy (IBE) food waste anaerobic digester facility in Chino, California, is an important example of a practical application that has been in operation since 2012, handling up to 300 tons per day of solid food waste. The biogas produced is filtered and compressed to transportation-grade renewable natural gas for utilization in refuse collection trucks, demonstrating the importance of biogas production from food waste in the circular economy (Jameel et al., 2024).

Even though biogas generation from food waste has promising outcomes, certain challenges and limitations exist. The irregular composition of food waste and supply of the feedstock for biogas production is the key challenge impacting on the biogas yield and process stability. Food waste from different sources like households, restaurants, food processing industries, and varying compositions requires careful characterization, pretreatment process, and biogas production optimization. The presence of plastic and metal contaminants, in food waste can affect the AD process, which needs effective pretreatment methods to remove such impurities. The technical complexity of the AD process involving continual monitoring and control of multiple parameters ensuring optimal performance is another limiting factor in widespread practical application. This complexity leads to the capital expenses to tackle the technical issues, restricting acceptance of this technology, especially in developing countries. The transportation of food waste at centralized biogas plants is one of the critical challenges in regions with poor waste management infrastructure.

To make the AD process more energy efficient and dependable, extensive research is requisite to integrate another form of renewable energy like solar and wind energies with the biogas process. The development of low-cost and effective pretreatment applicable to food waste with varied compositions can be beneficial. The innovative reactor design, which enhances the AD process, can significantly contribute to advancement in the field (Martí-Herrero et al., 2019).

In summary, biomethane production from food waste through AD is a sustainable solution for waste management and energy crisis. The investigations have proved that the high organic content of food waste is an excellent substrate for AD, and can be an alternative to energy crops in biofuel production. The digestion process is largely influenced by various parameters, like temperature, pH, and VFA concentration. The integration of the nanocatalyst with other substrates in the co-digestion process has shown affirmative output-enhancing biomethane production. Biogas production from food waste can be economically viable when other benefits like waste management fee savings and environmental advantages are taken into consideration. The technology alignment with the UN SDG is a promising factor in sustainable development. The existence of several challenges like economic viability, varied food composition, contaminants, and technical complexity is also a reality and needs to be addressed by societal acceptance of the technology. Future research should focus on addressing these limitations through the development of more efficient pretreatment methods, innovative reactor designs, and the amalgamation of biogas production with other renewable energy approaches. While challenges exist, the AD of food waste for biomethane generation offers a promising pathway toward a more sustainable future, transforming waste into energy and contributing to the circular economy.

3.3.3 Biohydrogen Production

The search for clean, renewable alternatives to fossil fuels has intensified owing to the global energy problem and ever-rising ecological concerns. Hydrogen, with its energy content (142 kJ/g) and carbon-free combustion products, offers a potentially viable solution (Rao and Basak, 2021). Meanwhile, food waste management presents a huge worldwide challenge. Approximately 1.3 billion tons of food, one-third of what is produced, gets wasted each year (Habashy et al., 2021). Converting that food into clean-burning hydrogen could address two massive, interrelated problems: energy production and waste management (Sahota et al., 2024).

The term "biohydrogen" comes from the Greek language. "Bio" means life, "hydro" means water, and "gen" means generator or producer. So, the combined term "biohydrogen" designates a product that comes from living sources that can generate nonbiodegradable fuel (Samrot et al., 2023). Because it comes from biological sources, it is considered a clean fuel. Hydrogen does not produce carbon when it burns; it produces only water vapor. So, if the H_2 is produced cleanly (as it can be by microbes in anaerobic conditions), it becomes a very appealing fuel. It burns very cleanly, very efficiently, and with a very high energy density. Still, there are big impediments to H_2 production using this method (Hallenbeck and Ghosh, 2009).

Food waste mainly contains carbohydrates in the form of starch, which must first be converted to glucose before H_2-producing microbes can effectively use it (Han et al., 2016). The overall complex and variable composition of food waste creates problems for biohydrogen production (Gbiete et al., 2024).). However, it also presents a large opportunity, given the amount of food waste generated daily. Soluble starch and other nutrients in food waste need to be converted into even simpler forms. This conversion process has emerged as a promising method because it can effectively reduce starch found in food waste down to glucose without generating unwanted by-products that could affect subsequent steps in the process. In one study, a commercial enzyme known as glucoamylase was used to convert a liquid portion of food waste into fermentable sugars with a ratio of 10% (w/v). They found that after seven hours of the reaction, liquid that could now be used as potential feedstock for biohydrogen production was produced in the form of a hydrolysate that contained 24 g/L glucose (Han et al., 2016). This is a clear demonstration of hydrolysis. Biohydrogen can be produced using different methods, like biophotolysis, which employs large-scale algae, or using microbes in microbial electrolysis, which is better suited for hard-to-recycle organic materials. These methods rely heavily on large quantities of pure water; biomaterials otherwise use their components for construction and metabolic activities. Even dark fermentation can be performed using water or wastewater, which must be pretreated to a certain degree so that the microbes can grow satisfactorily and perform the fermentation under anaerobic conditions (Rao and Basak, 2021).

Biohydrogen production from food waste is significantly influenced by the process parameters. Typically, the temperature is held at around 37°C so as to optimize the activity of mesophilic bacteria. Meanwhile, the pH is maintained above about 4.0 to prevent any inhibition of the hydrogen-producing microorganisms. Another very important parameter in continuous systems such as this is the hydraulic retention time (HRT). Studies suggest that an optimal HRT for such systems is about six hours, which results in the obtaining of a maximum H_2 production rate of 8.02 mmol/h/L. If the HRT is shorter than this, for instance, at about four hours, which is two-thirds of a six-hour HRT, then the production rate drops significantly to 0.76 mmol/h/L in part because of washout of insufficiently retained biomass. On the other hand, if the HRT were much longer, we might begin to obtain significant quantities of a metabolic end product such as acetic acid. Here, we also need to consider the rate at which we add more food waste, or the organic loading rate (OLR), which is also very fine-tuned for getting good production rates (Han et al., 2016).

Biohydrogen production performance is different in batch and continuous systems. As per the aforementioned literature, the following values can be used for simulating biohydrogen production. The maximum rate of hydrogen produced is 277.8 mL/h. The maximum potential for the production of hydrogen is 6437 mL with a lag Phase of 11.2 hours (Han et al., 2016). The highest rate of production of hydrogen of 8.02 mmol/h/L in the continuous agitated tank reactor (CSTR) is for HRT 6 hours. The yield of hydrogen in the batch system is higher than that of the continuous system. For instance, the following data shows that the 1 g of food residue is about 0.304 g of glucose, which results in the production of about 245.7 mL of hydrogen in batch operations and 205.8 mL in continuous operation. The inconsistency between the two values is due to glucose washout in continuous operations. The detailed investigation has shown that nearly 10% of the experiments of the biohydrogen production for the food residue are up to or greater than 15.55% COD/COD, up to or greater than 305.16 mg/L/d Productivity, and 64% v/v of hydrogen (Moussa et al., 2022).

The primary soluble microbial products yielded by food fermentation for waste-pressure methane fermentation of food waste are the production of ethanol, then acetate, and according to the literature, about 108.93 mmol and 79.4 mmol are produced from ethanol and acetate. The recovery of the carbon analysis shows that about 83.8% of carbon is recovered from the glucose absorbed during fermentation in the form of soluble, gaseous microbial products, and the remaining 16.2% shall be the leftover part, which would change into biomass. So, for about 32.5%, additional nondigested food residue is left, and the rest of the 67.5% carbon dioxide and soluble microbial products are produced. Ethanol (15.7%–18.52%) and CO_2 (13.74%–27.88%) are the major carbon consumed substrates, due to their solubility and gaseous nature, and are the final products of the fermentation procedure. Fermentation is the metabolic process that converts sugars into acids, gases, and alcohols. Han et al. reported that the major by-products of the fermentation process are carbon dioxide, hydrogen, and ethanol (Han et al., 2016).

Advanced methods of biohydrogen production commenced focusing on the enhancement of the yield and efficiency of the process. The methods found to be effective were used in the co-production of hydrogen using food waste with the co-digestion techniques. Using other waste streams to improve hydrogen production has shown a promising result. For instance, if the cattle manure is co-digested with food wastes at a mixing ratio of 47% to 51%, it results in higher hydrogen production of 30.00 mL/g/VS-added, and the rate of production is estimated to be 1.00 L/L/d of cattle waste (Liu et al., 2020). The co-fermentation of swine manure and molasses gave a hydrogen production rate of 31.9 L/d and a yield of 1.52 L/g of sugar (Wu et al., 2013). Innovative techniques used are the biohydrogen production using SSF in a dark process to boost the efficacy of the system. Other findings showed 1.00 L/L/d to be the hydrogen production of continuous mixed immobilized sludge reactors used in activated carbon as a carrier (Samrot et al., 2023). Table 3.3 provides a few examples of food waste utilized for biohydrogen production.

After the production of H_2, the solid residue left behind can be converted into biochar. Biochar is a useful element for soil enhancement. Soil biochar obtained from food waste is alkaline, with a pH being 7.88 ± 0.33. The biochar is rich in nutrients and can be efficiently used to improve soil productivity (Pradhan et al., 2024). The integrated approach utilized here helps in harnessing energy and treating wastewater over a broader spectrum. There is a proposal for integrating bio-thermo-chemical processes of food waste to recover the maximum amount of energy. Furthermore, food waste can become a useful energy source through a bio-thermo-chemical process that converts these wastes into energy vectors such as bio-H_2, bio-CH_4, and hydro-char. This process also includes compelling digestate and the production of biofertilizers (Sahota et al., 2024).

There are various challenges that hinder the propelling of the biohydrogen concept, such as a varied and complicated composition of food waste and continuously changing, unstable food waste that affects the unpredictable performance of the process, which leads to certain inhibitors in the process of hydrogen production. (Gbiete et al., 2024). The yield of the substrate greatly depends upon the extent of hydrolysis. If the hydrolysis is not a success, then the quantity of substrate fed to the hydrogen-producing microorganisms is so minimal that it acts as a limiting factor. The inhibitors generated during the pretreatment process sometimes have detrimental impacts on the hydrogen-producing microorganisms. AD has been later proven as a non-viable future clean energy source. Biohydrogen production, along with other bioprocesses in a biorefinery approach, may increase economic feasibility with multiple recoveries of products. In addition, a thorough techno-economic analysis and life cycle assessments are necessary to evaluate the sustainability and commercial feasibility of biohydrogen scale-up and production (Qu et al., 2022).

TABLE 3.3
Biohydrogen Generation Through Fermentation of Food Waste

Substrate	Reactor	Microorganism	Light/Dark	H_2 Production	References
Maize straw		*Rhodobacter sphaeroides*	4,000 ± 200 lx	4.6 mol-H_2/mol-reducing sugar	Wang et al. (2018)
Cornstalk pith		Mixed culture	2,000 lx by filament lamps	2.6 mol-H_2/mol-sugar consumed	Jiang et al. (2016)
Corn stover			3,500 lx	141.4 ml-H_2/g-TS	Zhang et al. (2018a)
Dark fermentation effluent of sugarcane bagasse	Batch	*Rhodopseudomonas*	8.5 W/m^2 by a tungsten lamp	755.0 ml-H_2/l-waste	Rai et al. (2014)
Sugar beet molasses		*Rhodopseudomonas palustris*	114 W/m^2 by a tungsten lamp	19.0 mol-H_2/mol-sucrose	Sagir et al. (2017)
Rotten apple		Mixed culture	3,030 lx	112.0 ml-H_2/g-TS	Lu et al. (2016)
Dark fermentation effluent of corn stover	Continuous		3,000 lx by solar and LED bundles	4.7 m^3-H_2/m^3/d	Zhang et al. (2018a)
Wheat waste hydrolysate			Dark	645.7 ml-H_2/g-TS	Karaosmanoglu Gorgec and Karapinar (2019)
Fruit and vegetable wastes		*Clostridium*, *Bifidobacterium*, and *Lactobacillus*	Dark	23.5 ml-H_2/g-VS	Abubackar et al. (2019)
Garden and food waste	Batch	*Caldicellulosiruptor saccharolyticus*	Dark	46.0 ± 1 ml-H_2/g-VS	Abreu et al. (2019)

Source: Habashy et al. (2021).

Various inhibitory factors limiting biohydrogen production with their controlling measures are described in Figure 3.3.

In conclusion, biohydrogen produced from food waste may be a promising pathway to sustainable waste management and clean energy production for the environment. The production of this method diverts organic waste from landfill sites, which leads to CA emissions and produces carbon-free energy carriers that may be overshadowed by many environmental conditions. Research continued development and increased fundamental knowledge related to both science and technology, and its efficiency and

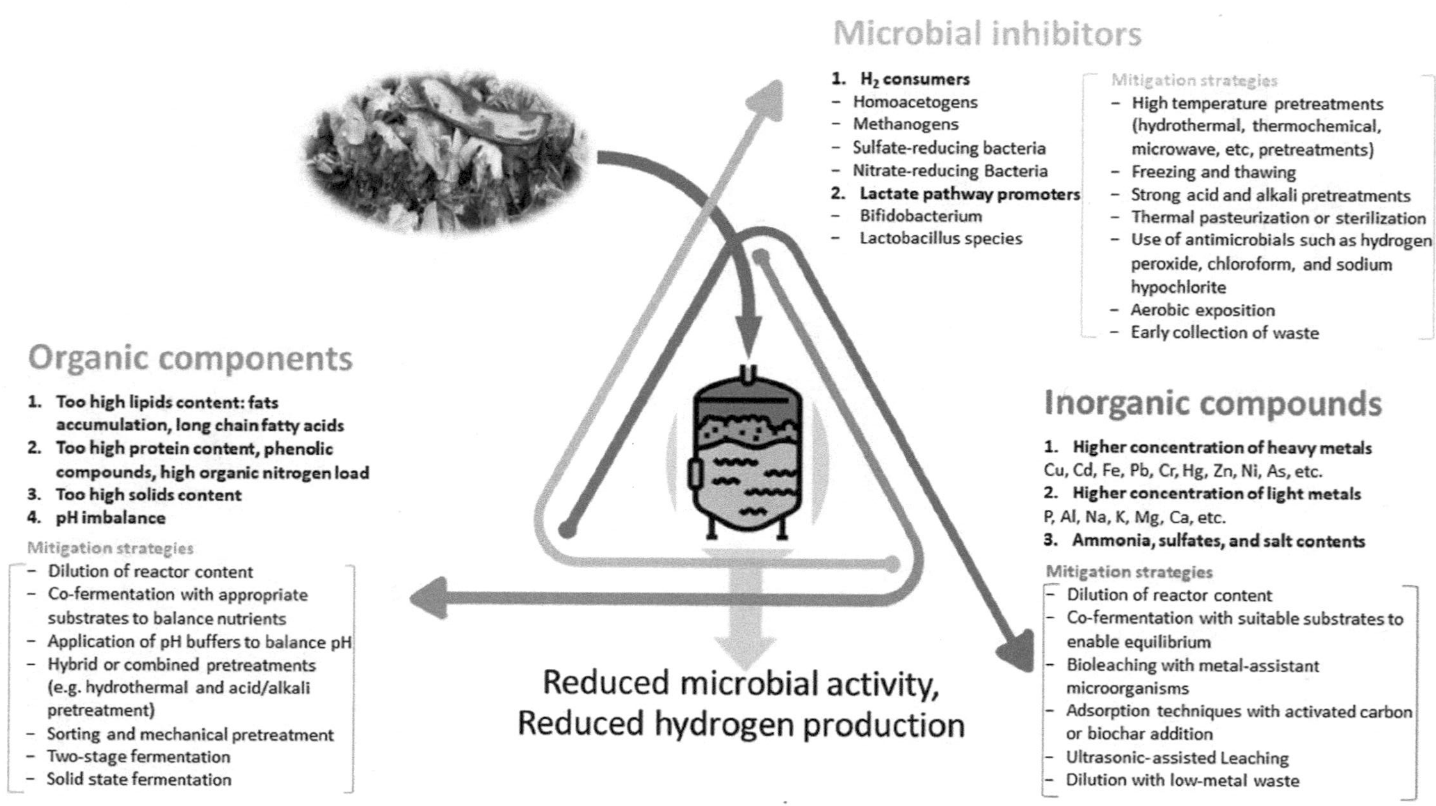

FIGURE 3.3 Native H_2 production using FW with inhibiting components.

viability can be enhanced. This production process serves as a crucial way to apply circular economy principles and to meet SDGs, and for that, technical, economic, and scaling up challenges are required to be addressed for commercial implementation.

3.3.4 Biodiesel Production

Biodiesel is one of the important types of biofuels, as it is considered the prominent alternative to diesel derived from fossil fuels mitigating the issues with hazardous gas emissions from petroleum diesel engines and burden on the conventional fuel sources. Biodiesel is made up of methyl esters of long-chain fatty acids, produced through the transesterification of oil feedstock with the use of alcohol and catalyst (acid or base). Transesterification is a chemical process of exchanging organic groups of ester with alcohols which is applied in the conversion of fats, oils, and greases (FOGs) from food waste to biofuel. Here the lipid reacts with alcohol like methanol in the presence of a catalyst (e.g. lipase enzyme) forming biodiesel and glycerol by-product (Demirbas, 2009). Significant studies have proved that biodiesel can be produced from renewable sources. The lubricant properties, energy density, and combustion efficiency are close to the petroleum based biodiesel, allowing the use of biodiesel to be applied with in exiting diesel engines with no alterations (Ajala et al., 2015). Also, the study has shown the low emission of particles, sulfur, and CO_2 from the biodiesel-based engine compared to petroleum diesel demonstrating as a clean energy source (Zhang et al., 2011).

Traditionally, vegetable oils, soybean oils, cottonseed oil, animal fats, and nonedible oils have been the feedstock for biofuel production where enzymes and microorganisms have been used as catalysts for the esterification process at temperature (Wang et al., 2021). Microbial-derived oil was also used as a medium for biodiesel production (Carmona-Cabello et al., 2021). For the last decades, the focus from the traditional feedstock has shifted toward domestic and industrial waste like dairy waste, kitchen waste, and sugarcane waste where the food waste directly or pretreated is used as the substrate for microbial growth (Mahish et al., 2024). The lipid content of the food waste is 5%–30%, which is further subjected to solvent extraction to remove lipid content from food waste (Pleissner et al., 2013). Hydrolysis was also performed in some food waste using fungal species to remove lipids from the waste, and further separation was done with centrifugation (Pleissner et al., 2014). The hydrolysate is heated to remove water from the lipid content, making it a suitable medium for microbial transesterification process. The processed lipid hydrolysates act as nutrient-rich substrates where microorganisms like bacteria, fungi, and algae convert food waste into valuable biodiesel.

The lipase enzyme derived from fungi has been proved to be an excellent choice as a catalyst over chemical catalysts as it shows excellent transesterification reaction efficiency. So, efforts were taken to utilize food waste for the conversion by filamentous fungal species. Gujjala et al. (2019) used laccase-treated lignocellulose waste liquor producing a lipid yield of 134 mg/g sugar consumed, which was further used as the substrate for the growth of *Aspergillus awamori* converting lipid into biodiesel in 48 h of incubation (Gujjala et al., 2019). The process has been optimized by response surface methodology to further utilize the obtained parameter for pilot scale production employing fungal species. The production of biodiesel was also demonstrated using the oleaginous yeast *Wickerhamomyces anomalus* strain EC28 grown on agro-industrial wastewaters and waste frying oils having 75% and 24% of lipid, respectively (Arous et al., 2017). Two fungal species, *Rhodosporidium toruloides* and *Cryptococcus laurentii*, were gown on orange peel waste for the biodiesel production of 36.9% and 31.9% biodiesel yields were obtained for *R. toruloides* and *C. laurentii* respectively (Carota et al., 2020).

Biodiesel production using microalgae cultivation using food waste as a feedstock is considered an emerging sustainable practice where food waste provides essential nutrients such as carbon, nitrogen, and phosphorus in optimum condition to promote algal growth. Several studies have shown the prominent lipid accumulation efficiency of the microalgae revealing their effective role in biodiesel production. The food waste hydrolysate grown *Chlorella vulgaris* microalgae has the potential to accumulate 40%–45% lipid in the cell producing biodiesel at very adequate environmental conditions (Kavitha et al., 2024). Anaerobically digested effluent from kitchen waste (ADE-KW) was used to cultivate two microalgae species viz *Chlorella sorokiniana* SDEC-18 and *Scenedesmus* SDEC-8 accumulating 30.27%–41.69% and 35.97%–47.39% lipid content under photoautotrophic condition, which was further demonstrated for biodiesel production (Zhang et al., 2018b). Several microalgae like *Botryococcus* sp. (Gani et al., 2023), *Chlamydomonas* sp. (Kothari et al., 2013), and *Desmodesmus* sp. (Das et al., 2022) have been demonstrated for biodiesel production from food waste feedstock. However, challenges such as optimizing nutrient composition, scaling up processes, and ensuring economic feasibility must be addressed for large-scale implementation. Additional examples of microbial biodiesel production from food waste are provided in Table 3.4.

TABLE 3.4
Biodiesel Production Using Food Waste by Microorganisms

Food Waste Type	Microorganism Used	Lipid Yield (%)	Biodiesel Yield (%)	Reference
Waste cooking oil	*Yarrowia lipolytica*	55	85	Mitrea et al. (2022)
Dairy waste	*Cryptococcus curvatus*	60	88	Krikigianni et al. (2022)
Waste cooking oil	*Endomelanconiopsis endophytica*	89	90	Rodrigues et al. (2023)
Coffee waste	*Rhodosporidium toruloides*	57	80	Giannakis et al. (2023)
Ricotta cheese whey (RCW)	*Oleaginous yeasts*	41	33	Carota et al. (2017)
Sugarcane bagasse	*Trichosporon oleaginosum*	55	86	Brar et al. (2017)
Sugar beet residues	*Lipomyces starkeyi*	47	—	Martani et al. (2020)
Vegetable waste	*Rhodococcus opacus*	57	90	Patnaik et al. (2021)
Saccharified liquid	*Rhodosporidium toruloides*	61	59	Ma et al. (2019)
Food waste	*Rhodococcus* sp. YHY01	69	39	Bhatia et al. (2019)

3.4 CIRCULAR ECONOMY

Voukkali et al. (2023) described that the circular economy of the food waste upscaling framework is based on four key principles: sustainability, energy efficiency, resource recovery, and waste prevention (Voukkali et al., 2023). The food loss can be minimized at the point of origin by preventing waste, optimally managing the food supply, and effective distribution of food. To recover resources, food waste must be systematically gathered and converted into valuable products like biofuel before pilling as waste. The biofuel conversion process makes use of energy-efficient microbial techniques, such as genetically modified strains and metabolic engineered microbial colonies. Furthermore, sustainability is accomplished through utilizing residual by-products like AD digestate, as biofertilizers in agricultural applications to set up a closed-loop system that boosts soil health and crop yield.

The implementation of a circular economy in upscaling food waste not only improve the economy but also provide advantage to the environment by reducing the environmental footprints and elevating resource efficiency. The substantial reduction in landfill methane emission, ease in dependency on fossil fuels, and promotes climate change mitigation are the major outcomes of circular economy adoption. The process also promotes the establishment of new bio-based industries, green employment generation, several income sources form waste valorization. Countries like India, European Union, and the United States have established circular economy policies assuring the upscaling of food waste-to-biofuel conversion. The European Green Deal, EPA's Food Recovery Hierarchy, and India's National Bio-Energy Mission have promoted the microbial conversion of food waste to biofuel as eco-friendly process.

The use of microbial systems in food waste conversion to valuable commodity is a revolutionary strategy under the circular economy providing a sustainable strategy for food waste management which indeed provides opportunity to reduce ecological crisis and increase bioenergy production. The amalgamation of the strategies of food waste management and microbial biotechnology with a bioenergy recovery system in food waste conversion can be the most fertile technology for future sustainable, and renewable energy. As governments, industries, and researchers innovate further, the circular economy for food waste management will be central to attaining global energy security and environmental sustainability.

3.5 CHALLENGES AND FUTURE PROSPECTS

The microbial conversion of food waste to biofuel has several critical challenges involving feedstock availability, pretreatment efficiency, enzyme costs, and scaling up for large-scale production. Overcoming these challenges has been the most critical aspect of the upscaling of food waste to biofuels resource.

i. **Feedstock availability**
   Though food waste is abundantly available, the consistent availability and composition of the food waste are largely affected by the seasonal variability. These fluctuations in the supply influence the microbial process for

biofuel production (Karmee and Lin, 2014). For the sustainable production and economic viability of continuous microbial conversion in biofuel production using food waste, a steady and uniform supply of feedstock is essential.

ii. **Pretreatment efficiency**

Food waste contains a lot of lignocellulosic biomass. This biomass needs good pretreatment to break down its complex polymers into sugars that can ferment. Old pretreatment methods, like acid hydrolysis, can create inhibitors such as furfural and hydroxymethylfurfural. These inhibitors make it hard for microbes to ferment. People have looked into other methods, like ammonia fiber expansion (AFEX), to boost efficiency and cut down on inhibitor formation.

iii. **Enzyme costs**

To turn cellulose into sugars that can ferment, we need cellulase enzymes. Right now, these enzymes cost more than the ones used to make first-generation biofuels. Some think that enzymes for cellulosic ethanol production might cost 20 to 40 times more than those for corn-based ethanol. This is because we need more of them, and they don't work as well. We need breakthroughs in enzyme engineering and production methods to lower these costs.

iv. **Scaling up for large-scale production**

Moving from small lab tests to big industrial biofuel production comes with several hurdles:

Nutrient Supply and Recycling: It's key to keep enough nutrients available and recycle them well to keep microbes active on a large scale.

Gas Transfer and Exchange: We need good systems to handle the movement and exchange of gases, like oxygen and carbon dioxide, to make sure microbes work at their best.

Light Delivery: Photoautotrophic microorganisms like some algae need even light spread to boost photosynthesis to its peak.

Cultural Integrity and Environmental Control: Keeping cultures stable and managing factors like temperature and pH plays a key role to avoid contamination and ensure steady biofuel output.

Land and Water Availability: Big operations need lots of land and water so careful planning helps to cut down on harm to the environment.

Harvesting Techniques: Coming up with cheap and quick ways to harvest is crucial to lower running costs.

Genetic and Metabolic Engineering: Genetically modified with improved metabolic process microbial species can improve biofuel yields improving biofuel conversion efficiency.

To address these challenges, the requirement of cross-disciplinary research and innovative technological breakthroughs in the development of sustainable and cost-effective methods are requisite for producing high yields of biofuel from food waste.

3.5.1 Future Prospects

With several procedural and technological constraints, the advancements in biofuel production from food waste using microorganisms have held significance in recent years, focusing on enhancing efficiency, sustainability, and scalability.

The advancement of the biotechnology and genetic engineering process has significantly improved the efficiency of converting food waste into biofuels. This includes the development of new microbial strains, which utilize a wide range of food waste verities and sustain adverse conditions during the conversion process to improve yield and reduce production costs (Wang et al., 2024). The utilization of food waste in biofuel production addresses both food waste management and biofuel production. The reduction of gas emissions lowered landfills burden are the most important aspects of food waste utilization which indeed contribute to future sustainable energy (Kurniawan et al., 2024). The advancement in technology will lead to a reduction in the biofuel production cost, thus lowering the reliance on fossil fuel. This could lead to increased investment and growth in the biofuel sector (Van Rooyen et al., 2024). The inclusive policies and regulations will have a significant role in advancement of the biofuel production. The government and organization are likely to establish frameworks and incentives to promote biofuel production using food waste.

In conclusion, the future prospects for upscaling food waste-to-biofuel production using microbial systems are bright, with significant potential for technological, environmental, and economic benefits (Dhiman and Mukherjee, 2021). The current hurdles involved in large-scale production of biofuel from food stock like feedstock sources, optimizing microbial processes, and developing efficient bioreactors can be overcome by continual contribution by the research community (Hafid et al., 2022).

3.6 CONCLUSION

The use of microbial systems in food waste upscaling represents a sustainable and efficient approach to addressing the global food waste management and energy crisis. Various bioprocesses involved in upscaling like fermentation, AD, and transesterification offer a promising solution for the conversion of organic complex molecules to biofuels (bioethanol, biomethane, biohydrogen, and biodiesel). These processes are not only solving the landfill accumulation issues but also converting waste into valuable energy sources contributing to a circular economy.

For effective food waste upscaling, the challenges such as process optimization, microbial strain efficiency, feedstock variability, and large-scale feasibility must be taken into consideration. The advancement in technologies of engineering and molecular biology will play a critical role in improving biofuel yield and process efficiency through bioreactor design, genetic engineering, and metabolic pathway optimization. Additionally, improved policies, financial support, and industrial investments are essential for the successful commercialization and social acceptance of microbial biofuel produced from food waste.

Global energy security, environmental conservation, and renewable energy solutions can be achieved through food waste management with innovative biotechnological approaches with sustainable waste management strategies.

ACKNOWLEDGMENT

SVO is thankful to the University Grand Commission (UGC), Government of India for awarding Dr. D. S. Kothari a Fellowship.

REFERENCES

Abreu AA, Tavares F, Alves MM, et al (2019) Garden and food waste co-fermentation for biohydrogen and biomethane production in a two-step hyperthermophilic-mesophilic process. *Bioresource Technology* 278:180–186. https://doi.org/10.1016/j.biortech.2019.01.085

Abubackar HN, Keskin T, Arslan K, et al (2019) Effects of size and autoclavation of fruit and vegetable wastes on biohydrogen production by dark dry anaerobic fermentation under mesophilic condition. *International Journal of Hydrogen Energy* 44:17767–17780. https://doi.org/10.1016/j.ijhydene.2019.05.106

Ajala OE, Aberuagba F, Odetoye TE, Ajala AM (2015) Biodiesel: Sustainable energy replacement to petroleum-based diesel fuel – A review. *Chemical and Biochemical Engineering Reviews* 2:145–156. https://doi.org/10.1002/cben.201400024

Alvarado-Ramírez L, Santiesteban-Romero B, Poss G, et al (2023) Sustainable production of biofuels and bioderivatives from aquaculture and marine waste. *Frontiers in Chemical Engineering* 4:1072761. https://doi.org/10.3389/fceng.2022.1072761

Al-Wahaibi A, Osman AI, Al-Muhtaseb AH, et al (2020) Techno-economic evaluation of biogas production from food waste via anaerobic digestion. *Scientific Reports* 10:15719. https://doi.org/10.1038/s41598-020-72897-5

Arous F, Atitallah IB, Nasri M, Mechichi T (2017) A sustainable use of low-cost raw substrates for biodiesel production by the oleaginous yeast Wickerhamomyces anomalus. *3 Biotech* 7:268. https://doi.org/10.1007/s13205-017-0903-6

Bharathi P, Dayana R, Rangaraju M, et al (2022) Biogas production from food waste using nanocatalyst. *Journal of Nanomaterials* 2022:7529036. https://doi.org/10.1155/2022/7529036

Bhatia SK, Gurav R, Choi T-R, et al (2019) Effect of synthetic and food waste-derived volatile fatty acids on lipid accumulation in Rhodococcus sp. YHY01 and the properties of produced biodiesel. *Energy Conversion and Management* 192:385–395. https://doi.org/10.1016/j.enconman.2019.03.081

Bibra M, Rathinam NK, Johnson GR, Sani RK (2020) Single pot biovalorization of food waste to ethanol by Geobacillus and Thermoanaerobacter spp. *Renewable Energy* 155:1032–1041. https://doi.org/10.1016/j.renene.2020.02.093

Bibra M, Samanta D, Sharma NK, et al (2022) Food waste to bioethanol: Opportunities and challenges. *Fermentation* 9:8. https://doi.org/10.3390/fermentation9010008

Brar KK, Sarma AK, Aslam M, et al (2017) Potential of oleaginous yeast Trichosporon sp., for conversion of sugarcane bagasse hydrolysate into biodiesel. *Bioresource Technology* 242:161–168. https://doi.org/10.1016/j.biortech.2017.03.155

Brexó RP, Sant'Ana AS (2017) Impact and significance of microbial contamination during fermentation for bioethanol production. *Renewable and Sustainable Energy Reviews* 73:423–434. https://doi.org/10.1016/j.rser.2017.01.151

Bux C, Cangialosi F, Amicarelli V (2023) Biomethane and compost production by anaerobic digestion of organic waste: Suggestions for rural communities in southern Italy. *Sustainability* 15:15644. https://doi.org/10.3390/su152115644

Carmona-Cabello M, García IL, Papadaki A, et al. (2021) Biodiesel production using microbial lipids derived from food waste discarded by catering services. *Bioresource Technology* 323:124597. https://doi.org/10.1016/j.biortech.2020.124597

Carota E, Crognale S, D'Annibale A, et al (2017) A sustainable use of Ricotta cheese Whey for microbial biodiesel production. *Science of the Total Environment* 584–585:554–560. https://doi.org/10.1016/j.scitotenv.2017.01.068

Carota E, Petruccioli M, D'Annibale A, et al (2020) Orange peel waste–based liquid medium for biodiesel production by oleaginous yeasts. *Applied Microbiology and Biotechnology* 104:4617–4628. https://doi.org/10.1007/s00253-020-10579-y

Chaudhary P, Garg S, George T, et al (2021) Underreporting and open burning – The two largest challenges for sustainable waste management in India. *Resources, Conservation and Recycling* 175:105865. https://doi.org/10.1016/j.resconrec.2021.105865

Chen Y (2011) Development and application of co-culture for ethanol production by co-fermentation of glucose and xylose: A systematic review. *Journal of Industrial Microbiology and Biotechnology* 38:581–597. https://doi.org/10.1007/s10295-010-0894-3

Chen Y, Cheng JJ, Creamer KS (2008) Inhibition of anaerobic digestion process: A review. *Bioresource Technology* 99:4044–4064. https://doi.org/10.1016/j.biortech.2007.01.057

Das V, Dunford N, Deka D (2022) Waste utilization and biodiesel production from Desmodesmus maximus grown in swine wastewater using waste eggshells as a catalyst. *Aquacultural Engineering* 99:102293. https://doi.org/10.1016/j.aquaeng.2022.102293

De Castro TR, De Macedo DC, De Genaro Chiroli DM, et al (2022) The potential of cleaner fermentation processes for bioplastic production: A narrative review of polyhydroxyalkanoates (PHA) and polylactic acid (PLA). *Journal of Polymers and the Environment* 30:810–832. https://doi.org/10.1007/s10924-021-02241-z

Demiray E, Açıkel E, Ertuğrul Karatay S, Dönmez G (2022) *Saccharomyces cerevisiae* and newly isolated *Candida boidinii* co-fermentation of industrial tea waste for improved bioethanol production. *Energy Sources, Part A: Recovery, Utilization, and Environmental Effects* 44:1160–1172. https://doi.org/10.1080/15567036.2022.2053763

Demiray E, González-Fernández C, Tomás-Pejó E (2024) Kluyveromyces marxianus and *Bacillus coagulans* co-cultivation: Efficient co-production of bioethanol and lactic acid from pomegranate peels. *Bioresource Technology Reports* 25:101808. https://doi.org/10.1016/j.biteb.2024.101808

Demirbas A (2009) Biodiesel from waste cooking oil via base-catalytic and supercritical methanol transesterification. *Energy Conversion and Management* 50:923–927. https://doi.org/10.1016/j.enconman.2008.12.023

Dhiman S, Mukherjee G (2021) Present scenario and future scope of food waste to biofuel production. *Journal of Food Process Engineering* 44:e13594. https://doi.org/10.1111/jfpe.13594

El-Araby R (2024) Biofuel production: Exploring renewable energy solutions for a greener future. *Biotechnol Biofuels* 17:129. https://doi.org/10.1186/s13068-024-02571-9

ElMekawy A, Diels L, De Wever H, Pant D (2013) Valorization of cereal based biorefinery byproducts: Reality and expectations. *Environmental Science & Technology* 47:9014–9027. https://doi.org/10.1021/es402395g

Emmanuel JK, Nganyira PD, Shao GN (2022) Evaluating the potential applications of brewers' spent grain in biogas generation, food and biotechnology industry: A review. *Heliyon* 8:e11140. https://doi.org/10.1016/j.heliyon.2022.e11140

Favaro L, Cagnin L, Basaglia M, et al (2017) Production of bioethanol from multiple waste streams of rice milling. *Bioresource Technology* 244:151–159. https://doi.org/10.1016/j.biortech.2017.07.108

Ferdeș M, Zăbavă B Ștefania, Paraschiv G, et al (2022) Food waste management for biogas production in the context of sustainable development. *Energies* 15:6268. https://doi.org/10.3390/en15176268

Gallego-García M, Moreno AD, Manzanares P, et al (2023) Recent advances on physical technologies for the pretreatment of food waste and lignocellulosic residues. *Bioresource Technology* 369:128397. https://doi.org/10.1016/j.biortech.2022.128397

Gani P, Apandi NM, Mohamed Sunar N, et al (2023) Characterisation of bio-oil extracted from microalgae Botryococcus sp. biomass grown in domestic and food processing wastewaters for valuable hydrocarbon production. *Biofuels* 14:433–444. https://doi.org/10.1080/17597269.2022.2147139

Gbiete D, Narra S, Mani Kongnine D, et al (2024) Insights into biohydrogen production through dark fermentation of food waste: Substrate properties, inocula, and pretreatment strategies. *Energies* 17:6350. https://doi.org/10.3390/en17246350

Giannakis N, Carmona-Cabello M, Makri A, et al (2023) Spent coffee grounds and orange peel residues based biorefinery for microbial oil and biodiesel conversion estimation. *Renewable Energy* 209:382–392. https://doi.org/10.1016/j.renene.2023.01.110

Gujjala LKS, Bandyopadhyay TK, Banerjee R (2019) Production of biodiesel utilizing laccase pretreated lignocellulosic waste liquor: An attempt towards cleaner production process. *Energy Conversion and Management* 196:979–987. https://doi.org/10.1016/j.enconman.2019.06.036

Habashy MM, Ong ES, Abdeldayem OM, et al (2021) Food waste: A promising source of sustainable biohydrogen fuel. *Trends in Biotechnology* 39:1274–1288. https://doi.org/10.1016/j.tibtech.2021.04.001

Hafid HS, Omar FN, Abdul Rahman N, Wakisaka M (2022) Innovative conversion of food waste into biofuel in integrated waste management system. *Critical Reviews in Environmental Science and Technology* 52:3453–3492. https://doi.org/10.1080/10643389.2021.1923976

Haldar D, Shabbirahmed AM, Singhania RR, et al (2022) Understanding the management of household food waste and its engineering for sustainable valorization – A state-of-the-art review. *Bioresource Technology* 358:127390. https://doi.org/10.1016/j.biortech.2022.127390

Hallenbeck PC, Ghosh D (2009) Advances in fermentative biohydrogen production: The way forward? *Trends in Biotechnology* 27:287–297. https://doi.org/10.1016/j.tibtech.2009.02.004

Han S (2004) Biohydrogen production by anaerobic fermentation of food waste. *International Journal of Hydrogen Energy* 29:569–577. https://doi.org/10.1016/j.ijhydene.2003.09.001

Han W, Yan Y, Shi Y, et al (2016) Biohydrogen production from enzymatic hydrolysis of food waste in batch and continuous systems. *Scientific Reports* 6:38395. https://doi.org/10.1038/srep38395

Hang YD, Lee CY, Woodams EE, Cooley HJ (1981) Production of alcohol from apple pomace. *Applied and Environmental Microbiology* 42:1128–1129. https://doi.org/10.1128/aem.42.6.1128-1129.1981

Hossain T, Miah AB, Mahmud SA, Mahin AA (2018) Enhanced bioethanol production from potato peel waste via consolidated bioprocessing with statistically optimized medium. *Applied Biochemistry and Biotechnology* 186:425–442. https://doi.org/10.1007/s12010-018-2747-x

Huang H, Qureshi N, Chen M-H, et al (2015) Ethanol production from food waste at high solids content with vacuum recovery technology. *The Journal of Agricultural and Food Chemistry* 63:2760–2766. https://doi.org/10.1021/jf5054029

Ingram LO, Doran JB (1995) Conversion of cellulosic materials to ethanol. *FEMS Microbiology Reviews* 16:235–241. https://doi.org/10.1111/j.1574-6976.1995.tb00170.x

Isono N, Hayakawa H, Usami A, et al. (2012) A comparative study of ethanol production by Issatchenkia orientalis strains under stress conditions. *Journal of Bioscience and Bioengineering* 113:76–78. https://doi.org/10.1016/j.jbiosc.2011.09.004

Jameel MK, Mustafa MA, Ahmed HS, et al (2024) Biogas: Production, properties, applications, economic and challenges: A review. *Results in Chemistry* 7:101549. https://doi.org/10.1016/j.rechem.2024.101549

Jiang D, Ge X, Zhang T, et al (2016) Photo-fermentative hydrogen production from enzymatic hydrolysate of corn stalk pith with a photosynthetic consortium. *International Journal of Hydrogen Energy* 41:16778–16785. https://doi.org/10.1016/j.ijhydene.2016.07.129

Karaosmanoglu Gorgec F, Karapinar I (2019) Biohydrogen production from hydrolyzed waste wheat by dark fermentation in a continuously operated packed bed reactor: The effect of hydraulic retention time. *International Journal of Hydrogen Energy* 44:136–143. https://doi.org/10.1016/j.ijhydene.2018.08.155

Karmee SK, Lin CSK (2014) Valorisation of food waste to biofuel: Current trends and technological challenges. *Sustainable Chemical Processes* 2:22. https://doi.org/10.1186/s40508-014-0022-1

Kavitha S, Ravi YK, Kumar G, et al (2024) Integrated microalgae cultivation and sustainable biofuel production from pretreated food waste. *Journal of Cleaner Production* 466:142829. https://doi.org/10.1016/j.jclepro.2024.142829

Kazemi Shariat Panahi H, Dehhaghi M, Guillemin GJ, et al (2022) Bioethanol production from food wastes rich in carbohydrates. *Current Opinion in Food Science* 43:71–81. https://doi.org/10.1016/j.cofs.2021.11.001

Khaswal A, Mishra SK, Chaturvedi N, et al (2024) Microbial enzyme production: Unlocking the potential of agricultural and food waste through solid-state fermentation. *Bioresource Technology Reports* 27:101880. https://doi.org/10.1016/j.biteb.2024.101880

Khedkar MA, Nimbalkar PR, Kamble SP, et al (2018) Process intensification strategies for enhanced holocellulose solubilization: Beneficiation of pineapple peel waste for cleaner butanol production. https://doi.org/10.48550/ARXIV.1810.05543

Kim S, Dale BE (2004) Global potential bioethanol production from wasted crops and crop residues. *Biomass and Bioenergy* 26:361–375. https://doi.org/10.1016/j.biombioe.2003.08.002

Kim YS, Jang JY, Park SJ, Um BH (2018) Dilute sulfuric acid fractionation of Korean food waste for ethanol and lactic acid production by yeast. *Waste Management* 74:231–240. https://doi.org/10.1016/j.wasman.2018.01.012

Kiran UE, Trzcinski AP, Ng WJ, Liu Y (2014) Bioconversion of food waste to energy: A review. *Fuel* 134:389–399. https://doi.org/10.1016/j.fuel.2014.05.074

Koike Y, An M-Z, Tang Y-Q, et al (2009) Production of fuel ethanol and methane from garbage by high-efficiency two-stage fermentation process. *Journal of Bioscience and Bioengineering* 108:508–512. https://doi.org/10.1016/j.jbiosc.2009.06.007

Kothari R, Prasad R, Kumar V, Singh DP (2013) Production of biodiesel from microalgae Chlamydomonas polypyrenoideum grown on dairy industry wastewater. *Bioresource Technology* 144:499–503. https://doi.org/10.1016/j.biortech.2013.06.116

Krikigianni E, Matsakas L, Rova U, et al (2022) Investigating the bioconversion potential of volatile fatty acids: Use of oleaginous yeasts rhodosporidium toruloides and cryptococcus curvatus towards the sustainable production of biodiesel and odd-chain Fatty acids. *Applied Sciences* 12:6541. https://doi.org/10.3390/app12136541

Kurniawan TA, Hassan GK, Al-Hazmi HE, et al (2024) Landfill mining: A step forward to reducing CH4 emissions and enhancing CO2 sequestration from landfill. *Journal of Hazardous Materials Advances* 100512. https://doi.org/10.1016/j.hazadv.2024.100512

Lin CSK, Pfaltzgraff LA, Herrero-Davila L, et al (2013) Food waste as a valuable resource for the production of chemicals, materials and fuels. Current situation and global perspective. *Energy Environmental Science & Technology* 6:426. https://doi.org/10.1039/c2ee23440h

Lin J, Zuo J, Gan L, et al (2011) Effects of mixture ratio on anaerobic co-digestion with fruit and vegetable waste and food waste of China. *Journal of Environmental Sciences* 23:1403–1408. https://doi.org/10.1016/S1001-0742(10)60572-4

Liu J, Dantoft SH, Würtz A, et al (2016) A novel cell factory for efficient production of ethanol from dairy waste. *Biotechnol Biofuels* 9:33. https://doi.org/10.1186/s13068-016-0448-7

Liu S, Li W, Zheng G, et al (2020) Optimization of Cattle Manure and Food Waste Co-Digestion for Biohydrogen Production in a Mesophilic Semi-Continuous Process. *Energies* 13:3848. https://doi.org/10.3390/en13153848

Lu C, Zhang Z, Ge X, et al (2016) Bio-hydrogen production from apple waste by photosynthetic bacteria HAU-M1. *International Journal of Hydrogen Energy* 41:13399–13407. https://doi.org/10.1016/j.ijhydene.2016.06.101

Ma H, Wang Q, Zhang W, et al (2008) Optimization of the medium and process parameters for ethanol production from kitchen garbage by zymomonas mobilis. *International Journal of Green Energy* 5:480–490. https://doi.org/10.1080/15435070802498176

Ma H, Wang Y, Lv P, et al (2024) Ethanol production from a mixture of waste tissue paper and food waste through saccharification and mixed-culture fermentation. *Fermentation* 10:194. https://doi.org/10.3390/fermentation10040194

Ma K, Ruan Z, Shui Z, et al (2016) Open fermentative production of fuel ethanol from food waste by an acid-tolerant mutant strain of Zymomonas mobilis. *Bioresource Technology* 203:295–302. https://doi.org/10.1016/j.biortech.2015.12.054

Ma X, Gao Z, Gao M, et al (2019) Microbial lipid production from food waste saccharified liquid under two-stage process. *Bioresource Technology* 289:121626. https://doi.org/10.1016/j.biortech.2019.121626

Mahish PK, Verma DK, Ghritlahare A, et al (2024) Microbial bioconversion of food waste to bio-fertilizers. *Sustainable Food Technology* 2:689–708. https://doi.org/10.1039/D3FB00041A

Mahmoodi P, Karimi K, Taherzadeh MJ (2018) Efficient conversion of municipal solid waste to biofuel by simultaneous dilute-acid hydrolysis of starch and pretreatment of lignocelluloses. *Energy Conversion and Management* 166:569–578. https://doi.org/10.1016/j.enconman.2018.04.067

Maleki F, Changizian M, Zolfaghari N, et al (2021) Consolidated bioprocessing for bioethanol production by metabolically engineered *Bacillus subtilis* strains. *Scientific Reports* 11:13731. https://doi.org/10.1038/s41598-021-92627-9

Martani F, Maestroni L, Torchio M, et al (2020) Conversion of sugar beet residues into lipids by Lipomyces starkeyi for biodiesel production. *Microbial Cell Factories* 19:204. https://doi.org/10.1186/s12934-020-01467-1

Martí-Herrero J, Soria-Castellón G, Diaz-de-Basurto A, et al (2019) Biogas from a full scale digester operated in psychrophilic conditions and fed only with fruit and vegetable waste. *Renewable Energy* 133:676–684. https://doi.org/10.1016/j.renene.2018.10.030

Matsakas L, Kekos D, Loizidou M, Christakopoulos P (2014) Utilization of household food waste for the production of ethanol at high dry material content. *Biotechnol Biofuels* 7:4. https://doi.org/10.1186/1754-6834-7-4

Mgeni ST, Mero HR, Mtashobya LA, Emmanuel JK (2024) The prospect of fruit wastes in bioethanol production: A review. *Heliyon* 10:e38776. https://doi.org/10.1016/j.heliyon.2024.e38776

Mihajlovski K, Radovanović Ž, Carević M, Dimitrijević-Branković S (2018) Valorization of damaged rice grains: Optimization of bioethanol production by waste brewer's yeast using an amylolytic potential from the *Paenibacillus chitinolyticus* CKS1. *Fuel* 224:591–599. https://doi.org/10.1016/j.fuel.2018.03.135

Mitrea L, Călinoiu L-F, Teleky B-E, et al (2022) Waste cooking oil and crude glycerol as efficient renewable biomass for the production of platform organic chemicals through oleophilic yeast strain of Yarrowia lipolytica. *Environmental Technology & Innovation* 28:102943. https://doi.org/10.1016/j.eti.2022.102943

Monika, Banga S, Pathak VV (2023) Biodiesel production from waste cooking oil: A comprehensive review on the application of heterogenous catalysts. *Energy Nexus* 10:100209. https://doi.org/10.1016/j.nexus.2023.100209

Moussa RN, Moussa N, Dionisi D (2022) Hydrogen production from biomass and organic waste using dark fermentation: An analysis of literature data on the effect of operating parameters on process performance. *Processes* 10:156. https://doi.org/10.3390/pr10010156

Mu D, Ma K, He L, Wei Z (2023) Effect of microbial pretreatment on degradation of food waste and humus structure. *Bioresource Technology* 385:129442. https://doi.org/10.1016/j.biortech.2023.129442

Mujtaba M, Fernandes Fraceto L, Fazeli M, et al (2023) Lignocellulosic biomass from agricultural waste to the circular economy: A review with focus on biofuels, biocomposites and bioplastics. *Journal of Cleaner Production* 402:136815. https://doi.org/10.1016/j.jclepro.2023.136815

Muñoz P, Cordero C, Tapia X, et al (2020) Assessment of anaerobic digestion of food waste at psychrophilic conditions and effluent post-treatment by microalgae cultivation. *Clean Technologies and Environmental Policy* 22:725–733. https://doi.org/10.1007/s10098-019-01803-z

Mussatto SI, Dragone G, Guimarães PMR, et al (2010) Technological trends, global market, and challenges of bio-ethanol production. *Biotechnology Advances* 28:817–830. https://doi.org/10.1016/j.biotechadv.2010.07.001

Narisetty V, Nagarajan S, Gadkari S, et al (2022) Process optimization for recycling of bread waste into bioethanol and biomethane: A circular economy approach. *Energy Conversion and Management* 266:115784. https://doi.org/10.1016/j.enconman.2022.115784

Negri C, Ricci M, Zilio M, et al (2020) Anaerobic digestion of food waste for bio-energy production in China and Southeast Asia: A review. *Renewable and Sustainable Energy Reviews* 133:110138. https://doi.org/10.1016/j.rser.2020.110138

Nigam PS, Singh A (2011) Production of liquid biofuels from renewable resources. *Progress in Energy and Combustion Science* 37:52–68. https://doi.org/10.1016/j.pecs.2010.01.003

Ohkouchi Y, Inoue Y (2007) Impact of chemical components of organic wastes on l(+)–lactic acid production. *Bioresource Technology* 98:546–553. https://doi.org/10.1016/j.biortech.2006.02.005

Onyeaka H, Mansa RF, Wong CMVL, Miri T (2022) Bioconversion of starch base food waste into bioethanol. *Sustainability* 14:11401. https://doi.org/10.3390/su141811401

Patnaik S, Saravanabhupathy S, Singh S, et al (2021) Multi-objective optimization for biomass and lipid production by oleaginous bacteria using vegetable waste as feedstock. *Environmental Engineering Research* 27:210061. https://doi.org/10.4491/eer.2021.061

Pilarska AA, Kulupa T, Kubiak A, et al (2023) Anaerobic digestion of food waste—A short review. *Energies* 16:5742. https://doi.org/10.3390/en16155742

Pleissner D, Kwan TH, Lin CSK (2014) Fungal hydrolysis in submerged fermentation for food waste treatment and fermentation feedstock preparation. *Bioresource Technology* 158:48–54. https://doi.org/10.1016/j.biortech.2014.01.139

Pleissner D, Lam WC, Sun Z, Lin CSK (2013) Food waste as nutrient source in heterotrophic microalgae cultivation. *Bioresource Technology* 137:139–146. https://doi.org/10.1016/j.biortech.2013.03.088

Pradhan S, Yuzer B, Bicer Y, et al (2024) Hydrogen gas and biochar production from kitchen food waste through dark fermentation and pyrolysis. *Frontiers in Chemical Engineering* 6:1450151. https://doi.org/10.3389/fceng.2024.1450151

Prasoulas G, Gentikis A, Konti A, et al (2020) Bioethanol production from food waste applying the multienzyme system produced on-site by *Fusarium oxysporum* F3 and mixed microbial cultures. *Fermentation* 6:39. https://doi.org/10.3390/fermentation6020039

Qu X, Zeng H, Gao Y, et al (2022) Bio-hydrogen production by dark anaerobic fermentation of organic wastewater. *Frontiers in Chemistry* 10:978907. https://doi.org/10.3389/fchem.2022.978907

Rai PK, Singh SP, Asthana RK, Singh S (2014) Biohydrogen production from sugarcane bagasse by integrating dark- and photo-fermentation. *Bioresource Technology* 152:140–146. https://doi.org/10.1016/j.biortech.2013.10.117

Raita M, Ibenegbu C, Champreda V, Leak DJ (2016) Production of ethanol by thermophilic oligosaccharide utilising Geobacillus thermoglucosidasius TM242 using palm kernel cake as a renewable feedstock. *Biomass and Bioenergy* 95:45–54. https://doi.org/10.1016/j.biombioe.2016.08.015

Ranganathan S, Dutta S, Moses JA, Anandharamakrishnan C (2020) Utilization of food waste streams for the production of biopolymers. *Heliyon* 6:e04891. https://doi.org/10.1016/j.heliyon.2020.e04891

Rao R, Basak N (2021) Development of novel strategies for higher fermentative biohydrogen recovery along with novel metabolites from organic wastes: The present state of the art. *Biotechnology and Applied Biochemistry* 68:421–444. https://doi.org/10.1002/bab.1964

Rodrigues JGC, Cardoso FV, Santos CCD, et al (2023) Biocatalyzed transesterification of waste cooking oil for biodiesel production using lipase from the Amazonian fungus endomelanconiopsis endophytica. *Energies* 16:6937. https://doi.org/10.3390/en16196937

Roslan MAM, Jefri NQUA, Ramlee N, et al (2021) Enhancing food waste biodegradation rate in a food waste biodigester with the synergistic action of hydrolase-producing Bacillus paralicheniformis GRA2 and Bacillus velezensis TAP5 co-culture inoculation. *Saudi Journal of Biological Sciences* 28:3001–3012. https://doi.org/10.1016/j.sjbs.2021.02.041

Sagir E, Ozgur E, Gunduz U, et al (2017) Single-stage photofermentative biohydrogen production from sugar beet molasses by different purple non-sulfur bacteria. *Bioprocess and Biosystems Engineering* 40:1589–1601. https://doi.org/10.1007/s00449-017-1815-x

Sahota S, Kumar S, Lombardi L (2024) Biohythane, biogas, and biohydrogen production from food waste: Recent advancements, technical bottlenecks, and prospects. *Energies* 17:666. https://doi.org/10.3390/en17030666

Samrot AV, Rajalakshmi D, Sathiyasree M, et al (2023) A review on biohydrogen sources, production routes, and its application as a fuel cell. *Sustainability* 15:12641. https://doi.org/10.3390/su151612641

Sar T, Harirchi S, Ramezani M, et al (2022) Potential utilization of dairy industries by-products and wastes through microbial processes: A critical review. *Science of the Total Environment* 810:152253. https://doi.org/10.1016/j.scitotenv.2021.152253

Sarrion A, Ipiales RP, De La Rubia MA, et al (2023) Chicken meat and bone meal valorization by hydrothermal treatment and anaerobic digestion: Biofuel production and nutrient recovery. *Renewable Energy* 204:652–660. https://doi.org/10.1016/j.renene.2023.01.005

Singh D, Tembhare M, Machhirake N, Kumar S (2023) Biogas generation potential of discarded food waste residue from ultra-processing activities at food manufacturing and packaging industry. *Energy* 263:126138. https://doi.org/10.1016/j.energy.2022.126138

Soleimani SS, Adiguzel A, Nadaroglu H (2017) Production of bioethanol by facultative anaerobic bacteria: Production of fermentative bioethanol. *Journal of the Institute of Brewing* 123:402–406. https://doi.org/10.1002/jib.437

Sugebo B (2022) A review on enhanced biofuel production from coffee by-products using different enhancement techniques. *Materials for Renewable and Sustainable Energy* 11:91–103. https://doi.org/10.1007/s40243-022-00209-0

Tiago GAO, Rodrigo NPB, Lourinho G, et al (2025) Systematic and bibliometric review of biomethane production from biomass-based residues: Technologies, economics and environmental impact. *Fuels* 6:8. https://doi.org/10.3390/fuels6010008

Tlais AZA, Fiorino GM, Polo A, et al (2020) High-value compounds in fruit, vegetable and cereal byproducts: An overview of potential sustainable reuse and exploitation. *Molecules* 25:2987. https://doi.org/10.3390/molecules25132987

Uncu ON, Cekmecelioglu D (2011) Cost-effective approach to ethanol production and optimization by response surface methodology. *Waste Management* 31:636–643. https://doi.org/10.1016/j.wasman.2010.12.007

Van Rooyen J, Mbella Teke G, Coetzee G, et al (2024) Enhancing bioethanol yield from food waste: Integrating decontamination strategies and enzyme dosage optimization for sustainable biofuel production. *Fuel* 378:133026. https://doi.org/10.1016/j.fuel.2024.133026

Voukkali I, Papamichael I, Loizia P, et al (2023) Waste metrics in the framework of circular economy. *Waste Management & Research* 41:1741–1753. https://doi.org/10.1177/0734242X231190794

Wu X, Lin H, Zhu J (2013) Optimization of continuous hydrogen production from co-fermenting molasses with liquid swine manure in an anaerobic sequencing batch reactor. *Bioresource Technology* 136:351–359. https://doi.org/10.1016/j.biortech.2013.02.109

Wang H, Peng X, Zhang H, et al (2021) Microorganisms-promoted biodiesel production from biomass: A review. *Energy Conversion and Management: X* 12:100137. https://doi.org/10.1016/j.ecmx.2021.100137

Wang M, Ye X, Bi H, Shen Z (2024) Microalgae biofuels: Illuminating the path to a sustainable future amidst challenges and opportunities. *Biotechnology for Biofuels* 17:10. https://doi.org/10.1186/s13068-024-02461-0

Wang X, Fang Y, Wang Y, et al (2018) Single-stage photo-fermentative hydrogen production from hydrolyzed straw biomass using Rhodobacter sphaeroides. *International Journal of Hydrogen Energy* 43:13810–13820. https://doi.org/10.1016/j.ijhydene.2018.01.057

Xia A, Cheng J, Song W, et al (2015) Fermentative hydrogen production using algal biomass as feedstock. *Renewable and Sustainable Energy Reviews* 51:209–230. https://doi.org/10.1016/j.rser.2015.05.076

Yuan H, Chen Y, Zhang H, et al (2006) Improved bioproduction of short-chain fatty acids (SCFAs) from excess sludge under alkaline conditions. *Environmental Science & Technology* 40:2025–2029. https://doi.org/10.1021/es052252b

Zeng J, Zeng H, Wang Z (2022) Review on technology of making biofuel from food waste. *International Journal of Energy Research* 46:10301–10319. https://doi.org/10.1002/er.7868

Zhang C, Su H, Baeyens J, Tan T (2014) Reviewing the anaerobic digestion of food waste for biogas production. *Renewable and Sustainable Energy Reviews* 38:383–392. https://doi.org/10.1016/j.rser.2014.05.038

Zhang J, He K, Shi X, Zhao Y (2011) Comparison of particle emissions from an engine operating on biodiesel and petroleum diesel. *Fuel* 90:2089–2097. https://doi.org/10.1016/j.fuel.2011.01.039

Zhang L, Cheng J, Pei H, et al (2018b) Cultivation of microalgae using anaerobically digested effluent from kitchen waste as a nutrient source for biodiesel production. *Renewable Energy* 115:276–287. https://doi.org/10.1016/j.renene.2017.08.034

Zhang Q, Zhang Z, Wang Y, et al (2018a) Sequential dark and photo fermentation hydrogen production from hydrolyzed corn stover: A pilot test using 11 m^3 reactor. *Bioresource Technology* 253:382–386. https://doi.org/10.1016/j.biortech.2018.01.017

Zwirzitz A, Alteio L, Sulzenbacher D, et al (2021) Ethanol production from wheat straw hydrolysate by issatchenkia orientalis isolated from waste cooking oil. *Journal of Fungi* 7:121. https://doi.org/10.3390/jof7020121

4 Resource Recovery from Wastewater and Microbial Biofuel Production

Suvendu Manna and Ajaya Kumar Behera

4.1 INTRODUCTION

The growing global population, industrial expansion, and increasing energy demands have placed immense pressure on natural resources, necessitating sustainable approaches for resource recovery and energy production. Wastewater, traditionally viewed as a waste product, is now recognized as a valuable source of water, nutrients, and bioenergy. The shift toward a circular economy in wastewater treatment has led to the development of advanced technologies for recovering essential resources while mitigating environmental pollution. Among these technologies, microbial biofuel production has emerged as a promising solution for sustainable energy generation, utilizing wastewater as both a substrate and a treatment medium (Li et al., 2023; Zhao et al., 2023). Table 4.1 provides a comparative analysis of wastewater production across major countries, highlighting annual wastewater generation, treatment rates, and key challenges in wastewater management. Developed nations like Germany, Japan, and the United Kingdom have high wastewater treatment rates (>90%), driven by strict environmental regulations and advanced treatment technologies. In contrast, developing countries like India, Indonesia, and South Africa struggle with low treatment rates (<50%) due to poor infrastructure, inadequate investment, and rapid urbanization, whereas countries like China and Brazil fall into the moderate treatment category (50%–75%), facing challenges such as industrial pollution and untreated discharge into natural water bodies. Water-scarce nations, including Australia and South Africa, focus on wastewater recycling and reuse but face infrastructural and financial barriers. This comparative assessment underscores the urgent need for investment, technological advancements, and policy reforms to enhance wastewater treatment and resource recovery worldwide.

Microbial biofuel production harnesses the metabolic capabilities of microorganisms to convert organic matter in wastewater into biofuels such as biogas (methane), biohydrogen, biodiesel, and bioethanol. Traditional methods such as anaerobic digestion (AD) have been widely used for methane production, whereas emerging

DOI: 10.1201/9781003633709-4

TABLE 4.1
Comparative Analysis of Wastewater Production by Country

Country	Total Wastewater Production (Billion m³/year)	Wastewater Treatment (%)	Key Challenges in Wastewater Management
China	70–80	~75	Rapid urbanization, high industrial discharge, need for advanced treatment technologies
India	40–50	~30	Poor wastewater infrastructure, lack of treatment facilities, groundwater contamination
United States	50–60	~85	Aging treatment plants, nutrient pollution, high energy consumption in treatment
Brazil	12–15	~50	Untreated discharge into rivers, lack of investment in wastewater treatment
Russia	15–20	~70	Cold climate impact on biological treatment, industrial wastewater pollution
Germany	10–12	~96	Stricter regulations, high operational costs, focus on circular economy
Japan	15–18	~95	High-tech treatment systems, challenges in sludge management and reuse
Indonesia	8–10	~20	Poor sanitation coverage, high untreated domestic wastewater
Mexico	12–14	~60	High industrial pollution, water reuse challenges
South Africa	6–8	~50	Water scarcity, infrastructure deterioration, industrial effluent discharge
Australia	5–7	~90	Water recycling for agriculture, high desalination dependence
United Kingdom	10–12	~98	Stringent wastewater regulations, microplastic pollution concerns

bioelectrochemical systems (BES), microbial electrolysis cells (MECs), and microbial fuel cells (MFCs) offer new pathways for bioelectricity and hydrogen recovery (Sharma et al., 2023; Wang et al., 2024). Additionally, microalgae-based systems have been extensively studied for biodiesel production, leveraging their ability to grow in nutrient-rich wastewater while simultaneously removing contaminants such as nitrogen and phosphorus (Patel et al., 2024). Recent advancements in metabolic engineering, microbial consortia optimization, and synthetic biology have significantly improved the efficiency of wastewater-based biofuel production. Researchers have developed engineered microbial strains with enhanced metabolic pathways to increase hydrogen and methane yields, while nanomaterials and innovative electrode designs have improved the performance of bioelectrochemical systems (Zhao et al., 2023; Gupta et al., 2025). Furthermore, integrated hybrid systems combining AD with BES have demonstrated superior energy-recovery efficiencies (Zhou et al., 2024).

The integration of wastewater treatment and biofuel production offers multiple benefits, including renewable energy generation, waste minimization, and economic feasibility. Conventional wastewater treatment plants (WWTPs) consume large amounts of energy, contributing to their high operational costs and carbon footprint. By implementing biofuel recovery technologies, WWTPs can transition from energy-intensive facilities to energy-positive systems, reducing reliance on fossil fuels while achieving sustainable waste management (Singh et al., 2023). Additionally, nutrient recovery from wastewater, particularly phosphorus and nitrogen, supports sustainable agriculture by providing alternative fertilizer sources, reducing dependency on chemical fertilizers (Li et al., 2023).

Despite these advancements, several challenges hinder large-scale implementation of wastewater-derived biofuels. One major issue is variability in wastewater composition, which affects microbial performance and process stability. Wastewater contains a diverse range of organic and inorganic compounds, requiring tailored microbial consortia and process optimization strategies to ensure consistent biofuel production (Gupta et al., 2025). Additionally, the high capital and operational costs of advanced bioelectrochemical and algal-based systems limit their commercial viability (Singh et al., 2023). Another significant challenge is the regulatory and policy landscape, as many countries lack clear guidelines and incentives for integrating wastewater treatment with biofuel production (Wang et al., 2024). To overcome these challenges, future research should focus on several key areas. First, the development of robust and adaptable microbial consortia is crucial for enhancing biofuel yields under variable wastewater conditions (Zhao et al., 2023). Second, artificial intelligence (AI) and machine learning-based process optimization can improve real-time monitoring and control of wastewater treatment and biofuel production systems (Zhou et al., 2024). Third, hybrid technologies combining multiple recovery pathways such as integrating AD with MFCs or using microalgae for both wastewater treatment and biodiesel production can maximize resource recovery efficiency (Sharma et al., 2023). Finally, strong policy support and financial incentives are essential for scaling up biofuel production from wastewater, enabling wider adoption and commercialization (Patel et al., 2024).

The transition toward wastewater-based biofuel production represents a significant step toward achieving a circular and sustainable bioeconomy. The synergistic benefits of wastewater treatment and biofuel generation make it a highly attractive approach for addressing environmental and energy challenges. With continuous advancements in microbial engineering, bioelectrochemical technologies, and policy frameworks, wastewater-derived biofuels have the potential to play a critical role in the future of sustainable energy. Future interdisciplinary efforts in biotechnology, environmental engineering, and policymaking will be essential for realizing the full potential of wastewater-based resource recovery (Li et al., 2023; Patel et al., 2024).

4.2 RESOURCE RECOVERY FROM WASTEWATER

Wastewater is no longer considered a waste but a valuable source of water, nutrients, and energy. With the increasing global emphasis on sustainability and circular economy principles, resource recovery from wastewater has gained significant attention. Traditional wastewater treatment methods focus on contaminant removal, but modern approaches integrate resource recovery strategies, including water reclamation, nutrient extraction (nitrogen and phosphorus), biofuel production, and energy recovery. Wastewater streams from municipal, industrial, and agricultural sources contain valuable organic and inorganic compounds that can be harnessed for sustainable applications. The composition of wastewater varies depending on its source, influencing the choice of treatment and recovery processes. Industrial wastewater often contains heavy metals and complex organic pollutants, while municipal wastewater is rich in biodegradable organic matter and nutrients. Table 4.2 presents a comparative composition of different wastewater types, emphasizing their potential for resource recovery.

The comparative Table 4.2, highlights the variability in wastewater composition across different sources, emphasizing their potential for resource recovery. Municipal wastewater, which has moderate levels of organic matter (biochemical oxygen demand (BOD) and chemical oxygen demand (COD)), nitrogen, and phosphorus, is widely used for biogas production, nutrient extraction, and water reclamation. However, its pathogen content requires additional treatment before reuse. Industrial

TABLE 4.2
Comparative Composition of Different Types of Wastewater

Wastewater Type	Major Components	Concentration Range (mg/L)	Resource Recovery Potential	References
Municipal Wastewater	Organic matter (BOD, COD), nitrogen, phosphorus, pathogens	BOD: 100–400, COD: 250–1,000, TN: 20–85, TP: 5–20	Biogas (methane), nutrient recovery (fertilizers), reclaimed water	Wang et al. (2024)
Industrial Wastewater	Heavy metals, organic solvents, hydrocarbons, high COD	COD: 500–10,000, Heavy Metals: 0.1–50	Metal recovery, biohydrogen, bioelectricity (MFCs)	Zhang et al. (2024)
Agricultural Wastewater	Pesticides, fertilizers, organic matter, nitrogen, phosphorus	COD: 500–3,000, TN: 20–200, TP: 2–50	Biogas, nutrient extraction (fertilizers), irrigation water reuse	Singh et al. (2023)
Dairy Wastewater	High organic load, fats, proteins, lactose, nitrogen	COD: 2,000–10,000, TN: 50–500, Fats: 100–1,000	Biogas, biodiesel from fats, protein recovery	Patel et al. (2024)
Textile Wastewater	Synthetic dyes, heavy metals, surfactants, high salinity	COD: 500–5000, Heavy Metals: 0.5–10, TDS: 1,000–10,000	Water reuse, biosorption of dyes, bioelectricity	Gupta et al. (2025)
Slaughterhouse Wastewater	Blood, fats, proteins, ammonia, phosphorus	COD: 2,000–8,000, TN: 50–400, TP: 10–100	Biogas, biodiesel, protein extraction	Zhao et al. (2023)

wastewater exhibits the highest COD concentrations (500–10,000 mg/L) due to complex organic compounds and heavy metals, making it challenging to treat but valuable for metal recovery and MFC. In contrast, agricultural wastewater contains fertilizers, pesticides, and organic matter, making it an ideal source for biogas and nutrient extraction. Dairy and slaughterhouse wastewater have high organic loads (COD: 2,000–10,000 mg/L), making them promising for biodiesel, protein recovery, and methane generation. Textile wastewater, characterized by synthetic dyes, surfactants, and heavy metals, requires specialized treatment but has potential for biosorption-based dye removal and bioelectricity generation. This analysis underscores the importance of customized treatment strategies to maximize resource recovery from diverse wastewater streams, contributing to a sustainable circular economy.

4.2.1 Nutrient Recovery

Phosphorus and nitrogen are essential nutrients in agriculture and industry. Their recovery from wastewater not only reduces eutrophication risks but also supports sustainable fertilizer production.

4.2.1.1 Phosphorus Recovery

Phosphorus is an essential element for all living organisms and plays a crucial role in agriculture, industry, and ecological balance. However, its excessive release into water bodies leads to eutrophication, causing algal blooms and oxygen depletion, which threaten aquatic life (Ye et al., 2017). At the same time, global phosphorus reserves are rapidly depleting, raising concerns about future food security and the need for sustainable phosphate recovery from wastewater (Zhang et al., 2013). Wastewater from municipal, industrial, and agricultural sources contains significant amounts of phosphorus, offering a valuable opportunity for resource recovery and recycling. Biological approaches, including microalgal and bacterial phosphate recovery, provide sustainable alternatives for phosphorus removal. Polyphosphate-accumulating organisms (PAOs) are used in enhanced biological phosphorus removal (EBPR) systems, where bacteria absorb and store phosphate intracellularly. These microbes can be harvested to extract phosphorus for reuse (Chrispim et al., 2019). Additionally, microalgae can absorb phosphorus from wastewater while simultaneously producing biomass for biofuels and biofertilizers, making this approach a promising circular economy solution (Ni et al., 2025).

4.2.1.2 Nitrogen Recovery

Nitrogen is a critical nutrient for plant growth and industrial applications, but excessive nitrogen discharge into water bodies leads to eutrophication, groundwater contamination, and greenhouse gas emissions (Zhang et al., 2024). Conventional wastewater treatment processes primarily focus on nitrogen removal via nitrification-denitrification, which results in nitrogen loss as nitrogen gas. However, with increasing concerns over environmental sustainability and resource recovery, nitrogen recovery technologies are gaining attention as a means to close the nitrogen cycle and reduce dependency on synthetic fertilizers (Wang et al., 2024). Biological nitrogen recovery processes leverage microalgae, nitrifying bacteria, and anaerobic ammonium oxidation (anammox) bacteria to convert nitrogen into biomass or useful by-products (Singh et al., 2023). Microalgae can uptake ammonium and nitrate, using them as nutrients for growth, while bacteria-based systems can produce ammonium-enriched effluents that can be used as biofertilizers. Despite its sustainability, low nitrogen removal rates and large land requirements limit widespread implementation.

4.3 ORGANIC MATTER AND ENERGY RECOVERY

Wastewater contains a significant amount of organic matter, primarily in the form of carbohydrates, proteins, fats, and other biodegradable compounds. Traditionally, wastewater treatment has focused on removing these substances to prevent environmental pollution. However, with the shift toward resource recovery and circular economy principles, organic matter is now being recognized as a valuable resource for bioenergy production. Advanced technologies such as AD and BEFs, hydrothermal liquefaction, and biogas recovery enable the efficient conversion of organic waste into useful energy forms, including biogas (methane), biohydrogen, and bioelectricity. These approaches not only help in reducing the carbon footprint of wastewater treatment plants but also contribute to sustainable energy generation. By integrating organic matter recovery with energy production, wastewater treatment facilities can transition from energy-consuming systems to energy-positive infrastructures, promoting environmental sustainability and economic viability.

4.3.1 AD for Biogas Production

AD is a biological process that converts organic matter from wastewater, agricultural residues, and food waste into biogas under oxygen-free conditions. This process is driven by a diverse microbial community that breaks down complex organic molecules into methane (CH_4) and carbon dioxide (CO_2) through a series of biochemical reactions (Rajendran et al., 2022). AD is a sustainable method for waste stabilization, energy recovery, and sludge volume reduction, making it a crucial part of the circular bioeconomy and renewable energy sector (Andronikou et al., 2025). Biogas produced from AD contains approximately 50%–70% methane, depending on feedstock composition, operating conditions, and reactor design. This methane-rich biogas can be utilized for electricity generation, heating, or as a vehicle fuel, significantly reducing reliance on fossil fuels (Feng et al., 2023). Additionally, AD produces a nutrient-rich digestate, which can be used as a biofertilizer for agricultural applications (Jiang et al., 2024).

4.3.1.1 Stages of Anaerobic Digestion

The AD process consists of four primary steps, *Hydrolysis*: Complex carbohydrates, proteins, and lipids are broken down into simpler molecules such as sugars, amino acids, and fatty acids by hydrolytic bacteria. This step is often the rate-limiting stage in AD (Feng et al., 2023). *Acidogenesis*: Fermentative bacteria convert the hydrolyzed products into volatile fatty acids (VFAs), hydrogen (H_2), ammonia (NH_3), and carbon dioxide (CO_2). This phase generates acidic intermediates, which can lower system pH if not properly balanced. *Acetogenesis*: VFAs and other by-products are further converted into acetate, hydrogen, and CO_2 by acetogenic bacteria, which are essential for methane production in the final stage. *Methanogenesis*: Methanogenic archaea utilize acetate, hydrogen, and CO_2 to produce methane (CH_4) and additional CO_2. This stage determines the overall biogas yield and is sensitive to environmental conditions such as pH, temperature, and ammonia levels (González et al., 2022).

4.3.1.2 Key Factors Influencing Biogas Production

The quality and quantity of biogas depend on the carbon-to-nitrogen (C/N) ratio of the feedstock. A C/N ratio between 20 and 30 ensures microbial stability and prevents ammonia inhibition. Feedstocks rich in lipids and carbohydrates yield higher methane content than protein-rich substrates (Duan et al., 2023). AD can be operated under mesophilic (30°C–40°C) or thermophilic (50°C–60°C) conditions. Thermophilic digestion enhances biogas production rates and pathogen removal but requires higher energy input and careful process control (El Mashad & Zhang, 2020). The amount of organic matter fed into the digester per unit volume per day affects system efficiency. A high OLR can increase methane production, but excessive loading may lead to acid accumulation and process failure (Ramos-Suarez et al., 2024). HRT is the time organic matter remains inside the digester. Longer HRT allows for complete degradation of complex compounds, but reducing HRT can improve reactor efficiency if microbial adaptation is maintained. Optimal HRT ranges from 10 to 30 days, depending on reactor type and feedstock. The ideal pH range for methanogenesis is 6.8–7.5. A drop in pH due to excess acid accumulation can inhibit methanogenic bacteria, while high alkalinity can slow microbial activity (Shi et al., 2022).

4.3.1.3 Advancements in Anaerobic Digestion for Biogas Optimization

Co-digestion strategies: Co-digestion involves mixing different substrates (e.g., municipal wastewater sludge with agricultural residues) to enhance nutrient balance and increase biogas yield. Studies have shown that co-digestion can improve methane production by up to 50% (González et al., 2022; Table 4.3).

Bioaugmentation and microbial engineering: The introduction of specialized microbial consortia can enhance hydrolysis and methanogenesis, improving overall digestion efficiency. Engineered microbial strains with higher tolerance to ammonia and VFAs have been developed to sustain methane production under stressful conditions (Feng et al., 2023).

Integration with microbial electrochemical systems (MES): Emerging technologies such as MECs and MFCs are being integrated with AD to enhance methane yield and recover bioelectricity from wastewater (El Mashad & Zhang, 2020).

4.3.2 Microbial Fuel Cells (MFCs) for Direct Electricity Generation

MFCs are a promising bioelectrochemical technology that directly converts organic waste into electricity using electroactive bacteria. This innovative approach utilizes microorganisms as biocatalysts to oxidize organic matter, transferring electrons to an anode, which are then conducted through an external circuit to generate electric current (Patil et al., 2021). Compared to traditional wastewater treatment processes, MFCs offer a dual advantage of waste degradation and energy recovery, making

TABLE 4.3
Comparison of AD Systems

AD Technology	Temperature (°C)	HRT (Days)	Methane Yield (m^3 CH_4/kg VS)	Advantages	Reference
Mesophilic Digestion	30–40	15–30	0.2–0.4	Stable, lower energy demand	Jiang et al. (2024)
Thermophilic Digestion	50–60	10–20	0.4–0.6	Higher methane yield, pathogen removal	González et al., 2022).
Two-Stage Digestion	35–55	10–25	0.5–0.7	Enhanced hydrolysis and methanogenesis	El Mashad and Zhang (2020)
Co-digestion	35–50	15–25	0.5–0.8	Improved nutrient balance, higher methane yield	Shi et al., (2022)

them an environmentally sustainable solution. MFCs have gained significant attention for their potential applications in wastewater treatment, biosensors, and remote power generation. Recent advances in electrode materials, microbial community engineering, and reactor designs have led to improved power output and enhanced long-term stability of MFCs (Huang et al., 2024). This subsection explores the working principles, influencing factors, and advancements in MFC technology for direct electricity generation.

4.3.2.1 Working Principle of MFCs

MFCs operate through the biocatalytic activity of electroactive microorganisms that break down organic matter in an anoxic environment, releasing electrons and protons. The electrons travel through an external circuit to the cathode, while protons diffuse through a proton exchange membrane (PEM) to the cathode, where they combine with an electron acceptor (e.g., oxygen, ferricyanide) to generate electricity (Rahimnejad, 2023). The net result is the conversion of organic matter into electricity, water, and CO_2, making MFCs an effective energy-recovery technology in wastewater treatment systems (Sharma et al., 2023).

4.3.2.2 Factors Influencing MFC Performance

Electrode material: Electrode materials significantly impact the efficiency of electron transfer. Carbon-based electrodes (graphene, carbon cloth, carbon nanotubes) are widely used due to their high conductivity and biocompatibility (Kumar et al., 2021). Conductive polymers, such as polyaniline and polypyrrole, have also been incorporated to improve biofilm formation and electron transfer rates.

Microbial communities: The performance of an MFC is directly influenced by the type of bacteria colonizing the anode. Geobacter and Shewanella species are commonly used as exoelectrogenic microbes due to their ability to transfer electrons extracellularly (Huang et al., 2024). Engineering microbial consortia by adding synthetic electron shuttles has been explored to enhance power density.

TABLE 4.4
Comparison of MFC Technologies

MFC Type	Power Output (mW/m²)	Key Features	References
Single-Chamber MFC	200–600	Simple design, low cost	Rahimnejad, 2023
Double-Chamber MFC	500–1,000	High efficiency, membrane-dependent	Huang et al. (2024)
Stacked MFCs	1,000–2,000	Higher scalability, complex maintenance	Sharma et al. (2023)

Substrate type: Different organic substrates, such as glucose, acetate, and wastewater-derived organic compounds, yield varying power densities. Complex wastewaters containing proteins and lipids require hydrolysis before microbial oxidation, affecting overall efficiency (Mustafa et al., 2021; Table 4.4).

Reactor configuration: The design of the MFC affects electron transfer efficiency and mass transport. Common configurations include (a) Single-Chamber MFCs: Compact, cost-effective but limited oxygen diffusion. (b) Double-Chamber MFCs: Higher efficiency but require a membrane. (c) Stacked MFCs: Multiple MFCs in series for increased power output (Saratale et al., 2022).

4.3.2.3 Recent Advances in MFC Technology

Bioelectrode engineering: Recent studies have focused on enhancing biofilm formation on electrodes using nanomaterials, conductive polymers, and 3D-structured electrodes to improve electron transfer efficiency (Patil et al., 2021).

Hybrid systems: Integrating MFCs with AD or constructed wetlands has shown promising results in maximizing electricity generation and wastewater treatment efficiency (Sharma et al., 2023).

Scaling up MFCs: Although lab-scale MFCs have demonstrated efficiency, scaling up remains a challenge due to electrode fouling, low power density, and system costs. Efforts to develop modular and stackable MFCs are ongoing to enable practical applications (Mustafa et al., 2021).

4.3.3 Microbial Biofuel Production from Wastewater

Microbes play a critical role in converting wastewater organic matter into biofuels such as biogas, biohydrogen, biodiesel, and bioethanol.

4.3.3.1 Methane Production

Methane production from wastewater treatment is a critical process in AD, where microorganisms break down organic matter in the absence of oxygen to generate biogas – a mixture primarily composed of methane and carbon dioxide. This bioenergy recovery strategy not only mitigates the environmental impact of organic waste but also contributes to renewable energy generation (Wainaina and Taherzadeh, 2021). Methane production from wastewater has gained significant attention due to its potential for carbon neutrality, reduction of greenhouse gas (GHG) emissions, and its role in sustainable wastewater management. Recent advancements in microbial consortia, reactor design, and co-digestion strategies have significantly enhanced the efficiency of methane production. Optimizing parameters such as substrate composition, temperature, and hydraulic retention time (HRT) is crucial to improving methane yields and process stability (Liu et al., 2024). This section explores the biochemical process, influencing factors, and technological advancements in methane recovery from wastewater.

4.3.3.1.1 Recent Advances in Methane Production Technologies

Bioaugmentation with methanogenic microorganisms: The addition of specialized microbial consortia enhances methane production efficiency. Engineered synthetic microbial communities have been explored to boost process resilience and methane yields (Zhou et al., 2024).

Advanced reactor designs: New reactor configurations, such as anaerobic membrane bioreactors (AnMBRs) and upflow anaerobic sludge blanket (UASB) reactors, have demonstrated higher methane recovery rates compared to conventional digesters (Dahiya and Mohan, 2022).

Electrochemical enhancement of methanogenesis: Applying an external voltage to anaerobic digesters has been shown to stimulate electron transfer and increase methane production. BESs accelerate methanogenesis by providing additional electron donors (Sun et al., 2022).

Biogas upgrading techniques: Methane-rich biogas often contains impurities like CO_2 and hydrogen sulfide (H_2S). Technologies such as membrane separation, water scrubbing, and chemical absorption are being developed to improve biogas purity for direct use in energy applications (Li et al., 2024).

4.3.3.2 Biodiesel Production from Microalgae

Microalgae-based wastewater treatment is an emerging, sustainable approach that integrates nutrient removal and biomass generation for biodiesel production. Microalgae efficiently assimilate nitrogen, phosphorus, and organic pollutants while generating valuable lipid-rich biomass, which can be converted into biodiesel (Wang et al., 2024). This dual-purpose system provides an eco-friendly alternative to conventional wastewater treatment while contributing to renewable energy production. The ability of microalgae to thrive in wastewater streams, utilizing nutrients as growth substrates, significantly reduces treatment costs and greenhouse gas (GHG) emissions. Moreover, recent advancements in genetic engineering, cultivation techniques, and harvesting technologies have improved biomass productivity and lipid accumulation, making microalgae-based wastewater treatment a viable strategy for sustainable energy generation (Singh, 2024).

4.3.3.2.1 Microalgae-based Wastewater Treatment

Microalgae treatment systems function through photosynthesis, where microalgae utilize CO_2 and inorganic nutrients (nitrate, ammonium, and phosphate) to grow and remove contaminants from wastewater. Commonly used microalgae for wastewater treatment include *Chlorella*, *Scenedesmus*, *Spirulina*, and *Nannochloropsis*, which exhibit high growth rates and lipid accumulation potential. Key treatment mechanisms include *Nutrient Assimilation*: Microalgae absorb nitrate (NO_3^-), ammonium (NH_4^+), and phosphate (PO_4^{3-}) from wastewater, reducing eutrophication potential (Zhou et al., 2024). *Oxygenation and Bacterial Synergy*: Microalgae release oxygen into the water, enhancing aerobic microbial activity, which further degrades organic pollutants (Chen et al., 2024). *Heavy Metal Removal*: Certain microalgae species exhibit biosorption capabilities, removing heavy metals such as lead, cadmium, and arsenic from wastewater (López-Sánchez et al., 2022). *Pathogen Reduction*: Microalgae produce antimicrobial compounds that inhibit pathogen growth, contributing to wastewater disinfection (Kumar et al., 2021).

4.3.3.2.2 Biodiesel Production from Microalgae Biomass

Microalgae-derived biodiesel is considered a promising alternative to fossil-based fuels due to its high lipid content and rapid growth rate (Table 4.5). The process of converting microalgae into biodiesel involves several key steps:

Biomass Cultivation: Microalgae are cultivated in various wastewater systems, including open ponds, raceway ponds, and photobioreactors. Recent studies highlight the use of high-rate algal ponds (HRAPs) for efficient wastewater treatment and biomass recovery (Yang et al., 2024).

Harvesting and Dewatering: Due to the dilute nature of microalgal cultures, efficient harvesting techniques such as flocculation, centrifugation, and membrane filtration are required. The integration of bioflocculants has been explored to enhance harvesting efficiency (Zhao et al., 2023).

Lipid Extraction and Transesterification: Microalgae lipids are extracted using solvent-based methods (hexane, chloroform-methanol) or green technologies like supercritical CO_2 extraction. Extracted lipids undergo transesterification with alcohol and a catalyst to produce biodiesel. Catalyst-free microwave-assisted transesterification has been explored to enhance biodiesel yield (Patel et al., 2024).

Despite its potential, several challenges must be addressed to improve the efficiency and scalability of microalgae-based wastewater treatment and biodiesel

TABLE 4.5
Comparison of Wastewater Treatment Methods Using Microalgae

Treatment System	Nutrient Removal Efficiency (%)	Lipid Content (%)	Energy-Recovery Potential	References
Open Ponds	Moderate (60–80)	20–30	Moderate	Yang et al. (2024)
Photobioreactors	High (80–95)	30–50	High	Zhou et al. (2024)
High-Rate Algal Ponds	High (85–90)	25–35	Moderate to High	Wang et al. (2024)

production: (1) High costs associated with biomass harvesting and lipid extraction remain a major limitation. (2) Developing low-cost harvesting techniques, such as self-flocculating microalgae, could enhance economic viability (López-Sánchez et al., 2022). (3) Outdoor systems are affected by temperature fluctuations, light intensity, and contamination. (4) Genetically modified microalgae with enhanced stress resistance are being explored (Singh, 2024). (5) Different wastewater compositions affect microalgal growth and lipid productivity. (6) Pretreatment of wastewater to remove toxic compounds can improve system performance (Zhao et al., 2023). (7) The co-production of biodiesel, biogas, and biohydrogen could enhance energy recovery and system efficiency. (8) AD of microalgal residues is being explored for biogas production (Patel et al., 2024).

4.3.3.3 Bioethanol and Biobutanol from Wastewater

The increasing demand for sustainable biofuels has led to extensive research on the production of bioethanol and biobutanol from various waste sources, including wastewater. Wastewater contains organic carbon, sugars, and other biodegradable compounds, which can serve as substrates for fermentative microorganisms to produce biofuels. Utilizing wastewater as a feedstock for bioethanol and biobutanol production not only provides a renewable energy source but also aids in waste treatment, reducing organic pollutants, and minimizing environmental impact (Kumar et al., 2021). Both bioethanol and biobutanol are considered viable alternatives to fossil fuels due to their higher energy content, reduced GHG emissions, and compatibility with existing fuel infrastructure. Compared to bioethanol, biobutanol exhibits superior energy density, lower volatility, and better blending properties with gasoline, making it a promising candidate for future biofuel applications (Patel & Singh, 2024). The integration of wastewater treatment with bioethanol and biobutanol production offers a circular bioeconomy approach, transforming waste into valuable energy resources while addressing wastewater pollution challenges.

4.3.3.3.1 Bioethanol Production from Wastewater

4.3.3.3.1.1 Microbial Fermentation for Bioethanol Synthesis Bioethanol production from wastewater relies on microbial fermentation, primarily using yeasts (*Saccharomyces cerevisiae*) and bacteria (*Zymomonas mobilis, Escherichia coli*). Studies indicate that wastewater from food processing industries, dairy plants, and municipal waste has high sugar content, making it an ideal substrate for bioethanol production. The use of engineered yeast strains and metabolic pathway optimization can further enhance ethanol yield (Mishra et al., 2023). These microorganisms convert fermentable sugars into ethanol under anaerobic or facultative anaerobic conditions.

The process involves (a) *Pretreatment of Wastewater*: Removal of toxic compounds and inhibitors to ensure microbial viability. Concentration of sugars and organic matter to optimize fermentation efficiency (Singh, 2024). (b) *Fermentation*: Microorganisms metabolize glucose, sucrose, and other sugars into ethanol and CO_2. Factors such as pH, temperature, and nutrient availability significantly impact ethanol yield (Kumar et al., 2021). (c) *Distillation and Purification*: Ethanol is separated from the fermentation broth and purified to fuel-grade standards.

4.3.3.3.2 *Biobutanol Production from Wastewater*

4.3.3.3.2.1 Clostridial Fermentation for Biobutanol Synthesis Biobutanol production is typically carried out using Clostridium species, such as *Clostridium acetobutylicum, Clostridium beijerinckii,* and *Clostridium saccharoperbutylacetonicum.* These bacteria perform Acetone-Butanol-Ethanol (ABE) fermentation, producing a mixture of butanol, acetone, and ethanol in a 3:6:1 ratio (Dutta et al., 2021).

The ABE fermentation process includes (a) *Substrate Utilization*: Organic compounds in wastewater, including carbohydrates, amino acids, and VFAs, serve as feedstock for Clostridium fermentation. Hydrolyzed lignocellulosic biomass in wastewater improves biobutanol yield (Kumar et al., 2021). (b) *Fermentation Conditions*: Butanol toxicity to microbial cells limits productivity; hence, adaptation strategies such as genetic modifications and in situ product recovery techniques are employed (Patel & Singh, 2024). (c) *Product Recovery*: Biobutanol is separated using distillation, gas stripping, or pervaporation, improving final biofuel purity.

4.4 CHALLENGES AND FUTURE PERSPECTIVES

4.4.1 Challenges in Resource Recovery and Microbial Biofuel Production

The integration of resource recovery from wastewater and microbial biofuel production presents a promising avenue for sustainable waste management and renewable energy generation. However, several challenges hinder large-scale implementation and economic feasibility.

Variability in wastewater composition: Wastewater characteristics vary significantly depending on source, location, season, and treatment processes. Industrial effluents, domestic sewage, and agricultural runoff contain different levels of organic matter, nutrients, and toxic compounds, impacting the efficiency of microbial biofuel production. The inconsistency in substrate availability makes process standardization difficult and affects microbial metabolic pathways, leading to fluctuations in biofuel yields.

Low biomass conversion efficiency: Microorganisms used in biofuel production, such as bacteria, fungi, and algae, often exhibit low substrate utilization efficiency due to inhibitory compounds in wastewater. For instance, heavy metals, antibiotics, and xenobiotics can disrupt microbial metabolism, reducing biofuel yields. Additionally, inefficient hydrolysis of complex organic matter into fermentable substrates limits microbial activity, necessitating pretreatment processes that add to operational costs.

Energy-intensive separation and purification processes: The recovery of biofuels like bioethanol, biobutanol, and biogas requires energy-intensive separation techniques, including distillation, membrane filtration, and gas stripping. These processes significantly impact the overall energy balance, often making biofuel production less energy-efficient than conventional fossil fuels. For instance, the low concentration of butanol in fermentation broth (typically $<2\%$) necessitates multi-stage distillation, increasing energy consumption and operational costs.

***Microbial inhibition and process stability*:** Microbial biofuel production systems, such as AD, MFCs, and fermentation-based bioethanol/butanol synthesis, face challenges related to process stability and microbial inhibition. Factors such as pH fluctuations, accumulation of toxic metabolites, and competition among microbial communities can lead to process failures and reduced productivity. Additionally, the sensitivity of microorganisms to environmental conditions demands real-time monitoring and control systems, increasing operational complexity.

***Economic and scale-up constraints*:** Despite technological advancements, scaling up biofuel production from wastewater remains challenging. High capital investment, infrastructure requirements, and maintenance costs hinder commercialization. Moreover, microbial biofuel systems often compete with other waste-to-energy technologies, such as incineration and AD for methane production, making their economic viability uncertain. Policy and regulatory frameworks also influence the adoption of waste-derived biofuels, impacting investment opportunities and industry growth.

4.4.2 Future Directions

***Advancing microbial engineering and synthetic biology*:** Genetic modifications and metabolic engineering of microbes can enhance substrate utilization, tolerance to inhibitors, and biofuel yields. Advanced techniques like CRISPR-based gene editing, adaptive evolution, and synthetic biology can create robust microbial strains tailored for wastewater biofuel production. Engineered yeasts, Clostridium species, and microalgae with improved fermentation capabilities can optimize ethanol, butanol, and biodiesel yields.

***Development of energy-efficient separation technologies*:** To improve the energy balance of biofuel production, future research should focus on the following: Membrane-based separation techniques (e.g., pervaporation, nanofiltration) for ethanol and butanol recovery. Integrated gas upgrading methods for biogas purification, reducing CO_2 and H_2S content. Hybrid separation strategies, combining adsorption, extraction, and distillation to enhance biofuel purity with reduced energy input. Integration of Circular Bioeconomy Approaches

***A circular bioeconomy model can maximize resource recovery from wastewater while minimizing waste generation*.**

- Co-production of biofuels and high-value bioproducts, such as bioplastics, biofertilizers, and biochar
- Utilization of wastewater-derived nutrients (phosphorus, nitrogen) for microalgae cultivation, improving biodiesel production
- Coupling MFCs with wastewater treatment, enabling simultaneous electricity generation and pollutant removal

***AI and process optimization*:** AI-driven process monitoring, predictive modeling, and automation can enhance the efficiency of wastewater-based biofuel systems. Machine learning algorithms can optimize fermentation parameters, microbial growth conditions, and biofuel separation techniques, improving overall yield and process stability.

Policy Support and Economic Incentives: Governments and policymakers can accelerate adoption through: Subsidies and tax incentives for wastewater-based biofuel projects. Carbon credit mechanisms to promote biofuel integration in energy markets. Standardization and certification of biofuels derived from wastewater to enhance market acceptance.

4.5 CONCLUSION

Resource recovery from wastewater and microbial biofuel production presents a sustainable and innovative approach to addressing environmental and energy challenges. By utilizing advanced microbial processes, valuable resources such as nutrients (phosphorus and nitrogen), organic matter, and biofuels (biogas, bioethanol, biobutanol, and biodiesel) can be effectively recovered from wastewater streams. Technologies like AD, MFCs, and microalgae-based biodiesel production offer promising solutions for energy generation while simultaneously improving wastewater treatment efficiency. Despite these advancements, several challenges, including process stability, energy-intensive biofuel separation, variability in wastewater composition, and economic constraints, hinder large-scale implementation. However, emerging technologies such as synthetic biology, AI-driven process optimization, and membrane-based separation techniques can significantly improve biofuel yields and energy efficiency. Furthermore, integrating circular bioeconomy approaches and implementing supportive policies and incentives will be crucial in promoting the adoption of these technologies. Looking ahead, continued research and development, coupled with industrial collaborations and policy interventions, can help scale up wastewater-based biofuel production. By overcoming current limitations, resource recovery from wastewater can contribute to a more sustainable, low-carbon future, reducing environmental pollution while generating renewable energy and valuable bioproducts.

BIBLIOGRAPHY

Abdalla, S.B., Moghazy, R.M., Hamed, A.A. et al., 2024 Strain selection and adaptation of a fungal-yeast-microalgae consortium for sustainable bioethanol production and wastewater treatment from livestock wastewater. *Microbial Cell Factories*, 23, 288. https://doi.org/10.1186/s12934-024-02537-4

Alam, M., Rahman, M. 2025. Waste-to-energy transition in wastewater management: A Case study of microbial fuel cells. *Environmental Science and Pollution Research*, 32, 1822–1840. https://doi.org/10.1007/s10311-025-01822-x

Alengebawy, A., Ran, Y., Osman, A.I. et al. 2024. Anaerobic digestion of agricultural waste for biogas production and sustainable bioenergy recovery: A review. *Environmental Chemistry Letters*, 22, 2641–2668. https://doi.org/10.1007/s10311-024-01789-1

Al-Juboori, R.A., Al-Shaeli, M., Johnson, D., Hilal, N., 2025. Ammonia recovery from wastewater. In Abu Zahrim Yaser, Junidah Lamaming, Daniel Johnson, Nidal Hilal (Eds.), *Nutrients and colored compounds in wastewater*, pp. 19–45. https://doi.org/10.1016/B978-0-443-21701-2.00004-0

Andronikou, M., Christoforou, P., Constantinou, D., Charalambous, P., Samanides, C.G., Karachaliou, P. and Vyrides, I., 2025. Critical role of bicarbonate in Zero-Valent iron for hydrogen generation and biogas upgrading in anaerobic digestion. *Bioresource Technology*, 426, p. 132236.

Apraku, E., Farmer, M., Lavallais, C., Soriano, D.A., Notestein, J., Tyo, K, Dunn, J., Tarpeh, W.A., Wells, G.F., 2025. Toward a circular nitrogen bioeconomy: Integrating nitrogen bioconcentration, separations, and high-value products for nitrogen recovery. *Current Opinion in Biotechnology*, 9, 103225. https://doi.org/10.1016/j.copbio.2024.103225

Archana, K., Visckram, A.S., Senthil Kumar, P., Manikandan, S., Saravanan, A., Natrayan, L. 2024. A review on recent technological breakthroughs in anaerobic digestion of organic biowaste for biogas generation: Challenges towards sustainable development goals. *Fuel* 358, 130298. https://doi.org/10.1016/j.fuel.2023.130298

Beschkov, V.N., Angelov, I.K. 2025. Volatile fatty acid production vs. methane and hydrogen in anaerobic digestion. *Fermentation*, 11(4):174. https://doi.org/10.3390/fermentation11040172

Bora, Abhispa, Rajan, A.S.T., Ponnuchamy, Kumar, Muthusamy, Govarthanan, Alagarsamy, Arun, 2024. Microalgae to bioenergy production: Recent advances, influencing parameters, utilization of wastewater – A critical review. *Science of the Total Environment*, 946, 174230. https://doi.org/10.1016/j.scitotenv.2024.174230

Brandon, Clark, Sharma, Neha, Apraku, Edward, Dong, Hang, Tarpeh, William A. 2024. Ligand exchange adsorbents for selective phosphate and total ammonia nitrogen recovery from wastewaters. *Accounts of Materials Research*, 5(4), 492–504. https://doi.org/10.1021/accountsmr.3c00290

Chakraborty, A., Roy, P., Das, S. 2025. Integration of AI-driven optimization in biofuel recovery from wastewater. *Applied Energy*, 348, 125202. https://doi.org/10.1016/j.apenergy.2024.125202

Chen, Q., Li, Z., Li, Y., Liu, M., Wu, Y., Chen, Z. and Zhu, B., 2024. Biodegradation of benzo [a] pyrene by a marine Chlorella vulgaris LH-1 with heterotrophic ability. *Marine Pollution Bulletin*, 198, 115848.

Chrispim, M.C., Scholz, M., Nolasco, M.A. 2019. Phosphorus recovery from municipal wastewater treatment: Critical review of challenges and opportunities for developing countries, *Journal of Environmental Management*, 248, 109268. https://doi.org/10.1016/j.jenvman.2019.109268

Condor, Billriz E., de Luna, Mark Daniel G., Lacson, Carl Francis Z., Acebu, Paula Isabel G., Abarca, Ralf Ruffel M., Nagarajan, Dillirani, Lee, Duu-Jong, Chang, Jo-Shu, 2024. Effects of carbon dioxide concentration and swine wastewater on the cultivation of Chlorella vulgaris FSP-E and bioethanol production from microalgae biomass. *Applied Energy*, 369, 123617. https://doi.org/10.1016/j.apenergy.2024.123617

Cuihong, Song, Jun-Jie, Zhu, Willis, John L., Moore, Daniel P., Zondlo, Mark A., Zhiyong, Jason Ren, 2023. Methane emissions from municipal wastewater collection and treatment systems. *Environmental Science & Technology*, 57(6), 2248–2261. https://doi.org/10.1021/acs.est.2c04388

Da Silva, C., Serra-Toro, A., Pelizzaro, V., Valentino, F., Astals, S, Mas, F., Dosta, J., 2025. Modeling nitrogen recovery and water transport in gas-permeable membranes, *Water Research*, 269, 122771. https://doi.org/10.1016/j.watres.2024.122771

Dahiya, S. and Mohan, S.V., 2022. Synergy of selective buffering, intermittent pH control and bioreactor configuration on acidogenic volatile fatty acid production from food waste. *Chemosphere*, *302*, p. 134755.

Devens, Kauanna Uyara, Ribeiro, Alexandre Rodrigues, Camargo, Franciele Pereira, Sakamoto, Isabel Kimiko, Varesche, Maria Bernadete Amâncio, Silva, Edson Luiz 2024. Unlocking potential: Exploring the methane production potential in anaerobic co-digestion of cassava wastewater and glycerol. *Journal of Environmental Chemical Engineering* 12(3), 113043. https://doi.org/10.1016/j.jece.2024.113043

Devens, Kauanna Uyara, Ribeiro, Alexandre Rodrigues, Camargo, Franciele Pereira, Sakamoto, Isabel Kimiko, Varesche, Maria Bernadete Amâncio, Silva, Edson Luiz, Obileke, KeChrist, Onyeaka, Helen, Meyer, Edson L, Nwokolo, Nwabunwanne, 2021.

Microbial fuel cells, a renewable energy technology for bio-electricity generation: A mini-review. *Electrochemistry Communications* 125, 107003. https://doi.org/10.1016/j.elecom.2021.107003

Du, J., Waite, T.D., Feng, J. et al. 2023. Coupled electrochemical methods for nitrogen and phosphorus recovery from wastewater: A review. *Environmental Chemistry Letters* 21, 885–909. https://doi.org/10.1007/s10311-023-01561-x

Duan, S., He, J., Xin, X., Li, L., Zou, X., Zhong, Y., Zhang, J. and Cui, X., 2023. Characteristics of digested sludge-derived biochar for promoting methane production during anaerobic digestion of waste activated sludge. *Bioresource Technology*, 384, p. 129245.

Dutta, S., Ghosh, S., Manna, D. and Chowdhury, R., 2021. Energy and environmental performance of a near-zero-effluent rice straw to butanol production plant. *Indian Chemical Engineer*, *63*(2), pp.139–151.

El Mashad, H. & Zhang, R. (2020). Biogas Energy from Organic Wastes. In Holden, N. M., Wolfe, M. L., Ogejo, J. A., & Cummins, E. J. (Ed.), *Introduction to Biosystems Engineering*. https://doi.org/10.21061/IntroBiosystemsEngineering/Biogas

El Mashad, H.M., Barzee, T.J., Franco, R.B., Zhang, R., Kaffka, S. and Mitloehner, F., 2023. Anaerobic digestion and alternative manure management technologies for methane emissions mitigation on californian dairies. *Atmosphere*, 14(1), p. 120.

Elhenawy, S., Khraisheh, M., AlMomani, F., Al-Ghouti, M, Hassan, M.K. 2022. From waste to watts: updates on key applications of microbial fuel cells in wastewater treatment and energy production. *Sustainability* 14(2), 955. https://doi.org/10.3390/su14020955

Elsayed, Ahmed, Kakar, Farokh Laqa, Marcus, Andrew, AlSayed, Ahmed, Mohamed Sherif Zagloul, C. M., Bell, K. Y., Santoro, D., Norton, John, Elbeshbishy, Elsayed, 2025. Application of additives for anaerobic digestion intensification: A comprehensive review on improving biogas production and methane yield. *Applied Energy*, 381, 125202. https://doi.org/10.1016/j.apenergy.2024.125202

El-Sheekh, M.M., Galal, H.R., Mousa, A.S.H. et al. 2023. Coupling wastewater treatment, biomass, lipids, and biodiesel production of some green microalgae. *Environmental Science and Pollution Research* 30, 35492–35504. https://doi.org/10.1007/s11356-023-25628-y

Esfandyari, M., Jafari, D., Azami, H. 2023. Microbial fuel cells for energy production in wastewater treatment plants—A review. *Biofuels* 15(6), 743–753. https://doi.org/10.1080/17597269.2023.2294227

Farghali, M., Chen, Z., Osman, A.I. et al. 2024. Strategies for ammonia recovery from wastewater: A review. *Environmental Chemistry Letters* 22, 2699–2751. https://doi.org/10.1007/s10311-024-01768-6

Feng, L., Hu, T., Ma, H., Gao, Z., Liu, Y., He, S., Ding, J., Jiang, J., Zhao, Q. and Wei, L., 2023. Impacts of biochar derived from oil sludge on anaerobic digestion of sewage sludge: performance and associated mechanisms. *Journal of Cleaner Production*, 425, p. 138838.

Gawali, Sushant, Alam, Shahrukh Nawaj, Tilva, Jeel, Jagadale, Krushna, Singh, Bhaskar, Guldhe, Abhishek, 2025. Development of scalable cultivation strategy for microalgal biomass generation using wastewater and integrated biorefinery approach for biofuel production. *Journal of Environmental Chemical Engineering* 13(2), 115441. https://doi.org/10.1016/j.jece.2025.115441

Gol, Negar, Taghavijeloudar, Mohsen, Jalilian, Neda, Rezania, Shahabaldin, 2025. Microalgae cultivation in semi-transparent photovoltaic bioreactor for sustainable power generation, wastewater treatment and biodiesel production. *Energy Conversion and Management* 325, 119417. https://doi.org/10.1016/j.enconman.2024.119417

González, R., Peña, D.C. and Gómez, X., 2022. Anaerobic co-digestion of wastes: reviewing current status and approaches for enhancing biogas production. *Applied Sciences*, *12*(17), p. 8884.

Guishi, Cheng, Yihao, Yang, Ercheng, Luo, Suthar, Diljeet Kumar, Ying, Zhao, Xue, Luan, Xiaoqiang, Wang, Changqing, Dong, 2024. Degradation and power generation performance of modified-anode microbial fuel cells for kitchen wastewater. *Journal of Power Sources* 591, 233841., https://doi.org/10.1016/j.jpowsour.2023.233841

Gupta, R., Singh, S., Kumar, P. 2025. Bioelectrochemical approaches for enhanced biofuel recovery from industrial wastewater. *Results in Engineering* 19, 104161. https://doi.org/10.1016/j.rineng.2025.104161

Hajizadeh, A., Mohamadi-Baghmolaei, M., Saady, N.M.C., Zendehboudi, S. 2022. Hydrogen production from biomass through integration of anaerobic digestion and biogas dry reforming. *Applied Energy* 309, 118442. https://doi.org/10.1016/j.apenergy.2021.118442

Han, F., Zhou, W., 2022. Nitrogen recovery from wastewater by microbial assimilation – A review. *Bioresource Technology* 363, 127933. https://doi.org/10.1016/j.biortech.2022.127933

Haozhe, Huang, Shaorong, Zhong, Siqi, Wen, Chao, Luo, Tianyu, Long, 2022. Improving the efficiency of wastewater treatment and microalgae production for biofuels. *Resources, Conservation and Recycling* 178, 106094. https://doi.org/10.1016/j.resconrec.2021.106094

Hermawan, E., Adiarso, A., Nelly, A., Wicaksana, D.E.P., Setiawan, H., Handayani, I.D., Setiadi, S., Wijono, A., Marsudi, A., Ferabiane, A.L. 2025. Techno-economic assessment of bioethanol wastewater from empty fruit bunch for biogas production. https://ssrn.com/abstract=5193586. https://doi.org/10.2139/ssrn.5193586

Hossain, M.S., Karim, T.U., Onik, M.H. et al. 2022. Impact of temperature, inoculum flow pattern, inoculum type, and their ratio on dry anaerobic digestion for biogas production. *Scientific Reports* 12, 6162. https://doi.org/10.1038/s41598-022-10025-1

Hu, Q., Zhou, J., Qiu, B. et al. 2021. Synergistically improved methane production from anaerobic wastewater treatment by iron/polyaniline composite. *Advanced Composites and Hybrid Materials* 4, 265–273. https://doi.org/10.1007/s42114-021-00236-x

Huang, Y., Liu, B., Chi, Y., Li, J., Chi, Y., Ji, M., Zhai, H., Wang, R., Yuan, T. and Yu, H., 2024. Study on the treatment of simulated dye wastewater containing FMPs using the CW-MFC system. *Journal of Water Process Engineering*, 66, p. 105810.

Iakovidou, G., Itziou, A., Tsiotsias, A., Lakioti, E., Samaras, P., Tsanaktsidis, C., Karayannis, V. 2024. Application of microalgae to wastewater bioremediation, with CO_2 biomitigation, health product and biofuel development, and environmental biomonitoring. *Applied Sciences* 14(15), 6727. https://doi.org/10.3390/app14156727

Induchoodan, T.G., Haq, I., Kalamdhad, A.S. 2022. Factors affecting anaerobic digestion for biogas production: A review. In Chaudhery Hussain, Subrata Hait (Eds.), *Advanced organic waste management*, Elsevier, pp. 223–233. https://doi.org/10.1016/B978-0-323-85792-5.00020-4

Javed, F., Rehman, Fahad, Khan, Asad Ullah, Fazal, Tahir, Hafeez, Ainy, Rashid, Naim, 2022. Real textile industrial wastewater treatment and biodiesel production using microalgae. *Biomass and Bioenergy* 165, 106559. https://doi.org/10.1016/j.biombioe.2022.106559

Jialing, Tang, Yunhui, Pu, Ting, Zeng, Yisong, Hu, Jin, Huang, Shengwang, Pan, Xiaochang, C. Wang, Yuyou Li, Abd El-Fatah Abomohra, 2022. Enhanced methane production coupled with livestock wastewater treatment using anaerobic membrane bioreactor: Performance and membrane filtration properties. *Bioresource Technology* 345, 126470. https://doi.org/10.1016/j.biortech.2021.126470

Jiang, Y., Zhang, X., An, L. and Liu, Y., 2024. A novel biochar-augmented enzymatic process for conversion of food waste to biofertilizers: Planting trial with leafy vegetable. *Bioresource Technology*, 399, p. 130554.

Jie, Zhou, Qiaoling, Lu, Wenlu, Li, Dezhi, Sun, Yan, Dang, Dan, Qu, Qiang, Wang, Bin, Qiu, 2024. Enhanced methane production from anaerobic wastewater treatment by carbon nanotubes: Surface defect mediated interfacial electron transfer. *Chemical Engineering Journal* 497, 154576. https://doi.org/10.1016/j.cej.2024.154576

Khabiri, B., Ferdowsi, M., Buelna, G., Jones, J. P., & Heitz, M. 2021. Bioelimination of low methane concentrations emitted from wastewater treatment plants: A review. *Critical Reviews in Biotechnology* 42(3), 450–467. https://doi.org/10.1080/07388551.2021.1940830

Kieft, B., Finke, N., McLaughlin, R.J. et al. 2023. Genome-resolved correlation mapping links microbial community structure to metabolic interactions driving methane production from wastewater. *Nature Communications* 14, 5380. https://doi.org/10.1038/s41467-023-40907-5

Kumar, K., Manju, P. and Gajalakshmi, S., 2023. Harnessing plant microbial fuel cells for resource recovery and methane emission reduction in paddy cultivation. *Energy Conversion and Management*, 294, p. 117545.

Kumar, N., Yadav, S. 2023. A review on sustainable wastewater treatment and biofuel production through hybrid bioprocesses. *Water* 15(22), 3967. https://doi.org/10.3390/w15223967

Kumar, Vinod, Krishna Kumar Jaiswal, Monu Verma, Mikhail S. Vlaskin, Manisha Nanda, Pankaj Kumar Chauhan, Ajay Singh, and Hyunook Kim. "Algae-based sustainable approach for simultaneous removal of micropollutants, and bacteria from urban wastewater and its real-time reuse for aquaculture." *Science of the Total Environment* 774 (2021): 145556.

Kyung, Jin Min, Doo, Young Oh, Ki, Young Park 2022. Field test of water-net based wastewater treatment for nutrient removal and bioethanol production. *Chemosphere* 301, 134791. https://doi.org/10.1016/j.chemosphere.2022.134791

Li, R., Jiang, Y., Huang, J., Luo, K., Fan, X., Guo, R., Liu, T., Zhang, Y. and Fu, S., 2024. Simultaneous biogas upgrading and single cell protein production using hydrogen oxidizing bacteria. *Chemical Engineering Journal*, *490*, p. 151576.

Li, Y., Zhao, X., Wang, C., Xu, J., Zheng, Y., Wang, X. 2023. Advances in microbial biofuel production from wastewater: Progress, challenges, and future prospects. *Science of the Total Environment* 899, 163758. https://doi.org/10.1016/j.scitotenv.2023.163758

Linxuan, Che, Bo, Yang, Qing, Tian, Hui, Xu, 2022. Iron-based biochar derived from waste-activated sludge enhances anaerobic digestion of synthetic salty organic wastewater for methane production. *Bioresource Technology* 345, 126465. https://doi.org/10.1016/j.biortech.2021.126465

Liu, X., Wang, J., Chen, H. 2024. Process optimization and sustainability assessment of wastewater-based biohydrogen production. *Fuel* 360, 131855. https://doi.org/10.1016/j.fuel.2024.131855

López-Sánchez, Anaid, Ana Laura Silva-Gálvez, Óscar Aguilar-Juárez, Carolina Senés-Guerrero, Danielle A. Orozco-Nunnelly, Danay Carrillo-Nieves, and Misael Sebastián Gradilla-Hernández. "Microalgae-based livestock wastewater treatment (MbWT) as a circular bioeconomy approach: Enhancement of biomass productivity, pollutant removal and high-value compound production." *Journal of environmental management* 308 (2022): 114612.

Mahto, K.U., Das, S. 2024. Electroactive biofilm communities in microbial fuel cells for the synergistic treatment of wastewater and bioelectricity generation. *Critical Reviews in Biotechnology* 45(2), 434–453. https://doi.org/10.1080/07388551.2024.2372070

Manigandan, S., Praveenkumar, T.R., Anderson, A., AlMallahi, M., Elgendi, M., 2023. Benefits of pretreated water hyacinth for enhanced anaerobic digestion and biogas production. *International Journal of Thermofluids* 19, 100369. https://doi.org/10.1016/j.ijft.2023.100369

Marangon, Bianca Barros, De Siqueira Castro, Jackeline, Calijuri, Maria Lúcia, 2024. Aviation fuel based on wastewater-grown microalgae: Challenges and opportunities of hydrothermal liquefaction and hydrotreatment. *Journal of Environmental Management*, 354, 120418. https://doi.org/10.1016/j.jenvman.2024.120418

Mehra, Kuber Singh, Abrar, Iyman, Bhatia, Ravi Kant, Goel, Varun, 2025. A comprehensive review of algae consortium for wastewater bioremediation and biodiesel production. *Energy Conversion and Management* 325, 119428. https://doi.org/10.1016/j.enconman.2024.119428

Mishra, M., Shukla, S., Mishra, A., Zohra, F., Singh, S., Kushwaha, A., Goswami, L. and Upadhyay, N., 2023. Biotechnological interventions in the valorization of the organic waste. In *Bio-based materials and waste for energy generation and resource management* (pp. 357–385). Elsevier.

Mishra, S., Kumar, R., Patel, D. 2024. Circular bioeconomy and wastewater-based biofuel generation. *Environmental Science and Pollution Research* 31, 1789–1805. https://doi.org/10.1007/s10311-024-01789-1

Mohsen, Askari, Saeid, Rajabzadeh, Leonard, Tijing, Ho, Kyong Shon, 2024. Advances in capacitive deionization (CDI) systems for nutrient recovery from wastewater: Paving the path towards a circular economy. *Desalination* 583, 117695, https://doi.org/10.1016/j.desal.2024.117695

Mondal, Subhadeep, Rakshit, Subham, Pal, Kalyanbrata, Halder, Suman Kumar, Mondal, Keshab Chandra, 2025. Valorization of wastewater for biobutanol production: A promising approach of wastewater biorefinery. In Maulin P. Shah, Angana Sarkar (Eds.), *Development in wastewater treatment research and processes, bio refinery of wastewater treatment*, Elsevier, pp. 79–92. https://doi.org/10.1016/B978-0-12-823764-9.00014-5

Moreno-Cruz, Carlos Francisco, Tzintzun-Camacho, Olivia, Gonzalez-Joaquin, Marlon C., Aguilar-Martinez, Xiomara E., González-Mendoza, Daniel, Martinez-Quiroz, Marisela, 2025. Livestock wastewater as a microalgae growth medium for potential production of biodiesel in arid areas of Mexico. *Algal Research* 86, 103957. https://doi.org/10.1016/j.algal.2025.103957

Mustafa, H.M., Hayder, G. and Jagaba, A.H., 2021. Microalgae: a renewable source for wastewater treatment and feedstock supply for biofuel generation. *Biointerface Research in Applied Chemistry*, 11(1), pp.7431–7444.

Nan, Jiang, Jialing, Song, Mengying, Yan, Yuan, Hu, Miaomiao, Wang, Yanbiao, Liu, Manhong, Huang, 2023. Iron cobalt-doped carbon nanofibers anode to simultaneously boost bioelectrocatalysis and direct electron transfer in microbial fuel cells: Characterization, performance, and mechanism. *Bioresource Technology* 367, 128230. https://doi.org/10.1016/j.biortech.2022.128230

Ni, M., Pan, Y., Gong, J., Chen, Z., Li, D., Huang, Y., Li, L., Ding, Y., Bi, Z., 2025. Glycogen-accumulating organisms promote phosphate recovery from wastewater by pilot-scale biofilm sequencing batch reactor: Performance and mechanism. *Bioresource Technology* 418, 131910. https://doi.org/10.1016/j.biortech.2024.131910

Nkuna, R., Roopnarain, A., Rashama, C., Adeleke, R. 2021. Insights into organic loading rates of anaerobic digestion for biogas production: A review. *Critical Reviews in Biotechnology* 42(4), 487–507. https://doi.org/10.1080/07388551.2021.1942778

Patel, K. and Singh, S.K., 2024. Sustainable waste management: A comprehensive life cycle assessment of bioethanol production from agricultural and municipal waste. *Environmental Science and Pollution Research, 31*(39), pp. 51431–51446.

Patel, R., Mehta, P., Joshi, N. 2024. Algal biofuel production from wastewater: A sustainable alternative for clean energy. *Fuel* 356, 134266. https://doi.org/10.1016/j.fuel.2024.134266

Patil, S. A., Bakliwal, A. A., Talele, S. G. 2021. Microbial Fuel Cell: A Green Microbial Approach for Production of Bioelectricity Using a Conducting Polymer. In Rawat, N. K., Volova, T. G., Haghi, A. K. (Eds.), Applied Biopolymer Technology and Bioplastics, vol-1, Apple Academic Press, New York, https://doi.org/10.1201/9781003045458

Peter, A.P., Khoo, K.S., Chew, K.W. et al. 2021. Microalgae for biofuels, wastewater treatment and environmental monitoring. *Environmental Chemistry Letters* 19, 2891–2904. https://doi.org/10.1007/s10311-021-01219-6

Pusphanathan, K., Mohd Zaini, Makhtar M., Shukor, H. et al. 2024. Effectiveness of Direct Sulfonated Polysulfone in Dual Chamber Microbial Fuel Cells Based Dewatered Sludge for Power Generation. *International Journal of Environmental Research* 18, 100. https://doi.org/10.1007/s41742-024-00650-1

Qian, S., Chen, L., Xu, S., Zeng, C., Lian, X., Xia, Z., Zou, J. 2025. Research on methane-rich biogas production technology by anaerobic digestion under carbon neutrality: A review. *Sustainability* 17(4), 1425. https://doi.org/10.3390/su17041425

Qing, Shao, Ying, Zhang, Zhe, Liu, Lizhi, Long, Zizheng, Liu, Yiqun, Chen, Xin-Ming, Hu, Mingming, Lu, Li-Zhi, Huang, 2022. Phosphorus and nitrogen recovery from wastewater by ceramsite: Adsorption mechanism, plant cultivation and sustainability analysis. *Science of the Total Environment* 805, 150288. https://doi.org/10.1016/j.scitotenv.2021.150288

Qiuyue, Guan, Yunhe, Qu, Yujia, Zhai, Wansheng, Shi, Mingxing, Zhao, Zhenxing, Huang, Wenquan, Ruan, 2023. Enhancement of methane production in anaerobic digestion of high salinity organic wastewater: The synergistic effect of nano-magnetite and potassium ions. *Chemosphere* 318, 137974. https://doi.org/10.1016/j.chemosphere.2023.137974

Rahayu, A., Hakika, D.C., Amrillah, N.A.Z., et al. 2025. The adsorption and kinetics studies of nitrate ions from bioethanol wastewater by ammonium-based ion exchange polymer. *Ecological Engineering & Environmental Technology* 26(3), 54–69. https://doi.org/10.12912/27197050/199560

Rahimnejad, M., 2023. MFC designing and performance. In *Biological Fuel Cells* (pp. 147–175). Elsevier.

Rajendran, N., Kang, D., Han, J. and Gurunathan, B., 2022. Process optimization, economic and environmental analysis of biodiesel production from food waste using a citrus fruit peel biochar catalyst. *Journal of Cleaner Production*, 365, p. 132712.

Ramos-Suarez, M., Zhang, Y. and Heaven, S., 2024. Acidogenic Fermentation of Organic Residual Solids: Effect of Different Alkaline Sources on pH, Alkalinity, and Fermentation Performance. *Fermentation*, 10(11), p.571.

Ren, J., Li, N., Du, M., Zhang, Y., Hao, C., & Hu, R. 2021. Study on the effect of synergy effect between the mixed cultures on the power generation of microbial fuel cells. *Bioengineered* 12(1), 844–854. https://doi.org/10.1080/21655979.2021.1883280

Rintala, Kati, Kinnunen, Viljami, Berg, Andreas, Rintala, Jukka, Kokko, Marika, 2024. Biogas production from piloted wood-based bioethanol process using high-rate anaerobic treatment – Focusing for inhibition and improving biodegradability. *Renewable Energy* 237, 121594. https://doi.org/10.1016/j.renene.2024.121594

Saratale, G.D., Banu, J.R., Nastro, R.A., Kadier, A., Ashokkumar, V., Lay, C.H., Jung, J.H., Shin, H.S., Saratale, R.G. and Chandrasekhar, K., 2022. Bioelectrochemical systems in aid of sustainable biorefineries for the production of value-added products and resource recovery from wastewater: A critical review and future perspectives. *Bioresource technology*, 359, p. 127435.

Sharma, R., Pandey, A., Kumar, D. 2023. Chapter 14: Microbial fuel cells and biohydrogen production from wastewater. In *Advances in microbial biotechnology for sustainable wastewater treatment and resource recovery*, pp. 315–340. Springer. https://doi.org/10.1007/978-981-97-1152-9_14

Shi, C., Ma, J., Wu, H., Luo, J., Liu, Y., Li, K., Zhou, Y. and Wang, K., 2022. Evaluation of pH regulation in carbohydrate-type municipal waste anaerobic co-fermentation: Roles of pH at acidic, neutral and alkaline conditions. Science of the Total Environment, 853, p. 158327.

Shuo, Wang, Qixin, Liu, Ji, Li, Zhiwu, Wang, 2021. Methane in wastewater treatment plants: status, characteristics, and bioconversion feasibility by methane oxidizing bacteria for high value-added chemicals production and wastewater treatment. *Water Research* 198, 117122. https://doi.org/10.1016/j.watres.2021.117122

Singh, V., Kumar, A., & Verma, P. 2023. Enhancing biofuel production from wastewater through synthetic biology approaches. *Bioresource Technology Reports* 20, 101738. https://doi.org/10.1016/j.biteb.2023.101738

Singh R. 2024 June. (Bio) electrochemical system: A systematic approach from agricultural waste to sewage wastewater treatment with nutrients and hydrogen recovery. *Journal of Cleaner Production*, 10(457), 142387.

Sun, Y., Ter Heijne, A., Rijnaarts, H. and Chen, W.S., 2022. The effect of anode potential on electrogenesis, methanogenesis and sulfidogenesis in a simulated sewer condition. *Water Research*, 226, p. 119229.

Thakur, Somil, Calay, Rajnish Kaur, Mustafa, Mohamad Y., Eregno, Fasil E., Patil, Ravindra R., 2025. Importance of substrate type and its constituents on overall performance of microbial fuel cells. *Current Research in Biotechnology* 9, 100272. https://doi.org/10.1016/j.crbiot.2025.100272

Wainaina, S. and Taherzadeh, M., 2021. Microbial conversion of food waste: volatile fatty acids platform. In *Current Developments in Biotechnology and Bioengineering* (pp. 205–233). Elsevier.

Wang, H., Li, Q., & Zhang, J. 2024. Wastewater-to-energy transition: A policy and technology review. *Journal of Environmental Management* 345, 125018. https://doi.org/10.1016/j.jenvman.2025.125018

Yang, M., Liu, Z., Wang, A., Nopens, I., Hu, H. and Chen, H., 2024. High biomass yields of Chlorella protinosa with efficient nitrogen removal from secondary effluent in a membrane photobioreactor. *Journal of Environmental Sciences*, 146, pp.272–282.

Ye, Y., Ngo, H.H., Guo, W., Liu, Y., Li, J., Liu, Y., Zhang, X., Jia, H., 2017. Insight into chemical phosphate recovery from municipal wastewater. *Science of the Total Environment* 576, 159–171. https://doi.org/10.1016/j.scitotenv.2016.10.078

Yolanda, Yustika Desti, Sangsik, Kim, Weonjung, Sohn, Ho, Kyong Shon, Euntae, Yang, Sungyun, Lee, 2025. Simultaneous nutrient-abundant hydroponic wastewater treatment, direct carbon capture, and bioenergy harvesting using microalgae–microbial fuel cells. *Desalination and Water Treatment* 321, 100941. https://doi.org/10.1016/j.dwt.2024.100941

Yonglan, Xi, Chengcheng, Wang, Xiaomei, Ye, Yang, Liu, Qingbo, Xiao, Zhaoyan, Jia, Chunhui, Cao, Ting, Han, Ting, Guo, Zhongbing, Chen, 2025. Methods for improving microbial electrolysis cell for swine wastewater treatment. *Renewable and Sustainable Energy Reviews* 207, 114947. https://doi.org/10.1016/j.rser.2024.114947

Yunfei, Li, Jia, Liu, Xuepeng, Chen, Xiaole, Yuan, Nan, Li, Weihua, He, Yujie, Feng, 2021. Tailoring spatial structure of electroactive biofilm for enhanced activity and direct electron transfer on iron phthalocyanine modified anode in microbial fuel cells. *Biosensors and Bioelectronics*, 191, 113410. https://doi.org/10.1016/j.bios.2021.113410

Yunjie, Song, Mengyu, Ma, Qingmuke, Su, Ying, Du, Dongyun, Du, 2024. The new strategy of using pine-cone biochar to enhance anaerobic digestion of liquor wastewater for methane production. *Journal of Cleaner Production* 438, 140748. https://doi.org/10.1016/j.jclepro.2024.140748

Zhang, F., Shen, C., Zhao, Y., Zhang, S., Wang, Y., Ji, B., Liu, R., Wong, M.H., Shan, S. and Zhang, J., 2024. An integrated constructed wetland-Microbial fuel cell system with sewage sludge-biochar to enhance treatment and energy recovery efficiencies. *Chemical Engineering Journal*, 486, p. 150431.

Zhang, Y., Desmidt, E., Van Looveren, A., Pinoy, L., Meesschaert, B., Van der Bruggen, B., 2013. Phosphate separation and recovery from wastewater by novel electrodialysis. *Environmental Science & Technology* 11, 5888–5895. https://doi.org/10.1021/es4004476

Zhao, Y., Liu, Z., Chen, G., Sun, W., & Zhou, J. 2023. Microbial consortia engineering for efficient biohydrogen production from wastewater. *Bioresource Technology* 385, 129250. https://doi.org/10.1016/j.biortech.2023.129250

Zhixuan, Ma, Lifang, Yu, Ze, Wang, Xin, Tian, Lanxiang, Zheng, Ran, Liu, 2025. Anaerobic co-digestion of winery wastewater with sewage sludge for methane production: Complementary feedstocks and potential direct interspecies electron transfer. *Fuel* 381, 133190. https://doi.org/10.1016/j.fuel.2024.133190

Zhou, L., Kim, H., Lee, S. (2024). Advances in microbial electrochemical systems for wastewater treatment and energy recovery. *ACS Sustainable Chemistry & Engineering* 12(5), 2789–2802. https://doi.org/10.1021/acssuschemeng.4c02779

Zhu, Liang, Wei, Ding, Xinhou, Zhang, Yun, Chen, Nan, Shen, 2025. Magnetite-mediating direct interspecies electron transfer is not pivotal role in directing electrons toward methane production from sulfate containing wastewater. *Bioresource Technology* 416, 131809. https://doi.org/10.1016/j.biortech.2024.131809

Zixuan, Deng, Chengde, Sun, Guangyi, Ma, Xinzheng, Zhang, Hongbo, Guo, Tao, Zhang, Xuefei, Zhou, Yalei, Zhang, Yong, Hu, Dapeng, Li, Yu-You, Li, Zhe, Kong, 2025. Anaerobic treatment of refractory industrial organic wastewater: A review of bioenergy recovery and low-carbon nitrogen removal toward carbon neutrality. *Journal of Water Process Engineering* 72, 107515. https://doi.org/10.1016/j.jwpe.2025.107515

5 Carbon Capture and Biofuel

A Sustainable Synergy for a Greener World

Zixin Zhou, Zehua Zhao, Xinchen Gong, and Honghong Lyu

5.1 INTRODUCTION

Since the Industrial Revolution, the world's economy and science and technology have developed rapidly, but the increase in the use of fossil fuels and a large amount of industrial smoke emissions have led to increasingly serious environmental problems, including carbon dioxide (CO_2), nitrogen oxides (NOx), hydrocarbons, carbon monoxide (CO), particulate matter, and sulfur dioxide (SO_2), which are almost all greenhouse gases, and thus have become an urgent global issue (Rahman et al., 2017). The sixth report released by the IPCC (United Nations Intergovernmental Panel on Climate Change) in 2021 shows that global temperatures have risen by 1.1°C since the Industrial Revolution and that CO_2 emissions have risen sharply to 36.2 billion tons in 2019. These figures reflect the changes in the global climate, including the increase in extreme temperatures and extreme meteorological phenomena, which threaten and seriously affect, among other things, natural ecology and human economic activities.

However, the main cause of global climate change is the "greenhouse effect" created by global carbon emissions since the Industrial Revolution in 1850. Carbon emissions are a global problem, and reducing them requires a coordinated response to the challenge by all countries. As one of the largest carbon emitters, China has responded positively by proposing a carbon peak and carbon neutral (dual-carbon) target, which is a leading and systematic approach to green and low-carbon development, and can bring about multiple effects in terms of environmental quality improvement and green transformation of the industrial structure, thus promoting the sustainable development of the economy and the environment (Holzinger et al., 2012).

In this context, some scholars have put forward the concept of "carbon balance" in which the left end of the balance is the "emission reduction project," which builds a low-carbon and efficient new energy system by means of clean or electric energy substitution and energy interconnection, and the right end of the balance is the

DOI: 10.1201/9781003633709-5

"carbon sink project," which carries out large-scale carbon absorption by means of carbon capture, utilization, sequestration and negative emission. At present, there are three main methods: (1) increase the energy use rate, to achieve better results with less energy consumption; (2) vigorously develop clean renewable energy, such as biofuels, reduce the consumption of traditional energy, and adjust the structure of energy development; (3) Carbon Capture, Utilization and Storage (CCUS).

CCUS is a new trend developed on the basis of the original Carbon Capture and Storage (CCS) technology with the addition of U (Utilization), which is an effective carbon treatment technology (Zahed et al., 2021). The principle is to capture and separate carbon dioxide from relevant emission sources; purify it, which can be put into new production processes for recycling; and achieve long-term isolation from the atmosphere, not just simple sequestration. The removal of excess carbon dioxide from the atmosphere can be accompanied by the provision of low-polluting sustainable energy sources, focusing on the development of efficient, clean, and economical energy systems (Thiyagarajan et al., 2019). This is considered a key means of reducing greenhouse gas emissions and addressing global climate change. The development of carbon capture and utilization technologies is therefore an important idea for carbon neutrality, driving the development of the downstream carbon utilization industry chain through the direct capture of carbon dioxide.

CCUS technology is a cluster of several technologies, which can be divided into carbon capture, carbon transportation, carbon utilization and storage according to the technical route, and the specific applications are shown in Figure 5.1.

Carbon capture is one of the steps, which mainly refers to the process of capturing carbon dioxide from different emission sources such as industrial production, energy utilization and atmosphere, and separating, collecting and compressing it. According to the different sequences of carbon source capture and combustion

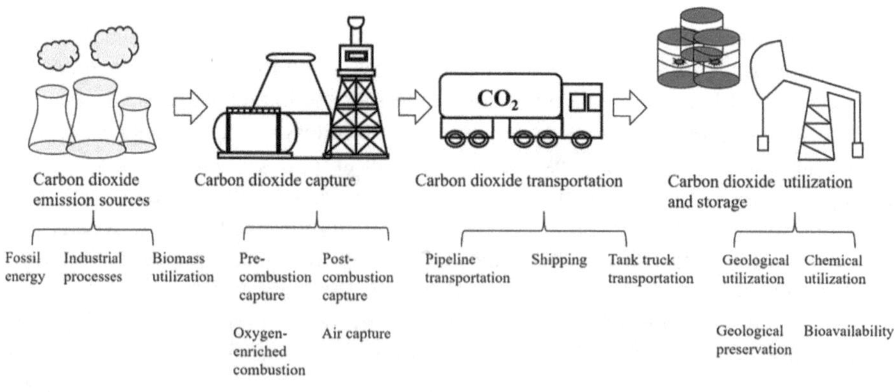

FIGURE 5.1 CCUS technology schematic.

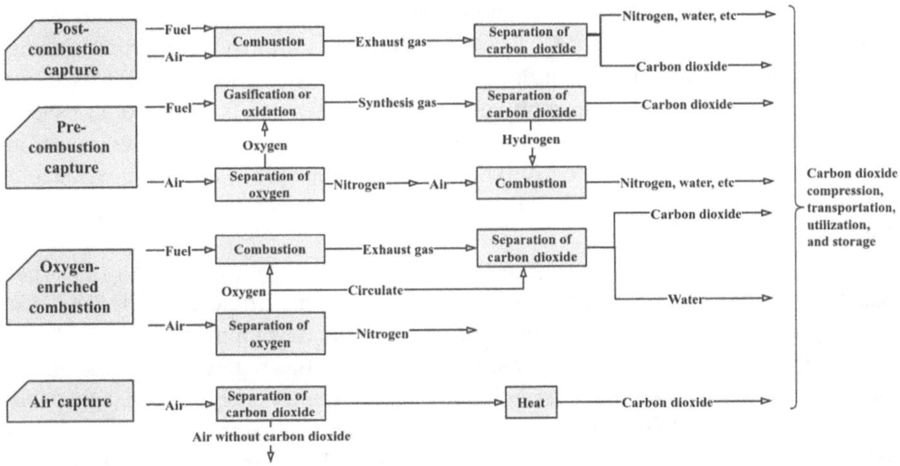

FIGURE 5.2 Classification and principles of carbon capture technology.

process, it can be categorized into four methods: post-combustion capture, pre-combustion capture, oxygen-enriched combustion, and air capture, and the specific principles are shown in Figure 5.2. These methods have their own advantages and disadvantages, and need to be applied in combination with specific gas sources and scenario equipment.

Carbon capture is the highest energy consumption and cost in the whole CCUS system, and its cost accounts for two-thirds or more of the total cost. For example, a "Future Power Generation" project in Illinois, United States, was a practical attempt to investigate the effect of CCS in real power generation, but the project was canceled in January 2018 due to the increase in the estimated cost from $830 million to $1.8 billion (Muratori et al., 2017). Therefore, the main research direction of carbon capture technology is to improve the capture efficiency and reduce the cost of capture (Hu et al., 2023). The 2021 report "Carbon Capture, Utilization and Storage (CCUS)" released by the United Nations Economic Commission for the European Union (UNECE) introduces popular carbon capture technologies, helps evaluate advantages, and promotes the deployment of technologies to achieve carbon reduction (Bakker et al., 2023).

The point source CCUS technology focuses on capturing carbon dioxide from industries such as cement and steel production, fossil fuel hydrogen production, waste incineration, and power generation. After compression, it is injected into deep underground rock layers for long-term preservation. Through different engineering methods, its capture efficiency exceeds 90%. Due to the advantage of storage capacity, the overall potential for carbon sequestration is great, but the cost is slightly higher, requiring policy guidance and large-scale facility construction.

In addition to traditional CCUS technology, there are also various new composite forms of CCUS technology, typical of which are biomass energy carbon capture and storage (BECCS) and direct air carbon capture and storage (DACCS; Lap et al., 2023). BECCS combines biomass utilization and CCUS technology to capture, utilize, or store carbon dioxide generated during biomass combustion or conversion, achieving negative emissions (NETs) throughout the entire process from biomass raw material generation to utilization (Johnson et al., 2014). The carbon released into the atmosphere is extracted through vegetation and then safely returned and stored underground in the form of combustion products during biomass combustion. DACCS directly captures carbon dioxide from the atmosphere and stores it. Due to differences in capture methods, the cost of DACCS technology is approximately twice that of BECCS.

As more and more countries choose CCUS technology as a means of carbon reduction in long-term climate policies, mainly through BECCS and DACCS for carbon removal, it is necessary to pay attention to the development of international certification and negative emission trading standards in this regard. In the sustainable development scenario of the International Energy Agency, it is expected that they will play a more significant role after 2050. Existing CCUS projects in Europe are mainly deployed in the North Sea region, including the Sleipner project in Norway (Edwards and Celia, 2018), which was commissioned in 1996, and the Snøhvit project in Norway, which was commissioned in 2008, and which have a storage capacity of 1.7 Mt/a of carbon dioxide (Shirazaki et al., 2024). As of 2021, there are 70 CCUS projects in operation, under construction and in planning in Europe. The role of CCUS technology is recognized by about 80% of countries in their long-term low-emission development strategies submitted to the United Nations Framework Convention on Climate Change (UNFCCC). According to IEA projections, carbon capture under CCUS technology could reach 7.6 billion t/year in 2050, a 190-fold increase over the 2020 global carbon capture of 40 Mt/year.

According to information from the Ministry of Ecology and Environment, China currently emits around 10 billion tons of carbon dioxide per year, but the carbon capture capacity is only around 3 million tons per year, with an application scale of less than one in 10,000 tons (Wu et al., 2023). According to the main energy structure of China, in order to achieve the goal of carbon emission control, it is very important to increase the research and application of carbon capture technology. The current application scenarios of carbon capture in China include iron and steel, industry, shipping, coal power, energy, automotive and petrochemicals.

Although research in China started late, with national policy support and the joint efforts of the government and enterprises, the technology has continued to develop and the construction of pilot demonstration projects is progressing steadily. There are about 40 existing CCUS projects in different stages of implementation in China, involving a variety of capture sources such as power plants and cement factories, most of which are in small-scale form. The report of the Global Carbon Capture and Storage Institute 2020 shows that there are currently 65 large-scale commercial CCS projects (industrial carbon source capture capacity ≥ 400,000t/a; power station carbon source capture capacity ≥ 800,000t/a) in the world, of which 26 are in operation; of these, China's Sinopec Zhongyuan Oilfield CCS project and PetroChina Jilin Oilfield CO_2-EOR project are two operating large-scale CCUS projects. This shows that China's capture sources and storage types are rich and diverse, and it has the ability to build large-scale facilities.

5.2 RESEARCH PROGRESS OF EXISTING CARBON CAPTURE TECHNOLOGIES

5.2.1 Chemical Absorption

Chemical absorption is a technique that separates carbon dioxide from combustion emissions through a chemical reaction. In the chemical absorption method, an absorber (such as ammonia, amine, alkaline solution) is usually used to contact the gas containing carbon dioxide, and the carbon dioxide is absorbed into the absorber to form a compound through a chemical reaction, and then the carbon dioxide is released from the absorber to achieve the capture and recovery of carbon dioxide (Figure 5.3; Jafri et al., 2022). The core of the method is alkaline absorbent. At present, organic alcohol amine solution, ammonia water and potassium carbonate aqueous solution are the most widely used chemical absorbent. Chemical absorption methods are subdivided into organic amines, mixed amines, ammonia, and hot potassium bases (Spigarelli and Kawatra, 2013).

Chemical absorption is one of the most widely used carbon capture technologies, and its advantages are as follows: (1) High efficiency: chemical absorption can efficiently separate carbon dioxide from the gas with high capture efficiency. (2) Mature: the technology of chemical absorption is relatively mature, and it has been widely used in the industrial field. (3) Controllable: by adjusting the parameters such as the type of absorber and its concentration, we can control the carbon dioxide absorption and release process (Aghel et al., 2022). Currently, the technology is widely applied to large-scale industrial projects with stationary sources of carbon emissions, such as coal-fired power plants and chemical plants (Singh and Dhar, 2019). However, there are many reasons that limit the development of chemical absorption: (1) High energy

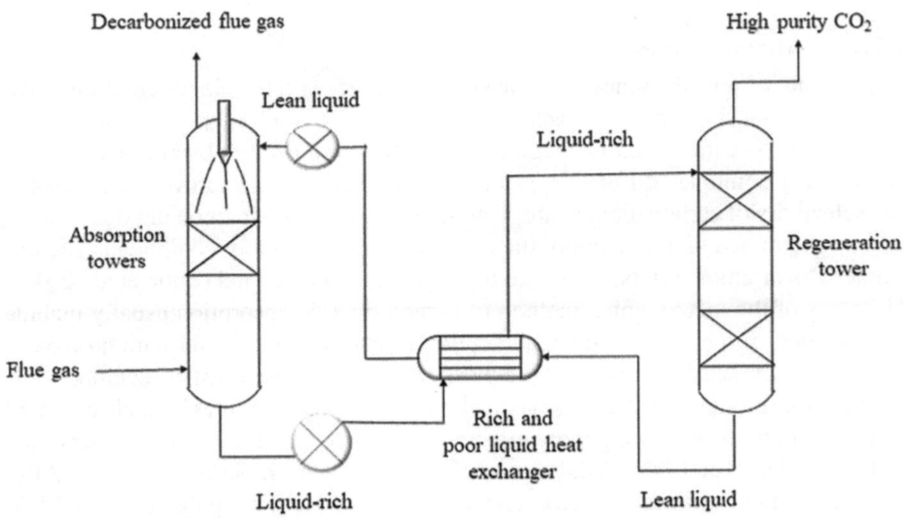

FIGURE 5.3 Process diagram of chemical solution absorption method.

consumption: chemical absorption requires a large amount of energy to achieve the absorption and release of carbon dioxide, resulting in high costs. (2) Difficulty in regeneration of absorbents: some absorbents need to be regenerated after the absorption of carbon dioxide, which is a complicated and energy-intensive regeneration process. (3) Potential environmental impacts: some absorbents may have a negative impact on the environment, such as ammonia leakage (Hernández et al., 2014).

5.2.1.1 Organic Amines

Organic amine method is a chemical absorption method in carbon capture technology, which mainly reacts with carbon dioxide through an organic amine solution to absorb carbon dioxide and form stable compounds (Zhao and Su, 2014). It is mainly divided into two stages: the absorption stage and the desorption stage. In the absorption stage, carbon dioxide gas passes through an organic amine solution in an absorber and reacts chemically with the organic amine to form compounds such as amine salts or amine esters. This step is the transfer of carbon dioxide from the gas phase to the liquid phase. The desorption stage involves the release of carbon dioxide from the absorbed liquid phase in the desorber by means of heating or depressurization, etc., to form pure carbon dioxide gas and a regenerated organic amine solution (Jafri et al., 2022).

The end products of CO_2 absorption by organic amines are usually compounds such as amine salts or amine esters. These compounds can be further transformed or utilized in subsequent treatment processes. For example, the absorbed compounds are separated and extracted from the absorption solution and purified, and the separated and purified compounds are introduced into a bioreactor to be converted into bioenergy, such as biodiesel and biomethane, by using microorganisms or enzyme catalysts. The biologically converted bioenergy can be collected, purified and used for power generation, fuel or other energy purposes (Lively et al., 2014).

5.2.1.2 Mixed Amines

The concept of mixed amines was introduced in 1985 with the aim of combining the advantages of a single amine to achieve better overall absorption performance (Song et al., 2018). The mixed amine method is a method used for the absorption of carbon dioxide that combines different types of amine solutions to improve the efficiency and selectivity of carbon dioxide absorption. In the process of carbon dioxide absorption by the mixed amine method, the end products are usually compounds such as amine salts or amine esters, similar to the organic amine method (Zhou et al., 2017). The steps of the mixed amine method of carbon dioxide absorption usually include the selection of the mixed amine, absorption of carbon dioxide, formation of products from the reaction, separation and purification, and subsequent utilization.

With the mixture of primary, secondary, and tertiary amines, which attracted much attention in developing more effective amine absorbers, hybrid absorbers such as MEA/MDEA and DEA/MDEA have been proposed (Tapia et al., 2018). After that, new organic amines such as cyclic amine piperazine (PZ), polyamines (AEEA, BDA, MAPA), spatial site-resistant amine (AMP), allylamine (DETA, TETA), HMPD and so on have been discovered, and the compound amine solution has been fully researched and widely used. For example, the CO_2 capture project at Snøhvit gas field in Norway and the Gorgon Carbon Dioxide Injection project in Australia used activated MDEA solution (aMDEA) as the absorbent, and the 150,000t/a CCUS demonstration project at Jinjie Power Plant in China used the MA series of absorbent from NANCI for post-combustion carbon capture (Han et al., 2019).

Although the mixed amine process allows for the absorption and conversion of carbon dioxide into compounds such as amine salts or amine esters, these products are not directly used for bioenergy production. However, the process of absorbing and converting carbon dioxide can offer some potential opportunities for bioenergy production. For example, carbon dioxide is captured and recovered through mixed amine methods and can be used in biomass fermentation to produce biofuels or biogas. In addition, through carbon recycling techniques, captured carbon dioxide is injected into the growth of plants to promote plant growth, and the plants are then used for bioenergy production.

5.2.1.3 Ammonia

In the ammonia method, carbon dioxide is absorbed by ammonia to form ammonium bicarbonate, and then the carbon dioxide is released from the solution by heating or depressurization, for example. In the ammonia method, carbon dioxide is absorbed by passing a gas containing carbon dioxide through ammonia, which reacts with the ammonia in the ammonia to form ammonium bicarbonate (Bian et al., 2020). The ammonium bicarbonate is decomposed by heating or depressurization, releasing carbon dioxide and ammonia. The released ammonia is recycled, while the carbon dioxide can be collected and stored or used for other purposes. It is also known as the three steps of absorption, desorption and regeneration (Aminu et al., 2017).

Conversion of carbon dioxide absorbed by the ammonia method into bioenergy is a viable method that can be achieved by providing carbon dioxide absorbed by the ammonia method to microorganisms or plants for photosynthesis or fermentation processes, which ultimately produce bioenergy, such as biofuels or biomass gases. A common method is to utilize microalgae for photosynthesis to convert carbon dioxide into biomass and further extract bio-oil or biomass gas as bioenergy (Cuéllar-Franca and Azapagic, 2015). Alternatively, an anaerobic fermentation process can be utilized to convert carbon dioxide into biomass gas, such as methane or hydrogen, along with other waste or organic matter. This process also faces some challenges, such as limitations in the efficiency of carbon dioxide utilization by microorganisms or plants, bioenergy extraction and utilization technologies (Spigarelli and Kawatra, 2013).

The ammonia method of carbon dioxide absorption is mainly suitable for small-scale industrial processes or for carbon dioxide capture in laboratories. In large-scale industrial applications, the ammonia method may be less suitable, as its equipment material requirements and ammonia handling issues add cost and operational complexity (Li et al., 2013).

5.2.1.4 Hot Potassium Bases

The hot-potassium-alkali method is a common carbon dioxide absorption technique typically used for the capture and treatment of carbon dioxide in industrial emissions gases. In this method, the gas passes through a solution containing potassium hydroxide (sodium or calcium hydroxide), and the carbon dioxide reacts with an alkaline substance to be absorbed and removed. Specific steps: When a gas passes through a solution containing an alkaline substance, the carbon dioxide reacts with the alkaline substance to form carbonates. The carbonate produced by the reaction will precipitate or otherwise separate from the solution, thus enabling the absorption and removal of carbon dioxide. Carbonates can be broken down by heating or other methods to release carbon dioxide, and the alkaline substance can be reused for carbon dioxide absorption (Leung et al., 2014). At present, when using potassium carbonate as an absorbent, activators such as diethanolamine (DEA), piperazine (PZ) and other alcohol amine organics are usually added to it to increase its absorption rate as well as its absorption capacity. Enid Fertilizer plant in the United States using piperazine as an activator of the Benfield method for CO_2 capture, the annual capture capacity of about 700,000 t (Fu et al., 2022).

The hot potash alkali method of carbon dioxide absorption is primarily used to capture and remove carbon dioxide from gases, rather than to convert it directly into bioenergy. However, after capturing and removing carbon dioxide, it can be used for bioenergy production. A common method is to use the captured carbon dioxide for photosynthesis, i.e., through organisms such as plants or algae to convert the carbon dioxide into organic matter for the eventual production of bioenergy such as biodiesel and bioethanol. This process requires light energy as a source of energy to facilitate the conversion of carbon dioxide and the synthesis of organic matter. Another method is to use catalysts to convert carbon dioxide into organic compounds such as methane and methanol, which are then used to produce bioenergy. This method requires efficient catalysts to facilitate the carbon dioxide conversion reaction (Patel et al., 2017).

The characteristics of different absorbents of chemical absorption method are shown in Table 5.1. In general, the chemical absorption method has higher energy consumption and more serious corrosion degradation problems, but its high CO_2 capture efficiency and high purity of separated CO_2 are suitable for application in large-scale CCUS projects, and it is the most promising method for arranging CCUS projects at this stage.

5.2.2 Physical Absorption

Physical absorption is another commonly used CO_2 capture technology, which, in contrast to chemical absorption, focuses on the absorption of CO_2 from a gas into a solution through a physical process. Physical absorption involves the dissolution of CO_2 in an absorbent based on electrostatic interactions or weak interactions of van der Waals forces between molecules, and the solubility is dependent on the temperature and pressure of the system, as well as the nature and concentration of the absorbent. Physical absorption methods usually utilize the physical properties of the solvent (e.g., solubility, absorption rate) to achieve separation and capture of CO_2 (Creamer and Gao, 2016).

Usually, low temperature and high pressure are favorable to the absorption of CO_2. The key to this method lies in the selection of the absorbent, which should have good solubility for CO_2, as well as a high boiling point, good stability, and

TABLE 5.1
Characteristics of Different Absorbents in Chemical Absorption Method

Methodologies	Advantages	Drawbacks
Organic amines	Fast absorption, high absorption efficiency, capable of handling high concentrations of carbon dioxide gas	Higher energy consumption, loss of organic amines, regeneration and disposal of solutions need to be addressed
Mixed amines	Fast absorption, high absorption efficiency, high load resistance, relatively low energy consumption, reduced use of volatile amines	High equipment complexity, difficult solution treatment, difficult selection of amines
Ammonia	Large absorption capacity, relatively low regeneration energy consumption, synergistic removal of other acid gases	Extremely slow absorption rate, high ammonia escape loss, difficult to control
Hot potassium bases	Low absorption cost, relatively small regeneration energy consumption, good stability	Slow absorption rate, need to add activator, small absorption capacity

nontoxicity and harmlessness. Low temperature and high pressure are the best operating conditions for the physical absorption method (Lau et al., 2021). Among the many physical absorbents, the better developed and widely used absorbents include methanol (low-temperature methanol wash Restisol method), polyethylene glycol dimethyl ether (Seloxol method), N-methyl pyrrolidone (Purisol method), and propylene carbonate (Flou method). Currently, the physical absorption methods used in CCUS projects in operation around the world are mainly the Restisol method and the Seloxol method. The methanol absorbent used in the Restisol method is not easy to be degraded; it has low viscosity and high stability, but must be operated at low temperatures, and the process is more complicated. The CO_2 separation of the Seloxol method consumes less energy, and is chemically stable, with the disadvantages of high viscosity, high stability, and low energy consumption. Physical absorption is currently used in the field of natural gas extraction and processing (Dutcher et al., 2015).

The advantages of carbon dioxide absorption by physical absorption method are (1) low energy consumption: compared with the chemical absorption method, the physical absorption method can usually realize lower energy consumption and reduce the operation cost; (2) simple operation: the operation of physical absorption method is relatively simple, and it does not involve complex chemical reactions; and (3) chemicals required: compared with the chemical absorption method which requires the use of chemicals, the physical absorption method does not require the addition of chemicals, which reduces the operation risk (Dziejarski et al., 2023). But there are also many reasons to limit the development of physical absorption: (1) poor selectivity: physical absorption of carbon dioxide selectivity is poor, may be accompanied by the absorption of other gases, and may require the need for follow-up treatment; (2) high cost of solvent regeneration: solvent regeneration process may require the consumption of more energy and cost; and (3) the operating temperature and pressure effects: physical absorption of the operating temperature and pressure requirements are higher, and there need to be strict control of operating conditions (Spigarelli and Kawatra, 2013).

5.2.3 CO_2 Capture by Adsorption

CO_2 enrichment by adsorption is now a relatively mature technology, which has obtained very considerable utilization and a lot of markets, and it accounts for a very important part in carbon capture. Adsorption method for CO_2 enrichment mainly uses porous solid materials with surface selectivity. After CO_2 is captured, external conditions such as temperature and pressure are adjusted to release CO_2, such as amino-supported adsorbents or PEI. The metal oxide is directly used for adsorption (Silveira et al., 2023). The physical adsorption method is preferentially adsorbed by van der Waals force and weak intermolecular force, and then desorption is performed by changing temperature or pressure. In the capture and removal of CO_2, adsorption separation technology uses the interaction between adsorbent and CO_2 to achieve the separation and recovery of CO_2. Common adsorption methods include Pressure Swing Adsorption (PSA) and Temperature Swing Adsorption (TSA; Dutcher et al., 2015).

As shown in Figure 5.4, the basic principle of PSA is that the gas components are different on the solid material, and the adsorption amount will change with the change of pressure so as to achieve the separation and purification of gas based on the periodic pressure change.

Table 5.2 will compare the advantages and disadvantages of the PSA method and TSA method. In the process of CO_2 removal by the PSA method, the air is pretreated, dedusted and cooled, and then sent to the adsorption tower by the compressor, where the CO_2 is purified. TSA is the same as PSA, using the difference in the adsorption

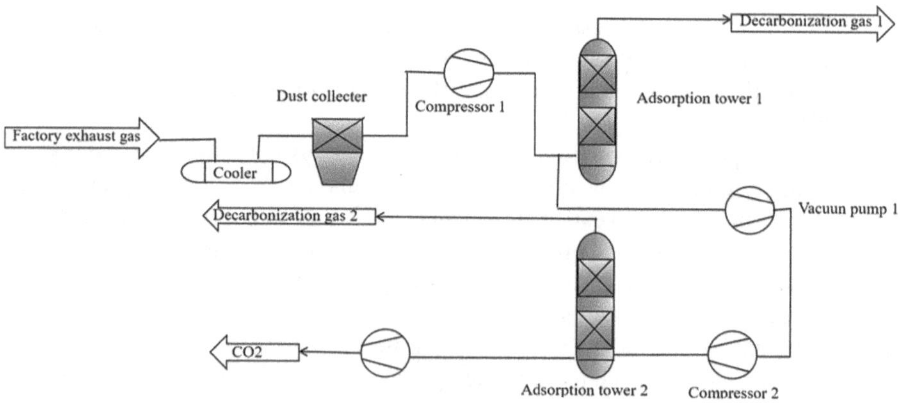

FIGURE 5.4 Process flow of PSA separation of CO_2.

TABLE 5.2
Comparison of Advantages and Disadvantages of PSA Method and TSA Method

CO_2 Trapping technique	Advantage	Disadvantage
PSA	The process cycle is short, the generated product is high purity, the equipment involved is simple to operate and easy to maintain, and if the first-level PSA cannot achieve the separation effect, it can also be connected to multiple towers and continuous cycle operation.	The adsorbent is difficult to desorption and the product is easy to damage.
TSA	The process is simple and the product recovery rate is high.	The process cycle is long, additional heating is required, and the cost is high.

performance of each component of the gas on the solid material for gas adsorption purification, the difference between TSA and PSA is that it will be affected by temperature changes, the use of periodic temperature change to achieve gas separation and purification.

Nowadays, the physical adsorption method is widely popular because of its small energy consumption, simple process, it is environmentally friendly, and will not cause instrument corrosion, but there are still many problems affecting the development of physical adsorption: (1) The capacity of the adsorbent is small, cannot carry out large-scale adsorption, so it needs to be improved to be used as biofuels. (2) The cost of the new adsorbent is too high and is not suitable for large-scale use (Luberti and Ahn, 2022). (3) Unable to deal with the gas with more water, the adsorbent will preferentially adsorb water molecules, resulting in a decline in the ability to adsorb CO_2, resulting in unsatisfactory adsorption. In the future, the research direction of CO_2 enrichment by adsorption method is mainly to develop adsorbents with high adsorption capacity, fast adsorption kinetics, good chemical stability and low cost, as well as gas drying technology that can deal with wet target flue gas. In the construction industry, some achievements have been made in capturing carbon through adsorption to achieve carbon recycling (Bahrun et al., 2022).

The solid adsorbent technology was put into use in 2014. The technology captures CO_2 80%–90% efficiently, and in 2020, the Portland Cement plant was supported by the US Department of Energy's National Energy Technology Laboratory, which has the potential to become the largest carbon capture plant in the cement industry. Based on the aforementioned practical applications, it can be seen that carbon collection by adsorption method is feasible, and it is completely feasible to use this method to capture carbon to produce biofuels, and after certain development, the cost is controllable, and it is very competitive in the market (Chai et al., 2022).

5.2.4 Membrane Separation Method

The working principle of the membrane separation method is as follows (Figure 5.5): the dissolution rate and diffusion rate of different components of gas in the membrane are different, and due to the pressure difference on both sides of the membrane, the gas with high permeability will pass through the membrane to the side with low pressure, while the gas with low permeability cannot pass through the membrane to the high pressure side (Mubashir et al., 2017).

Membrane treatment is mainly used to treat water gas composed of CO_2 and H_2 before combustion (Solangi et al., 2021). Generally, the membranes used to treat water gas are divided into two categories, CO_2 preferential permeation membrane and H_2 preferential permeation membrane. Among these two membranes, CO_2 preferential permeation membrane is relatively more advantageous, because the CO_2 in water gas is lower than the membrane area required for H_2 separation. It can save the cost of carbon collection. At the same time, because H_2 is high pressure and CO_2 is low pressure, the pressure loss is small, which is conducive to the subsequent steps. At the same time, the enrichment of H_2 has low requirements for membrane separation factors (Shabbani et al., 2024).

The range of membrane materials is very wide, organic membrane including polysulfone ether, polyoxyethylene, polyphenyl ether, inorganic membrane, including zeolite, MoS_2, MOFs, oxygen fossil, and ink, among which the stability of organic film is poor, the use is subject to a lot of restrictions, but the advantages of inorganic film are more obvious, inorganic film has good chemical stability, but

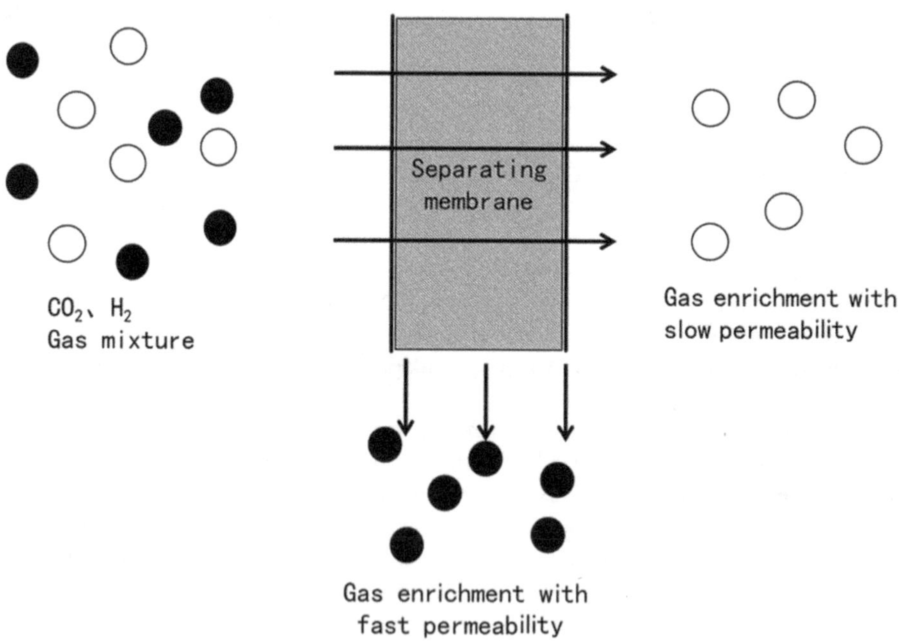

FIGURE 5.5 Schematic diagram of membrane separation technology.

inorganic film also has certain problems: inorganic film production costs are high; many times, it cannot be put into large-scale use (Shen et al., 2023). Mixed matrix membrane combines the two, which not only has good stability but also has excellent permeability and selectivity, but the mixed matrix membrane technology is still immature and needs further research.

Compared with other removal methods, membrane separation has its own advantages: (1) it uses less energy; (2) it is highly efficient in enrichment; (3) it causes less pollution to the environment; 4) it is easy to operate and does not require too much technical training, and efficiently captured carbon is suitable for use as biofuels (Dai et al., 2016), but membrane separation also has many problems to be solved: (1) the technology is still in the laboratory research, lack of large-scale application; (2) the capture cost is very high, not suitable for large-scale application (Ng et al., 2022). The limitations and advantages of the membrane separation method are very obvious, and many institutions have invested a lot of energy in the study of the membrane method, such as Polaris membrane of MTR Company in the United States, PVAm membrane of NUST in Norway, PolyActive membrane of HZG in Germany, and amine Xugen mass transfer membrane of Ohio State University (Han et al., 2019).

In the future, if the membrane separation method is to be put into the production of biofuels, on the one hand, it is necessary to produce and research and develop high-performance membranes, and more importantly, it is to combine the membrane separation method with other carbon capture methods, such as the membrane separation and chemical absorption method have been combined (Zhang, 2016). Membrane separation method, does not cause pollution, the required energy consumption is very small, the enrichment effect is good, and the good use can be well invested in the production of biomass fuel, but before obtaining the market and competitiveness, it also needs to invest more human and material resources to produce high-performance membranes that can be recycled, which takes a certain amount of time (Feng et al., 2023).

5.2.5 Cryogenic Method

For some gas sources with strong pressure and high CO_2 concentration (CO_2 volume fraction > 60%), if CO_2 separation is to be carried out in the flue gas after oxygen-rich combustion, or exploitation in natural gas Wells with high CO_2 content, the cryogenic method can be used (Song et al., 2019). The cryogenic method uses the difference of boiling points of different components in the gas source to condense the gas before low temperature and then separates it one by one by distillation according to the different boiling points of various components. Now the more well-known cryogenic method has the CCC process (as seen in Figure 5.6).

Although the cryogenic method can deal with strong pressure, and high CO_2 concentration of gas, there is still a lack of industrial application for this method; there is not enough large-scale application experience to learn from; in addition, the refrigeration process will consume a lot of energy, will bring a lot of cost consumption. If we want to put the future cryogenic method into large-scale application, we need to develop a smaller energy consumption of the refrigeration process. This reduces

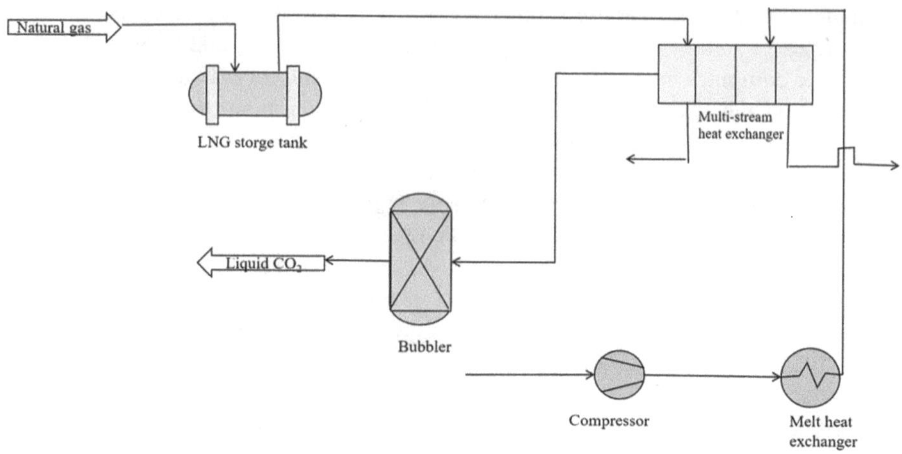

FIGURE 5.6 CCC process simplification diagram.

costs and reduces the footprint of the equipment (Katare et al., 2023). Although there are a lot of problems, the cryogenic method can deal with strong pressure and high CO_2 concentration of gases, which is not available in other methods. So although the cost is high because of the need for certain personnel technology, there are still quite a lot of researchers invested in it; its market is very broad (Hu et al., 2016).

5.2.6 Biocapture Technology

CO_2 biocapture is a technology that utilizes organisms such as microorganisms or plants to convert CO_2 into organic matter. This technology helps to reduce the concentration of CO_2 in the atmosphere, thereby combating climate change and mitigating the effects of global warming. Biocapture focuses on the fixation of CO_2 by plants and microorganisms.

Algae, as the main raw material for the production of third-generation fuel technologies, can utilize CO_2 for photosynthesis, and has the following advantages in the production of biofuels: high photosynthetic efficiency, fast growth rate, good environmental adaptation, rich lipid content, and high carbon sequestration capacity. Algae is considered an attractive biomaterial for sequestration of CO_2 and also for the production of renewable biofuels, food, animal and aquaculture feed products, and other value-added products such as cosmetics, nutrients, pharmaceuticals, biofertilizers, and bioactives. Nexant believes that algae can yield more oil from the same area compared to other oilseed crops. Therefore, there is an increasing interest in technologies that utilize algae for oil production. Figure 5.7 summarizes the process of using algal organisms for diesel production (Singh and Dhar, 2019).

Microorganisms are a group of tiny organisms, including bacteria, fungi, and protozoa, which have a variety of metabolic pathways and biological functions that allow them to capture and convert CO_2 in different environments. For example, some bacteria and fungi can utilize CO_2 to carry out smack oxygen respiration and convert CO_2 into organic matter. Some algal microorganisms and bacteria utilize the energy

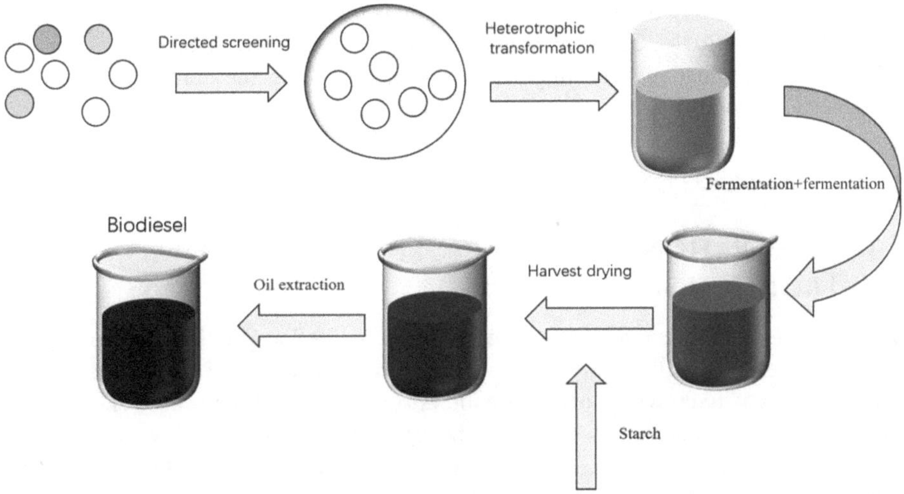

FIGURE 5.7 Process flow chart of algal biodiesel preparation.

of sunlight to convert CO_2 into organic matter and oxygen through the process of photosynthesis. Methane-producing bacteria can use CO_2 and hydrogen to produce methane, converting CO_2 into another carbon source. Some bacteria can convert CO_2 into nitrogen or other compounds through the processes of nitrification and denitrification (Saravanan et al., 2022).

Biogenic methane is CH_4 produced by the reduction of CO_2 through biological processes, and biogenic methane is produced mainly through the metabolic activities of some specific microorganisms, which are called methanogenic bacteria. Methanogenic bacteria can use CO_2, H_2 and other substrates to produce methane, a process known as methane fermentation or methane synthesis. Methanogenic bacteria include methanogens and some bacteria. Methanogenic bacteria are capable of multi-electron reduction of inert molecules in the presence of ATP-dependent nitrogen-fixing enzymes. Naturally occurring nitrogenases are unable to perform CO_2 reduction because several amino acids in close proximity to the ferromolybdenum cofactor also play a role in substrate binding and reduction. Nitrogenase was modified by replacing Gln with α-195 and Ala with α-70 to enable it to reduce CO_2 to CH_4 (Bhatia et al., 2019). There are also enzymes that act on both plants and microorganisms to facilitate CO_2 capture by plants and microorganisms. One of the most common enzymes in photosynthesis is Rubisco, an enzyme found in plants, algae, and some bacteria. The function of Rubisco is divided into two parts: carboxylation and oxygenation. In carboxylation, Rubisco combines CO_2 with RuBP to produce 3-PGA, which is a key step in photosynthesis and promotes carbon fixation. In contrast, in oxidation, Rubisco binds oxygen to RuBP, leading to energy loss and a decrease in the efficiency of photosynthesis, which is known as photorespiration. Although Rubisco plays a crucial role in photosynthesis, it is also considered one of the less efficient enzymes because of its relatively slow catalytic rate and the energy loss it causes during oxidation. Scientists are working to find ways to improve the efficiency of Rubisco in order to increase the efficiency of photosynthesis in plants and potentially help increase crop yields (Lin et al., 2014). Carbonic anhydrases are a class of enzymes widely found in living organisms that play an important role in regulating the conversion reaction between CO_2 and H_2O. Carbonic anhydrase accelerates the reaction between CO_2 and water, hydrates CO_2 into carbonate ions (HCO_3^-), and promotes Rubisco activity, thereby facilitating photosynthesis (Aspatwar et al., 2018).

5.3 INTEGRATION OF CO_2 CAPTURE WITH BIOFUEL PRODUCTION

Because of technological and storage constraints, as well as political and ecological concerns, we need to find ways to use carbon dioxide in a more sustainable way. One such way is to convert carbon dioxide into valuable and renewable energy. The abundance of raw materials will bridge the gap between the world's energy supply and demand and alleviate the impoverishment of the energy supply. In addition, due to the finite nature of fossil fuels, finding alternative fuels such as biofuels is imperative (Hawrot-Paw et al., 2020). Therefore, utilizing carbon dioxide as a biofuel will not only help to slow down global warming, but will also provide an inexhaustible source of carbon for the future.

Compared to traditional fossil fuels, biofuels have greater advantages in terms of energy stability, ecology, and political economy. In addition to this, by incorporating agriculture, new income generation opportunities can be created for the people, along with innovative development of new projects (Putra et al., 2019). The large-scale application of energy fuels in various industries promotes the development of alternative and renewable energy sources. Thus in recent years, there has been a growing trend toward the conversion and utilization of biofuels globally.

Although the use of biofuels based on crop production has the potential to reduce greenhouse gas emissions and mitigate the carbon effect, it also poses certain challenges. On the one hand, the emission of flue gases from machinery and equipment used in production activities may lead to an increase in greenhouse gas emissions, and on the other hand, the impact of biofuels on economic, social, and industrial structural factors, due to the fact that the production of biofuels may ultimately be in direct competition with the production of food, as well as through indirect effects of land utilization changes, carbon emissions and crop yields, leading to increased CO_2 emissions (Öztürk and Türkay, 2016).

Part of carbon utilization in CCUS is the production of biofuels from carbon dioxide emissions, using microorganisms to efficiently produce biofuels from animal and plant feedstocks, including liquid and gaseous biofuels. The liquid biofuels that are widely used today are bioethanol and biodiesel, with bioethanol fermentation being the largest microbial fermentation process available. Since the microbial method of fermentation route can convert carbohydrates into biofuels such as alcohols, fatty acid esters, straight-chain hydrocarbons and cycloalkanes by utilizing everyday crops such as straw, grass and shredded wood as agricultural wastes or non-food crops, it has attracted a lot of attention, and many countries around the world have set national-level targets for the development of bioethanol (Jin et al., 2014). However, there are many problems in microbial fermentation production, and the main problems in production are pretreatment of biomass, simultaneous saccharification fermentation (SSF) and co-fermentation (CBP), separation of biofuel and fermentation broth. Among the pretreatment methods are ionic liquid method and liquid ammonia isomerization treatment, which affects the enzymatic reaction and fermentation pattern by controlling the relationship between cellulase and trichoderma. SSF research focuses on the appropriate temperature of action due to the high requirement of efficient stability of genetically engineered bacteria for SSF technology, while CBP requires thermophilic microorganisms such as *Clostridium perfringens* and *Clostridium thermofibrillarum*. Most of its feedstock is converted to proteins, and the rest is used for conversion to ethanol and growth of the organisms. In terms of industrialization, SSF and CBP are most expected to be the first to be used on a large scale. The products are separated and purified by osmotic evaporation, permeation, distillation, molecular sieve adsorption, organic solvent extraction, etc. The fermentation method for the production of acetone and butanol by *Clostridium byssidianum* as a starter strain (acetone-butanol-ethanol), referred to as ABE fermentation, uses oleyl alcohol to extract butanol from the fermentation broth and then distill it. Due to the toxicity of the by-products n-butanol and isobutanol to microorganisms, it is necessary to enhance the adaptability of the strain or use solvent pumps during the fermentation process to increase the yield, etc. ABE fermentation is one of the larger-scale biochemical industries other than bioethanol fermentation (Garg et al., 2023).

Whereas gaseous biofuels include biogas and hydrogen, large-scale plants have a better future in producing biogas and carbon dioxide from biomass derived from organisms and industrial wastes while cleaning up municipal waste. Biogas is usually plant fiber fermented by anaerobic or partially anaerobic microorganisms through hydrolysis or acidification and is produced in stages, requiring control of pH and temperature at different stages to increase the yield. Hydrogen is considered to be the most desirable fuel of the future in terms of transport use and economic and environmental considerations. Biofuels are highly likely to become an important part of the solution to greenhouse gas emissions (Gao et al., 2017). As fossil fuels are depleted and oil prices are expected to continue to rise, both conserving resources and producing alternatives such as biofuels are important strategies (DeCicco, 2018). Future biofuels will combine physical, chemical, and biological fermentation methods, such as the synthesis of ethyl tertiary butyl ether (ETBE). Biofuels will be a very important part of carbon capture technology in achieving carbon reduction and carbon neutrality in the future, and its development will be carried out under the science of sustainability, with functional and diversified systems through the integration of resources and socio-environmental factors.

5.4 CONCLUSIONS

At present, China has entered a period of ecological civilization construction focusing on pollution reduction, carbon reduction and synergies. CCUS technology has the potential to reduce overall emission reduction costs and increase emission reduction flexibility, and combined with the green and sustainable characteristics of biofuels, it is gradually becoming an important technological means at the current stage of carbon dioxide emission reduction. Carbon capture will be used as a powerful tool to promote the adjustment of industrial structure and energy structure, and will be combined with a variety of technological means to achieve the dual-carbon goal, contributing to China's strength in combating global climate change.

In addition, there are many uncertainties and unknowns in the development and utilization of carbon capture technology in terms of cost, life cycle effects, storage capacity and storage durability. The electricity situation, stability of long-term policy frameworks, and implementation costs may be challenges when applying the technology. At the same time, it offers many potential opportunities, such as the development of value-added carbon dioxide storage, industrial utilization, and carbon trading markets.

REFERENCES

Aghel, B., Janati, S., Wongwises, S., Shadloo, M.S. 2022. Review on CO_2 capture by blended amine solutions. *International Journal of Greenhouse Gas Control*, **119**, 103715.

Aminu, M.D., Nabavi, S.A., Rochelle, C.A., Manovic, V. 2017. A review of developments in carbon dioxide storage. *Applied Energy*, **208**, 1389–1419.

Aspatwar, A., Haapanen, S., Parkkila, S. 2018. An Update on the metabolic roles of carbonic anhydrases in the model alga Chlamydomonas reinhardtii. *Metabolites*, **8**(1), 22.

Bahrun, M.H.V., Bono, A., Othman, N., Zaini, M.A.A. 2022. Carbon dioxide removal from biogas through pressure swing adsorption – A review. *Chemical Engineering Research and Design*, **183**, 285–306.

Bakker, S., Moorman, S., Knoope, M., Terwindt, M. 2023. Land use of energy supply for carbon neutral mobility: a well-to-wheel analysis. *European Transport Research Review*, **15**(1), 26.

Bhatia, S.K., Bhatia, R.K., Jeon, J.-M., Kumar, G., Yang, Y.-H. 2019. Carbon dioxide capture and bioenergy production using biological system – A review. *Renewable and Sustainable Energy Reviews*, **110**, 143–158.

Bian, B., Bajracharya, S., Xu, J., Pant, D., Saikaly, P.E. 2020. Microbial electrosynthesis from CO_2: Challenges, opportunities and perspectives in the context of circular bioeconomy. *Bioresource Technology*, **302**.

Chai, S.Y.W., Ngu, L.H., How, B.S., Chin, M.Y., Abdouka, K., Adini, M.J.B.A., Kassim, A.M. 2022. Review of CO_2 capture in construction-related industry and their utilization. *International Journal of Greenhouse Gas Control*, **119**, 103727.

Creamer, A.E., Gao, B. 2016. Carbon-Based Adsorbents for Postcombustion CO_2 Capture: A Critical Review. *Environmental Science & Technology*, **50**(14), 7276–7289.

Cuéllar-Franca, R.M., Azapagic, A. 2015. Carbon capture, storage and utilisation technologies: A critical analysis and comparison of their life cycle environmental impacts. *Journal of CO_2 Utilization*, **9**, 82–102.

Dai, Z., Ansaloni, L., Deng, L. 2016. Recent advances in multi-layer composite polymeric membranes for CO_2 separation: A review. *Green Energy & Environment*, **1**(2), 102–128.

DeCicco, J. M. 2018. Methodological Issues Regarding Biofuels and Carbon Uptake. *Sustainability* **10** (5), 1581.

Dutcher, B., Fan, M., Russell, A.G. 2015. Amine-Based CO_2 Capture Technology Development from the Beginning of 2013—A Review. *ACS Applied Materials & Interfaces*, **7**(4), 2137–2148.

Dziejarski, B., Krzyżyńska, R., Andersson, K. 2023. Current status of carbon capture, utilization, and storage technologies in the global economy: A survey of technical assessment. *Fuel*, **342**, 127776.

Edwards, R.W.J., Celia, M.A. 2018. Infrastructure to enable deployment of carbon capture, utilization, and storage in the United States. *Proceedings of the National Academy of Sciences*, **115**(38), E8815–E8824.

Feng, S., Du, X., Luo, J., Zhuang, Y., Wang, J., Wan, Y. 2023. A review on facilitated transport membranes based on π-complexation for carbon dioxide separation. *Separation and Purification Technology*, **309**, 122972.

Fu, L., Ren, Z., Si, W., Ma, Q., Huang, W., Liao, K., Huang, Z., Wang, Y., Li, J., Xu, P. 2022. Research progress on CO_2 capture and utilization technology. *Journal of CO_2 Utilization*, **66**.

Gao, G., Clare, A., Rose, C., Caldwell, G., 2017. Ulva rigida in the future ocean: potential for carbon capture, bioremediation and biomethane production. *GCB Bioenergy* **10**(1): 39–51.

Garg, D., Samota, M.K., Kontis, N., Patel, N., Bala, S., Rosado, A.S. 2023. Revolutionizing biofuel generation: Unleashing the power of CRISPR-Cas mediated gene editing of extremophiles. *Microbiological Research*, **274**.

Han, Y., Salim, W., Chen, K.K., Wu, D., Ho, W.S.W. 2019. Field trial of spiral-wound facilitated transport membrane module for CO_2 capture from flue gas. *Journal of Membrane Science*, **575**, 242–251.

Hawrot-Paw, M., Koniuszy, A., Sędłak, P., Seń, D. 2020. Functional properties and microbiological stability of fatty acid methyl esters (FAME) under different storage conditions. *Energies*, **13**(21).

Hernández, C.E.L.R., Leite, A.L., Dantas, P.V., Ramos, S.P., Perez, M.d.l.A., Takaki, G.M.d.C. 2014. Hybrid microbial-photosynthetic biofuel cells for simultaneous bacterial glycerol biotransformation and algal carbon dioxide capture. *Journal of the Brazilian Chemical Society*, **25**, 560–571.

Holzinger, M., Le Goff, A., Cosnier, S. 2012. Carbon nanotube/enzyme biofuel cells. *Electrochimica Acta*, **82**, 179–190.

Hu, G., Nicholas, N.J., Smith, K.H., Mumford, K.A., Kentish, S.E., Stevens, G.W. 2016. Carbon dioxide absorption into promoted potassium carbonate solutions: A review. *International Journal of Greenhouse Gas Control*, **53**, 28–40.

Hu, J., Jåstad, E.O., Bolkesjø, T.F., Rørstad, P.K. 2023. Impact of large-scale Bio-CCS deployment on forest biomass competition and forest industry production. *Biomass and Bioenergy*, **175**, 106896.

Jafri, Y., Ahlström, J.M., Furusjö, E., Harvey, S., Pettersson, K., Svensson, E., Wetterlund, E. 2022. Double yields and negative emissions? Resource, climate and cost efficiencies in biofuels with carbon capture, storage and utilization. *Frontiers in Energy Research*, **10**, 797529.

Jin, H., Chen, L., Wang, J., Zhang, W. 2014. Engineering biofuel tolerance in non-native producing microorganisms. *Biotechnology Advances*, **32**(2), 541–548.

Johnson, N., Parker, N., Ogden, J. 2014. How negative can biofuels with CCS take us and at what cost? Refining the economic potential of biofuel production with CCS using spatially-explicit modeling. *Energy Procedia*, **63**, 6770–6791.

Katare, A., Kumar, S., Kundu, S., Sharma, S., Kundu, L.M., Mandal, B. 2023. Mixed matrix membranes for carbon capture and sequestration: Challenges and scope. *ACS Omega*, **8**(20), 17511–17522.

Lap, T., Benders, R., van der Hilst, F., Faaij, A. 2023. BECCS as climate mitigation option in a Brazilian low carbon energy system: Estimating potential and effect of gigatonne scale CO_2 storage. *International Journal of Greenhouse Gas Control*, **128**, 103945.

Lau, H.C., Ramakrishna, S., Zhang, K., Radhamani, A.V. 2021. The role of carbon capture and storage in the energy transition. *Energy & Fuels*, **35**(9), 7364–7386.

Leung, D.Y.C., Caramanna, G., Maroto-Valer, M.M. 2014. An overview of current status of carbon dioxide capture and storage technologies. *Renewable and Sustainable Energy Reviews*, **39**, 426–443.

Li, L., Zhao, N., Wei, W., Sun, Y. 2013. A review of research progress on CO_2 capture, storage, and utilization in Chinese Academy of Sciences. *Fuel*, **108**, 112–130.

Lin, M.T., Occhialini, A., Andralojc, P.J., Parry, M.A.J., Hanson, M.R. 2014. A faster Rubisco with potential to increase photosynthesis in crops. *Nature*, **513**(7519), 547–550.

Lively, R.P., Sharma, P., McCool, B.A., Beaudry-Losique, J., Luo, D., Thomas, V.M., Realff, M., Chance, R.R. 2014. Anthropogenic CO_2 as a feedstock for the production of algal-based biofuels. *Biofuels, Bioproducts and Biorefining*, **9**(1), 72–81.

Luberti, M., Ahn, H. 2022. Review of polybed pressure swing adsorption for hydrogen purification. *International Journal of Hydrogen Energy*, **47**(20), 10911–10933.

Mubashir, M., Fong, Y.Y., Leng, C.T., Keong, L.K. 2017. Issues and current trends of hollow-fiber mixed-matrix membranes for CO_2 separation from N_2 and CH_4. *Chemical Engineering & Technology*, **41**(2), 235–252.

Muratori, M., Kheshgi, H., Mignone, B., Clarke, L., McJeon, H., Edmonds, J. 2017. Carbon capture and storage across fuels and sectors in energy system transformation pathways. *International Journal of Greenhouse Gas Control*, **57**, 34–41.

Ng, T.Y.S., Viriya, V., Chew, T.L., Yeong, Y.F., Ahmad, A.L., Ho, C.-D., Jawad, Z.A. 2022. Optimization of CO_2/H_2 separation over Ba-SAPO-34 zeolite membrane synthesized by microwave heating. *Membranes*, **12**(9), 850.

Öztürk, A., Türkay, M. 2016. Bicriteria optimization approach to analyze incorporation of biofuel and carbon capture technologies. *AIChE Journal* **62**(9): 3473–3483.

Patel, H.A., Byun, J., Yavuz, C.T. 2017. Carbon dioxide capture adsorbents: Chemistry and methods. *ChemSusChem*, **10**(7), 1303–1317.

Putra, Z., Kurnia J., Sasmito, A., Muraza, O. 2019. Process design and techno-economic analysis of ethyl levulinate production from carbon dioxide and 1,4-butanediol as an alternative biofuel and fuel additive. *International Journal of Energy Research*, **43**(11), 5932–5945.

Rahman, F.A., Aziz, M.M.A., Saidur, R., Bakar, W.A.W.A., Hainin, M.R., Putrajaya, R., Hassan, N.A. 2017. Pollution to solution: Capture and sequestration of carbon dioxide (CO_2) and its utilization as a renewable energy source for a sustainable future. *Renewable and Sustainable Energy Reviews*, **71**, 112–126.

Saravanan, A., Deivayanai, V.C., Senthil Kumar, P., Rangasamy, G., Varjani, S. 2022. CO_2 bio-mitigation using genetically modified algae and biofuel production towards a carbon net-zero society. *Bioresource Technology*, **363**, 127982.

Shabbani, H.J.K., Othman, M.R., Al- Janabi, S.K., Barron, A.R., Helwani, Z. 2024. H_2 purification employing pressure swing adsorption process: Parametric and bibliometric review. *International Journal of Hydrogen Energy*, **50**, 674–699.

Shen, J., Wang, X., Chen, Y. 2023. Adsorbents for adsorption separation of CO_2 and CH_4: A literature review. *The Canadian Journal of Chemical Engineering*, **101**(12), 7115–7133.

Shirazaki, S., Pishvaee, M.S., Sobati, M.A. 2024. Integrated supply chain network design and superstructure optimization problem: A case study of microalgae biofuel supply chain. *Computers & Chemical Engineering*, **180**, 108468.

Silveira, B.H.M., Costa, H.K.M., Santos, E.M. 2023. Bioenergy with carbon capture and storage (BECCS) in Brazil: A review. *Energies*, **16**(4), 16042021.

Singh, J., Dhar, D.W. 2019. Overview of carbon capture technology: Microalgal biorefinery concept and state-of-the-art. *Frontiers in Marine Science*, **6**.

Solangi, N.H., Anjum, A., Tanjung, F.A., Mazari, S.A., Mubarak, N.M. 2021. A review of recent trends and emerging perspectives of ionic liquid membranes for CO_2 separation. *Journal of Environmental Chemical Engineering*, **9**(5), 105860.

Song, C., Liu, Q., Ji, N., Deng, S., Zhao, J., Li, Y., Song, Y., Li, H. 2018. Alternative pathways for efficient CO_2 capture by hybrid processes—A review. *Renewable and Sustainable Energy Reviews*, **82**, 215–231.

Song, C., Liu, Q., Qi, Y., Chen, G., Song, Y., Kansha, Y., Kitamura, Y. 2019. Absorption-microalgae hybrid CO_2 capture and biotransformation strategy—A review. *International Journal of Greenhouse Gas Control*, **88**, 109–117.

Spigarelli, B.P., Kawatra, S.K. 2013. Opportunities and challenges in carbon dioxide capture. *Journal of CO_2 Utilization*, **1**, 69–87.

Tapia, J.F.D., Lee, J.-Y., Ooi, R.E.H., Foo, D.C.Y., Tan, R.R. 2018. A review of optimization and decision-making models for the planning of CO_2 capture, utilization and storage (CCUS) systems. *Sustainable Production and Consumption*, **13**, 1–15.

Thiyagarajan, S., Varuvel, E.G., Martin, L.J., Beddhannan, N. 2019. Mitigation of carbon footprints through a blend of biofuels and oxygenates, combined with post-combustion capture system in a single cylinder CI engine. *Renewable Energy*, **130**, 1067–1081.

Wu, B., Lin, R., Bose, A., Huerta, J.D., Kang, X., Deng, C., Murphy, J.D. 2023. Economic and environmental viability of biofuel production from organic wastes: A pathway towards competitive carbon neutrality. *Energy*, **285**, 129322.

Zahed, M.A., Movahed, E., Khodayari, A., Zanganeh, S., Badamaki, M. 2021. Biotechnology for carbon capture and fixation: Critical review and future directions. *Journal of Environmental Management*, **293**, 112830.

Zhang, Z. 2016. Comparisons of various absorbent effects on carbon dioxide capture in membrane gas absorption (MGA) process. *Journal of Natural Gas Science and Engineering*, **31**, 589–595.

Zhao, B., Su, Y. 2014. Process effect of microalgal-carbon dioxide fixation and biomass production: A review. *Renewable and Sustainable Energy Reviews*, **31**, 121–132.

Zhou, W., Wang, J., Chen, P., Ji, C., Kang, Q., Lu, B., Li, K., Liu, J., Ruan, R. 2017. Bio-mitigation of carbon dioxide using microalgal systems: Advances and perspectives. *Renewable and Sustainable Energy Reviews*, **76**, 1163–1175.

6 Microbial Enzymes Involved in Biofuel Production

Sahar Akhavan-Mahdavi

6.1 INTRODUCTION

The quest for sustainable and eco-friendly energy sources has placed biofuel production at the forefront of scientific research and industrial application. Among the various methods employed for biofuel production – ranging from thermochemical and mechanical to enzymatic and microemulsion processes – the enzymatic approach stands out for its safety, cleanliness, cost-effectiveness, and minimal environmental footprint. Enzymes, as natural catalysts, accelerate the chemical reactions essential for transforming biomass into biofuels, without being consumed in the process, thus offering a remarkable specificity, efficiency, and the advantage of operating under mild conditions (Bangaru et al., 2022; Soni et al., 2023).

The enzymatic method of biofuel production is increasingly gaining global acceptance, particularly for the production of biodiesel and biohydrogen, which are now widely used across various industries. This preference stems from the method's ability to overcome the limitations associated with conventional chemical catalysts, such as improving yield and efficiency, reducing environmental impact, and enhancing the quality of biofuels. The potential for the repeated use of enzymes further underscores their economic advantage, especially given the lower utility costs of enzymatic hydrolysis compared to acidic hydrolysis, which requires harsh conditions and subsequent treatment steps (Himmel et al., 2010). Figure 6.1 represents a flowchart that provides a visual representation of the biofuel production process, highlighting the pivotal role of microbial enzymes in the conversion of biomass into sustainable energy.

At the heart of biofuel production lies the challenge of breaking down complex biomass, such as cellulosic biomass found in agricultural residues, grasses, and wood chips, into simpler, fermentable sugars. Enzymes such as cellulases, hemicellulases, and ligninases play pivotal roles in this process, hydrolyzing cellulose, hemicellulose, and degrading lignin, respectively. Such enzymatic action not only facilitates the conversion of these sugars into bioalcohols like ethanol but also enhances the efficiency of producing other biofuels through processes like the transesterification of fats and oils into biodiesel by lipases, and the conversion of organic substrates into biohydrogen by dehydrogenases (Chaurasia et al., 2020).

 DOI: 10.1201/9781003633709-6

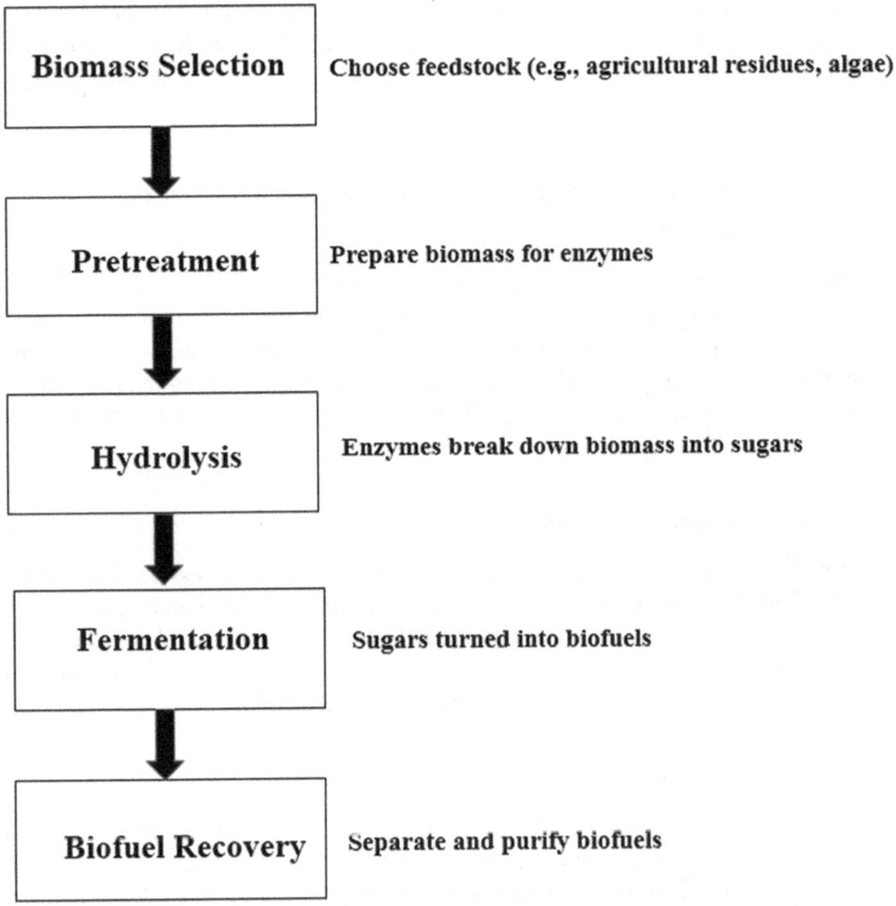

FIGURE 6.1 Overview of biofuel production with microbial enzymes.

Despite the evident advantages of enzymatic processes, including energy efficiency, substrate specificity, and reduced environmental impact, challenges such as the high cost of enzyme production, the need for enzyme recovery and recycling, and the development of enzymes that are stable and active under industrial conditions persist. Addressing these challenges has spurred research into genetic engineering, enzyme immobilization, and the development of novel enzymes, promising new avenues for more sustainable and efficient biofuel production (Prasad et al., 2019). The historical evolution of enzyme utilization, from ancient practices of fermentation to modern biotechnological innovations, underscores the significant role enzymes have played and continue to play in biofuel production (Kumar and Kumar, 2017). The advent of recombinant DNA technology, enzyme engineering, and advancements in bioprocessing techniques have revolutionized the field, enhancing the economic viability and environmental sustainability of biofuel production. As we look toward the future, the ongoing research and technological advancements in enzyme technology hold the promise of further optimizing biofuel production processes, aligning with global sustainability goals, and paving the way for a more energy-efficient and eco-friendly industrial landscape (Soni et al., 2023).

6.2 THE FUNCTION OF ENZYMES IN THE PRODUCTION OF BIOFUELS

Enzymes play a crucial role in the production of biofuels, serving as catalysts that speed up the biochemical reactions needed to convert biomass into usable fuels. Their function is pivotal across various stages of biofuel production, from the breakdown of complex organic materials to the final conversion into bioethanol, biodiesel, and other forms of renewable energy. Here's a detailed overview of how enzymes function in the biofuel production process (Soni et al., 2023).

6.2.1 Breakdown of Biomass

Cellulolytic and hemicellulolytic enzymes: These enzymes are essential in breaking down cellulose and hemicellulose, the main components of plant cell walls, into simpler sugars. Cellulases, for example, catalyze the hydrolysis of cellulose into glucose, which can then be fermented to produce ethanol. Similarly, hemicellulases break down hemicellulose into its constituent sugars (Adegboye et al., 2021).

6.2.2 Lignin Removal

Lignin-modifying enzymes: Lignin is a complex polymer that provides structural support to plant cell walls and acts as a barrier to the enzymatic breakdown of cellulose and hemicellulose. Lignin-modifying enzymes, such as laccases and peroxidases, help in delignification, thereby improving the accessibility of cellulose and hemicellulose for further enzymatic action (Tamoor et al., 2021).

6.2.3 Sugar Fermentation

Fermentative enzymes: Once the complex polymers are broken down into simpler sugars, yeast, and bacteria use fermentative enzymes to convert these sugars into ethanol and other biofuels. For instance, the enzyme zymase, found in yeast, is crucial for the fermentation process that converts glucose into ethanol and carbon dioxide (Love, 2022).

6.2.4 Biodiesel Production

Lipases: In the production of biodiesel, lipases catalyze the transesterification of fats or oils, converting them into fatty acid methyl esters (FAMEs), which are the chemical constituents of biodiesel. This process involves the reaction of triglycerides (from fats or oils) with alcohol in the presence of a lipase catalyst (Bhandari et al., 2021).

6.2.5 Enhancing Efficiency and Reducing Costs

Enzymes are being engineered to be more efficient, stable, and cost-effective for industrial biofuel production. Advances in biotechnology and genetic engineering have enabled the development of enzyme mixtures optimized for specific substrates and process conditions, improving the overall efficiency and sustainability of biofuel production (Banu et al., 2021).

6.2.6 Environmental Benefits

The use of enzymes in biofuel production offers significant environmental benefits. It allows for the conversion of waste biomass into energy, reducing reliance on fossil fuels and lowering greenhouse gas emissions. Moreover, enzymatic processes often operate under milder conditions than chemical processes, reducing energy consumption and minimizing the production of harmful by-products. Enzymes are indispensable in the biofuel production process, facilitating the conversion of biomass into renewable energy in an efficient and environmentally friendly manner. Ongoing research and development in enzyme technology continue to improve the viability and sustainability of biofuels as an alternative energy source (Raghavendra et al., 2016).

6.3 TYPES OF MICROBIAL ENZYMES IN BIOFUEL PRODUCTION

Microbial enzymes, sourced from bacteria, fungi, and yeasts, are instrumental in the sustainable and efficient conversion of biomass into biofuels, serving as biological catalysts that break down complex organic materials into simpler molecules. These enzymes facilitate the production of various biofuels, including bioethanol, biodiesel, and biohydrogen, by specifically acting on different biomass components. The advancement of biotechnological methods, such as genetic engineering and enzyme optimization, has significantly improved the effectiveness and reduced the costs of these enzymes in biofuel production processes. Unlike chemical processes that rely on chemical catalysts and heat, biochemical processes utilize these biological catalysts to transform raw materials into fuel, offering a solution to the energy-intensive and environmentally detrimental drawbacks of conventional chemical catalysts (Sharmila et al., 2024). Here is an overview of the main types of microbial enzymes involved in biofuel production (Table 6.1).

6.3.1 Cellulases and Hemicellulases

Cellulases and hemicellulases are crucial microbial enzymes in the production of biofuels, playing a pivotal role in the biochemical conversion of biomass to bioenergy. These enzymes break down cellulose and hemicellulose, the primary components of plant cell walls, into fermentable sugars, which can then be converted into biofuels such as bioethanol. Cellulases are a group of enzymes that hydrolyze cellulose into glucose units. They include endoglucanases, which cut internal bonds to disrupt the crystalline structure of cellulose and create new chain ends; exoglucanases, which

TABLE 6.1
Overview of Key Enzymes in Biofuel Production

Enzyme	Source	Function	Biofuel Application
Cellulases	Fungi (e.g., *Trichoderma reesei*) bacteria (e.g., *Clostridium thermocellum*)	Break down cellulose into glucose units. Cellulases are a complex of enzymes including endoglucanases, exoglucanases, and β-glucosidases, which work synergistically to hydrolyze cellulose, a major component of plant cell walls	Essential for the production of cellulosic ethanol by converting agricultural residues, grasses, and woody plants into fermentable sugars
Hemicellulases	Fungi (e.g., *Aspergillus niger*), bacteria (e.g., *Bacillus species*)	Decompose hemicellulose, a heteropolymer found in plant cell walls, into its component sugars, including xylose and arabinose. Key enzymes include xylanases, β-xylosidases, and mannanases	Used alongside cellulases to enhance the breakdown of biomass, improving the yield of fermentable sugars for biofuel production
Lipases	Fungi (e.g., *Candida antarctica*), bacteria (e.g., *Pseudomonas cepacia*)	Catalyze the transesterification reaction of triglycerides (fats and oils) to produce biodiesel (fatty acid methyl esters, FAMEs) and glycerol, using methanol or ethanol as the alcohol source	Crucial for biodiesel production from vegetable oils, animal fats, and recycled greases
Amylases	Bacteria (e.g., *Bacillus licheniformis*), fungi (e.g., *Aspergillus oryzae*)	Convert starches into sugars. Amylases include α-amylases, which break down long-chain starch molecules into maltose and dextrins, and glucoamylases, which further hydrolyze these products into glucose	Vital for the production of bioethanol from starchy crops like corn and wheat
Dehydrogenases	Various microorganisms	Involved in the fermentation processes, converting organic substrates into hydrogen through anaerobic digestion	Play a role in biohydrogen production, an emerging biofuel with potential for clean energy applications
Ligninases	White-rot fungi (e.g., *Phanerochaete chrysosporium*)	Degrade lignin, a complex aromatic polymer in plant cell walls that provides structural support and resistance to microbial attack. Key enzymes include laccases and peroxidases	Facilitate the breakdown of lignocellulosic biomass, improving access to cellulose and hemicellulose for further enzymatic conversion to biofuels

(*Continued*)

TABLE 6.1 (Continued)
Overview of Key Enzymes in Biofuel Production

Enzyme	Source	Function	Biofuel Application
Laccase	Fungi (e.g., basidiomycetes, deuteromycetes, ascomycetes), bacteria (e.g., *Marinomonas mediterranea, Streptomyces lavendulae, Bacillus subtilis, Geobacillus thermocatenulatus, Aquisalibacillus elongatus*)	Degradation of lignin by oxidizing lignin components, contributing to the breakdown of this complex aromatic polymer	Used in the pretreatment of lignocellulosic biomass for the removal of lignin, facilitating the hydrolysis of polysaccharides into fermentable sugars for biofuel production
Pectinases	Fungi (e.g., *Aspergillus* sp., *Penicillium* sp.), bacteria (e.g., *Bacillus* sp.)	Hydrolysis of pectin, breaking down pectic substances in plant cell walls	Aid in the breakdown of plant cell walls to enhance the accessibility of cellulose and other polysaccharides for conversion into bioethanol, especially in biomass where lignin content is low
Proteases	Fungi (e.g., *Aspergillus oryzae*), bacteria (e.g., *Bacillus subtilis, Penicillium roqueforti*)	Breakdown of proteins into amino acids and peptides through proteolysis	Improve the efficiency of microorganisms in converting sugars to biofuels by making nitrogen sources available and facilitating the liberation of higher alcohols for biofuel conversion

cleave two to four units from the ends of the exposed chains produced by endoglucanase action; and β-glucosidases, which hydrolyze the cellobiose and other short oligosaccharides into glucose. The efficient breakdown of cellulose into glucose is essential for the subsequent fermentation process to produce bioethanol (Singhania et al., 2021).

Hemicellulases, on the other hand, target hemicellulose, a heterogeneous polymer consisting of various sugars including xylose, mannose, galactose, rhamnose, and arabinose, among others. Hemicellulases include a variety of enzymes such as xylanases, which degrade xylan, the most abundant hemicellulosic polysaccharide, into xylose units. Other hemicellulases work synergistically to break down the complex sugar polymers into their constituent sugars, which are then fermented into biofuels. Cellulase, an enzyme of considerable industrial importance, contributes to about 20% of the global enzyme market. Its demand is expected to surge, particularly from the biofuel production sector. Thus, enhancing the production and efficiency of cellulase at an industrial scale has become a focal point (Srivastava et al., 2018). Microorganisms, especially certain bacteria and fungi, are efficient cellulase producers. Among these, fungi are preferred for their superior substrate utilization and penetration capabilities, despite the rarity of a single fungus possessing a complete cellulase system for effective biomass hydrolysis. Commercial cellulase production often involves using different fungal strains to develop the three key enzymes (exo, endo, and β-glucosidase) through submerged fermentation (SmF), a process known for its high costs due to low-end product concentration and the need for additional purification steps. Some companies are exploring the production of cellulase from genetically modified organisms (GMOs), but this is still in the experimental stage due to high costs. Therefore, finding cost-effective production methods is crucial for making biofuel production economically feasible. Recent studies have focused on using extremophilic fungi for their stable and active enzymes, which could make enzyme production more cost-effective(Narisetty et al., 2023).

Cellulases are classified into three main types based on their action on cellulose: endoglucanases, exoglucanases (also known as cellobiohydrolases), and β-glucosidases. These enzymes are part of the O-glycoside oxidases family and are involved in breaking the glycosidic bonds found between carbohydrates or between carbohydrate and non-carbohydrate molecules (Ranganathan et al., 2022). Endoglucanases: These enzymes hydrolyze internal β-1,4-glycosidic bonds within the cellulose chain, leading to a decrease in polymerization and an increase in the number of chain ends. They are primarily associated with the GH families 5–8, 12, 16, 44, 45, 48, 51, 64, 71, 74, 81, 87, 124, and 128. Endoglucanases, which have a cleft-shaped active site, are considered endo-acting enzymes due to their activity on the amorphous regions of cellulose, making them crucial for initiating cellulose breakdown (Prajapati et al., 2018).

Exoglucanases or Cellobiohydrolases: These enzymes work by cleaving β-1, 4-glycosidic bonds at the ends of the cellulose chain, effectively releasing cellobiose units from the crystalline structure of cellulose. They are mainly found in GH families 5–7 and 48. Their tunnel-shaped active site helps prevent readhesion of separated molecules, facilitating the solubilization of cellulose. The structural understanding of exoglucanases, such as the cellobiohydrolase from Phanerochaete chrysosporium (Cel7A), provides insights into their function and role in cellulose hydrolysis (Wilson, 2009).

β-glucosidases: These enzymes catalyze the conversion of cello-oligosaccharides into glucose by breaking β-glucosidic linkages. They are considered the rate-limiting step in the hydrolysis reaction and are essential for the final step in converting cellulose into fermentable sugars. β-glucosidases belong to GH families 1, 3, 4, 17, 30, and 116. Their action is crucial in biofuel production processes, where they work alongside other cellulases to produce reducing sugars(Areeshi, 2022).

Trichoderma reesei, *Clostridium thermocellum*, and *Aspergillus niger* are key microbial producers of enzymes pivotal for biofuel production. *T. reesei* is renowned for its highly efficient cellulase system, capable of converting cellulose into fermentable sugars, as highlighted in a landmark study by Singhania (2011) which showcased its capability for efficient biomass hydrolysis (Singhania, 2011). *C. thermocellum*, a thermophilic bacterium, produces a highly active cellulase complex known as the cellulosome, demonstrating exceptional efficiency in cellulose degradation and potential for direct conversion of lignocellulosic biomass into bioethanol, according to research by Arai et al. (2022) (Arai et al., 2022). *A. niger*, a filamentous fungus, produces a broad spectrum of cellulolytic enzymes, including cellulases and hemicellulases, used effectively in hydrolyzing lignocellulosic materials into simple sugars for biofuel production, as shown in studies by Juturu and Wu (2014). These microbes and their enzyme systems are integral to advancing biofuel research and development, underscoring the importance of microbial enzymes in the biofuel production process (Juturu and Wu, 2014).

Research on biomass hydrolysis using microbial enzymes, particularly cellulases, has significantly advanced biofuel production. Li et al. (2016) have demonstrated the efficiency of *Trichoderma reesei* cellulases in hydrolyzing pretreated corn stover to yield high amounts of fermentable sugars, underscoring the potential of these enzymes in biofuel processes. Similarly (Li et al., 2016), Arai et al. (2022) highlighted the capabilities of *Clostridium thermocellum* in consolidated bioprocessing (CBP), showing how this bacterium could potentially streamline biofuel production by merging enzyme production, biomass hydrolysis, and fermentation into a single step(Arai et al., 2022). The effectiveness of cellulase enzymes varies considerably among microbial sources, prompting comparative studies to identify the most active enzymes for industrial biofuel production. Zhang et al. (2019) found that cellulase enzymes from *Trichoderma reesei* displayed the highest activity under certain conditions, suggesting the importance of selecting the appropriate enzyme based on the biomass and process conditions to optimize biofuel production (Zhang et al., 2019). Moreover, a comprehensive review by Kuhad et al. (2011) discussed the variability of cellulase systems across different microbes, indicating that while no single cellulase enzyme is universally the best, enzymes from Trichoderma species often exhibit superior performance across various scenarios (Kuhad et al., 2011).

6.3.2 Lipases

Lipases, or triacylglycerol ester hydrolases (EC 3.1.1.3), are widespread enzymes present in animals, plants, and both natural and genetically modified microorganisms. They are crucial for fat metabolism, breaking down fats and oils into free fatty acids, diacylglycerols, monoglycerols, and glycerol through hydrolysis. Their ability to catalyze reactions with water-insoluble substrates is notable, with the efficiency of catalysis often requiring a water/lipid interface. Furthermore, in conditions with varying water content, lipases can facilitate synthetic reactions, including esterification, transesterification, and acidolysis, even in organic solvents.

Owing to their remarkable adaptability, lipases are deemed essential catalysts in biotechnological applications, contributing significantly to sectors such as food and dairy, pharmaceuticals, agrochemicals, oleochemicals, cosmetics, and detergents. The enzymatic transesterification process, turning vegetable oil and animal fat into biodiesel, has emerged as a particularly vibrant area of research. This surge in interest is attributed to biodiesel's potential in mitigating environmental degradation through its capacity to lower overall carbon dioxide emissions across its lifecycle, sourced from renewable origins. Additionally, biodiesel offers reductions in hydrocarbons, carbon monoxide, and particulate matter emissions. Its compatibility with existing engine technology, either as a pure form or in blends, without necessitating modifications, alongside its global availability as a renewable resource, presents a sustainable solution to the challenges of fossil fuel depletion and the geopolitical tensions in oil-rich regions (Fatma et al., 2021).

One of the key benefits of employing lipase-catalyzed reactions is the ability to operate under mild conditions of temperature and pressure, which significantly reduces energy consumption and minimizes the risk of generating unwanted by-products. Lipases are distinguished by their remarkable substrate specificity, enabling the selective conversion of feedstocks into biodiesel. This specificity not only minimizes the need for extensive purification processes but also enhances the overall yield of biodiesel production. Furthermore, lipases are capable of efficiently processing low-quality feedstocks, including waste cooking oils and fats. This capability not only reduces the cost associated with raw materials but also addresses critical waste management issues (Foukis et al., 2017). Lipases are versatile enzymes that play a critical role in the biodiesel production process, catalyzing the transesterification of oils and fats into FAMEs, the primary constituents of biodiesel. Several microbial sources are renowned for their lipase production capabilities, contributing significantly to biodiesel technology. For instance, the fungi *Aspergillus niger* and *Rhizopus oryzae* are prominent lipase producers, with their enzymes being widely applied in biodiesel synthesis due to their high specificity and efficiency under varying process conditions. Similarly, bacterial species such as Pseudomonas fluorescens and Bacillus subtilis have been identified as potent lipase producers, offering advantages in terms of enzyme stability and activity (Foukis et al., 2017).

The choice of oil feedstock is pivotal in biodiesel production, influencing both the economic viability and sustainability of the production process. Various oil resources have been utilized for biodiesel production, encompassing a wide range of vegetable oils and animal fats. Commonly used vegetable oils include soybean oil, palm oil, and canola oil, known for their availability and relatively high oil content (Rao et al., 2020). Waste cooking oil and nonedible oils, such as jatropha oil and neem oil, have also gained attention as cost-effective and environmentally friendly alternatives, promoting waste utilization and reducing competition with food resources. Furthermore, microalgae have emerged as a promising source of lipids for biodiesel production, offering high lipid productivity and the ability to grow in nonarable land, thus not competing with food crops (Costantini and Califano, 2021).

The potential for reusing immobilized lipases multiple times in the transesterification process presents a significant opportunity to lower the production costs associated with biodiesel. The immobilization technique also improves the stability and activity of lipases under operational conditions, making the biodiesel production process more economically viable (Zambare et al., 2021). Although numerous lipase immobilization technologies have been explored, only a handful have reached industrial applications due to the expensive nature of the immobilization process. Therefore, there is a pressing need for innovation in the development of cost-effective immobilization techniques. These advancements should aim for reduced reaction times, enhanced enzyme stability and activity, and the use of affordable supports, all of which are critical for the scalability of enzymatic biodiesel production at an industrial level (Rao et al., 2020). Lipase immobilization, a process that attaches enzyme molecules to insoluble support while retaining their catalytic activity, is crucial for their application in biodiesel production and other biotechnological fields. This process utilizes various techniques, mainly adsorption, entrapment, and covalent coupling, each with its specific advantages and limitations. Adsorption, characterized by weak interactions between the enzyme and support, is simple and preserves enzyme activity but risks enzyme leaching. Entrapment restricts mass transfer, limiting its practical use despite its potential for reusability and enhanced enzyme characteristics like thermal and pH stability (Rao et al., 2020). Covalent immobilization offers increased stability and recyclability, yet may not always retain lipase activity. Hydrophobic supports are preferred for lipase immobilization due to their ability to maintain enzyme activity by mimicking the natural water-oil interface environment, essential for the enzyme's interfacial activation (Abdulmalek and Yan, 2022). Moreover, innovations in mesoporous materials, particularly silica, have shown promise for lipase immobilization, offering tailored pore sizes and surface modifications to enhance enzyme stability and activity. These advancements highlight the ongoing efforts to develop cost-effective, efficient immobilization techniques crucial for scaling up enzymatic biodiesel production while addressing the challenges of enzyme costs and operational stability (Rao et al., 2020).

Recent advancements in biotechnology, particularly in the realms of enzyme engineering and immobilization strategies, have led to the creation of lipase variants that are not only more effective but also more resilient. These innovations specifically aim to tackle significant obstacles such as the deactivation of enzymes caused by methanol, which is frequently used in the transesterification process. This, in turn, enhances the cost-effectiveness of producing biodiesel. Furthermore, lipases have a vital role beyond biodiesel production; they are pivotal in other biofuel generation methods, like transforming oil-rich waste or grease into biogas via anaerobic digestion. The remarkable ability of lipases to facilitate lipid-based reactions highlights their essential contribution to the biofuel sector, marking a significant stride toward the realization of sustainable and renewable energy solutions. This emphasizes the impact of lipase enzymes in propelling the biofuel industry forward, as detailed by Fatma and colleagues in 2021 (Fatma et al., 2021).

6.3.3 Amylases

Amylases, including both α-amylases and β-amylases, are indispensable enzymes in the biofuel industry, playing a vital role primarily in the production of bioethanol from starch-based biomass. These enzymes are pivotal in catalyzing the breakdown of starch – a polysaccharide abundantly found in agricultural products such as corn, wheat, barley, and other grains – into simpler sugars. These sugars are then fermented into ethanol, showcasing the transformation of renewable plant materials into sustainable energy alternatives to fossil fuels.

For example, in the bioethanol production process, α-amylases initially break down long starch molecules into shorter chains called dextrins, which are then further hydrolyzed by β-amylases into maltose and ultimately glucose. These simple sugars are essential for the fermentation process, serving as substrates for yeast or bacteria to produce ethanol. The use of microbial amylases in this process significantly improves efficiency, ensuring a thorough and swift conversion of starch into fermentable sugars. This enhancement not only increases the yield of ethanol but also reduces the time required for production, as outlined by Vaidya et al. in 2015. Through such applications, amylases demonstrate their indispensable role in advancing sustainable and renewable energy solutions within the biofuel industry (Vaidya et al., 2015).

Moreover, the enzymatic hydrolysis of starch by amylases occurs at lower temperatures compared to chemical hydrolysis methods. This attribute leads to considerable energy savings and minimizes the environmental impact of the production process. The cost-effectiveness of using amylases is further highlighted by the reduced need for expensive chemical additives and harsh processing conditions, thereby lowering overall production costs. Additionally, advancements in enzyme engineering and biotechnology have led to the development of amylase variants with improved activity, stability, and substrate specificity. Such advancements enable the effective utilization of a broader range of starch-rich feedstocks, including nontraditional agricultural residues and waste materials, enhancing both the sustainability and cost-effectiveness of bioethanol production (de Castro and de Castro, 2012).

The implications of amylase application extend beyond bioethanol production, with potential uses in other biofuel processes like biobutanol production, where starch is a key feedstock. The efficiency of amylases in converting starch into simple sugars underscores their value across various biofuel technologies, contributing significantly to the development of renewable energy systems and reducing reliance on fossil fuels.

Amylases underscore the integration of biotechnological innovations into renewable energy production, playing a critical role in the efficient and sustainable generation of bioethanol from starch-based biomass. Their application not only exemplifies the potential of green energy solutions but also marks a significant stride toward the advancement of renewable energy sources(Arya et al., 2022).

6.3.4 Dehydrogenases

Dehydrogenases represent a crucial class of enzymes in the biofuel industry, significantly impacting the production of biohydrogen and bioalcohols such as bioethanol and biobutanol. These enzymes are central to the oxidation-reduction reactions within microbial metabolic pathways, enabling the transfer of electrons from organic substrates to acceptors like NAD^+ or $NADP^+$, which are then reduced to NADH or NADPH. This electron transfer is essential for the metabolic processes that yield biofuels, highlighting the pivotal role of dehydrogenases in bioenergy production (Tasca et al., 2008).

In the realm of biohydrogen production, dehydrogenases are instrumental in converting organic substrates into hydrogen gas through dark fermentation processes. Certain bacteria utilize these enzymes to oxidize organic compounds, thereby generating electrons. These electrons are subsequently employed by hydrogenase enzymes to produce hydrogen gas. The enzymatic reactions facilitated by dehydrogenases occur under ambient conditions, which makes biohydrogen production not only energy-efficient but also environmentally friendlier compared to conventional thermochemical hydrogen production methods (Aquino Neto and De Andrade, 2013).

Dehydrogenases also play a critical role in the production of bioethanol and biobutanol, where they are involved in the metabolic pathways of fermenting microorganisms. By oxidizing intermediates, these enzymes facilitate the conversion of sugars into alcohols, producing the necessary NADH for the reduction steps that follow in the pathway. Furthermore, advancements in genetic and metabolic engineering have leveraged the optimization of dehydrogenase activity within microbial hosts, thereby enhancing the yield and efficiency of bioalcohol production. Through the overexpression of specific dehydrogenases or their engineering for improved activity, researchers can significantly boost the conversion rates of substrates into biofuels.

Moreover, dehydrogenases contribute to the flexibility and optimization of biofuel production processes. Their ability to utilize a broad spectrum of substrates, including sugars, glycerol, and organic acids, allows for the employment of various feedstocks. This versatility includes agricultural wastes and industrial by-products, thereby enhancing the sustainability and economic viability of biofuel production. Optimizing the activity and stability of dehydrogenases through enzyme engineering can lead to increased biofuel yields and more efficient substrate conversion, further improving the economic aspects of biofuel production.

Dehydrogenases are key to the biofuel industry, enabling the efficient and sustainable conversion of organic substrates into valuable biohydrogen and bioalcohols. Their role in metabolic pathways underscores the significant potential of microbial enzymes in advancing sustainable and efficient energy solutions. This not only contributes to the development of renewable energy sources but also aids in reducing the reliance on fossil fuels, marking a significant step forward in the pursuit of green energy alternatives (Scheiblbrandner et al., 2022).

6.3.5 Proteases

Proteases, enzymes that catalyze the breakdown of proteins into peptides and amino acids, hold significant value in the biofuel production sector. Their applications span from enhancing the efficiency of biofuel production processes to optimizing the utilization of various biomass feedstocks, showcasing their versatility and utility in renewable energy technologies.

For instance, in the realm of algal biofuel production, proteases play a crucial role by hydrolyzing the protein components of microalgal cell walls, facilitating lipid release for biodiesel conversion. This enzymatic pretreatment significantly improves the accessibility of carbohydrates and lipids within the biomass, aiding their transformation into biofuels. Several microbes are renowned for their protease production capabilities, contributing to biofuel processes. Bacillus species, for example, are widely reported for their proteolytic enzyme production. Specifically, Bacillus subtilis has been studied for its ability to produce proteases that can efficiently break down proteinaceous materials in biomass, making it a valuable resource for biofuel production. Similarly, fungi such as Aspergillus oryzae and Aspergillus niger are recognized for their protease production, offering potential applications in biomass pretreatment processes to enhance biofuel yield. Through the exploitation of these microbial proteases, the biofuel industry can optimize the conversion of biomass into renewable energy sources, underscoring the enzymes' importance in advancing sustainable energy technologies (Arya et al., 2022).

Furthermore, proteases significantly enhance microbial fermentation processes, which are essential for the production of bioethanol or biobutanol. By breaking down proteins present in biomass or fermentation media, proteases release amino acids that act as nitrogen sources for fermenting microorganisms. This supports microbial growth and boosts fermentation efficiency. Additionally, proteases play a role in reducing process inhibitors by hydrolyzing proteins in biomass feedstocks that could otherwise impede enzyme activity or microbial fermentation, thereby improving overall biofuel yields (Wang et al., 2019).

Proteases also contribute to the valorization of waste streams generated during biofuel production. By hydrolyzing protein-rich waste into amino acids and peptides, these enzymes enable the conversion of waste into valuable products such as animal feed supplements or nutrients for further microbial fermentation processes. This not only supports the principles of a circular economy but also lessens the environmental footprint of biofuel production (Greene et al., 2015).

Recent advancements in biotechnology have facilitated the development of proteases that possess improved qualities such as heightened activity, enhanced stability, and greater specificity, which are particularly beneficial for the unique demands of biofuel production processes. For instance, genetic engineering techniques have been utilized to modify proteases from Bacillus species, resulting in enzymes with increased tolerance to the high temperatures often encountered in biofuel production. A notable example includes a genetically modified version of a Bacillus subtilis protease, engineered for improved thermal stability and activity, making it more suitable for the breakdown of proteinaceous biomass at elevated temperatures.

Proteases offer a myriad of applications in biofuel production, highlighting their critical role from biomass pretreatment to waste valorization. Their ability to efficiently break down proteins into utilizable forms and mitigate process inhibitors underscores their importance in the development and optimization of sustainable biofuel production technologies (Eijsink et al., 2008).

6.3.6 Pectinases

Pectinases, a vital group of enzymes that specialize in the breakdown of pectin – a complex polysaccharide present in plant cell walls – are gaining importance in the biofuel industry for their pivotal role in biomass processing. These enzymes enhance the production of biofuels, such as bioethanol, by degrading pectin to improve the accessibility of cellulose and hemicellulose for further enzymatic hydrolysis, thus facilitating a more efficient conversion of biomass into biofuels (Amin et al., 2019).

In biofuel production, pectinases play multiple roles. They are crucial in the biomass pretreatment stage, especially for lignocellulosic and pectin-rich biomass like fruit wastes, breaking down pectin barriers to expose cellulose and hemicellulose. This enzymatic pretreatment aids in the hydrolysis of these polysaccharides into fermentable sugars, setting the stage for their conversion into biofuels. Additionally, by reducing the viscosity of the biomass slurry and breaking down pectin, pectinases enhance fermentation efficiency and increase the yield of biofuels. They also enable the valorization of agro-industrial wastes, transforming pectin-rich materials such as citrus peels and apple pomace into valuable biofuels, thereby contributing to the sustainability of biofuel production and addressing waste management issues (Singhania et al., 2021). Pectinases often work in synergy with other biomass-degrading enzymes, like cellulases and hemicellulases, to improve biomass conversion efficiency. This synergistic action ensures a more comprehensive degradation of plant cell wall components, resulting in higher yields of fermentable sugars. The technological and economic benefits of utilizing pectinases in biofuel production are significant. They help reduce production costs by diminishing the need for harsh chemical pretreatments and enabling the use of low-cost, pectin-rich agricultural wastes. The enzymatic pretreatment process with pectinases is environmentally friendly, avoiding toxic chemicals and minimizing waste and pollutants, thus aligning with global efforts toward more sustainable biofuel production (Djuma'ali et al., 2012).

Advancements in enzyme technology, including the development of more efficient and thermostable pectinase variants, are enhancing the viability of enzymatic pretreatment for industrial-scale biofuel production. Genetic engineering and immobilization techniques are being explored to further improve the performance and reusability of pectinases, optimizing biofuel production processes.

In summary, pectinases offer a broad spectrum of applications in the biofuel industry, from improving biomass pretreatment to enhancing fermentation processes. Their contribution to the sustainable conversion of biomass into energy, coupled with their role in promoting the economic viability and environmental sustainability of biofuel production, underscores the significance of these enzymes in advancing renewable energy solutions (Xiao and Anderson, 2013).

6.4 MECHANISMS OF ACTION OF MICROBIAL ENZYMES

Microbial enzymes catalyze biochemical reactions essential for converting biomass into biofuels. These enzymes, derived from bacteria, fungi, and other microorganisms, work by breaking down complex organic molecules into simpler forms that can be further processed into biofuels. The mechanisms of action of these enzymes involve a series of steps that enable them to bind to their specific substrates, catalyze the conversion, and release the product. Understanding these mechanisms is crucial for optimizing biofuel production processes (Srivastava et al., 2019; Table 6.2).

6.4.1 Enzyme-Substrate Binding

Lock and Key Model: Enzymes have specific active sites that fit their substrate molecules in a manner similar to a lock and key. This specificity ensures that enzymes catalyze only specific reactions (Dutta and Wu, 2014).

Induced Fit Model: This model suggests that the active site of an enzyme is flexible and molds itself around the substrate upon binding. This change in shape facilitates the catalytic action of the enzyme (Eijsink et al., 2008).

TABLE 6.2
Mechanisms of Action and Roles of Microbial Enzymes in Biofuel Production Processes

Enzyme Category	Enzyme Examples	Substrate	Mechanism of Action	Biofuel Production Step
Cellulolytic enzymes	Cellulases, Hemicellulases	Cellulose, hemicellulose	Hydrolyze β-1,4-glycosidic bonds in cellulose and various bonds in hemicellulose, breaking them down into fermentable sugars.	Biomass breakdown
Lignin-degrading enzymes	Laccases, Peroxidases	Lignin	Oxidize lignin components, facilitating the degradation of lignin to enhance accessibility to cellulose and hemicellulose.	Biomass breakdown
Pectin-degrading enzymes	Pectinases	Pectin	Degrade pectin in plant cell walls, facilitating the disassembly of the cell wall structure.	Biomass breakdown
Starch-degrading enzymes	Amylases (α-Amylases, β-Amylases)	Starch	Convert starch into simple sugars by cleaving α-1,4-glycosidic bonds.	Sugar fermentation
Saccharifying snzymes	Glucoamylases, β-glucosidases	Oligosaccharides, disaccharides	Convert oligosaccharides and disaccharides into glucose for fermentation.	Sugar fermentation
Fermentative microorganisms	Yeasts (e.g., *Saccharomyces cerevisiae*)	Simple sugars (glucose)	Ferment simple sugars into bioethanol and other biofuels.	Biofuel synthesis
Lipid-transforming enzymes	Lipases	Triglycerides (fats and oils)	Catalyze the transesterification of triglycerides into FAMEs and glycerol (biodiesel).	Biofuel synthesis
Redox enzymes	Dehydrogenases, Hydrogenases	Organic substrates	Dehydrogenases oxidize organic substrates to generate electrons and protons for hydrogenase-catalyzed biohydrogen production.	Biofuel synthesis

6.4.2 Catalysis

Transition State Formation: Enzymes stabilize the transition state of the substrate, reducing the activation energy required for the reaction to proceed. This is achieved through various mechanisms, such as straining the substrate bonds, providing a favorable microenvironment, or directly participating in the chemical reaction (Christopher et al., 2014).

Chemical Reaction: Enzymes catalyze the conversion of substrates into products through mechanisms like hydrolysis (adding water to break bonds), oxidation-reduction (transfer of electrons), and group transfer (transferring functional groups from one molecule to another; Guldhe et al., 2015).

6.4.3 Product Release

Once the reaction is complete, the product has a lower affinity for the enzyme's active site and is released into the surrounding environment. The enzyme then returns to its original state, ready to catalyze another reaction cycle (Noraini et al., 2014).

In the realm of biofuel production, enzymatic actions play a pivotal role in converting biomass into sustainable energy sources. Cellulases, for instance, are instrumental in cellulosic ethanol production, where they attach to cellulose fibers and break down the β-1,4-glycosidic bonds, releasing glucose units. This process is facilitated by a combination of endoglucanases, exoglucanases, and β-glucosidases, ensuring cellulose is thoroughly converted into fermentable sugars. Similarly, lipases are key in biodiesel production, catalyzing the transesterification of fats or oils with alcohols like methanol or ethanol to produce fatty acid alkyl esters (FAAEs, commonly known as biodiesel) and glycerol, by transferring acyl groups from the triglyceride to the alcohol. Furthermore, amylases play a critical role in starch-based bioethanol production by hydrolyzing starch molecules into smaller sugars; α-amylase randomly cleaves the internal α-1,4-glycosidic bonds in starch to produce shorter polysaccharides, which are then broken down into glucose by glucoamylase for ethanol fermentation (Dodd and Cann, 2009).

The effectiveness of these microbial enzymes in the biofuel sector is influenced by several factors, including enzyme concentration, temperature, pH levels, and the presence of potential inhibitors or activators. Fine-tuning these conditions can significantly improve enzyme performance, leading to increased yields and efficiencies in biofuel production processes. Recent advancements in enzyme technology, particularly the development of engineered enzymes with enhanced features, offer promising prospects for the biofuel industry. Such innovations have the potential to not only increase the economic viability of biofuel production but also contribute to a more sustainable and energy-secure future (Dutta and Wu, 2014).

6.5 ENZYME IMMOBILIZATION TECHNIQUES

Enzyme immobilization techniques have become pivotal in enhancing the efficiency and sustainability of microbial enzymes used in biofuel production processes. These techniques involve fixing enzymes onto a solid support or within a matrix to restrict their movement but still retain their catalytic activity. Immobilization not only improves the stability and reusability of enzymes but also facilitates easier separation from the reaction mixture, making the process more cost-effective and environmentally friendly. Here are several commonly used immobilization techniques and their application in biofuel production (Verma et al., 2016).

Adsorption: This is one of the simplest and most widely used methods, where enzymes are physically adsorbed onto the surface of carriers such as activated carbon, silica gel, or alumina. Although this method is easy and inexpensive, it might lead to enzyme leaching due to weak enzyme-support interactions. Enzymes immobilized through the adsorption method have found substantial application in biofuel production due to the simplicity and cost-effectiveness of the process. For example, lipases from Candida antarctica (CALB) and Rhizopus oryzae have been adsorbed onto activated carbon and silica gel, respectively, for use in biodiesel production through the transesterification of oils. Another notable instance includes immobilizing cellulase on alumina for the hydrolysis of cellulose into fermentable sugars, a critical step in bioethanol production from lignocellulosic biomass (Puri et al., 2013).

Among various carriers, activated carbon stands out as a highly preferred matrix for enzyme immobilization by adsorption, thanks to its large surface area, high porosity, and the presence of functional groups that can interact with the enzyme molecules. This facilitates a higher loading of enzymes compared to other supports like silica gel or alumina (Puri et al., 2013).

The main advantages of the adsorption method include its operational simplicity, the non-requirement of complex chemical reagents for enzyme attachment, and the potential for enzyme recovery and support reusability, making it an economically appealing option. However, this method has its drawbacks, chiefly the risk of enzyme leaching. The weak interactions between the enzyme and the support can lead to enzyme detachment during the reaction process, especially under rigorous operational conditions, which might result in reduced catalytic efficiency and process instability. Additionally, the possibility of enzyme deactivation due to the physical adsorption onto hard surfaces poses a challenge to maintaining the desired activity levels throughout the process (Zanuso et al., 2021).

Despite these disadvantages, the adsorption method remains a valuable tool in the biofuel industry, especially when the operational conditions are carefully managed to minimize enzyme leaching and deactivation, thereby maximizing the economic and sustainable production of biofuels (Dutta and Wu, 2014). Covalent Bonding: Enzyme immobilization via covalent bonding has proven to be a significant advancement in biofuel production, offering enhanced stability and reusability of enzymes. This method involves the formation of strong covalent bonds between the enzyme molecules and support materials, which effectively reduces the risk of enzyme leakage during the biofuel production process. Examples of enzymes immobilized through covalent bonding for biofuel applications include lipases from Candida antarctica (CALB) and Burkholderia cepacia, which have been covalently attached to magnetic nanoparticles and agarose with epoxy groups, respectively, for biodiesel production through the transesterification of oils and fats (Almeida et al., 2022).

The choice of support materials in covalent bonding is crucial. Agarose with epoxy groups is favored for its ability to form stable covalent links with the enzyme, thereby ensuring the enzyme's activity over multiple uses. Magnetic nanoparticles offer the added advantage of easy separation from the reaction mixture using an external magnetic field, facilitating enzyme recovery and reuse, which is particularly advantageous in continuous flow systems for biodiesel production (Yang et al., 2012). Advantages of covalent bonding include the high stability of the immobilized enzyme, which can withstand harsh process conditions without significant loss of activity or leakage, and the potential for enzyme reusability, which contributes to cost savings in biofuel production. Moreover, the immobilized enzymes often exhibit altered substrate specificity and improved resistance to inhibitors, further enhancing the process efficiency (Zhang et al., 2012). However, there are disadvantages to consider. The covalent bonding process can be complex and time-consuming, requiring specific chemical modifications of the support material and careful control of reaction conditions to ensure effective enzyme attachment without losing catalytic activity. Additionally, the strong bond between the enzyme and the support may hinder the enzyme's flexibility and access to substrates, potentially affecting its catalytic efficiency. Furthermore, the irreversible nature of covalent bonding means that once an enzyme is deactivated, it cannot be easily replaced, necessitating the preparation of a new immobilized enzyme batch (Grimaldi et al., 2014).

Despite these challenges, covalent bonding remains a valuable method for enzyme immobilization in biofuel production, offering a balance between enzyme stability, activity, and operational feasibility, making it an integral part of advancing sustainable and efficient biofuel production technologies (Christopher et al., 2014).

Encapsulation: Enzymes are enclosed within a semi-permeable membrane or gel matrix, such as alginate beads or polyacrylamide gel. Encapsulation protects enzymes from harsh process conditions while allowing substrates and products to diffuse in and out. This technique is suitable for enzymes like cellulases and amylases in bioethanol production, providing a controlled environment that maintains enzyme activity over extended periods(Betancor and Luckarift, 2008). For instance, cellulases have been successfully encapsulated in alginate beads for use in the hydrolysis of lignocellulosic biomass. The encapsulation allows these enzymes to retain their activity over extended periods, enabling the continuous breakdown of cellulose into fermentable sugars. Similarly, amylases have been encapsulated in polyacrylamide gel to facilitate the conversion of starch-rich agricultural waste into bioethanol, showcasing the versatility of this immobilization method across various biomass types (Park et al., 2010).

Advantages of encapsulation include enhanced enzyme stability, as the encapsulating material protects enzymes from degradation under harsh process conditions such as extreme pH and temperature. Additionally, encapsulation allows for the easy recovery and reuse of enzymes, reducing the overall cost of biofuel production. The semi-permeable nature of the encapsulating material ensures that substrates can reach the enzyme and products can exit, maintaining efficient catalysis (Sirisha et al., 2016). However, there are also disadvantages to consider. The diffusion of substrates and products through the encapsulating material might be limited, potentially reducing the reaction rate compared to free enzymes. The preparation of encapsulated enzymes can be more complex and costly than other immobilization methods, involving additional steps and materials. Moreover, the physical size of the encapsulated enzyme systems might restrict their application in certain types of reactors or require specific reactor designs to accommodate them (Minteer, 2017)Despite these challenges, encapsulation offers a promising approach for enzyme immobilization in biofuel production, providing a balance between protecting enzymes and maintaining their catalytic efficiency. The method's ability to extend enzyme life and reduce operational costs makes it an attractive option for sustainable and economical biofuel production processes (Park et al., 2010).

Cross-Linking: Enzymes are cross-linked with each other using bi-functional reagents, forming enzyme aggregates without the need for a carrier. Cross-linked enzyme aggregates (CLEAs) enhance stability and can be easily recovered from the reaction mixture. This method has shown promise for various biofuel enzymes due to its simplicity and effectiveness in improving enzyme stability and activity (Shamsudin et al., 2021).

Immobilization on Nanomaterials: The use of nanomaterials as supports for enzyme immobilization offers advantages such as high surface area, enhanced enzyme loading, and improved mass transfer. Nanomaterials like nanoparticles, nanofibers, and nanotubes can significantly enhance the catalytic efficiency of enzymes like dehydrogenases in biohydrogen production (Yadav et al., 2021).

Nanoparticles, including gold and silver nanoparticles, offer a substantial surface-to-volume ratio, which facilitates an increase in enzyme attachment. This, in turn, can lead to higher catalytic activity. Dehydrogenases immobilized on gold nanoparticles, for example, have demonstrated improved biohydrogen production efficiencies due to the optimal enzyme orientation and stabilization on the nanoparticles' surfaces. The advantages of using nanoparticles include enhanced enzyme loading and improved electron transfer rates, crucial for reactions involving dehydrogenases. However, challenges such as the potential for nanoparticle aggregation and the high cost of synthesizing metal nanoparticles present limitations to this approach (Puri et al., 2013). Nanofibers, produced through techniques like electrospinning, provide a fibrous and highly porous structure that offers an increased surface area for enzyme immobilization. The immobilization of cellulases on electrospun nanofibers has shown to prolong enzyme stability and maintain sustained catalytic activity, attributed to the protective and supportive environment afforded by the nanofiber matrix. While nanofibers offer a stable support system that enhances enzyme activity and stability, controlling the fiber diameter and orientation during production can be challenging, which may influence the efficiency of enzyme immobilization and subsequent mass transfer (Shuai et al., 2017).

Carbon nanotubes (CNTs) are recognized for their mechanical strength, electrical conductivity, and substantial surface area. These properties make CNTs an excellent support for enzymes like hydrogenases, leading to an increase in biohydrogen production. The electrical properties of CNTs, in particular, can facilitate electron transfer in enzymatic reactions, crucial for the efficient production of biofuels. Despite their benefits, CNTs also pose challenges, including the potential for aggregation and difficulties in achieving even dispersion for optimal enzyme immobilization. Furthermore, the production and purification of CNTs can be costly and involve the use of toxic chemicals (Ansari and Husain, 2012).

The utilization of nanomaterials for enzyme immobilization offers a pathway to enhancing the catalytic efficiency of enzymes in biofuel production. Each type of nanomaterial – nanoparticles, nanofibers, and nanotubes – brings distinct advantages, including increased surface area, enhanced enzyme loading, and improved mass transfer. However, these benefits are accompanied by challenges such as potential aggregation, high production costs, and difficulties in control during synthesis. The choice of nanomaterial for enzyme immobilization will depend on the specific needs of the biofuel production process and the desired enzymatic reaction outcomes, balancing the advantages against the limitations to achieve optimal biofuel production efficiency (Gupta et al., 2011).

The choice of immobilization technique depends on the type of enzyme, the nature of the biofuel production process, and the specific requirements of the process, such as the need for high stability, resistance to inhibitors, or reusability. Advances in materials science and biotechnology continue to develop new immobilization strategies and materials, further improving the performance of microbial enzymes in biofuel production and making the processes more sustainable and economically viable (Zanuso et al., 2021).

6.6 OPTIMIZING ENZYME PERFORMANCE

Optimizing the performance of microbial enzymes in biofuel production is crucial for enhancing the efficiency, sustainability, and cost-effectiveness of these processes. Various strategies can be employed to improve enzyme activity, stability, and specificity, thereby increasing the yield of biofuels (Ge and Li, 2018). Here are key approaches to optimizing enzyme performance in biofuel production.

Enzyme engineering: Enzyme engineering has become a cornerstone in enhancing the efficiency of biofuel production by tailoring enzymes to better fit the demanding conditions of industrial processes. Through techniques such as site-directed mutagenesis, directed evolution, and rational design, enzymes like cellulases and lipases have been fine-tuned for optimal performance. For example, site-directed mutagenesis was employed to engineer a variant of the cellulase enzyme from Trichoderma reesei, resulting in a mutant with a significantly improved ability to withstand high temperatures, a common condition in biofuel production processes. This engineered enzyme showed a notable increase in cellulolytic activity at temperatures where the wild-type enzyme would typically denature, thereby enhancing the overall efficiency of biomass conversion to sugars for bioethanol production (Bilal et al., 2018). Directed evolution, another powerful tool in enzyme engineering, has been applied to lipases to increase their resistance to methanol, a common solvent in biodiesel production through the transesterification process. By subjecting lipase genes to iterative rounds of mutagenesis and selection, researchers were able to develop variants with enhanced tolerance to methanol, allowing for more efficient biodiesel production from FAMEs without significant enzyme deactivation (Huang et al., 2020).

Rational design has led to the creation of enzyme variants with altered substrate specificity, enabling the use of nontraditional feedstocks for biofuel production. For instance, a lipase initially specific to certain oil substrates was engineered to also process waste cooking oil, a cheaper and more sustainable feedstock, with high efficiency. This adjustment not only expanded the range of usable raw materials but also contributed to the economic viability and sustainability of the biodiesel production process (Güven et al., 2010).

By increasing activity, enhancing stability under industrial conditions, and broadening substrate specificity, engineered enzymes are pivotal in overcoming some of the most significant challenges in biofuel production. However, it's important to note that while enzyme engineering offers substantial benefits, it also involves complex and often costly research and development processes, necessitating a balance between innovation and practical application feasibility (Fu et al., 2020).

Process conditions optimization: The efficiency of enzymatic reactions is highly dependent on process parameters such as temperature, pH, enzyme concentration, and the presence of cofactors or inhibitors. Systematic optimization of these conditions through techniques like response surface methodology (RSM) can significantly enhance enzyme performance (Table 6.3; Zhao et al., 2015). Temperature significantly influences enzyme activity. Enzymes like cellulases, used in bioethanol production from lignocellulosic biomass, exhibit optimal activity at specific temperature ranges. Operating below or above this range can lead to reduced enzyme efficiency or denaturation. An instance of this can be seen in the engineering of thermostable variants of cellulases that maintain high activity levels at elevated temperatures, thereby improving the hydrolysis of cellulose into fermentable sugars under industrial conditions (Vaid et al., 2018). Similarly, pH affects enzyme structure and function. Lipases, crucial for biodiesel production through the transesterification of fats and oils, have specific pH requirements for optimal activity. By adjusting the reaction medium to the optimal pH, enzyme activity can be maximized, leading to higher biodiesel yields. Directed evolution techniques have been applied to modify lipases to be more active and stable across a broader pH range, accommodating the variability

TABLE 6.3
Operational Parameters of Key Enzymes in Biofuel Production

Enzyme Name	Optimal pH	Optimal Temperature	Stability Factors	Substrate Specificity
Cellulase	4.5–5.5	50°C–60°C	Halotolerant, resistant to ionic liquids	Cellulose, breaking down β-1,4-glycosidic bonds
Amylase (AmyPL)	5.6	50°C	Optimal activity in fecal metagenomic environments	Starch, hydrolyzing to glucose
Xylanase (Xyn22)	5.5–7.0	Up to 90°C	Thermostable, halotolerant (up to 5 M NaCl)	Xylan, degrading into xylose
Laccase (LacM)	3.0–7.0	30°C–70°C	Tolerant to organic solvents and salts, azo and triphenylmethane dye decolorization	Lignin, polyphenols, and related compounds
Lipase (Lip906)	7.8	74°C	Resistant to industrial detergents and low temperatures (up to 4°C)	Triglycerides, facilitating transesterification for biodiesel
Feruloyl Esterase (FAE-Xuan)	5.0	30°C	Stable across a broad pH range (3.0 to 10.0), high activity toward methyl ferulate	Hydroxycinnamic acids (e.g., ferulic acid), improving biomass digestibility
β-Glucosidase (MeBglD2)	5.5–6.0	60°C–70°C	Not inhibited by saccharification products	Cellulose, enhancing saccharification yields alongside other cellulases

in feedstock acidity and enhancing process flexibility (Gama et al., 2015). Enzyme concentration is another critical factor, where too little may result in inefficient reaction rates, while too much could lead to unnecessary cost increases without proportional gains in productivity. Optimizing enzyme concentration for specific processes, such as the saccharification step in bioethanol production, is vital for achieving high sugar yields without incurring excessive costs (Alam et al., 2019). The presence of cofactors or inhibitors also affects enzyme activity. For enzymes requiring cofactors, such as certain dehydrogenases in biohydrogen production, the addition of necessary cofactors can significantly enhance activity. Conversely, inhibitors, which might be present in raw biomass feedstocks, can adversely affect enzyme performance. Strategies to remove or neutralize these inhibitors prior to enzymatic processing have been developed to prevent enzyme inhibition and ensure efficient biofuel production (Knocke and Vogt, 2009).

The optimization of process conditions like temperature, pH, enzyme concentration, and the management of cofactors or inhibitors are paramount for maximizing enzyme activity and, by extension, the efficiency of biofuel production processes. Through the application of techniques such as RSM, these conditions can be fine-tuned to match the optimal requirements of specific enzymes, leading to improved biofuel yields and process sustainability (Antczak et al., 2009).

Immobilization: Immobilizing enzymes on solid supports or within matrices can enhance their stability, facilitate their reuse, and simplify their separation from the reaction products. Immobilization techniques such as adsorption, covalent attachment, and encapsulation play a crucial role in enhancing the efficiency and sustainability of biofuel production by improving enzyme stability, facilitating reuse, and simplifying separation from reaction products. Adsorption, exemplified by lipase immobilization on silica for biodiesel production, offers simplicity and cost-effectiveness but risks enzyme leaching. Covalent attachment, as seen with cellulases bound to magnetic nanoparticles, provides a stable immobilization method that reduces enzyme leaching and can enhance catalytic efficiency, particularly beneficial for the hydrolysis process in bioethanol production. Encapsulation, demonstrated by encapsulated amylases for starch hydrolysis, protects enzymes from harsh conditions and extends their operational life, though it may restrict substrate access. Each method has its advantages – such as cost-effectiveness, enhanced stability, and protection of the enzyme – and disadvantages, including potential for enzyme leaching, complexity, and limited substrate access. Choosing the appropriate immobilization technique tailored to specific biofuel production needs can significantly contribute to more effective and sustainable biofuel production processes (Dutta and Wu, 2014).

Coenzyme regeneration: For enzymatic reactions that require cofactors, developing efficient coenzyme regeneration systems is crucial for maintaining high catalytic efficiency. Implementing strategies for the in situ regeneration of cofactors like NADH or ATP can make processes like the enzymatic production of biohydrogen or bioalcohols more economically viable (Gama et al., 2015). In the enzymatic production of biohydrogen, hydrogenase enzymes often require the cofactor NADH to facilitate the conversion of organic substrates into hydrogen. One effective strategy for NADH regeneration involves coupling hydrogenase reactions with other enzymatic systems that can replenish NADH levels within the reaction environment. An instance of this approach is the use of formate dehydrogenase to regenerate NADH from formate, thereby maintaining the required NADH levels for continuous hydrogen production. Similarly, in the production of bioalcohols such as ethanol, the cofactor ATP is crucial for the glycolysis pathway. An innovative method for ATP regeneration is the use of substrate-level phosphorylation reactions, where enzymes like acetate kinase can catalyze the conversion of acetyl phosphate into acetate, simultaneously generating ATP from ADP. This approach ensures a consistent supply of ATP for the glycolytic enzymes, enhancing the overall yield of ethanol from glucose or other substrates.

Efficient coenzyme regeneration systems not only ensure sustained catalytic activity but also contribute to the economic feasibility of biofuel production by minimizing the need for external cofactor additions. Through the implementation of such strategies, the enzymatic production of biohydrogen, bioalcohols, and other biofuels can be made more economically viable, advancing the development of sustainable and efficient bioenergy technologies.

Enzyme cocktails: Developing enzyme mixtures or cocktails that combine multiple enzymes with complementary activities can enhance the breakdown of complex biomass into fermentable sugars. By carefully selecting and balancing the components of the cocktail, synergistic effects can be achieved, leading to more efficient degradation of lignocellulosic biomass. The degradation of lignocellulosic biomass, such as agricultural residues or forestry waste into fermentable sugars requires the concerted action of cellulases, hemicellulases, and ligninases. Cellulases break down cellulose into glucose, hemicellulases hydrolyze hemicellulose into sugars like xylose and arabinose, and ligninases help modify lignin, which is not a sugar but a complex polymer that shields polysaccharides from enzymatic attack. A well-designed enzyme cocktail that includes these three types of enzymes can effectively disassemble lignocellulosic biomass, making the sugars available for fermentation into bioethanol (Adsul et al., 2020).

Another example involves the production of biofuels from algae. Algal biomass contains proteins, carbohydrates, and lipids, requiring a combination of proteases, carbohydrases, and lipases for efficient breakdown. An enzyme cocktail containing these enzymes can facilitate the release of lipids for biodiesel production, as well as sugars and amino acids for other biofuel pathways (Dodd and Cann, 2009). The development of effective enzyme cocktails involves not just selecting the right enzymes but also optimizing their ratios to maximize the synergistic effects. For instance, studies have shown that certain ratios of cellulases and hemicellulases can significantly increase the release of fermentable sugars from specific types of biomass, compared to the use of either enzyme alone. However, challenges remain in identifying the optimal composition of enzyme cocktails for different biomass feedstocks, as the structure and composition of biomass can vary widely. Additionally, the cost of enzyme production and the need to maintain enzyme activity over the course of the biomass breakdown process are important considerations in the development and application of enzyme cocktails. Despite these challenges, enzyme cocktails represent a promising approach to unlocking the potential of various types of biomass for biofuel production, offering a path toward more sustainable and efficient bioenergy solutions (Lopes et al., 2018).

Use of Surfactants or Additives: The addition of surfactants or other additives can improve enzyme stability and activity by preventing enzyme denaturation or by enhancing substrate accessibility. This approach can be particularly beneficial in the enzymatic hydrolysis of lignocellulosic biomass, where surfactants can help disrupt the biomass structure and increase enzyme access to cellulose and hemicellulose (Fischer et al., 2008). The strategic incorporation of surfactants or additives into enzymatic reactions enhances enzyme stability and activity, proving especially effective in the enzymatic hydrolysis of lignocellulosic biomass. Surfactants, by disrupting biomass structure, improve the accessibility of enzymes like cellulases and hemicellulases to their substrates, facilitating efficient breakdown into fermentable sugars (Attaphong et al., 2012). Various types of surfactants are utilized, including non-ionic surfactants (e.g., Tween 80, Triton X-100) that prevent non-specific enzyme adsorption to lignin, ionic surfactants (e.g., SDS) used cautiously due to potential enzyme inactivation, biosurfactants (e.g., rhamnolipids, sophorolipids) that offer an eco-friendly option with effective hydrolysis enhancement, and zwitterionic surfactants (e.g., cocamidopropyl betaine) known for their mild impact on enzymes. While surfactants offer significant benefits in improving hydrolysis efficiency, considerations around optimal concentrations, cost implications, and environmental impact must be carefully managed to fully leverage their advantages in biofuel production processes (Trindade, 2011).

By employing these optimization strategies, the performance of microbial enzymes in biofuel production can be significantly enhanced, leading to processes that are more efficient, cost-effective, and environmentally friendly. Continuous research and development in enzyme technology and process engineering are key to unlocking the full potential of enzymes in the biofuel industry (Eckert and Trinh, 2016).

6.7 APPLICATIONS OF ENZYMES IN BIODIESEL AND BIOHYDROGEN PRODUCTION

The rapid advancements in technology, coupled with economic growth and a burgeoning global population, have precipitated a critical energy crisis. Traditional energy sources are becoming increasingly scarce and are accompanied by the emission of harmful gases, necessitating the shift toward more sustainable and environmentally friendly alternatives. Biofuels such as biogas, biodiesel, biohydrogen, and bioethanol emerge as viable substitutes, offering a less harmful impact on the environment (Awogbemi and Von Kallon, 2024). Microorganisms and their enzymes, particularly those involved in the production of biodiesel and biohydrogen, are at the forefront of research due to their potential for green and efficient biofuel production. Enzymes like cellulases, lipases, and proteases play pivotal roles in converting biomass into biofuel through various biochemical steps, with enzymatic hydrolysis standing out as a more economical and higher-yielding method compared to acid hydrolysis (Table 6.4) (Srivastava et al., 2020).

The biological production of hydrogen presents an eco-friendly alternative, with both photosynthetic and dark-fermentative pathways being extensively explored for H_2 generation. Studies have shown that certain bacterial species can produce significant yields of hydrogen through these methods. The enhancement of these processes, alongside the metabolic optimization of bacteria using biowaste as feedstock, aims to increase the efficiency of hydrogen production. This approach not only holds promise for meeting the escalating energy demands but also for integrating the production

TABLE 6.4
Comparative Analysis of Biofuel Production Processes

Process Step	Traditional Method	Enzyme-Assisted Method	Efficiency/Output Comparison
Biomass pretreatment	Uses harsh chemicals and high temperatures to break down biomass.	Utilizes specific enzymes to target and break down biomass components under milder conditions.	Enzyme-assisted methods are more energy-efficient and have lower environmental impact.
Saccharification	Relies on acids or high temperatures to hydrolyze biomass into sugars.	Employs a cocktail of enzymes like cellulases and hemicellulases to convert biomass into fermentable sugars.	Enzyme-assisted saccharification is more specific, leading to higher yields of fermentable sugars.
Fermentation	Ferments sugars into biofuels using yeast or bacteria.	Similar process, but the higher purity and specificity of sugars from enzyme-assisted saccharification can improve fermentation efficiency.	Enzyme-assisted methods can lead to faster fermentation rates and higher biofuel yields due to higher sugar purity.
Biofuel Extraction	Requires energy-intensive distillation for bioethanol or mechanical pressing and chemical reactions for biodiesel.	Enzymatic methods can enhance the separation and purification of biofuels, reducing the need for harsh chemicals.	Enzyme-assisted extraction can be more efficient and environmentally friendly, though technology is still emerging.
Waste Management	Traditional methods may produce hazardous waste that requires treatment.	Enzyme-assisted processes result in less toxic by-products and can utilize waste biomass more effectively.	Enzyme-assisted processes offer a more sustainable approach to waste management, reducing environmental impact.

of other bio-based products like polyhydroxyalkanoate and methane, thereby augmenting the overall feasibility of biofuel production (Mumtaz et al., 2022).

Enzyme immobilization on nanomaterials has opened new avenues in biofuel research, offering enhanced stability and efficiency for industrial applications. Techniques such as the use of calcium phosphate mineralized chitosan beads have improved anti-swelling characteristics, storage capacity, and heat stability, facilitating batch catalysis in various processes. The immobilization of enzymes like β-galactosidase and lipase on chitosan-based carriers has demonstrated significant improvements in catalytic efficiency and thermal stability. Such advancements underscore the potential of immobilized enzymes in making biofuel processing more versatile and efficient, particularly through the use of thermophilic or thermostable enzymes derived from resilient microorganisms (Singh et al., 2022).

However, the field of nanotechnology in biofuel production faces challenges such as the high cost of materials synthesis, scale-up limitations, and the environmental impact of nanoparticle dispersion. Despite these hurdles, nanotechnology has significantly benefited biofuel production in terms of conversion efficiency and yield. Future research efforts are directed toward the discovery of novel lignocellulolytic enzymes, engineering enzymes with improved industrial applicability, and overcoming the challenges of hydrogen purification and storage. The development of integrated strategies and the use of modified microbial fuel cells are among the innovative approaches being explored to enhance biohydrogen yield and address the cost-efficiency of biofuel production at a commercial scale (Osman et al., 2023).

The industrial application of microbial enzymes in biofuel production represents a rapidly evolving field, driven by the need for sustainable energy solutions. Through various case studies, the practical benefits and challenges of implementing enzyme-based processes on an industrial scale become evident. These case studies not only highlight the technological advancements but also underscore the economic and environmental impacts of enzyme applications in biofuel production (Table 6.5). These case studies exemplify the diverse applications of microbial enzymes in

TABLE 6.5
Case Studies on the Application of Microbial Enzymes in Biofuel Production

Case Study	Company/ Project	Location	Overview	Achievements	Challenges
Cellulosic ethanol production	Poet-DSM Advanced Biofuels, Project LIBERTY	Emmetsburg, Iowa, USA	Utilizes corn crop residues to produce ethanol through enzymatic hydrolysis using a cocktail of cellulases and hemicellulases.	Demonstrated the feasibility of converting agricultural waste into renewable fuel at a commercial scale.	High costs of enzyme production and the need for process optimization for cost-competitiveness.
Biodiesel production using immobilized lipases	Novozymes	Global operations	Developed a biodiesel production process using immobilized lipases for transesterification, suitable for low-quality feedstocks with high free fatty acid content.	Operates under milder conditions than chemical processes, reducing environmental impact and allowing the use of a wider range of feedstocks.	Costs associated with immobilized enzymes and the need for their recovery and reuse.
Biohydrogen production via dark fermentation	LanzaTech		Produces biohydrogen and valuable chemicals from industrial waste gases through gas fermentation using genetically engineered microorganisms.	Innovative use of waste as a resource, contributing to circular economy principles and reducing greenhouse gas emissions. Successfully scaled up with global partnerships.	Complexity of gas fermentation and ensuring the economic viability of biohydrogen production.

biofuel production, from liquid fuels like ethanol and biodiesel to gaseous fuels like biohydrogen. They highlight the importance of continued research and development to overcome challenges related to cost, efficiency, and scalability. Future advancements in enzyme technology, process integration, and the development of novel feedstocks are expected to further enhance the sustainability and economic viability of biofuel production, contributing to global energy security and environmental sustainability. Collaboration between industry, academia, and government will be key to realizing these advancements and fostering the growth of the biofuel sector.

6.8 CHALLENGES AND LIMITATIONS IN ENZYME USE

The adoption of enzymes in biofuel production presents numerous advantages, including enhanced efficiency and environmental sustainability. However, this approach faces several challenges that hinder its integration into the commercial biofuel sector. The high cost of producing specialized enzymes is a significant barrier, as industrial-scale production entails expenses related to fermentation, purification, and formulation. Additionally, the operational conditions within biofuel production, such as extreme temperatures and pH levels, can adversely affect enzyme stability and activity, leading to reduced efficacy and the need for frequent enzyme replenishment (Rodionova et al., 2017). This issue is compounded by the complexity of biomass, which requires a diverse cocktail of enzymes to achieve efficient conversion to fermentable sugars, further escalating costs and complicating the production process (Velvizhi et al., 2020).

Addressing the challenge of enzyme recovery and reuse is vital for the economic viability of enzymatic biofuel production. Although immobilization techniques offer a solution, they introduce additional complexity and expenses, and enzymes typically exhibit diminished activity with each reuse cycle. Moreover, scaling enzymatic processes from the laboratory to industrial levels involves overcoming obstacles related to mass transfer, enzyme loading, and precise process control, alongside the requisite substantial capital investment in specialized infrastructure (Ebaid et al., 2019).

Regulatory and environmental considerations also pose challenges, particularly concerning the use of GMOs for enzyme production and biofuel fermentation, which may face regulatory hurdles and public skepticism in certain regions. Furthermore, the environmental footprint of enzyme production, encompassing energy and resource consumption, necessitates careful management to uphold the sustainability of enzymatic biofuel processes. Enzymatic methods must also prove their competitive edge over traditional chemical processes, which are typically more established, economically optimized, and scalable (Goria et al., 2022).

To surmount these hurdles, a concerted effort in biotechnology, process engineering, and materials science is essential. Ongoing research aims to refine enzyme production techniques, improve enzyme stability through genetic and protein engineering, and develop effective systems for enzyme recovery and recycling. Collaboration among academic institutions, industry, and regulatory bodies is crucial to navigate regulatory landscapes and to foster the adoption of enzymatic processes. With these challenges addressed, enzymatic biofuel production stands to significantly contribute to sustainable energy solutions, underscoring the importance of continued innovation and cooperation in this field (Mumtaz et al., 2022).

6.9 FUTURE TRENDS AND INNOVATIONS IN ENZYME TECHNOLOGY

The landscape of enzyme technology in biofuel production is poised for transformative advancements, promising to address current challenges and open new avenues for more sustainable and efficient biofuel generation. The forefront of these advancements is marked by progress in genetic and protein engineering, including the refinement of CRISPR/Cas9 and similar genome editing technologies. These innovations offer precise alterations of microbial genomes to produce enzymes with improved functionalities specifically suited for biofuel applications. Furthermore, synthetic biology is set to redefine biofuel production by designing novel metabolic pathways and enzymes capable of converting a broader array of substrates into biofuels, thereby diversifying usable feedstocks (Pandey et al., 2019).

Another promising area of development is the creation of multifunctional enzymes, which aims to streamline biofuel production by reducing reliance on enzyme cocktails. This involves engineering enzymes that possess the ability to catalyze multiple reactions or fusing different enzymatic activities into a single protein. Additionally, advancements in enzyme immobilization and stabilization, particularly through the use of nanomaterials and smart materials, are expected to enhance enzyme stability, activity, and recyclability. These materials not only provide optimal environments for enzyme action but also facilitate easier enzyme recovery, contributing to process efficiency and control (Yadav et al., 2021).

The integration of machine learning and computational tools in enzyme design and process optimization represents a significant leap forward. These technologies enable the prediction of enzyme structure-function relationships and the rational design of targeted enzymes, while also optimizing biofuel production processes for maximum efficiency. Moreover, the concept of integrated bioprocessing, which combines enzyme production, biomass conversion, and biofuel purification into a seamless operation, is anticipated to reduce operational costs and boost overall process efficiency. This integrated approach will benefit immensely from metabolic engineering and systems biology advances (Punia et al., 2021).

To ensure the environmental sustainability of enzyme-based biofuel production, exploring renewable energy sources for enzyme production and adopting circular economy principles are crucial. Additionally, developing clear regulatory frameworks and fostering public acceptance through transparent communication about the benefits and safety of genetically engineered enzymes are essential for commercialization. As enzyme technology advances, collaboration across the scientific community, industry, and policymakers will be vital in harnessing these innovations for a sustainable energy future, making biofuel production more economically viable and environmentally friendly.

6.10 ENVIRONMENTAL IMPACT AND SUSTAINABILITY CONSIDERATIONS

The integration of microbial enzymes into biofuel production signifies a pivotal shift toward environmental sustainability and plays a crucial role in global efforts to combat climate change. By facilitating a more efficient conversion of biomass into biofuels, enzyme technology heralds the advent of renewable energy sources that boast significantly lower environmental footprints compared to traditional fossil fuels. This transition, however, necessitates a comprehensive assessment of the environmental impact and sustainability across the entire biofuel lifecycle – from biomass cultivation and enzyme production to biofuel generation and its ultimate utilization. Key to this endeavor is the realization that biofuels, especially those produced with microbial enzymes, have the potential to achieve near carbon neutrality. This is due to the cyclical nature of biofuels' carbon emissions, where CO_2 absorbed by biomass during growth approximately balances out the CO_2 emitted during biofuel combustion. Moreover, enzymatic processes, by virtue of operating under less stringent conditions than their chemical counterparts, contribute to lower energy consumption and a consequent reduction in greenhouse gas emissions (Mandotra et al., 2021).

Sustainable biomass utilization stands as another cornerstone of enzyme-assisted biofuel production. By targeting agricultural and forest residues, along with municipal solid waste, enzymatic conversion processes aim to minimize waste and promote sustainable land usage practices. The development of advanced enzymes, capable of processing diverse biomass types, further supports the use of nonfood crops and the exploitation of marginal lands – thereby alleviating competition for food resources and aiding in biodiversity preservation. Water usage and quality also benefit from enzyme technology, as these processes typically demand less water and mitigate water pollution by circumventing the need for harsh chemicals and extensive use of fertilizers and pesticides in biomass cultivation (Kumar et al., 2016).

The production and utilization of enzymes themselves must adhere to sustainability principles, calling for the minimization of environmental impacts through renewable energy use and the recycling of nutrients within enzyme production facilities. The biodegradability of enzymes presents an advantage over chemical catalysts, easing disposal concerns and environmental contamination risks. Comprehensive lifecycle assessments (LCAs) are indispensable in evaluating the overall environmental footprint of enzyme-assisted biofuel production, providing insights into the full spectrum of environmental benefits and identifying potential areas for improvement (Noraini et al., 2014).

Looking ahead, the enzyme technology sector is ripe for further innovations that promise to bolster the environmental benefits of biofuel production. Research focused on enhancing enzyme efficiency and robustness, coupled with advancements in biomass processing, is essential for optimizing biofuel production's sustainability. Policy frameworks that incentivize sustainable agricultural and production practices, alongside standards and certifications for sustainably produced biofuels, will play a critical role in fostering the biofuel market. As the field progresses, interdisciplinary collaboration will be key to maximizing the sustainability and minimizing the environmental impacts of biofuel production, ensuring that enzyme-assisted biofuel production aligns with broader carbon neutrality and sustainability objectives (Samrot et al., 2023).

6.11 ADVANCEMENTS IN MICROBIAL ENZYME TECHNOLOGY FOR SUSTAINABLE BIOFUEL PRODUCTION

In the quest for sustainable and renewable energy sources, the exploration of microbial enzymes in biofuel production has emerged as a promising path toward meeting the world's growing energy demands while addressing environmental concerns. Enzymes play a crucial role in biofuel production, facilitating the breakdown of complex biomass into fermentable sugars and enabling the conversion of various substrates into biofuels through processes such as the enzymatic hydrolysis of lignocellulosic biomass, transesterification for biodiesel production, and photobiological water splitting for biohydrogen production. However, the cost of biofuel production remains high, and achieving viable alternatives to fossil fuels necessitates the development of enzymes with enhanced activity, broad substrate specificity, and resilience to harsh processing conditions (Singh and Rathore, 2017).

Advancements in metagenomics have opened new avenues for identifying novel enzymes from diverse microbial communities inhabiting extreme environments. These efforts have led to the discovery of enzymes with unique properties, such as high halotolerance and resistance to ionic liquids, making them suitable for industrial biofuel production. For example, novel cellulases and amylases with optimal activity under specific conditions have been identified from metagenomic libraries, demonstrating the potential for improved biofuel production processes. Similarly, enzymes like xylanases and laccases have been discovered, offering enhanced efficiency in the degradation of lignocellulosic biomass, a critical step in converting biomass into biofuels (Li et al., 2022).

Fungal and algal systems also present opportunities for improving biofuel synthesis through genetic engineering and metabolic engineering approaches. Fungi, responsible for producing a significant portion of commercially available enzymes, have been engineered to increase enzyme production and biofuel synthesis efficiency. Strategies include the heterologous expression of xylanase genes and the metabolic engineering of strains for enhanced saccharification efficiency and lipid accumulation, crucial for biodiesel production. Algae, with their high lipid content and potential for biohydrogen production, have been genetically modified to increase lipid synthesis and overcome challenges in biohydrogen generation, showcasing the versatility of microbial systems in biofuel production (Arya et al., 2021).

The termite microbiome has been explored for novel enzymes capable of efficient lignocellulose degradation, highlighting the symbiotic relationship between termites and their gut microbiota. Novel endoglucanases, β-glucosidases, and xylanases identified from termite gut microbiomes have shown promise in enhancing the digestion of plant biomass, offering pathways to more efficient and cost-effective biofuel production. These discoveries, coupled with advancements in microbial enzyme technology and genetic engineering, underscore the potential of microbial enzymes in revolutionizing the biofuel industry, paving the way for sustainable and environmentally friendly energy solutions (Mona et al., 2020).

6.12 FUTURE PERSPECTIVES

The exploration of microbial enzymes in the realm of biofuel production heralds a new era for sustainable and renewable energy solutions, positioning enzyme technology as a pivotal force in transforming the biofuel industry. The progression from understanding enzymes' basic actions to leveraging advanced genetic engineering and immobilization techniques marks a path of continuous innovation aimed at optimizing enzyme performance in terms of efficiency, stability, and cost-effectiveness. Technological advancements in genetic and protein engineering have led to the development of more robust enzymes, tailored to withstand the harsh conditions of industrial biofuel production processes, thereby enhancing the economic feasibility of these sustainable energy solutions.

The environmental benefits of enzyme-assisted biofuel production, as a cleaner alternative to fossil fuels, are significant, offering a method to cut greenhouse gas emissions while utilizing renewable biomass. Despite these advances, the industry faces challenges, including the high costs associated with enzyme production, maintaining enzyme stability, and scaling processes for broader applications. These hurdles must be overcome to ensure the broader adoption of enzyme-based biofuel technologies. Looking ahead, the discovery of novel enzymes from unique environments such as extremophiles presents opportunities for breakthroughs in biofuel process enhancement. Moreover, the future success of biofuel production will likely hinge on the development of integrated bioprocesses that consolidate enzyme production, biomass conversion, and biofuel purification, thereby streamlining operations to reduce costs and increase efficiency.

A comprehensive approach to sustainability, considering the entire lifecycle of biofuel production from biomass cultivation through to the end-use, is imperative. This perspective necessitates continuous lifecycle assessments to confirm that biofuel production aligns with environmental, economic, and social sustainability goals. Furthermore, the growth of the biofuel sector will be significantly influenced by supportive policies and market development, including incentives for adopting sustainable practices and investing in research and development to foster market readiness for biofuels.

6.13 CONCLUSION

In conclusion, the evolving role of microbial enzymes in biofuel production stands on the brink of considerable growth, driven by relentless innovation and a dedicated focus on sustainability. The synergistic efforts of researchers, industry professionals, policymakers, and the broader community are crucial to harnessing the full potential of enzyme technologies, paving the way for biofuel production to become a key component of the global renewable energy mix. This collaborative approach promises not only to advance the biofuel industry but also to contribute significantly to achieving a sustainable energy future.

REFERENCES

Abdulmalek SA and Yan Y. (2022) Recent developments of lipase immobilization technology and application of immobilized lipase mixtures for biodiesel production. *Biofuels, Bioproducts and Biorefining*, 16: 1062–1094.

Adegboye MF, Ojuederie OB, Talia PM, et al. (2021) Bioprospecting of microbial strains for biofuel production: Metabolic engineering, applications, and challenges. *Biotechnology for Biofuels*, 14: 1–21.

Adsul M, Sandhu SK, Singhania RR, et al. (2020) Designing a cellulolytic enzyme cocktail for the efficient and economical conversion of lignocellulosic biomass to biofuels. *Enzyme and Microbial Technology*, 133: 109442.

Alam MA, Yuan T, Xiong W, et al. (2019) Process optimization for the production of high-concentration ethanol with Scenedesmus raciborskii biomass. *Bioresource Technology*, 294: 122219.

Almeida FL, Prata AS and Forte MB. (2022) Enzyme immobilization: What have we learned in the past five years? *Biofuels, Bioproducts and Biorefining*, 16: 587–608.

Amin F, Bhatti HN and Bilal M. (2019) Recent advances in the production strategies of microbial pectinases—A review. *International Journal of Biological Macromolecules*, 122: 1017–1026.

Ansari SA and Husain Q. (2012) Potential applications of enzymes immobilized on/in nano materials: A review. *Biotechnology Advances*, 30: 512–523.

Antczak MS, Kubiak A, Antczak T, et al. (2009) Enzymatic biodiesel synthesis–Key factors affecting efficiency of the process. *Renewable Energy*, 34: 1185–1194.

Aquino Neto S and De Andrade AR. (2013) New energy sources: The enzymatic biofuel cell. *Journal of the Brazilian Chemical Society*, 24: 1891–1912.

Arai T, Aikawa S, Sudesh K, et al. (2022) Electrotransformation of thermophilic bacterium Caldimonas manganoxidans. *Journal of Microbiological Methods*, 192: 106375.

Areeshi MY. (2022) Microbial cellulase production using fruit wastes and its applications in biofuels production. *International Journal of Food Microbiology*, 378: 109814.

Arya I, Poona A, Dikshit PK, et al. (2021) Current trends and future prospects of nanotechnology in biofuel production. *Catalysts*, 11: 1308.

Arya PS, Yagnik SM, Rajput KN, et al. (2022) Valorization of agro-food wastes: Ease of concomitant-enzymes production with application in food and biofuel industries. *Bioresource Technology*, 361: 127738.

Attaphong C, Do L and Sabatini DA. (2012) Vegetable oil-based microemulsions using carboxylate-based extended surfactants and their potential as an alternative renewable biofuel. *Fuel*, 94: 606–613.

Awogbemi O and Von Kallon DV. (2024) Recent advances in the application of nanomaterials for improved biodiesel, biogas, biohydrogen, and bioethanol production. *Fuel*, 358: 130261.

Bangaru A, Sree KA, Kruthiventi C, et al. (2022) Role of enzymes in biofuel production: Recent developments and challenges. *Bio-Clean Energy Technologies*, 1: 81–112.

Banu JR, Kumar G and Chattopadhyay I. (2021) Management of microbial enzymes for biofuels and biogas production by using metagenomic and genome editing approaches. *3 Biotech*, 11: 1–15.

Betancor L and Luckarift HR. (2008) Bioinspired enzyme encapsulation for biocatalysis. *Trends in Biotechnology*, 26: 566–572.

Bhandari S, Poudel DK, Marahatha R, et al. (2021) Microbial enzymes used in bioremediation. *Journal of Chemistry*, 2021: 1–17.

Bilal M, Iqbal HM, Hu H, et al. (2018) Metabolic engineering and enzyme-mediated processing: A biotechnological venture towards biofuel production–a review. *Renewable and Sustainable Energy Reviews*, 82: 436–447.

Chaurasia D, Boudh S, Singh JS, et al. (2020) Microbes and microbial enzymes as a sustainable energy source for biofuel production. *Microbial Services in Restoration Ecology*, 361: 207–217.

Christopher LP, Kumar H and Zambare VP. (2014) Enzymatic biodiesel: Challenges and opportunities. *Applied Energy*, 119: 497–520.

Costantini A and Califano V. (2021) Lipase immobilization in mesoporous silica nanoparticles for biofuel production. *Catalysts*, 11: 629.

de Castro SM and de Castro AM. (2012) Assessment of the Brazilian potential for the production of enzymes for biofuels from agroindustrial materials. *Biomass Conversion and Biorefinery*, 2: 87–107.

Djuma'ali Da, Soewarno N, Sumarno S, et al. (2012) Cassava pulp as a biofuel feedstock of an enzymatic hydrolysis proces. *Makara Journal of Technology*, 15: 14.

Dodd D and Cann IK. (2009) Enzymatic deconstruction of xylan for biofuel production. *GCB Bioenergy*, 1: 2–17.

Dutta S and Wu KC-W. (2014) Enzymatic breakdown of biomass: Enzyme active sites, immobilization, and biofuel production. *Green Chemistry*, 16: 4615–4626.

Ebaid R, Wang H, Sha C, et al. (2019) Recent trends in hyperthermophilic enzymes production and future perspectives for biofuel industry: A critical review. *Journal of Cleaner Production*, 238: 117925.

Eckert CA and Trinh CT. (2016) *Biotechnology for biofuel production and optimization* Elsevier.

Eijsink VG, Vaaje-Kolstad G, Vårum KM, et al. (2008) Towards new enzymes for biofuels: Lessons from chitinase research. *Trends in Biotechnology*, 26: 228–235.

Fatma T, Zafar Z, Fatima S, et al. (2021) Computational assessment of Botrytis cinerea lipase for biofuel production. *Catalysts*, 11: 1319.

Fischer CR, Klein-Marcuschamer D and Stephanopoulos G. (2008) Selection and optimization of microbial hosts for biofuels production. *Metabolic Engineering*, 10: 295–304.

Foukis A, Gkini OA, Stergiou P-Y, et al. (2017) Sustainable production of a new generation biofuel by lipase-catalyzed esterification of fatty acids from liquid industrial waste biomass. *Bioresource Technology*, 238: 122–128.

Fu L, Zhang J and Si T. (2020) Recent advances in high-throughput mass spectrometry that accelerates enzyme engineering for biofuel research. *BMC Energy*, 2: 1.

Gama R, Van Dyk JS and Pletschke BI. (2015) Optimisation of enzymatic hydrolysis of apple pomace for production of biofuel and biorefinery chemicals using commercial enzymes. *3 Biotech*, 5: 1075–1087.

Ge Y and Li L. (2018) System-level energy consumption modeling and optimization for cellulosic biofuel production. *Applied Energy*, 226: 935–946.

Goria K, Kothari R, Singh HM, et al. (2022) Biohydrogen: Potential applications, approaches, and hurdles to overcome. *Handbook of biofuels*. Elsevier, 399–418.

Greene ER, Himmel ME, Beckham GT, et al. (2015) Glycosylation of cellulases: Engineering better enzymes for biofuels. *Advances in Carbohydrate Chemistry and Biochemistry*, 72: 63–112.

Grimaldi J, Collins C and Belfort G. (2014) Toward cell-free biofuel production: Stable immobilization of oligomeric enzymes. *Biotechnology Progress*, 30: 324–331.

Guldhe A, Singh B, Mutanda T, et al. (2015) Advances in synthesis of biodiesel via enzyme catalysis: Novel and sustainable approaches. *Renewable and Sustainable Energy Reviews*, 41: 1447–1464.

Gupta MN, Kaloti M, Kapoor M, et al. (2011) Nanomaterials as matrices for enzyme immobilization. *Artificial Cells, Blood Substitutes, and Biotechnology*, 39: 98–109.

Güven G, Prodanovic R and Schwaneberg U. (2010) Protein engineering—An option for enzymatic biofuel cell design. *Electroanalysis: An International Journal Devoted to Fundamental and Practical Aspects of Electroanalysis*, 22: 765–775.

Himmel ME, Xu Q, Luo Y, et al. (2010) Microbial enzyme systems for biomass conversion: Emerging paradigms. *Biofuels*, 1: 323–341.

Huang J, Zhao P, Jin X, et al. (2020) Enzymatic biofuel cells based on protein engineering: Recent advances and future prospects. *Biomaterials Science*, 8: 5230–5240.

Juturu V and Wu JC. (2014) Microbial cellulases: Engineering, production and applications. *Renewable and Sustainable Energy Reviews*, 33: 188–203.

Knocke C and Vogt J. (2009) Biofuels–challenges & chances: How biofuel development can benefit from advanced process technology. *Engineering in Life Sciences*, 9: 96–99.

Kuhad RC, Chandna P, Lata N et al (2011) Composting of lignocellulosic waste material for soil amendment. *Bioaugmentation, Biostimulation and Biocontrol*, 107–128.

Kumar K, Ghosh S, Angelidaki I, et al. (2016) Recent developments on biofuels production from microalgae and macroalgae. *Renewable and Sustainable Energy Reviews*, 65: 235–249.

Kumar R and Kumar P. (2017) Future microbial applications for bioenergy production: A perspective. *Frontiers in Microbiology*, 8: 450.

Li C, Lin F, Li Y, et al. (2016) A β-glucosidase hyper-production Trichoderma reesei mutant reveals a potential role of cel3D in cellulase production. *Microbial Cell Factories*, 15: 1–13.

Li S, Li F, Zhu X, et al. (2022) Biohydrogen production from microalgae for environmental sustainability. *Chemosphere*, 291: 132717.

Lopes AdM, Ferreira Filho E and Moreira L. (2018) An update on enzymatic cocktails for lignocellulose breakdown. *Journal of Applied Microbiology*, 125: 632–645.

Love J. (2022) Microbial pathways for advanced biofuel production. *Biochemical Society Transactions*, 50: 987–1001.

Mandotra S, Sharma C, Srivastava N, et al. (2021) Current prospects and future developments in algal bio-hydrogen production: A review. *Biomass Conversion and Biorefinery*, 13: 1–18.

Minteer SD. (2017) *Enzyme stabilization and immobilization*. Springer.

Mona S, Kumar SS, Kumar V, et al. (2020) Green technology for sustainable biohydrogen production (waste to energy): A review. *Science of the Total Environment*, 728: 138481.

Mumtaz M, Baqar Z, Hussain N, et al. (2022) Application of nanomaterials for enhanced production of biodiesel, biooil, biogas, bioethanol, and biohydrogen via lignocellulosic biomass transformation. *Fuel*, 315: 122840.

Narisetty V, Tarafdar A, Bachan N, et al. (2023) An overview of cellulase immobilization strategies for biofuel production. *Bioenergy Research*, 16: 4–15.

Noraini M, Ong HC, Badrul MJ, et al. (2014) A review on potential enzymatic reaction for biofuel production from algae. *Renewable and Sustainable Energy Reviews*, 39: 24–34.

Osman AI, Lai ZY, Farghali M, et al. (2023) Optimizing biomass pathways to bioenergy and biochar application in electricity generation, biodiesel production, and biohydrogen production. *Environmental Chemistry Letters*, 21: 2639–2705.

Pandey A, Mohan SV, Chang J-S, et al. (2019) *Biomass, biofuels, biochemicals: Biohydrogen*. Elsevier.

Park B-W, Yoon D-Y and Kim D-S. (2010) Recent progress in bio-sensing techniques with encapsulated enzymes. *Biosensors and Bioelectronics*, 26: 1–10.

Prajapati AS, Panchal KJ, Pawar VA, et al. (2018) Review on cellulase and xylanase engineering for biofuel production. *Industrial Biotechnology*, 14: 38–44.

Prasad RK, Chatterjee S, Mazumder PB, et al. (2019) Bioethanol production from waste lignocelluloses: A review on microbial degradation potential. *Chemosphere*, 231: 588–606.

Punia H, Tokas J, Malik A, et al. (2021) Enzymes as nanoadditives: A promising alternative for biofuel production. *Nanomaterials*. Elsevier, 631–662.

Puri M, Barrow CJ and Verma ML. (2013) Enzyme immobilization on nanomaterials for biofuel production. *Trends in Biotechnology*, 31: 215–216.

Raghavendra M, Nayaka SC and Gupta VK. (2016) Microbial enzymes for conversion of biomass to bioenergy. *Microbial Enzymes in Bioconversions of Biomass*, 1–26, Springer Nature.

Ranganathan S, Mahesh S, Suresh S, et al. (2022) Experimental and computational studies of cellulases as bioethanol enzymes. *Bioengineered*, 13: 14028–14046.

Rao A, Sathiavelu A and Mythili S. (2020) Mini review on nanoimmobilization of lipase and cellulase for biofuel production. *Biofuels*, 11: 191–200.

Rodionova MV, Poudyal RS, Tiwari I, et al. (2017) Biofuel production: Challenges and opportunities. *International Journal of Hydrogen Energy*, 42: 8450–8461.

Samrot AV, Rajalakshmi D, Sathiyasree M, et al. (2023) A review on biohydrogen sources, production routes, and its application as a fuel cell. *Sustainability*, 15: 12641.

Scheiblbrandner S, Csarman F and Ludwig R. (2022) Cellobiose dehydrogenase in biofuel cells. *Current Opinion in Biotechnology*, 73: 205–212.

Shamsudin M, Tan L and Tsuji T. (2021) Enzyme immobilization technology in biofuel production: A review. *IOP Conference Series: Materials Science and Engineering*. IOP Publishing, 012056.

Sharmila VG, Shanmugavel SP and Banu JR. (2024) A review on emerging technologies and machine learning approaches for sustainable production of biofuel from biomass waste. *Biomass and Bioenergy*, 180: 106997.

Shuai W, Das RK, Naghdi M, et al. (2017) A review on the important aspects of lipase immobilization on nanomaterials. *Biotechnology and Applied Biochemistry*, 64: 496–508.

Singh A and Rathore D. (2017) *Biohydrogen production: Sustainability of current technology and future perspective*. Springer.

Singh R, Langyan S, Rohtagi B, et al. (2022) Production of biofuels options by contribution of effective and suitable enzymes: Technological developments and challenges. *Materials Science for Energy Technologies*, 5: 294–310.

Singhania RR. (2011) Production of celluloytic enzymes for the hydrolysis of lignocellulosic biomass. *Biofuels*. Elsevier, 177–201.

Singhania RR, Ruiz HA, Awasthi MK, et al. (2021) Challenges in cellulase bioprocess for biofuel applications. *Renewable and Sustainable Energy Reviews*, 151: 111622.

Sirisha VL, Jain A and Jain A. (2016) Enzyme immobilization: An overview on methods, support material, and applications of immobilized enzymes. *Advances in Food and Nutrition Research*, 79: 179–211.

Soni SK, Sharma A and Soni R. (2023) Microbial enzyme systems in the production of second generation bioethanol. *Sustainability*, 15: 3590.

Srivastava N, Rathour R, Jha S, et al. (2019) Microbial beta glucosidase enzymes: Recent advances in biomass conversation for biofuels application. *Biomolecules*, 9: 220.

Srivastava N, Srivastava M, Mishra PK, et al. (2018) Applications of fungal cellulases in biofuel production: Advances and limitations. *Renewable and Sustainable Energy Reviews*, 82: 2379–2386.

Srivastava RK, Shetti NP, Reddy KR, et al. (2020) Biofuels, biodiesel and biohydrogen production using bioprocesses. A review. *Environmental Chemistry Letters*, 18: 1049–1072.

Tamoor M, Samak NA, Jia Y, et al. (2021) Potential use of microbial enzymes for the conversion of plastic waste into value-added products: A viable solution. *Frontiers in Microbiology*, 12: 777727.

Tasca F, Gorton L, Harreither W, et al. (2008) Direct electron transfer at cellobiose dehydrogenase modified anodes for biofuel cells. *The Journal of Physical Chemistry C*, 112: 9956–9961.

Trindade SC. (2011) Nanotech biofuels and fuel additives. *Biofuel's Engineering Process Technology*. *InTech*, 103–114.

Vaid S, Nargotra P and Bajaj BK. (2018) Consolidated bioprocessing for biofuel-ethanol production from pine needle biomass. *Environmental Progress & Sustainable Energy*, 37: 546–552.

Vaidya S, Srivastava P, Rathore P, et al. (2015) Amylases: A prospective enzyme in the field of biotechnology. *Journal of Applied Biosciences*, 41: 1–18.

Velvizhi G, Ranjitha J, Vijayalakshmi S, et al. (2020) For bioenergy and biofuel production. *Nanobiotechnology for Sustainable Bioenergy and Biofuel Production*, 124.

Verma ML, Puri M and Barrow CJ. (2016) Recent trends in nanomaterials immobilised enzymes for biofuel production. *Critical Reviews in Biotechnology*, 36: 108–119.

Wang H, Hart DJ and An Y. (2019) Functional metagenomic technologies for the discovery of novel enzymes for biomass degradation and biofuel production. *Bioenergy Research*, 12: 457–470.

Wilson DB. (2009) Cellulases and biofuels. *Current Opinion in Biotechnology*, 20: 295–299.

Xiao C and Anderson CT. (2013) Roles of pectin in biomass yield and processing for biofuels. *Frontiers in Plant Science*, 4: 67.

Yadav AK, Pandey S, Tripathi AD, et al. (2021) Role of enzymes in biofuel production. *Bioenergy Research: Evaluating Strategies for Commercialization and Sustainability*, 1–18, Wiley.

Yang X-Y, Tian G, Jiang N, et al. (2012) Immobilization technology: A sustainable solution for biofuel cell design. *Energy & Environmental Science*, 5: 5540–5563.

Zambare V, Patankar R, Bhusare B, et al. (2021) Recent advances in feedstock an d lipase research and development towards commercialization of enzymatic biodiesel. *Processes*, 9: 1743.

Zanuso E, Gomes DG, Ruiz HA, et al. (2021) Enzyme immobilization as a strategy towards efficient and sustainable lignocellulosic biomass conversion into chemicals and biofuels: Current status and perspectives. *Sustainable Energy & Fuels*, 5: 4233–4247.

Zhang B, Weng Y, Xu H, et al. (2012) Enzyme immobilization for biodiesel production. *Applied Microbiology and Biotechnology*, 93: 61–70.

Zhang F, Bunterngsook B, Li J-X, et al. (2019) Regulation and production of lignocellulolytic enzymes from Trichoderma reesei for biofuels production. *Advances in Bioenergy*, 4: 79–119.

Zhao X, Qi F, Yuan C, et al. (2015) Lipase-catalyzed process for biodiesel production: Enzyme immobilization, process simulation and optimization. *Renewable and Sustainable Energy Reviews*, 44: 182–197.

7 Advances in Immobilization Technologies for Biofuel Production

Mayur R. Ladole, Sujata S. Patil, Pravin B. Pokle, Lalit M. Gantayet, and Steven Ferguson

7.1 INTRODUCTION

The significant increase in the global population has led to a rise in energy demand. Fossil fuels from non-renewable sources are being depleted at a record rate (Abdulmalek and Yan, 2022; Paraskar et al., 2023). It has been estimated that, by the end of 2050, the oil reserves may be exhausted with the current fuel consumption rate (Harun et al., 2010; Ho et al., 2011; Kim et al., 2018). Hence, petroleum-based fuels have shown a considerable difference in price fluctuations. Since then, continuous research on fuel production from renewable sources has been ongoing (Lee et al., 2015). Production of biofuels from locally available renewable sources or biomass can be seen as a potential alternative to meet the future energy demand (Abraham and Puri, 2020; Najjar et al., 2022; Pugh et al., 2011; Reshmy et al., 2023). Bioethanol, biohydrogen, biodiesel, and biomethane are a few examples of biofuels generated from different biomass (Kralova and Sjöblom, 2010; Sheikh et al., 2023).

From these biofuels, bioethanol is produced using plant biomass or waste biomass. Different crop products, such as corn, wheat, potatoes, and sugarcane, were potentially used to produce bioethanol. However, this has resulted in a negative impact on world grain prices. Hence, to address moral and ethical concerns of using food grains to produce biofuel, the plant biomass left over after the harvesting of crops, generally known as lignocellulosic biomass, has received significant attention as a promising renewable source to produce biofuels as a replacement to the petroleum-based fuels. During the process of biofuel production, enzymes play an essential role. The use of enzymes during this step is preferred due to the high selectivity of enzymes, reaction at mild temperature conditions, and being environmentally benign (Du et al., 2005; Kumari et al., 2007; Munde et al., 2017; Verma and Barrow, 2016). Two enzymes, majorly lipases and cellulase, play a pivotal role in the large-scale production of biofuels.

 DOI: 10.1201/9781003633709-7

The enzyme lipases (E.C. 3.1.1.3) with high enantioselectivity catalyzes the reaction at the hydrophilic or hydrophobic interface, such as the synthesis of oils and fats and hydrolysis reaction (Verma and Barrow, 2016). Due to their application in asymmetric synthesis, lipases have found a wide range of applications, such as the synthesis of biodiesel and fine chemicals for energy applications (Abdulmalek and Yan, 2022; Adlercreutz, 2013; Kanwar et al., 2008; Parandi et al., 2023a; Wei et al., 2023), in the food industry to produce structured lipids and oil hydrolysis (Goswami et al., 2013; Kralovec et al., 2010). Lipases are extensively studied and explored to carry out the transesterification reaction in which long-chain triglycerides are converted into fatty acid methyl esters (FAMEs). Biodiesel has been accepted as an alternate fuel due to its nontoxic nature, production from renewable sources, and biodegradability. It also provides environmental benefits, as it emits less NO, CO, particulate matter, and sulfur, thus minimizing the greenhouse effect (Abbaszaadeh et al., 2012; Calero et al., 2015). Moreover, biodiesel has received quite a good attraction due to its higher-octane number, flash point, lubricant efficiency, and combustion efficiency compared to diesel from the petroleum industry (Kim et al., 2018).

The lipase-synthesized biodiesel can be prepared using different renewable feedstocks and mainly consists of edible, nonedible, and edible oil after use, generally considered as waste edible oil. The high cost associated with edible oil restricts its use for biodiesel production, as there is a high demand for edible oil as a food source (Gharat and Rathod, 2013; Kim et al., 2018; Maroju et al., 2022). Considering this, edible oil is being replaced with nonedible oil, microalgae oil, and waste oil for biodiesel production (Aghabeigi et al., 2023; Parandi et al., 2022, 2023b; Wang et al., 2023; Wei et al., 2023). Nonedible oils such as castor, mahua, karanja, rubber, and Jatropha are cost-effective substrates for biodiesel production (Hama and Kondo, 2013).

The used oil after cooking (waste oil), soap stock, and grease are also used as a cheaper option for biodiesel production. It was estimated that every year, more than 16 million tons of waste cooking oil is generated (Azócar et al., 2010). Apart from using different oils for biodiesel production, microalgae are also considered a potential feedstock or nonedible crop that contains more than 25 times more oil than other plants (Aghabeigi et al., 2023). Microalgae have recently been gaining tremendous attention due to some of the benefits of using them as a feedstock. Apart from their contribution to protecting the environment by carbon dioxide fixation, they can grow relatively fast. Some microalgae species were found to have almost 50% oil by dry biomass weight (Ahmed et al., 2018; Lee et al., 2015).

Apart from lipids or oils as a source for biofuel production or specifically for biodiesel production, biofuel or bioethanol can also be produced from waste biomass, generally known as lignocellulosic biomass. Three typical steps are involved in converting lignocellulosic biomass into biofuel: pre-treatment of biomass, saccharification, and fermentation (Figure 7.1) (Abraham et al., 2014).

The pre-treatment step involves the removal of lignin, which is otherwise a nonvaluable by-product for the next step. The saccharification step is essential for degrading cellulosic biomass using cellulase (E.C. 3.2.1.4). The enzyme cellulase is ideally a mixture of three different enzymes: endoglucanase (E.C. 3.2.1.4), exoglucanase (E.C. 3.2.1.91), and β-glucosidase (E.C. 3.2.1.21). These enzymes act

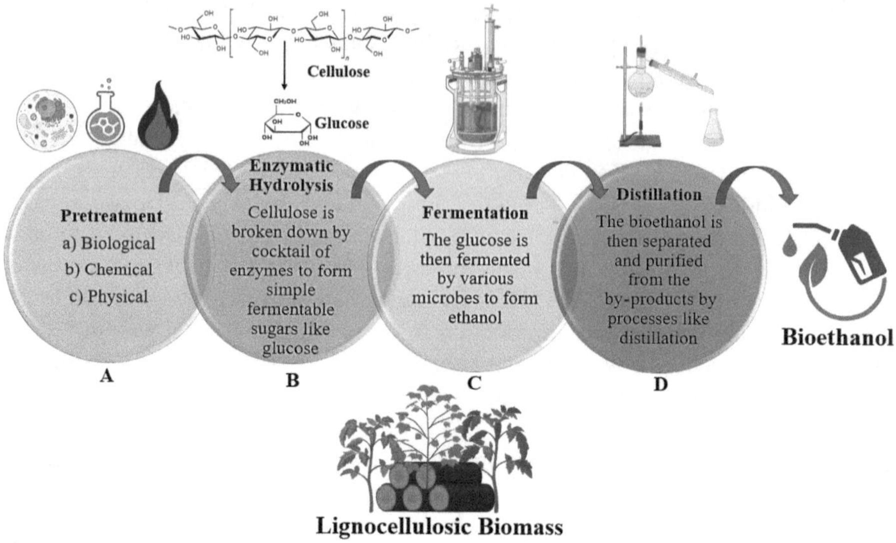

FIGURE 7.1 An overview of bioethanol production from the lignocellulosic biomass. **(Adapted from Rajnish et al., 2021.)**

simultaneously for cellulose degradation (Ladole et al., 2017; Rajnish et al., 2021; Reshmy et al., 2023). The formed product of this reaction, glucose, is then utilized during fermentation, which is the final step of bioethanol production (Abraham et al., 2014; Abraham and Puri, 2020; Lee et al., 2015).

Despite several advantages offered by the enzymatic production of biofuels, some limitations, such as less stability, storage issues, difficulty of removal from the reaction medium, and limited or non-reusability, need to be addressed. The nonreusability of enzymes leads to their wastage after reaction completion, which adds an extra cost. Hence, the use of enzymes in their free form is usually restricted as the process becomes more costly, resulting in the higher cost of end products. To address this issue, the immobilization of enzymes offers certain benefits such as more excellent stability of enzymes over a range of temperatures and pH, ease of separation from the reaction mixture, and, more importantly, reusability of enzymes, which helps to reduce the process cost. The immobilized enzymes can be easily separated after the completion of the reaction; hence, it is easier to purify the final or end product in the soluble form. Several enzyme immobilization technologies or techniques have been explored to immobilize several enzymes. From simple adsorption to covalent cross-linking and, most recently, covalent organic frameworks, the research on different enzyme immobilization strategies is continuously ongoing.

This chapter overviews the overall development of enzyme immobilization techniques. It also covers the traditional methods of enzyme immobilization to the most recent methods, with a special emphasis on applying immobilized enzymes in biofuel production.

7.2 ENZYME IMMOBILIZATION

The concept of immobilization of enzymes was discovered in 1971 with a suggestion by Ephraim Katchalski-Katzir at the Enzyme Engineering Conference in the USA (Katchalski-Katzir, 1993). The word "immobilization" means restricted, localized, or physically bound to a static support. In this concept of immobilization, the enzymes are bound to a stationary support, which allows its use multiple times to carry out a similar reaction without hampering its activity. The enzyme immobilization process may inhibit enzyme activity; hence, it is imperative to select an appropriate technique and optimize enzyme immobilization to retain maximum activity of enzymes in the immobilized form. As mentioned earlier, several enzyme immobilization methods are available based on the interaction between the enzyme and the carrier support used during immobilization. The immobilization methods such as adsorption, covalent bonding, entrapment, encapsulation, carrier-free immobilization, cross-linking etc., are widely used for enzyme immobilization (Kim et al., 2018; Salwan et al., 2020; Santos et al., 2020). A lot of efforts have been then made in the development of enzyme immobilization supports. Irrespective of several advantages offered by immobilized enzymes, such as kinetic and thermodynamic stability, stabilization against harsh conditions, high selectivity, capacity to work on multi-step enzyme cascade reactions, ease of separation, reduction in the formation of undesired products, and providing greater surface area-to-volume ratio compared to free enzyme (Fernandez-Lafuente, 2009; Iyer and Ananthanarayan, 2008; Polizzi et al., 2007). The limitations offered by conventional immobilization techniques have been carefully studied, and new developments are being explored for enzyme immobilization (Hernandez and Fernandez-Lafuente, 2011; Mateo et al., 2006).

It has also been reported that the activity of enzymes after immobilization may increase or decrease in some cases (Ottolina et al., 1992). This is attributed to the fact that the materials used for enzyme immobilization may hamper its activity by reducing the exposure of enzymes' active sites to the substrates. This eventually decreases the enzyme-substrate interaction, resulting in less product formation. Thus, many research activities have taken place to develop various immobilization supports that may not hamper enzymatic activity. Some of the materials, such as mesoporous materials with pore size in the range of 2–50 nm, macro-/micro-surfaces, and nanomaterials, are explored for immobilization for maximum activity recovery after immobilization, enhanced stability, and multiple reusability (Drechsler et al., 2004; Palocci et al., 2007; Yiu and Wright, 2005).

The immobilization of enzymes on micro or macro surfaces has been found to have potential usage in environmental remediation, biosensors, and the synthesis of value-added products. Different immobilization methods, such as selective immobilization of enzymes at specific sites and its applications in synthetic biology innovations, have emerged as essential and efficient covalent linking methods at specific sites. Moreover, various alternative enzyme immobilization methods, such as the development of active protein microarrays, have evolved. Along with this, the enzymes can be immobilized continuously in a flow system such as PASS flow system (Dräger et al., 2007). This system has been developed to immobilize and synthesize his-6-tagged recombinant proteins, which have been used efficiently to perform different biocatalytic reactions (Wong et al., 2009). The continuous improvement of enzyme immobilization has gained much attention and popularity for various potential applications. Enzymes in their free form make the reaction homogenous, making it difficult to recover and reuse, unviable after product formation. Hence, immobilized enzymes provide an appropriate solution as they increase enzyme stability and reusability.

Recently, the application of nanotechnology in biotechnology, generally called nano-biotechnology, has received a great deal of attention in various applications such as targeted drug delivery, imaging, cell labeling, and enzyme immobilization. Due to their high efficacy, biocompatibility, excellent stability, and reusability, they are the preferred supports for enzyme immobilization (Bosio et al., 2016; de Jesús Rostro-Alanis et al., 2016; Meryam Sardar, 2015). For biochemical applications, nanoparticles have been explored for enzyme loading or immobilization. Due to their exceptional surface area-to-volume ratio, nanoparticles provide high enzyme loading and higher stability than other supports (Ansari and Husain, 2012). The enzymes can be either adsorbed or cross-linked on nanoparticles. Nanoparticles need to be functionalized prior to cross-linking. Taking advantage of magnetic properties, magnetic nanoparticles are widely explored for enzyme immobilization (Ladole et al., 2021). These magnetic nano-biocatalysts can be quickly separated after the reaction completion using an external magnetic field. This allows the easier separation of enzymes and increases the repeatability of enzymes over successive reactions. These immobilized enzymes on magnetic nanoparticles also make product separation relatively easier. Several enzyme immobilization methods and their potential applications in biofuel production are described in the following section.

7.3 ENZYME IMMOBILIZATION METHODS

Typically, the enzymes are immobilized on the carrier support via different techniques. It is broadly categorized into two techniques: the physical and chemical methods of enzyme immobilization (Figure 7.2).

The physical method of enzyme immobilization consists of enzyme immobilization on a solid support. The immobilization can be done via simple adsorption, entrapment, and enzyme encapsulation. The chemical method of enzyme immobilization involves carrier-free or carrier-based techniques for immobilization. In this method, the enzymes are bonded to the support or carrier using a chemical reagent called a cross-linker. The cross-linker is covalently bound to the enzyme and the support matrix. The enzymes can also be immobilized without a carrier matrix by the process of enzyme precipitation followed by covalent cross-linking to form cross-linked enzyme aggregates (CLEAs). A detailed discussion of these immobilization methods and applications in biofuel production is given in the following section.

7.3.1 Physical Adsorption

Physical adsorption is well known for its simple method of immobilization. This immobilization technique is one of the oldest methods in which the enzymes are adsorbed on the surface of a stationary or insoluble support. In the process of

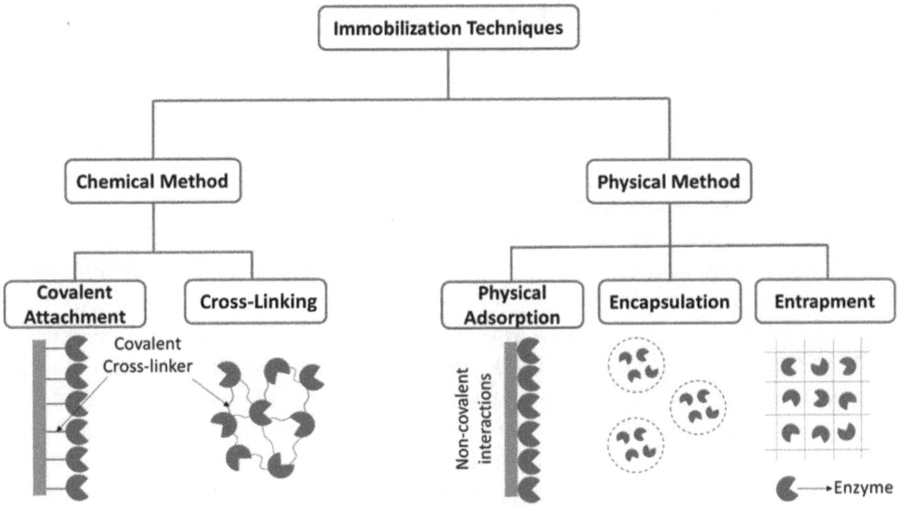

FIGURE 7.2 Different techniques and classification of enzyme immobilization.

immobilization using simple adsorption, the enzyme is combined with an appropriate adsorbent, which acts as a carrier under appropriate pH and ionic conditions. The unbound enzyme is then washed, leaving only the immobilized enzyme to be used. The weak forces such as hydrogen, Van der Waals forces, hydrophobic, and ionic form an interaction between the exterior surface of the adsorbent and enzymes (Brady and Jordaan, 2009; Santos et al., 2020; Villeneuve et al., 2000; Zdarta et al., 2018).

The adsorption process typically involves the adsorption of proteins and binding sites at multiple points on the immobilization surface, which may change the proteins' microenvironment (Johnson et al., 1996; Palocci et al., 2007). The amount of enzyme leakage is dependent on the type of adsorbent used. Furthermore, surface characteristics of support are responsible for preventing any chemical changes and adverse effects on enzyme activity. A significant drawback of this technique is the ease with which enzymes desorb (Rao et al., 1998; Salwan et al., 2020)

Using the simple adsorption technique of enzyme immobilization, MCM-41 molecular sieves are used as efficient support for lipase immobilization. At optimum immobilization conditions and pH 6, maximum lipase immobilization of around 250 mg/g was observed. The immobilized lipases were further explored for biodiesel synthesis from cotton seed oil using a transesterification reaction and methanol as the solvent. The transesterification reaction, when carried out at all optimized conditions such as pH 8, 12:1 methanol to oil molar ratio, and 40°C, a maximum yield of more than 98% was achieved for the FAME (Katiyar and Ali, 2012).

Biodiesel was also produced from nonedible oil extracted from *Jatropha curcas* using immobilized lipase extracted from *Pseudomonas cepacian*. The lipase was adsorbed on celite using a simple adsorption approach. The transesterification reaction was performed for 8 h by alcoholysis reaction at 50°C and in the presence of

4%–5% (w/w) water. The maximum yield of around 98% (w/w) was achieved at these reaction conditions. The immobilized biocatalyst was also examined for reusability studies, and the immobilized catalyst can be used at least four times without hampering enzymatic activity (Shah and Gupta, 2007).

As discussed in the introduction, waste cooking oil can also be used as a potential feedstock for biodiesel production using enzymatic transesterification. Four types of zeolites, namely FM-8, 5A, AW-300, 13-X, and hydrotalcite, were used for lipase immobilization using a simple adsorption technique. Lipase on hydrotalcite showed maximum loading at 4°C and pH 8.5. The immobilized lipase was then used in the transesterification reaction. It retained 36% of initial activity at 45°C and 14% when the reaction was performed at 55°C, even after the seventh cycle of its reuse (Yagiz et al., 2007). The reaction, when compared with the free enzymes for a period of 105 h, had a higher yield of about 95% compared to immobilized enzymes, which yielded 92.8%. However, the activity study indicated more activity for immobilized lipase than free lipase when the amount of lipase was taken into consideration. This is attributed to the active site for transesterification reaction shown by hydrotalcite. Thus, the synergistic effect of lipase and hydrotalcite as a heterogeneous catalyst shows more activity than free lipase (Yagiz et al., 2007).

An ion-exchange microporous hydrolytic resin was used to immobilize lipase extracted from *Rhizomocur miehei*. The experimental model was developed and compared with the data generated after experimental work. A correlation was observed between the theoretical and practical yield values, indicating 90% yield after 8 h of transesterification reaction using immobilized enzymes (De Paola et al., 2009). Cellulase was immobilized on silica nanoparticles using the physical adsorption approach. The reaction was carried out for 12 hours. In the concurrent saccharification and fermentation reaction of cellulose, immobilized cellulase was employed to enhance the ethanol yields. Compared to free enzymes, immobilized enzymes yielded a higher amount of glucose. When the enzyme is immobilized onto silica nanoparticles, it can stabilize and increase the activity under conditions where the reaction is not ideal, including high ethanol concentrations and temperatures (Lupoi and Smith, 2011).

The properties of mesoporous silica materials, including their large surface area, tuneable pore size, and ease of functionalization, present a unique opportunity for enzyme immobilization. This still poses a challenge as a significant cellulase activity gets trapped inside the silica pores. Hence, wrinkled silica nanoparticles (WSNs) were synthesized and explored for cellulase immobilization to produce a stable and functional biocatalyst. The cellulase immobilization in the prepared nanoparticles was enhanced using pentanol as a co-solvent, which helped to increase inter-wrinkle distance while synthesizing WSNs (Costantini et al., 2020). Results indicated WSNs have provided a favorable and protective microenvironment for the enzyme cellulase. The morphology, size, and pores distribution of the support were helpful during the entire process of cellulase immobilization. Moreover, this process allows secondary structure retention, which helps to maintain enzymes' catalytic activity. The immobilized catalyst showed a similar yield when reused over a successful reaction of over five times (Costantini et al., 2020).

7.3.2 Entrapment

One of the popular methods of enzyme immobilization in earlier days was entrapment. This process restricts the movement of enzymes as they are trapped inside fibers or porous gel. Despite entrapment, enzymes remain a free catalyst for carrying out the reaction. Enzyme entrapment is a relatively simple and easy-to-follow protocol where the substrate and end product possess low molecular weight, and the enzyme remains inside the polymeric matrix (Kuan et al., 2013; Salwan et al., 2020). The entrapped enzymes can be separated easily from the reaction mixture by filtration or centrifugation. Some other polymers, such as polyacrylamide (Deshpande et al., 1987), and polyvinyl alcohol hydrogel (Grosová et al., 2008) was also explored for enzyme entrapment. If natural polymers such as agar, agarose, and gelatin are used, reverse thermal polymerization can be employed. In the case of ionotropic gelation, carrageenan and alginates are used as polymers (Datta et al., 2013).

Biodiesel production was also carried out using an enzyme entrapment process. The process of lipase entrapment is usually performed in situ in which a monomeric phase solution of lipase is prepared, and with the formation of polymerization, the enzymes get entrapped inside the polymeric network (Santos et al., 2020; Souza et al., 2017). The enzyme entrapment provides more stability and protects enzymes from the external environment or surroundings, such as agitation, bubbles, and solvents like hydrophobic solvents. Transesterification of soybean oil was carried out using nine different lipases. Out of the nine selected lipases, the one extracted from *Pseudomonas cepacian* with a maximum yield of fatty acid alkyl esters (FAAE) was further entrapped in hydrophobic and chemically inert sol-gel support. The reaction with immobilized lipase was further studied, and it was found that after 1 h of transesterification, the yield of FAEE and FAME was 65% and 67%, respectively (Noureddini et al., 2005).

Similarly, polyallylamine-mediated biometric silica was used as a potential support for the entrapment of lipase extracted from *Pseudomonas cepacian*. The immobilized biocatalyst was used for the transesterification of waste cooking and soyabean oil into biodiesel. The optimal conditions for the transesterification reaction were found at 38% (w/w) n-hexane, substrate of molar ratio of 5:1, and temperature of 43.3°C. The theoretically predicted yield of 67% was obtained, while the experimental value of 68% has been reported (Kuan et al., 2013).

7.3.3 Covalent Bonding

The following method of enzyme immobilization is via covalent bonding. It is one of the popular and widely used methods for enzyme immobilization. Unlike enzyme immobilization via simple adsorption, the enzymes are linked to the supporting matrix via a covalent bond. The critical factor to be considered here is the immobilization of enzymes without affecting their activity. To do so, a suitable environment is provided that does not hamper the enzymatic activity. The occurrence of amino acids on the side chain, such as arginine, aspartic acid, and histidine, are involved in the covalent binding process. A high specific activity can also be achieved using specific carriers such as peptides for enzyme binding covalently. The enzymes can be directly bonded to the support matrix or with the help of specific linkages or linkers known as spacer or spacer arms (Salwan et al., 2020).

These spacer arms or linker molecules help with better flexibility to the enzymes bonded on the support matrix. This not only increases the enzyme activity directly but also increases the product yield. The support matrix is activated via specific functional groups to assist the formation of covalent linkage. Some of the chemical moieties, such as cyanogen bromide agarose, cyanogen bromide–activated Sepharose, glutaraldehyde, etc., are used to hold the carbohydrates and help to increase the thermal stability of enzymes bonded covalently (Guisan et al., n.d.; Salwan et al., 2020). For lipase, the cross-linking involves the covalent bonding of lipases on the support matrix in an irreversible manner. The reaction involves bond formation between active groups on the support matrix and lipase's active catalytic site on the exterior of active amino acid. The active sites usually present on the exterior of amino acids are the amine and thiol groups (Santos et al., 2020).

The covalent interaction between the support matrix and enzymes forms a stable immobilized structure, and the leaching of enzymes from the support matrix will be minimal. This is attributed to the direct interaction of immobilized enzymes and support matrix characteristics such as binding affinity toward enzymes, pore size, and chemical stability, making the immobilized enzyme relatively stable and active. The covalent attachment at multiple points for the efficient immobilization of enzymes compared with the operational and thermal stability of lipase using simple adsorption.

Lipases from heterologous *Rhizopus oryzae* were covalently immobilized on various support matrices and were tested for transesterification reaction. The biodiesel production was carried out using methanol as the solvent and Jatropha oil as the substrate. The reaction was performed at an optimal temperature of 30°C after consecutive addition of methanol seven times. When used in the reaction at 30°C for 4 h, the covalently immobilized lipase has shown an average FAME yield of around 51% to 65% (Rodrigues et al., 2016).

In another study, the lipase from *Thermomynces lanuginosus* was covalently attached to olive pomace. The transesterification of pomace oil was carried out using immobilized lipase and methanol as the solvent. A conversion of 93% was obtained when the reaction was performed for 24 h at the temperature of 25°C. The reusability study revealed that immobilized lipase retains its activity even after 10 cycles of its reuse (Yücel, 2011). Lipases from *Candida cylindracea* were covalently immobilized on glutamic acid-modified alkaline Ca-bentonite. The modified lipases exhibited higher biocatalytic activity. At a higher temperature of about 50°C for an 8 h reaction, the immobilized lipase maintained 50% of its initial activity; however, when the reaction was performed with free lipase for about 6 h, it only retained 6% of its initial activity. The comparison study with free and immobilized enzymes for the transesterification reaction revealed that immobilized and free lipase showed 99% and 53% conversion, respectively. This indicated the retention of higher enzymatic activity after immobilization. When checked for its reusability, the immobilized lipase showed more than 56% retention in its activity when used five times (Tang et al., 2019).

A multipoint covalent attachment method was employed for lipase immobilization extracted from *Pseudomonas fluorescens* and *Thermomynces lanuginosus* on Toyopearl AF-amino-650M resin. The obtained lipase-immobilized support received was thermally stable and showed promising activity recovery. A transesterification reaction using immobilized lipase was performed with palm and babassu oil. The reaction was performed using ethanol as the solvent. The prepared immobilized biocatalyst showed excellent thermal stability and hydrolytic activity compared to free lipase. The stability of immobilized lipase was found to be more than 27 times that of free lipase. The highest transesterification rate of palm oil, more than 93%, was obtained using both immobilized lipases (Mendes et al., 2011).

7.3.4 Cross-linking

Moving to the subsequent immobilization methods, cross-linking of enzymes has been introduced as one of the carrier-free immobilization methods. Cross-linking of enzymes is reasonably straightforward, and a crude mixture of aqueous enzymes can also be used. The process consists of enzyme precipitation using a (chilled) solvent, nonionic polymers, or salts such as a saturated ammonium sulfate solution. The precipitated enzymes later cross-linked in the next stage using covalent cross-linking agents to form a three-dimensional network of enzymes termed CLEAs (Sheldon, 2007; Talekar et al., 2013). This makes the enzymes in the insoluble form easier to separate from the reaction mixture either by centrifugation or filtration. Out of the many available cross-linkers, glutaraldehyde is one of the widely used and studied cross-linkers. This bifunctional agent showed excellent stability and high efficacy and is relatively cheaper than other agents. The formed CLEAs performed well during the reaction with much more stability, effectivity, and reusability (Talekar et al., 2013).

Due to the bifunctional nature of glutaraldehyde, it cross-links enzymes' amino groups on both ends (Sahu et al., 2016). However, the cross-linking process leads to a change in protein conformation, which ultimately results in the loss of catalytic activity. In the case of enzymes with low lysine groups/residues, the enzymes are supplemented with other proteins such as albumin, collagen, and gelatin, improving the overall immobilization efficiency (Talekar et al., 2013, 2012).

Lipase immobilization using a cross-linking technique comprises enzyme precipitation followed by cross-linking with one another using a suitable cross-linking agent. A detailed illustration of lipase CLEAs preparation is given in Figure 7.3. CLEAs lipase produced from *Pseudomonas cepacian* was used to transesterify *Madhuca indica oil.* Compared to free lipases, which showed 98% conversion after the reaction time of 6 h, the same quantity of lipase CLEAs showed a conversion of 92% in just 2.5 h reaction time. This indicates the overall efficacy of using lipase CLEAs over free lipases. Moreover, centrifugation can easily separate the lipase CLEAs after the reaction (Kumari et al., 2007).

Lipases were cross-linked using a new method in hydrophilic polyurethane foams using polyglutaraldehyde as a cross-linking agent. The transesterification of *canola oil* has been carried out using immobilized lipases and methanol. Results indicated

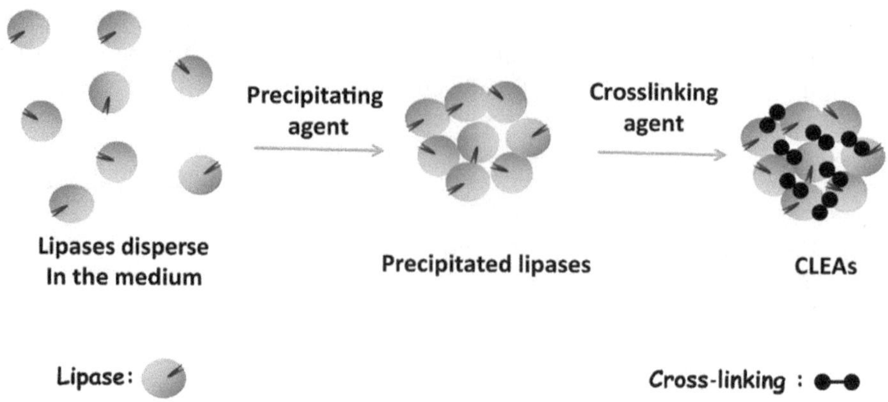

FIGURE 7.3 Preparation of cross-linked enzyme aggregate of lipase enzyme.

(Adapted from Sampaio et al., 2022.)

retention in the enzymatic activity of lipase even after ten cycles of reuse with only minimal loss of the original activity. The FAME yield of around 90% was obtained, and the immobilized lipases exhibited superior stability than free lipase (Dizge and Keskinler, 2008). Different immobilization techniques for enzyme immobilization and application in biofuel production are summarized in Table 7.1.

TABLE 7.1
Summary of Different Lipase Immobilization Techniques and Application of Immobilized Lipase in Biofuel Production

Immobilization Methods	Substrate	Solvent Used	Reaction Conditions	Yield (%)	References
Adsorption	Waste cooking oil	Methanol	Reaction temp. 24°C; 4% immob. lipase (wt%); Oil/MeOH molar ratio 1:4; reaction time 5 h	95	Yagiz et al. (2007)
Adsorption	*Jatropha curcas oil*	Ethanol	Reaction temp. 40°C; 10% immob. lipase; Oil/EtOH 1:4; reaction time 8 h	98	Shah and Gupta (2007)
Adsorption	Cotton seed oil	Methanol	Reaction temp. 40°C; 5% immob. lipase; Oil/MeOH molar ratio 1:12; reaction time 48 h	98.3	Katiyar and Ali (2012)
Cross-linking	*Jatropha curcas oil*	Ethanol	Reaction temp. 35°C; 52.5% immob. Lipase; Oil/EtOH molar ratio 1:10; reaction time 24 h	100	Abdulla and Ravindra (2013)
Cross-linking	*Madhuca indica oil*	Ethanol	Reaction temp. 40°C; 10% immob. lipase; Oil/EtOH molar ratio 1:4; reaction time 2.5 h	92	Kumari et al. (2007)

(Continued)

TABLE 7.1 (Continued)
Summary of Different Lipase Immobilization Techniques and Application of Immobilized Lipase in Biofuel Production

Immobilization Methods	Substrate	Solvent Used	Reaction Conditions	Yield (%)	References
Cross-linking	Olive oil	Methanol	Reaction temp. 40°C; Oil/MeOH molar ratio 1:4; reaction time 12 h	64	Han and Kim (2011)
Cross-linking	Canola oil	Methanol	Reaction temp. 40°C; Oil/MeOH molar ratio 1:6; reaction time 24 h	90	Dizge and Keskinler (2008)
Entrapment	Tallow and grease	Ethanol	Reaction temp. 50°C; Oil/EtOH molar ratio 1:4; reaction time 24 h	95	Hsu et al. (2001)
Entrapment	Palm oil	Ethanol or Methanol	Reaction temp. 35°C; 1% immob. lipase; Oil/EtOH molar ratio 1:15; reaction time 1 h **or** Reaction temp. 37°C; 1% immob. lipase; Oil/MeOH molar ratio 1:14; reaction time 1 h	70 or 85	Moreno-Pirajàn and Giraldo (2011)
Entrapment	Soybean oil	Methanol/Ethanol	Reaction temp. 35°C; 4.75% immob. lipase; Oil/ MeOH molar ratio 1:7.5, reaction time 1 h **or** Reaction temp. 35°C; 4.75% immob. lipase; Oil/EtOH molar ratio 1: 15.2; reaction time 1 h	67 or 65	Noureddini et al. (2005)
Entrapment	Soybean oil or waste cooking oil	Methanol	Reaction temp. 43.3°C; 1.0% immob. lipase; Oil/MeOH molar ratio 1:5; reaction time 36 h	68 or 67	Kuan et al. (2013)
Covalent Bonding	Camellia oil	Methanol	Reaction temp. 40°C; 40% immob. lipase; reaction time 48 h	99	Tang et al. (2019)
Covalent Bonding	Olive pomace oil	Methanol	Reaction temp. 25°C; Oil/MeOH molar ratio 1:6; reaction time 24 h	93	Yücel (2011)
Covalent Bonding	Palm or Babassu oils	Ethanol	Reaction temp. 45°C; Oil/EtOH molar ratio 1:18; reaction time 48 h or; Oil/EtOH molar ratio 1:9; reaction time 48 h	93	Mendes et al. (2011)
Covalent Bonding	Crude *Jatropha curcas oil*	Methanol	Reaction temp. 30°C, reaction time 4 h	51–65	Rodrigues et al. (2016)

7.3.5 Nanosupports

The use of various carrier-based immobilization nanosupports provides excellent stability and efficiency to the enzymes. The support also helps enzyme recycling, which ultimately decreases the overall cost of its potential use in industrial applications. Some of these supports, such as mesoporous materials, gel, and polymers (synthetic and natural), have already been discussed. With the evolution of nanotechnology, there has been a significant improvement in applying nanoparticles in enzyme immobilization (Figure 7.4).

Nanoparticles as a support for enzyme immobilization are quite exciting and one of the promising research areas in which robust nanomaterials are developed as carriers for enzyme immobilization. The continuous development and advancement in nanomaterials make it a favorable carrier for enzyme immobilization. The characteristic properties of nanoparticles, such as their nano-size (nm), high surface area-to-volume ratio, ease of functionalization, and modifications, led to their

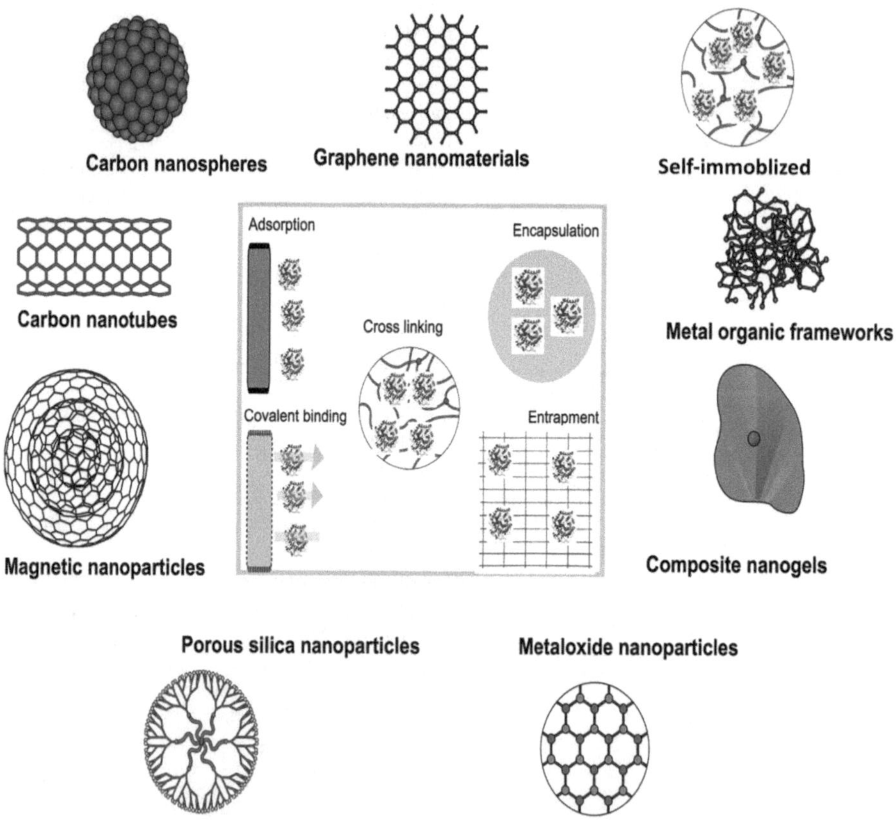

FIGURE 7.4 Different methods of enzyme immobilization using various magnetic and non-magnetic nanosupports.

(Adapted from Rajnish et al., 2021.)

applications in broad areas of biotechnology. The nanoparticles' high-loading capacity shows high enzyme activity retention compared to the conventional immobilization materials (Verma et al., 2013, 2016; Verma and Barrow, 2016).

The prepared nano-biocatalyst exhibits excellent stability over a broad range of temperatures and pH compared to the biocatalyst in its free form. The uniformity in the shape and size of nanoparticles, along with their well-structured core shells, makes it an ideal support for immobilization applications. Most nanoparticle synthesis protocols are simple and can be prepared with toxic chemicals and surfactants. One of the significant advantages of nanoparticles is immobilizing multiple enzymes, termed the co-immobilization process (Ladole et al., 2018; Patil and Rathod, 2022). Typically, nanomaterials are classified based on their magnetic and nonmagnetic properties (Ansari and Husain, 2012).

7.3.5.1 Nonmagnetic

The nanomaterials that do not possess magnetic properties are nonmagnetic nanomaterials or nanoparticles. Some typical nonmagnetic nanomaterials, polystyrene, zirconia, polylactic acid, silica, and chitosan, are explored for immobilization. Among these, silica nanoparticles, an inorganic material, are widely studied and suitable for various applications due to their rigid mechanical properties, chemical resistance, nontoxicity, good compatibility, and high thermal stability (Zhang et al., 2022). Due to the high porosity and surface area, the silica supports show excellent adsorption properties, resulting in efficient immobilization and lower diffusion (Adlercreutz, 2013; Kim et al., 2006; Liese and Hilterhaus, 2013). The activated immobilization support helps in enzyme reuse and retains essential biochemical properties (Iyer and Ananthanarayan, 2008; Mateo et al., 2007). Some of the widely used nanomaterials, such as nanofibers, nanotubes, nanoporous materials, and nanoparticles, have been studied extensively for biocatalytic applications (Salwan et al., 2020; Wang, 2006). Polymethyl methacrylate (PMAA), as a core nonmagnetic nanomaterial, was explored for cellulase immobilization (Ho et al., 2008). The immobilized cellulase showed high residual activity and thermal stability compared to the native cellulase solution. The use of PMAA nanomaterials has helped to gain thermally stable enzyme solutions, which in turn can be used for industrial applications (Ho et al., 2008).

The enzyme shells of these nanoparticles are thick and constantly dispersed. The epoxy group of these nanoparticles could provide covalent bonding with hydroxyl, amino, and thiol groups on the enzyme's surface. This research found that more time is required to build and interact with the enzyme and epoxy group. This can be addressed with the help of an external functionalization or physical and chemical interaction of the enzyme's surface. Thus, it helps in the process of formation of an intramolecular covalent linkage between the epoxy functional group and enzymes. This process helps strengthen and initiate immobilization regions for the formation of multiple covalent linkage formation (Mateo et al., 2007). Canola oil was used to synthesize biodiesel by methanolysis using lipase immobilized on mesoporous SBA-15 nanoparticles. The nanoparticles were functionalized using 3-glycidiloxypropyl trimethysilane prior to immobilization. The functionalization process helps in the formation of covalent bonds between enzymes and SBA nanoparticles. The immobilization helps to improve the thermal stability of lipase and showed an improved catalytic activity even at the high methanol concentration. The immobilized lipases showed excellent operational stability and did not affect the change in activity even after repeated use for several cycles (Babaki et al., 2016).

The porous inorganic silica nanoparticles were explored for lipase immobilization and application for biodiesel production using transesterification of triolein and methanol as the solvent. A conversion efficiency of 98% was obtained using this immobilization support, along with a biodiesel yield of 89% in 3 h and a temperature of 37°C. The ratio of methanol to oil was 6:1 (MacArio et al., 2013). Nonmagnetic metal nanoparticles, which include alumina, zirconia, titania, zinc, silver, gold, etc., are explored for enzyme immobilization. Some of the nanoparticles' critical properties include high loading, resistance to mechanical forces, and high stability over a wide range of pH and temperature. Due to the inert properties of these materials, they can be used at different reaction conditions and hence receive a good attraction and applications in the immobilization of different enzymes (Cipolatti et al., 2014; Santos et al., 2020).

Polydopamine-functionalized silver nanoparticles were used to support lipase immobilization and explored to produce biodiesel (Dumri and Hung Anh, 2014). The results indicated a biodiesel yield of about 95% and were examined using gas chromatography. The transesterification was carried out at 40°C for about 6 h. The immobilized lipase showed excellent reusability, remained active even after 7 cycles of reuse, and showed a reduced conversion efficiency of up to 27%. Zirconia nanoparticles modified with carboxylic acid containing a long alkyl chain have been explored to support lipase immobilization. Modifying zirconia nanoparticles improved the catalytic activity of lipases almost double compared to free lipases and lipase immobilized on unmodified zirconia nanoparticles. Moreover, with the best possible modification of zirconia nanoparticles, the activity enhancement was nearly 16 times more than native lipases (Chen et al., 2009). Different nonmagnetic supports for enzyme immobilization are given in Table 7.2.

7.3.5.2 Magnetic

The advantage of using nanoparticles for enzyme immobilization is that it helps in several ways, such as high surface area, greater loading capacity, improved stability, and activity over a range of pH and temperature. However, the separation of these nonmagnetic nanoparticles remains a topic of discussion. Energy-intensive unit operations such as high-speed centrifugation are required to separate these nanosupports or particles (Ghadi et al., 2015). The problem of separating these nanoparticles can be solved by synthesizing magnetic nanoparticles (MNPs). The unique property of these nanoparticles, being magnetic (in some cases the super-magnetism), size in nanometres, high surface area-to-volume ratio, less toxicity, and biocompatibility makes it an ideal support for enzyme immobilization (Ladole et al., 2021; Xie and Wang, 2012). The magnetic property helps for the ease in separation of these nanoparticles just by applying an external magnetic field (Johnson et al., 2011; Liese and Hilterhaus, 2013). The magnetic field helps to separate nanoparticles from the reaction mixture, and the reaction mixture or product can be easily separated. The separated nanoparticles can be reused by adding fresh reactant and removing the magnetic field. The supermagnetic property helps in the dispersion of magnetic nanoparticles without forming enzyme aggregates.

TABLE 7.2
A Summary of Various Nanosupports (Nonmagnetic) Used for Lipase Immobilization and Their Application in Biofuel Production

Immobilization Nanosupport	Substrate	Solvent	Reaction Conditions	Yield (%)	Ref.
Porous inorganic silica	Triolein	Methanol	Reaction temp. 37°C; 10% immob. lipase; Oil/MeOH molar ratio 1:6; reaction time 3 h	89	MacArio et al. (2013)
SBA-15 (mesoporous) functionalized by 3-glycidiloxypro pyltrimethylsilane	Canola oil	Methanol, t-butanol and triethyl-amine	Reaction temp. 50°C; 3.8 wt % immob. lipase; Oil/MeOH molar ratio 1:3; reaction time 72 h	24/33/34	Babaki et al. (2016)
Polydopamine-functionalized silver nanoparticles	Soybean Oil	Methanol	Reaction temp. 40°C; Oil/ MeOH molar ratio 1:3; reaction time 6 h	95	Dumri and Hung Anh (2014)
Single-walled carbon nanotubes (SWCNTs)	Sunfloweroil	Ethanol	Reaction temp. 35°C, 15 wt% immob. lipase; 3.9 L of ethanol; reaction time 4 h	83.4	Bencze et al. (2016)
Magnetic multiwalled carbon nanotubes grafted with polyamidoamine (mMWCNTs)	Waste cooking oil	t-butanol	Reaction temp. 50°C; 6 wt% immob. lipase; 20 wt% tert-butanol; reaction time 10 h	94	Tan et al. (2012)
Nanofibrous membrane of polyacrylate	Soybean oil	Methanol	Reaction temp. 30°C; 0.35 wt% immob. lipase; 51 wt% MeOH; reaction time 24 h	90	Zhao et al. (2007)
Electro spun poly acrylonitrile nanofibers	Rapeseed Oil	n-butanol	Reaction temp. 40°C; Oil/n-butanol molar ratio 1:3; reaction time 48 h	94	Fan et al. (2016)

Recently, magnetic nanoparticles have been widely used as support for enzyme immobilization and different applications (Dyal et al., 2003; Gupta et al., 2011, 2011). The magnetic nanoparticles can be synthesized using various methods such as chemical co-precipitation, hydrothermal, sol-gel, and ultrasound method (Dyal et al., 2003). Additionally, MNPs were initially extracted from the magneto tactic bacteria (Springer-Verlag et al., 1987). These bacteria are ubiquitous and usually found in freshwater sediments marine and biomineralize magnetosomes. These bacteria can be easily grown at the lab scale (Springer-Verlag et al., 1987). However, the quantity of MNPs obtained from these bacteria is limited and can't be scaled up easily.

To tackle this, the MNPs are synthesized using a chemical co-precipitation technique. Typically, the MNPs are synthesized using the chemical co-precipitation method, which is widely explored as it is quite simple and quick. The synthesis involves the precipitation of iron from the iron salts (ferric and ferrous) using an alkaline solution at a higher pH of around ten. The reaction mixture is further stirred at a higher temperature (80°C–90°C) under the nitrogen or argon atmosphere for 30 min. The addition of alkaline or ammonia solution while stirring the solution forms a black precipitate. After completion of the reaction, the mixture is cooled down. The formed black precipitate is then separated using an external magnet and washed with deionized water, followed by ethanol and 0.1M HCL unless the solution becomes neutral. The black precipitate is then kept for vacuum drying and stored in an airtight container for further use.

The synthesized magnetic nanoparticles (Fe_3O_4 or Fe_2O_3) possess amphoteric hydroxyl groups (OH) on the surface. The occurrence of these OH groups helps either in the direct adsorption of enzymes on MNPs or the surface can be modified using an amino functional group (Figure 7.5). The amino-functionalization of MNPs

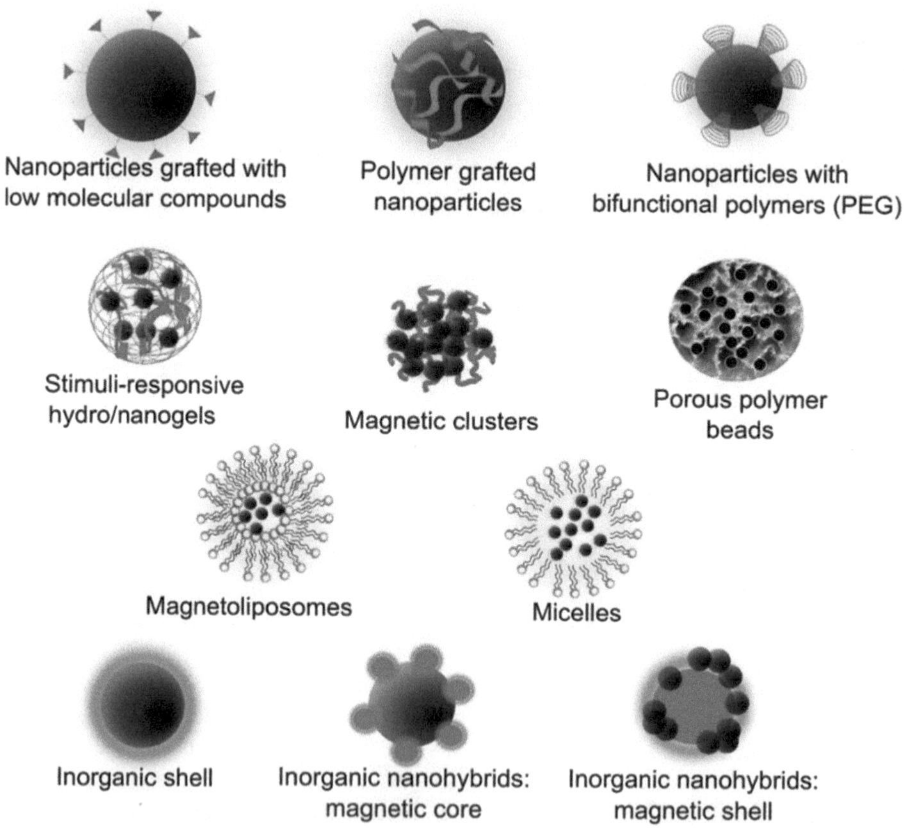

FIGURE 7.5 Represents the arrangement of magnetic nanoparticles using polymers, chemical molecules, and different inorganic nanoparticles.

(Adapted from Hola et al., 2015.)

helps in the covalent cross-linking of enzymes on MNPs using a suitable cross-linker. A variety of proteins and enzymes, such as bovine serum albumin (BSA), cellulases, lipases, streptokinase (Brady and Jordaan, 2009; Fernandez-Lafuente, 2009), glucose oxidase, trypsin, chymotrypsin, etc. using various coupling or cross-linking agents.

From various magnetic nanoparticles, magnetite (Fe_3O_4) is widely used due to its strong magnetic properties and quicker attraction to the magnetic field (Hola et al., 2015). The silica paramagnetic nanoparticles impregnated with nickel have also been explored for enzyme immobilization (Prakasham et al., 2007). The enzyme-immobilized silica paramagnetic nanoparticles possess excellent stability and activity at maximum temperature and pH conditions. This is attributed to an alteration in the enzyme's configuration during immobilization.

In the case of organic nanomaterials, the nanoparticles are synthesized from natural or renewable sources. Some widely explored or used organic polysaccharides are chitosan, cellulose, pectin, etc. Chitosan, a biopolymer, is extracted from prawns or shrimp shells (Ladole et al., 2021; Zhi et al., 2006). These nanoparticles are widely explored for enzyme immobilization applications owing to their harmless bioactivity, antimicrobial and antifungal activity, and outstanding chemical and physical activity (Xie and Wang, 2012). The low cost of chitosan makes it a favorable organic support for various commercial or industrial applications. The chitosan can also be used as a coating agent for the amino-functionalization of MNPs (Ghadi et al., 2015; Ladole et al., 2021).

The applications of chitosan for amino-functionalization of MNPs have been widely studied (Kamel Ariffin and Idris, 2022; Ladole et al., 2021). The transesterification of soybean oil into biodiesel was carried out using lipase immobilized on core-shell magnetic nanoparticles coated with chitosan as the immobilization support. There was an improvement in the activity of immobilized lipase compared to the free one. The improvement in the catalytic activity of enzymes after immobilization was attributed to the hydrophilic and polymeric properties of chitosan (Ghadi et al., 2015).

In another study, magnetic nanoparticles coated with chitosan were used to effectively convert soybean oil into biodiesel. Under ideal circumstances, with a methanol/oil molar ratio of 4:1 and a three-step injection of methanol at 35°C for 30 hours, this immobilized lipase achieved a FAME conversion rate of 87%. The immobilized lipases did not lose their catalytic activity and showed excellent reusability up to four times (Xie and Wang, 2012).

A one-step process has been developed for the co-immobilization of two lipases (extracted from *Aspergillus oryzae* (AOL) and *Rhizomucor miehei* (RML)). These lipases were rapidly encapsulated in chitosan-magnetic microcapsules (Wei et al., 2023). The immobilization was done via a self-assembly approach in which citrate-modified MNPs carrying negative charge cross-linked with lipase-containing chitosan-citrate aggregates. The prepared co-immobilized lipases were used to synthesize biodiesel from the waste cooking oil. When the lipases were bio-imprinted with tween-80, followed by co-immobilization, the formed biocatalyst showed an enhanced activity at optimized conditions and biodiesel yield of up to 98.5%. This yield was almost 1.3 and 1.6 times higher compared to biodiesel synthesis using a single lipase AOL and RML, respectively. The co-immobilized lipase showed excellent reusability and maintained more than 96% activity after three cycles of reuse for biodiesel production (Wei et al., 2023).

Similarly, another polysaccharide, cellulose, has been explored to prepare Fe_3O_{4}-cellulose nanocomposite for lipase immobilization. Multipoint covalent sites such as epoxy (C-O-C) and carboxyl (C=O) have been revealed after characterizing the prepared Fe_3O_4-cellulose nanocomposite. The occurrence of these multicovalent binding sites was helpful in forming a more stable, robust, and immobile structure between lipase and Fe_3O_4-cellulose nanocomposite. It showed high thermal stability by retaining over 70% residual activity when used for 30 min at 80°C. The prepared lipase-immobilized nanocomposite was used for the transesterification of waste cooking oil to biodiesel. The reusability study revealed that the prepared lipase nanocomposite can be reused and only showed a slight decline in activity when reused up to four times (Najjar et al., 2022; Shalini et al., 2023).

In another report, magnetic nanoparticles were aminofunctionalized using 3-aminopropyltriethoxysilane (APTES) as a functionalizing agent. The prepared AMNPs were used for biodiesel production by transesterification of fungal lipids extracted from *Aspergillus niger* (Jambulingam et al., 2019). Both free and immobilized lipases were evaluated to check the esterase activity at different reaction conditions, and results indicated improved catalytic activity of immobilized lipase compared to a free one. Optimization of several parameters was carried out for FAME production. Under optimized conditions, the FAME yield was achieved up to 84%. The reusability and stability of immobilized lipase were performed, and it was found that the nano-biocatalyst could be used for up to five cycles of reuse. The storage stability study was also conducted, and it was found that there was no loss of immobilized lipase activity for 60 days.

On a similar line of work, lipases produced from gamma-irradiated *Aspergillus niger* were immobilized on APTES-coated magnetic nanoparticles (APTES-Fe_3O_4) via a covalent cross-linking technique. The functionalization of MNPs was confirmed using different characterization techniques such as FTIR, XRD, and SEM. The optimization of pH and temperature for both free and immobilized lipase was carried out. The lipase-immobilized APTES-Fe_3O_4 nanoparticles were used for the transesterification of lipids extracted from *Chlorella pyrenoidosa.* Under all optimized conditions, such as a 1:4 M ratio of oil to methanol, 300 mg of APTES-Fe_3O_4-lipase loading, and a temperature of 45°C for 6 h of reaction, the biodiesel yield of around 90% was obtained. The immobilized lipase loses 15% of its initial activity after five cycles of reuse without any treatment (Shalini et al., 2023).

The enzyme co-immobilization approach can also be utilized for the transesterification of waste oil to biodiesel. To improve the biocatalytic yield, lipases with different advantages and specificities can be immobilized together to form a highly active and specific co-immobilized biocatalyst. To achieve this, lipases from two different sources such as *Thermomyces lanuginosus* (1,3 specificity) and *Burkholderia cepacian* (nonspecific) were co-immobilized on magnetic nanoparticles (Wang et al., 2023). The iron oxide nanoparticles (Fe_3O_4) were initially functionalized with 3-glycidyloxypropyltrimethoxysilane (3-GPTMS), followed by lipase co-immobilization. The co-immobilization process was optimized using response surface methodology (RSM). Compared to a single immobilized lipase system, the prepared co-immobilized lipases have shown an excellent enhancement in the reaction rate and biocatalytic activity with almost 93% product yield at optimized reaction conditions after 6 h. This yield was almost 30% higher than using only a single immobilized lipase. Moreover, when the co-immobilized lipases were used to carry out transesterification reactions with different feedstocks, the biodiesel yield was in the range of 90%–98%. This higher yield demonstrated the combined effect of both lipases when co-immobilized together on the magnetic nanoparticles. The reusability of co-immobilized lipases can be reused for several cycles. It can retain more than 77% of initial activity even after the ninth cycle of reuse after removing glycerol and methanol from the biocatalyst surface using t-butanol wash (Wang et al., 2023).

A multicomponent immobilization method was utilized to immobilize lipases extracted from different microbes. The immobilization was done using cyclohexyl isocyanide as a cross-linking agent on the aminofunctionalized magnetic nanoparticles. This newly developed method is very fast and can be performed in a single step under milder reaction conditions at pH 7.0, a temperature of 25°C, and water as a solvent (Figure 7.6). When immobilization was performed for 2 h, the average immobilization yield of different lipases was found to be around 99%. The prepared nano-biocatalyst was further utilized for biodiesel production via transesterification of waste cooking oil

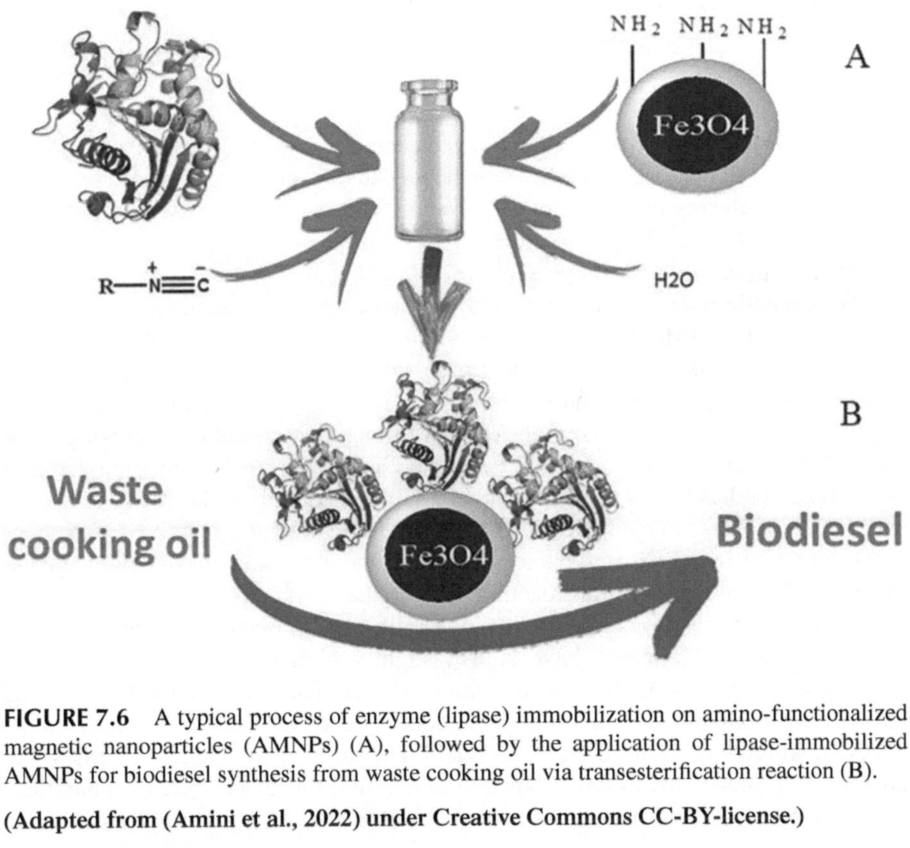

FIGURE 7.6 A typical process of enzyme (lipase) immobilization on amino-functionalized magnetic nanoparticles (AMNPs) (A), followed by the application of lipase-immobilized AMNPs for biodiesel synthesis from waste cooking oil via transesterification reaction (B).

(Adapted from (Amini et al., 2022) under Creative Commons CC-BY-license.)

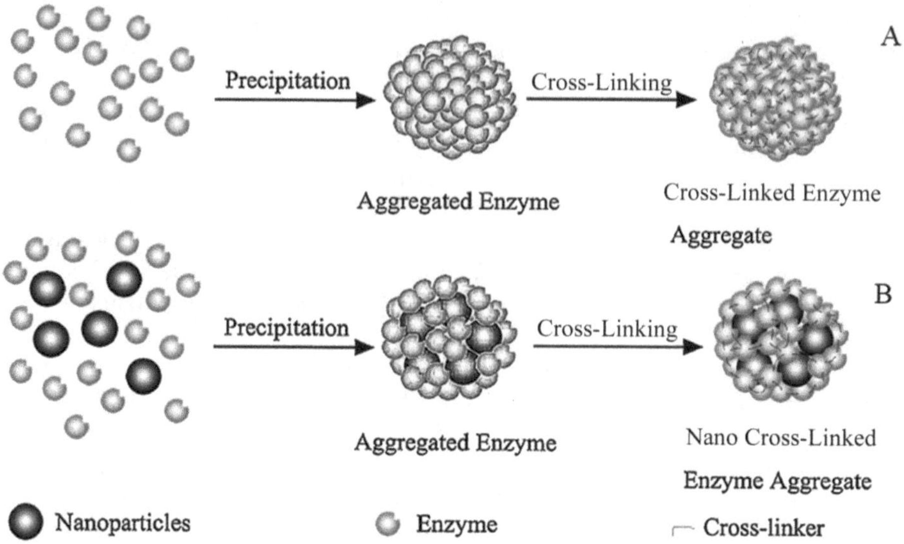

FIGURE 7.7 Schematic representation for cross-linked enzyme aggregate preparation of enzymes (A) followed by preparation of magnetic cross-linked enzyme aggregate preparation using magnetic nanoparticles (B).

(Adapted from Wang et al., 2017.)

using the RSM approach. Under all optimized conditions, an average yield of 78% was achieved using all lipases immobilized on the prepared magnetic support. When reused multiple times, the immobilized lipases showed a decrease in biocatalytic activity after the fifth cycle of reuse. The immobilized lipases were washed with n-hexane and dried before being used for the next cycle of reuse (Amini et al., 2022).

The magnetic nanoparticles can also be used to prepare magnetic cross-linked enzyme aggregates (M-CLEAs) of lipase (Talekar et al., 2012; Wang et al., 2017). A comparison was also made between lipase immobilized on aminofunctionalized magnetic nanoparticles (AMNPs) and magnetic CLEAs (Sampaio et al., 2022). For this, lipase was immobilized on AMNPs through a covalent cross-linking technique. For the preparation of M-CLEAs, the enzymes were first precipitated and cross-linked with AMNPs to form magnetic CLEAs (López et al., 2014). A typical preparation of lipase magnetic CLEAs is shown in Figure 7.7.

The immobilized lipase prepared using both these methods has been evaluated for olive oil transesterification reaction. The transesterification was carried out at 40°C and solvent (2 propanol) to oil molar ratio of 6:1. Results indicated that the magnetic CLEAs of lipases have shown slightly higher transesterification than lipase immobilized on AMNPs. Regarding biodiesel yield, the M-CLEAs of lipases have shown higher yields of up to 30% compared to the lipase immobilized on AMNPs, which have a more than 20% yield. Due to their magnetic properties, the prepared nano-biocatalysts have shown excellent reusability and ease of separation. However, biodiesel conversion has been reduced to over 50% after ten reuse cycles

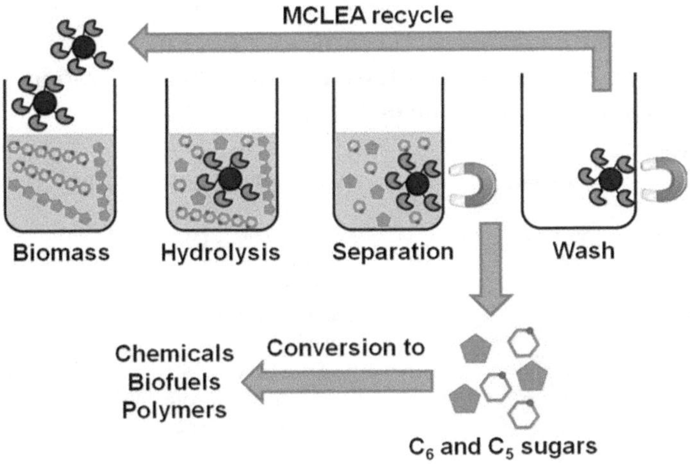

FIGURE 7.8 Schematic of application of magnetic cross-linked enzyme aggregates (M-CLEAs) in biomass hydrolysis for the conversion of biomass into biofuels.

(Adapted from Lucena et al., 2019.)

(López et al., 2014). The M-CLEAs have been explored for different applications, including synthesizing chemicals, biofuels, and polymers. Taking advantage of the magnetic property, the M-CLEAs can be easily separated from the reaction medium using an external magnet, washed, and reused to carry the next reaction. The application of M-CLEAs for biomass hydrolysis and its reusability is illustrated in Figure 7.8.

Table 7.3 Summarizes different magnetic nanosupports for lipase immobilization.

Similar studies were also conducted in which cellulase was immobilized on magnetic nanosupports and used for cellulose hydrolysis and other applications (Lucena et al., 2019). The study indicated high enzyme loading, higher hydrolytic yield, and easier separation from the reaction mixture. Table 7.4 includes some different supports for cellulase immobilization on magnetic nanoparticles.

7.3.6 Metal-organic Frameworks

The advantages of magnetic nanoparticles for enzyme immobilization for providing excellent stability, high-loading capacity, ease of separation, and reusability make it an ideal support. The magnetic nanoparticles have been explored for a wide range of applications, and almost all enzymes are immobilized on this support. However, there are certain limitations that are observed while using MNPs as the support for enzyme immobilization. The formation of aggregates is one of the issues that must be addressed. The MNPs tend to form aggregates when a magnetic field is applied, if the size of MNPs is above 50 nm, the chance of aggregate formation is relatively higher. This leads to the degradation of enzyme activity as the active sites of enzymes gets hindered and there is less mass transfer between the enzymes and substrate, which gives lower yield. Therefore, it is recommended to use supermagnetic nanoparticles with size less than 30 nm, which does not retain the magnetic property once the

TABLE 7.3
Different Magnetic Nanosupports Used for Lipase Immobilization and Application in Biofuel Production

Immobilization Nanosupport	Substrate	Solvent	Reaction Conditions	Yield (%)	Ref.
Chitosan-magnetic core-shell nanoparticles	Soybean oil	Methanol	Reaction temp. 37°C; 0.9% immob. lipase; Oil/MeOH molar ratio 1:3; reaction time 72 h	18.66	Zhi et al. (2006)
Magnetic cross-linked enzyme aggregates (M-CLEAs)	Olive oil	2-propanol	Reaction temp. 40°C; 1 wt% immob. lipase; Oil/2-Propanol molar ratio 1:6	30.3	Cruz-Izquierdo et al. (2014)
APTES-coated Fe_3O_4 nanoparticles	Fungal lipid	Methanol	Reaction temp. 45°C; 6 wt% immob. lipase; Oil/MeOH molar ratio 14; reaction time 4 h	85.3	Jambulingam et al. (2019)
Magnetic (Fe_3O_4) core coated with silica shell treated with [3-(trimethoxysilyl) propyl] ammonium chloride	Olive oil	Methanol	Reaction temp. 40°C; 11 wt% immob. lipase; Oil/MeOH molar ratio 1:4, reaction time 30 h	90	Tran et al. (2012)
Magnetic (Fe_3O_4) core coated with silica shell functionalized with (3-glycidoxypropyl) trimethoxy silane)	Waste cooking oil	Methanol	Reaction temp. 50°C; 4.5 wt% immb. lipase; 70% molecular sieve; Oil/MeOH molar ratio 1:3; reaction time 96 h	96	Sen et al. (2010)
Magnetic nanoparticle ($MgFe_2O$@CaO) catalyst coated by CaO	Soyabean oil	Methanol	Reaction temp. 70°C, immob. lipase 1 wt.%, Oil to methanol ratio 1:12, reaction time 3 h	98.3	Liu et al., (2016)

magnetic field is removed. Also, while cross-linking enzymes on AMNPs, enzymes' active sites may get involved, which decreases enzyme activity recovery. To overcome this, immobilization of enzymes in porous support material having affinity of the porous materials toward enzyme. One such material, metal-organic frameworks (MOFs), has been explored and widely studied recently (Giraldo et al., 2023; Ladole et al., 2020; Martínez Gil et al., 2022; Shomal et al., 2022). The MOFs are made up of a 3D-shaped matrix that is formed by an organic-inorganic complex. This enables the creation of a porous structure and high surface area to accommodate more enzymes while retaining greater stability and physicochemical properties (Zou et al., 2023).

TABLE 7.4
Cellulase Immobilization on Different Magnetic Nanoparticles and Applications in Biomass Hydrolysis for Biofuel Production

Enzyme Source	Immobilization Support	Percentage Immobilization (%)	Percentage Reusability (%)	Applications	Reference
Trichoderma reesei	Fe_3O_4.Chitosan NP	72.4	80	Production of glucose	Abigail et al. (2015)
Commercial cellulase	MNP-PMMA NP	49	31	Cellulose Hydrolysis	Needhidasan et al. (2014)
Aspergillus niger	Magnetic HNTs	Not given	68.2	Cellulose hydrolysis	Sillu and Agnihotri (2020)
Aspergillus niger	Functionalized MWCNTs	85	75	Cellulose hydrolysis	Ahmad and Khare (2018)
Commercial cellulase	Fe_3O_4/GO/CS	90	80	Bioethanol Production	Li et al. (2015)
Commercial cellulase	Magnetic CLEAs (M-CLEAs)	176 mg/g	74	Conversion of lignocellulosic Biomass	Jia et al. (2017)
Commercial cellulase	Magnetic nanoparticles coated with silica	0.43 g/g	20% (after 10 cycles)	Cellulose hydrolysis	Roth et al. (2016)
Trichoderma viride	GO@ Fe_3O_4 @4arm-PEG-NH_2	132 mg/g	45	Cellulose hydrolysis	Han et al. (2018)
Commercial cellulase	Mt K-10/Fe_3O_4	~861 mg/g	Not given	Lignocellulose degradation	Wu et al. (2019)
Commercial cellulase	Cellulose@ Fe_3O_4@CS	Not given	50%	Glucose Production	Tan et al. (2018)

The biodiesel was also produced from African Palm Oil (APO) by transesterification reaction and lipase immobilized on MOFs. For this, ZIF-8 and MOF-199 were used as immobilization support (Giraldo et al., 2023). The maximum adsorption capacity of nearly 30 mg/g of lipase was achieved at 45°C using ZIF-8 as the support. The lipase adsorption equilibrium study indicated that lipase was physically adsorbed on the surface of ZIF-8; however, chemical adsorption occurred between MOF-199 and lipase. Both ZIF-8 and MOF-199 immobilized lipases showed high efficacy for the transesterification reaction for biodiesel production. Methanol was used as the solvent. A conversion rate of 95% was observed with lipase immobilized on ZIF-8; however, the biodiesel yield was 90% with lipase immobilized on MOF-199. Due to the chemical bonding on lipases with MOF-199, it showed better reusability than ZIF-8-lipase (Giraldo et al., 2023).

The same research group also conducted a similar study in which the ZIF-8 and MOF-199, were modified with furfuryl alcohol (FA) to immobilize lipases efficiently.

These modified lipases were further used for the transesterification of APO. The FA-modified MOF-199 showed maximum immobilization of 80.3% with a specific activity of 125 (%) U/g. On the other hand, the results showed that the enzyme loading capacity of FA-modified ZIF-8 was around 91%, with a specific activity of 142.5 U/g of protein. The biodiesel from palm oil's transesterification reaction was greater than 90% using methanol, and both the immobilized lipases compared to only 70% using free lipases (Martínez Gil et al., 2022).

The UiO-66-NH-MOF functionalized with NH_2 has been used for cellulase immobilization (Ahmed et al., 2018). The prepared cellulase-immobilized MOFs with high stability and reusability were used for cellulose hydrolysis. Due to the larger surface area and high porosity, the prepared MOFs showed excellent loading capacity and accommodated around 350 mg/g of cellulase due to the availability of free NH_2 groups. Moreover, the outer layer of MOFs and extra NH_2 holding sites protect cellulase against a wide range of pH and make it thermally stable. The cellulase at optimal pH 5 and temperature of 55°C showed the highest activity, around 85%. The reusability study revealed retention of almost 72% of activity, even after ten cycles of reuse. When stored at 4°C for 30 days, the cellulase MOFs retain around 65% activity (Ahmed et al., 2018).

Like various nonmagnetic immobilization supports, MOFs can be reused from the reaction mixture by filtration or centrifugation. This makes MOF separation quite tedious and causes a loss in enzyme activity during separation. To address this limitation, magnetic nanoparticles can be incorporated into MOFs to create magnetic metal-organic frameworks (MMOFs; Ladole et al., 2020). The presence of a magnetic core shell in MMOFs makes the separation relatively easier with an external magnetic field. The preparation of MMOFs is similar to that of MOFs, with one additional step involved during its synthesis. In MMOFs preparation, the nanoparticles are modified to organic frameworks, keeping the magnetic nanoparticles at the center surrounded by 3D frameworks.

The cellulase inhibitors such as vanillin and formic acid, which are commonly found in lignocellulosic biomass, can be tolerated by immobilizing cellulase in magnetic UiO-66-NH magnetic metal-organic frameworks (UiO-66-NH-MMOFs). The prepared magnetic nano-biocatalyst enhances the cellulose hydrolytic yield by almost 19%. It also helps to protect cellulase over a wide range of pH and temperature conditions and is thermotolerant. The UiO-66-NH-MMOFs can be recycled several times without affecting cellulase activity (Qi et al., 2018).

MOFs of different materials, such as UIO-66-NH_2, PCN-250, ZIF-8, and 2MIL-100-Fe were also explored for cellulase immobilization. The cellulase enzyme was adsorbed inside the prepared MOFs. These MOFs were prepared and activated using the solvent replacement method, after which they were dried in a vacuum oven. The enzyme-loading capacity of each MOFs were evaluated, and it was found that ZIF-8 showed the highest enzyme loading (176.16 mg/g), followed by UIO-66-NH_2 (146.8 mg/g), MIL-100-Fe (136.45), and PCN-250 (119.88 mg/g), respectively.

The presence of 1-ethyl-3-methylimida-zolium diethylphosphate in ZIF-8-MOFs enhances cellulase activity. This is attributed to the affinity of cellulase toward ZIF-8. Moreover, the ZIF-8-cellulase-MOFs exhibited high tolerance to ionic liquids, and cellulase activity was enhanced almost by 112.5%. Regarding hydrolytic efficiency,

TABLE 7.5
Application of Metal and MMOFs for Cellulase Immobilization and Applications in Hydrolysis of Lignocellulosic Biomass for Biofuel Production

MOFs	Enzyme Loading (mg/g)	Residual Activity After Reusability (%)	Usage	References
Fe_3O_4@ UiO-66-NH_2	~126	70	Hydrolysis of cellulosic biomass	Qi et al. (2018)
UiO-66-NH_2	350	85	Hydrolysis of cellulosic biomass	Ahmed et al. (2018)
ZIF-8	~176	~93	Hydrolysis of lignocellulose	Zhou et al. (2021)
Cu-MOF	~163	90	Hydrolysis of cellulosic biomass	Wang et al. (2019)

cellulase-immobilized ZIF-8-MOFs showed enhanced activity by 92% compared to nonimmobilized cellulase (Zhou et al., 2019). The MOFs were also explored for the immobilization of β-glucosidase in Cu(PABA) biocomposite MOFs (Wang et al., 2019). The immobilization support exhibited an excellent loading capacity of around 163 mg/g. The prepared enzyme-embedded MOFs showed outstanding reusability, retaining more than 90% residual activity for more than ten cycles. The prepared *β*-glucoside-MOFs were co-immobilized with cellulase enzyme to form a co-immobilized mixture of cellulase-*β*-glucoside-Cu(PABA) MOFs. This co-immobilized biocatalyst has shown a cellulose hydrolytic yield of around 70% at pH 5. Cellulase, when immobilized alone in the prepared MOFs, showed a higher hydrolyzing yield of 98% compared to free cellulase (Wang et al., 2019). The summary of different MOFs used for biofuel production is given in Table 7.5.

7.4 FUTURE PROSPECTIVES

The increase in industrialization, modern societies, population, and high standard of living has increased high energy demand. Moreover, increased vehicles and transportation to meet the demand for goods and public transport have resulted in elevated fuel consumption. With an increase in fuel demand, a continuous depletion of fossil fuels has been observed. To tackle this and meet the high fuel demand, there is an urgent need to find alternative sources from which fuels can be produced much more efficiently. This can be done by producing biodiesel from waste oil and nonedible oil sources or bioethanol production from the waste biomass. Chemical processes are utilized to convert these waste sources into useful biofuel. However, considering the environmental impact of these methods, more focus was created on the bioprocesses in which the wastes can be converted into fuels via biocatalytic reactions. For the conversion of oil or lipids into biodiesel, lipases are used. The reaction involves the transesterification in the presence of a suitable solvent. Bioethanol production involves the biocatalytic conversion of cellulosic biomass into simple sugars via saccharification, followed by fermentation of simple sugars to bioethanol. The continuous evaluation and modification of biocatalysts helped to perform the biocatalytic process for various applications much more efficiently. The reactions can also be performed in aqueous solutions or organic media. The emergence and continuous development of immobilization methods from conventional methods such as adsorption, entrapment to recent methods such as cross-linking, covalent cross-linking on nanosupports helped to explore enzymes for a variety of applications. Also, with the emergence of newer fields of biotechnology, such as nano-biotechnology or nano-biology, nanosupports can be used to create nano-biocatalyst and explored in synthesis applications. The kinetic and biochemical properties provided by the immobilization supports helped the enzymes to achieve excellent stability, shelf-life, ease of separation from the reaction and reusability for multiple times.

One of the main advantages of using immobilized enzymes is their stability under different reaction conditions and, more importantly, their reusability for several cycles. The reusability of immobilized enzymes helps in lowering biofuel production costs. As industries always look for economical production processes, immobilized enzymes help compensate for the production cost of different products. Moreover, the immobilized enzymes are easy to handle and can be separated easily from the reaction mixture, making the product separation much easier. The selection of appropriate immobilization support is essential as the properties of immobilized enzymes greatly depend on the carrier or support used and the properties of enzymes that need to be immobilized.

Despite the advantages of immobilized enzymes, the conventional immobilization methods have some limitations, such as steric hindrance, enzyme leaching, limited reusability, difficulty in enzyme recycling, changing the enzyme configuration, and difficulty in mass transfer. Along with this, for the application of biofuel production, the limitations of immobilized enzymes include enzyme inactivation in the solvent media, high costs associated with the enzyme purchase, and scale-up of the biocatalytic reaction. These limitations can be overcome to a certain extent with the help of more efficient immobilization support, such as nanomaterials. The use of nanosupports can certainly provide an additional benefit to the enzyme immobilization process. Some of the key advantages of using nanomaterials, such as high-loading capacity, greater surface area, high activity recovery, and high-mass transfer, make it an ideal support for enzyme immobilization. Apart from this, it is critical to study and understand the interaction between the enzymes and the surface of nanomaterials, which helps to form a much more efficient enzyme-immobilized assembly. Some of the limitations associated with nonmagnetic nanoparticles, such as difficulty in separation, limited reusability, and lower yield, can be overcome by replacing them with MNPs. The magnetic property helps to quickly separate enzymes from the reaction mixture with the help of an external magnet field. The problem of aggregate formation associated with MNPs can also be overcome by synthesizing supermagnetic nanoparticles that get easily dispersed in the reaction mixture after removing the magnetic field without forming any aggregates.

Along with MNPs, research was also conducted in developing MOFs with additional benefits such as high enzyme loading, high activity retention, increased enzyme activity, protection from the external environment, and high reusability.

The magnetic property can also be introduced during MOFs preparation to form the MMOFs. This enables quick separation of MMOFs compared to MOFs that often require filtration or centrifugation. Irrespective of all these benefits and advances of using immobilized enzymes on MNPs, MOFs, and MMOFs, some limitations need to be addressed, which are given as follows.

1. The initial cost of enzymes: This is one of the most crucial factors to be considered for biocatalytic synthesis of biofuels. The initial cost of enzymes plays an important role in the overall product cost. Enzymes are quite expensive due to the high cost associated with the downstream process and low product yield. To overcome this, identifying or modifying the microbial strain that can produce the enzyme of interest in bulk is very important. It is also important to develop a downstream process with fewer enzyme purification stages. By doing this, the initial cost of enzymes can be reduced, and hence, more industries are attracted toward biofuel production using a biocatalytic process.
2. Process scale-up: Although the enzymatic processes, more specifically using free enzymes as the catalyst, are easy to scale up, at the same time, there is a limitation or difficulty in scaling up the process using immobilized enzymes. This is because the quantity of enzymes required during the biocatalytic process will be much higher in treating lignocellulosic biomass. Packed bed reactors can be utilized to use the immobilized enzyme at a larger scale to achieve significant product yield. Biofuel production using a stirrer tank reactor will be challenging as the agitation and mixing process will damage the immobilized enzyme, and the chances of enzyme leaching will be much higher. Also, the substrate used should be in the liquid phase to get an efficient product yield using immobilized enzymes. If the substrate used for the process is solid, it is challenging for the reaction to proceed, as both the substrate and enzymes will be in the same phase. This results in mass transfer limitations and less enzyme interaction with the substrate, hence a lesser product yield.
3. Development of newer supports for immobilization: The continuous development and synthesis of different carrier supports for enzyme immobilization helps to overcome the limitations of conventional immobilization supports. From simple adsorption techniques to nanosupports, these supports are explored for various applications, including biofuel production. Although recent developments and novelty in different supports have addressed most of the limitations associated with immobilized enzymes, it is necessary to develop an efficient immobilization support such as covalent organic frameworks. The continuous development in enzyme immobilization will help to develop an efficient support with high loading, greater immobilization efficiency, stability over a wide range of pH, temperature, and different solvent conditions, and, most importantly, multi-time reusability. It may also be possible to use the newly developed support with ease in scale-up for biofuel production.

7.5 CONCLUSION

In conclusion, the surge in energy demand driven by industrialization, urbanization, and modern lifestyles has intensified the consumption of fossil fuels, leading to their depletion. To mitigate this, the focus has shifted toward alternative fuel sources, particularly biofuels derived from waste materials through biocatalytic processes. Immobilized enzymes have emerged as critical tools in this area, offering stability, reusability, and ease of handling. Despite several advantages, some challenges persist, including enzyme cost, scale-up issues, and the need for innovative immobilization supports. However, ongoing research into microbial strain modification, process scaling techniques, and the development of novel supports promises to overcome these obstacles, flooring the way for a more sustainable and economically viable biofuel production process.

REFERENCES

Abbaszaadeh, A., Ghobadian, B., Omidkhah, M.R., Najafi, G., 2012. Current biodiesel production technologies: A comparative review. *Energy Conversion and Management*, 138–148. https://doi.org/10.1016/j.enconman.2012.02.027

Abdulla, R., Ravindra, P., 2013. Immobilized Burkholderia cepacia lipase for biodiesel production from crude Jatropha curcas L. oil. *Biomass and Bioenergy* 56, 8–13. https://doi.org/10.1016/j.biombioe.2013.04.010

Abdulmalek, S.A., Yan, Y., 2022. Recent developments of lipase immobilization technology and application of immobilized lipase mixtures for biodiesel production. *Biofuels, Bioproducts and Biorefining*. https://doi.org/10.1002/bbb.2349

Abigail, M.E.A., Samuel, S.M., Ramalingam, C., 2015. Addressing the environmental impacts of butachlor and the available remediation strategies: A systematic review. *International Journal of Environmental Science and Technology* https://doi.org/10.1007/s13762-015-0866-2

Abraham, R.E., Puri, M., 2020. Nano-immobilized cellulases for biomass processing with application in biofuel production. *Methods in Enzymology* (pp. 327–346). Academic Press Inc. https://doi.org/10.1016/bs.mie.2019.09.006

Abraham, R.E., Verma, M.L., Barrow, C.J., Puri, M., 2014. Suitability of magnetic nanoparticle immobilised cellulases in enhancing enzymatic saccharification of pretreated hemp biomass. *Biotechnology for Biofuels* 7. https://doi.org/10.1186/1754-6834-7-90

Adlercreutz, P., 2013. Immobilisation and application of lipases in organic media. *Chemical Society Reviews* 42, 6406–6436. https://doi.org/10.1039/c3cs35446f

Aghabeigi, F., Nikkhah, H., Zilouei, H., Bazarganipour, M., 2023. Immobilization of lipase on the graphene oxides magnetized with NiFe2O4 nanoparticles for biodiesel production from microalgae lipids. *Process Biochemistry* 126, 171–185. https://doi.org/10.1016/j.procbio.2023.01.012

Ahmad, R., Khare, S.K., 2018. Immobilization of *Aspergillus niger* cellulase on multiwall carbon nanotubes for cellulose hydrolysis. *Bioresource Technology* 252, 72–75. https://doi.org/10.1016/j.biortech.2017.12.082

Ahmed, I.N., Yang, X.L., Dubale, A.A., Li, R.F., Ma, Y.M., Wang, L.M., Hou, G.H., Guan, R.F., Xie, M.H., 2018. Hydrolysis of cellulose using cellulase physically immobilized on highly stable zirconium based metal-organic frameworks. *Bioresource Technology* 270, 377–382. https://doi.org/10.1016/j.biortech.2018.09.077

Amini, Y., Shahedi, M., Habibi, Z., Yousefi, M., Ashjari, M., Mohammadi, M., 2022. A multicomponent reaction for covalent immobilization of lipases on amine-functionalized magnetic nanoparticles: Production of biodiesel from waste cooking oil. *Bioresour Bioprocess* 9. https://doi.org/10.1186/s40643-022-00552-0

Ansari, S.A., Husain, Q., 2012. Potential applications of enzymes immobilized on/in nano materials: A review. *Biotechnology Advances* https://doi.org/10.1016/j.biotechadv.2011.09.005

Azócar, L., Ciudad, G., Heipieper, H.J., Navia, R., 2010. Biotechnological processes for biodiesel production using alternative oils. *Applied Microbiology and Biotechnology* https://doi.org/10.1007/s00253-010-2804-z

Babaki, M., Yousefi, M., Habibi, Z., Mohammadi, M., Yousefi, P., Mohammadi, J., Brask, J., 2016. Enzymatic production of biodiesel using lipases immobilized on silica nanoparticles as highly reusable biocatalysts: Effect of water, t-butanol and blue silica gel contents. *Renewable Energy* 91, 196–206. https://doi.org/10.1016/j.renene.2016.01.053

Bencze, L.C., Bartha-Vári, J.H., Katona, G., Toşa, M.I., Paizs, C., Irimie, F.D., 2016. Nanobioconjugates of Candida antarctica lipase B and single-walled carbon nanotubes in biodiesel production. *Bioresource Technology* 200, 853–860. https://doi.org/10.1016/j.biortech.2015.10.072

Bosio, V.E., Islan, G.A., Martínez, Y.N., Durán, N., Castro, G.R., 2016. Nanodevices for the immobilization of therapeutic enzymes. *Critical Reviews in Biotechnology*. https://doi.org/10.3109/07388551.2014.990414

Brady, D., Jordaan, J., 2009. Advances in enzyme immobilisation. *Biotechnology Letters*. https://doi.org/10.1007/s10529-009-0076-4

Calero, J., Luna, D., Sancho, E.D., Luna, C., Bautista, F.M., Romero, A.A., Posadillo, A., Berbel, J., Verdugo-Escamilla, C., 2015. An overview on glycerol-free processes for the production of renewable liquid biofuels, applicable in diesel engines. *Renewable and Sustainable Energy Reviews*. https://doi.org/10.1016/j.rser.2014.11.007

Chen, Y.Z., Ching, C.B., Xu, R., 2009. Lipase immobilization on modified zirconia nanoparticles: Studies on the effects of modifiers. *Process Biochemistry* 44, 1245–1251. https://doi.org/10.1016/j.procbio.2009.06.023

Cipolatti, E.P., Silva, M.J.A., Klein, M., Feddern, V., Feltes, M.M.C., Oliveira, J.V., Ninow, J.L., De Oliveira, D., 2014. Current status and trends in enzymatic nanoimmobilization. *Journal of Molecular Catalysis B: Enzymatic*. https://doi.org/10.1016/j.molcatb.2013.10.019

Costantini, A., Venezia, V., Pota, G., Bifulco, A., Califano, V., Sannino, F., 2020. Adsorption of cellulase on wrinkled silica nanoparticles with enhanced inter-wrinkle distance. *Nanomaterials* 10, 1–14. https://doi.org/10.3390/nano10091799

Cruz-Izquierdo, Á., Picó, E.A., López, C., Serra, J.L., Llama, M.J., 2014. Magnetic cross-linked enzyme aggregates (mCLEAs) of Candida antarctica lipase: An efficient and stable biocatalyst for biodiesel synthesis. *PLoS One* 9. https://doi.org/10.1371/journal.pone.0115202

Datta, S., Christena, L.R., Rajaram, Y.R.S., 2013. Enzyme immobilization: An overview on techniques and support materials. *3 Biotech* 3, 1–9. https://doi.org/10.1007/s13205-012-0071-7

de Jesús Rostro-Alanis, M., Mancera-Andrade, E.I., Patiño, M.B.G., Arrieta-Baez, D., Cardenas, B., Martinez-Chapa, S.O., Saldívar, R.P., 2016. Nanobiocatalysis: Nanostructured materials – A minireview. *Biocatalysis* 2, 1–24. https://doi.org/10.1515/boca-2016-0001

De Paola, M.G., Ricca, E., Calabrò, V., Curcio, S., Iorio, G., 2009. Factor analysis of transesterification reaction of waste oil for biodiesel production. *Bioresource Technology* 100, 5126–5131. https://doi.org/10.1016/j.biortech.2009.05.027

Deshpande, A., D'souza, S.F., Nadkarni, G.B., 1987. Coimmobilization of D-amino acid oxidase and catalase by entrapment of Trigonopsis variabilis in radiation polymerised Polyacrylamide beads. *Journal of Biosciences*, 11, 137–144.

Dizge, N., Keskinler, B., 2008. Enzymatic production of biodiesel from canola oil using immobilized lipase. *Biomass and Bioenergy* 32, 1274–1278. https://doi.org/10.1016/j.biombioe.2008.03.005

Dräger, G., Kiss, C., Kunz, U., Kirschning, A., 2007. Enzyme-purification and catalytic transformations in a microstructured PASSflow reactor using a new tyrosine-based Ni-NTA linker system attached to a polyvinylpyrrolidinone-based matrix. *Organic & Biomolecular Chemistry* 5, 3657–3664. https://doi.org/10.1039/b712804e

Drechsler, U., Fischer, N.O., Frankamp, B.L., Rotello, V.M., 2004. Highly efficient biocatalysts via covalent immobilization of Candida rugosa lipase on ethylene glycol-modified gold-silica nanocomposites. *Advanced Materials* 16, 271–274. https://doi.org/10.1002/adma.200306022

Du, W., Xu, Y.Y., Liu, D.H., Li, Z.B., 2005. Study on acyl migration in immobilized lipozyme TL-catalyzed transesterification of soybean oil for biodiesel production. *Journal of Molecular Catalysis B: Enzymatic* 37, 68–71. https://doi.org/10.1016/j.molcatb.2005.09.008

Dumri, K., Hung Anh, D., 2014. Immobilization of lipase on silver nanoparticles via adhesive polydopamine for biodiesel production. *Enzyme Research* 2014. https://doi.org/10.1155/2014/389739

Dyal, A., Loos, K., Noto, M., Chang, S.W., Spagnoli, C., Shafi, K.V.P.M., Ulman, A., Cowman, M., Gross, R.A., 2003. Activity of Candida rugosa lipase immobilized on γ-Fe2O3 magnetic nanoparticles. *Journal of the American Chemical Society* 125, 1684–1685. https://doi.org/10.1021/ja021223n

Fan, Y., Wu, G., Su, F., Li, K., Xu, L., Han, X., Yan, Y., 2016. Lipase oriented-immobilized on dendrimer-coated magnetic multi-walled carbon nanotubes toward catalyzing biodiesel production from waste vegetable oil. *Fuel* 178, 172–178. https://doi.org/10.1016/j.fuel.2016.03.071

Fernandez-Lafuente, R., 2009. Stabilization of multimeric enzymes: Strategies to prevent subunit dissociation. *Enzyme and Microbial Technology*. https://doi.org/10.1016/j.enzmictec.2009.08.009

Ghadi, A., Tabandeh, F., Mahjoub, S., Mohsenifar, A., Roshan, F.T., Alavije, R.S., 2015. Fabrication and characterization of core-shell magnetic chitosan nanoparticles as a novel carrier for immobilization of Burkholderia cepacia Lipase. *Journal of Oleo Science* 64, 423–430. https://doi.org/10.5650/jos.ess14236

Gharat, N., Rathod, V.K., 2013. Ultrasound assisted enzyme catalyzed transesterification of waste cooking oil with dimethyl carbonate. *Ultrasonics Sonochemistry* 20, 900–905. https://doi.org/10.1016/j.ultsonch.2012.10.011

Giraldo, L., Gómez-Granados, F., Moreno-Piraján, J.C., 2023. Biodiesel production using palm oil with a MOF-lipase B biocatalyst from candida antarctica: A kinetic and thermodynamic study. *International Journal of Molecular Sciences* 24. https://doi.org/10.3390/ijms241310741

Goswami, D., Basu, J.K., De, S., 2013. Lipase applications in oil hydrolysis with a case study on castor oil: A review. *Critical Reviews in Biotechnology*. https://doi.org/10.3109/07388551.2012.672319

Grosová, Z., Rosenberg, M., Rebroš, M., Šipocz, M., Sedláčková, B., 2008. Entrapment of β-galactosidase in polyvinylalcohol hydrogel. *Biotechnology Letters* 30, 763–767. https://doi.org/10.1007/s10529-007-9606-0

Guisan, J.M., Bolivar, J.M., López-Gallego, F., Rocha-Martín Editors, J., 2020. Correction to: immobilization of enzymes and cells. Immobilization of Enzymes and Cells: Methods and Protocols, C1-C1. https://link.springer.com/protocol/10.1007/978-1-0716-0215-7_33

Gupta, M.N., Kaloti, M., Kapoor, M., Solanki, K., 2011. Nanomaterials as matrices for enzyme immobilization. *Artificial Cells, Blood Substitutes, and Biotechnology* 39, 98–109. https://doi.org/10.3109/10731199.2010.516259

Hama, S., Kondo, A., 2013. Enzymatic biodiesel production: An overview of potential feedstocks and process development. *Bioresource Technology* 135, 386–395. https://doi.org/10.1016/j.biortech.2012.08.014

Han, J., Luo, P., Wang, Y., Wang, L., Li, C., Zhang, W., Dong, J., Ni, L., 2018. The development of nanobiocatalysis via the immobilization of cellulase on composite magnetic nanomaterial for enhanced loading capacity and catalytic activity. *International Journal of Biological Macromolecules* 119, 692–700. https://doi.org/10.1016/j.ijbiomac.2018.07.176

Han, J.Y., Kim, H.K., 2011. Transesterification using the cross-linked enzyme aggregate of Photobacterium lipolyticum Lipase M37. *Journal of Microbiology and Biotechnology* 21, 1159–1165. https://doi.org/10.4014/jmb.1106.06048

Harun, R., Danquah, M.K., Forde, G.M., 2010. Microalgal biomass as a fermentation feedstock for bioethanol production. *Journal of Chemical Technology and Biotechnology* 85, 199–203. https://doi.org/10.1002/jctb.2287

Hernandez, K., Fernandez-Lafuente, R., 2011. Control of protein immobilization: Coupling immobilization and site-directed mutagenesis to improve biocatalyst or biosensor performance. *Enzyme and Microbial Technology* https://doi.org/10.1016/j.enzmictec.2010.10.003

Ho, K.M., Mao, X., Gu, L., Li, P., 2008. Facile route to enzyme immobilization: Core-shell nanoenzyme particles consisting of well-defined poly(methyl methacrylate) cores and cellulase shells. *Langmuir* 24, 11036–11042. https://doi.org/10.1021/la8016529

Ho, S.H., Chen, C.Y., Lee, D.J., Chang, J.S., 2011. Perspectives on microalgal CO_2 emission mitigation systems – A review. *Biotechnology Advances.* https://doi.org/10.1016/j.biotechadv.2010.11.001

Hola, K., Markova, Z., Zoppellaro, G., Tucek, J., Zboril, R., 2015. Tailored functionalization of iron oxide nanoparticles for MRI, drug delivery, magnetic separation and immobilization of biosubstances. *Biotechnology Advances* 33(6), 1162–1176. https://doi.org/10.1016/j.biotechadv.2015.02.003

Hsu, A.F., Jones, K., Marmer, W.N., Foglia, T.A., 2001. Production of alkyl esters from tallow and grease using lipase immobilized in a phyllosilicate sol-gel. *JAOCS: Journal of the American Oil Chemists' Society* 78, 585–588. https://doi.org/10.1007/s11746-001-0308-x

Iyer, P. V., Ananthanarayan, L., 2008. Enzyme stability and stabilization – Aqueous and non-aqueous environment. *Process Biochemistry* 43(10), 1019–1032. https://doi.org/10.1016/j.procbio.2008.06.004

Jambulingam, R., Shalma, M., Shankar, V., 2019. Biodiesel production using lipase immobilised functionalized magnetic nanocatalyst from oleaginous fungal lipid. *Journal of Cleaner Production* 215, 245–258. https://doi.org/10.1016/j.jclepro.2018.12.146

Jia, J., Zhang, W., Yang, Z., Yang, X., Wang, N., Yu, X., 2017. Novel magnetic cross-linked cellulase aggregates with a potential application in lignocellulosic biomass bioconversion. *Molecules* 22. https://doi.org/10.3390/molecules22020269

Johnson, P.A., Park, H.J., Driscoll, A.J., 2011. Enzyme nanoparticle fabrication: Magnetic nanoparticle synthesis and enzyme immobilization. *Methods in Molecular Biology* 679, 183–191. https://doi.org/10.1007/978-1-60761-895-9_15

Johnson, R.D., Wang, Z.-G., Arnold, F.H., 1996. Surface site heterogeneity and lateral interactions in multipoint protein adsorption. *The Journal of Physical Chemistry* 100, 5134–5139.

Kamel Ariffin, M.F., Idris, A., 2022. Fe2O3/Chitosan coated superparamagnetic nanoparticles supporting lipase enzyme from Candida Antarctica for microwave assisted biodiesel production. *Renewable Energy* 185, 1362–1375. https://doi.org/10.1016/j.renene.2021.11.077

Kanwar, S.S., Gehlot, S., Verma, M.L., Gupta, R., Kumar, Y., Chauhan, G.S., 2008. Synthesis of geranyl butyrate with the poly(acrylic acid-co-hydroxy propyl methacrylate-cl-ethylene glycol dimethacrylate) hydrogel immobilized lipase of Pseudomonas aeruginosa MTCC-4713. *Journal of Applied Polymer Science* 110, 2681–2692. https://doi.org/10.1002/app.28241

Katchalski-Katzir, E., 1993. Immobilized enzymes-learning from past successes and failures. *Trends in Biotechnology* 11, 471–478.

Katiyar, M., Ali, A., 2012. Immobilization of Candida rugosa Lipase on MCM-41 for the transesterification of cotton seed oil. *Journal of Oleo Science*. https://doi.org/10.5650/jos.61.469

Kim, J., Grate, J.W., Wang, P., 2006. Nanostructures for enzyme stabilization. *Chemical Engineering Science* 61, 1017–1026. https://doi.org/10.1016/j.ces.2005.05.067

Kim, K.H., Lee, O.K., Lee, E.Y., 2018. Nano-immobilized biocatalysts for biodiesel production from renewable and sustainable resources. *Catalysts* 8(2), 68. https://doi.org/10.3390/catal8020068

Kralova, I., Sjöblom, J., 2010. Biofuels-renewable energy sources: A review. *Journal of Dispersion Science and Technology* 31, 409–425. https://doi.org/10.1080/01932690903119674

Kralovec, J.A., Wang, W., Barrow, C.J., 2010. Production of omega-3 triacylglycerol concentrates using a new food grade immobilized Candida antarctica lipase B. *Australian Journal of Chemistry* 63, 922–928. https://doi.org/10.1071/CH10087

Kuan, I.C., Lee, C.C., Tsai, B.H., Lee, S.L., Lee, W.T., Yu, C.Y., 2013. Optimizing the production of biodiesel using lipase entrapped in biomimetic silica. *Energies (Basel)* 6, 2052–2064. https://doi.org/10.3390/en6042052

Kumari, V., Shah, S., Gupta, M.N., 2007. Preparation of biodiesel by lipase-catalyzed transesterification of high free fatty acid containing oil from Madhuca indica. *Energy and Fuels* 21, 368–372. https://doi.org/10.1021/ef0602168

Ladole, M.R., Mevada, J.S., Pandit, A.B., 2017. Ultrasonic hyperactivation of cellulase immobilized on magnetic nanoparticles. *Bioresource Technology* 239, 117–126. https://doi.org/10.1016/j.biortech.2017.04.096

Ladole, M.R., Nair, R.R., Bhutada, Y.D., Amritkar, V.D., Pandit, A.B., 2018. Synergistic effect of ultrasonication and co-immobilized enzymes on tomato peels for lycopene extraction. *Ultrasonics Sonochemistry* 48, 453–462. https://doi.org/10.1016/j.ultsonch.2018.06.013

Ladole, M.R., Pokale, P.B., Patil, S.S., Belokar, P.G., Pandit, A.B., 2020. Laccase immobilized peroxidase mimicking magnetic metal organic frameworks for industrial dye degradation. *Bioresource Technology* 317. https://doi.org/10.1016/j.biortech.2020.124035

Ladole, M.R., Pokale, P.B., Varude, V.R., Belokar, P.G., Pandit, A.B., 2021. One pot clarification and debittering of grapefruit juice using co-immobilized enzymes@chitosanMNPs. *International Journal of Biological Macromolecules* 167, 1297–1307. https://doi.org/10.1016/j.ijbiomac.2020.11.084

Lee, O.K., Seong, D.H., Lee, C.G., Lee, E.Y., 2015. Sustainable production of liquid biofuels from renewable microalgae biomass. *Journal of Industrial and Engineering Chemistry*. https://doi.org/10.1016/j.jiec.2015.04.016

Li, Y., Wang, X.Y., Jiang, X.P., Ye, J.J., Zhang, Y.W., Zhang, X.Y., 2015. Fabrication of graphene oxide decorated with Fe3O4@SiO2 for immobilization of cellulase. *Journal of Nanoparticle Research* 17. https://doi.org/10.1007/s11051-014-2826-z

Liese, A., Hilterhaus, L., 2013. Evaluation of immobilized enzymes for industrial applications. *Chemical Society Reviews* 42, 6236–6249. https://doi.org/10.1039/c3cs35511j

Liu, Y., Zhang, P., Fan, M., Jiang, P., 2016. Biodiesel production from soybean oil catalyzed by magnetic nanoparticle MgFe2O4@CaO. *Fuel* 164, 314–321. https://doi.org/10.1016/j.fuel.2015.10.008

López, C., Cruz-Izquierdo, Á., Picó, E.A., García-Bárcena, T., Villarroel, N., Llama, M.J., Serra, J.L., 2014. Magnetic biocatalysts and their uses to obtain biodiesel and biosurfactants. *Frontiers in Chemistry* 2. https://doi.org/10.3389/fchem.2014.00072

Lucena, G.N., dos Santos, C.C., Pinto, G.C., da Rocha, C.O., Brandt, J.V., de Paula, A.V., Jafelicci Júnior, M., Marques, R.F.C., 2019. Magnetic cross-linked enzyme aggregates (MCLEAs) applied to biomass conversion. *Journal of Solid State Chemistry* 270, 58–70. https://doi.org/10.1016/j.jssc.2018.11.003

Lupoi, J.S., Smith, E.A., 2011. Evaluation of nanoparticle-immobilized cellulase for improved ethanol yield in simultaneous saccharification and fermentation reactions. *Biotechnology and Bioengineering* 108, 2835–2843. https://doi.org/10.1002/bit.23246

MacArio, A., Verri, F., Diaz, U., Corma, A., Giordano, G., 2013. Pure silica nanoparticles for liposome/lipase system encapsulation: Application in biodiesel production. *Catalysis Today* 204, 148–155. https://doi.org/10.1016/j.cattod.2012.07.014

Maroju, P. A., Ganesan, R., & Dutta, J. R. (2022). Efficient biodiesel production from rice bran oil using magnetite immobilized-recombinant lipase from probiotic Bacillus licheniformis. *Green Chemistry*, 24(22), 8800–8811.

Martínez Gil, J.M., Reyes, R.V., Bastidas-Barranco, M., Giraldo, L., Moreno-Piraján, J.C., 2022. Biodiesel production from transesterification with lipase from Pseudomonas cepacia immobilized on modified structured metal organic materials. *ACS Omega* 7, 41882–41904. https://doi.org/10.1021/acsomega.2c02873

Mateo, C., Fernandes, B., Van Rantwijk, F., Stolz, A., Sheldon, R.A., 2006. Stabilisation of oxygen-labile nitrilases via co-aggregation with poly(ethyleneimine). *Journal of Molecular Catalysis B: Enzymatic* 38, 154–157. https://doi.org/10.1016/j.molcatb.2005.12.007

Mateo, C., Grazu, V., Palomo, J.M., Lopez-Gallego, F., Fernandez-Lafuente, R., Guisan, J.M., 2007. Immobilization of enzymes on heterofunctional epoxy supports. *Nature Protocols* 2, 1022–1033. https://doi.org/10.1038/nprot.2007.133

Mendes, A.A., Giordano, R.C., Giordano, R.D.L.C., De Castro, H.F., 2011. Immobilization and stabilization of microbial lipases by multipoint covalent attachment on aldehyde-resin affinity: Application of the biocatalysts in biodiesel synthesis. *Journal of Molecular Catalysis B: Enzymatic* 68, 109–115. https://doi.org/10.1016/j.molcatb.2010.10.002

Meryam Sardar, R.A., 2015. Enzyme immobilization: An overview on nanoparticles as immobilization matrix. *Biochemistry & Analytical Biochemistry* 04. https://doi.org/10.4172/2161-1009.1000178

Moreno-Pirajàn, J.C., Giraldo, L., 2011. Study of immobilized candida rugosa lipase for biodiesel fuel production from palm oil by flow microcalorimetry. *Arabian Journal of Chemistry* 4, 55–62. https://doi.org/10.1016/j.arabjc.2010.06.019

Munde, P.J., Muley, A.B., Ladole, M.R., Pawar, A. V., Talib, M.I., Parate, V.R., 2017. Optimization of pectinase-assisted and tri-solvent-mediated extraction and recovery of lycopene from waste tomato peels. *3 Biotech* 7. https://doi.org/10.1007/s13205-017-0825-3

Najjar, A.A., Hassan, E.A., Zabermawi, N.M., Almasaudi, S.B., Moulay, M., Harakeh, S., Abd El-Aal, M., 2022. Efficacy of the immobilized Kocuria flava lipase on Fe3O4/cellulose nanocomposite for biodiesel production from cooking oil wastes. *Catalysts* 12. https://doi.org/10.3390/catal12090977

Needhidasan, S., Samuel, M., Chidambaram, R., 2014. Electronic waste – An emerging threat to the environment of urban India. *Journal of Environmental Health Science and Engineering* https://doi.org/10.1186/2052-336X-12-36

Noureddini, H., Gao, X., Philkana, R.S., 2005. Immobilized Pseudomonas cepacia lipase for biodiesel fuel production from soybean oil. *Bioresource Technology* 96, 769–777. https://doi.org/10.1016/j.biortech.2004.05.029

Ottolina, G., Carrea, G., Riva, S., Sartore, L., Veronese, F.M., 1992. Effect of the enzyme form on the activity, stability and enantioselectivity of lipoprotein lipase in toluene. *Biotechnology Letters*, 14, 947–952.

Palocci, C., Chronopoulou, L., Venditti, I., Cernia, E., Diociaiuti, M., Fratoddi, I., Russo, M.V., 2007. Lipolytic enzymes with improved activity and selectivity upon adsorption on polymeric nanoparticles. *Biomacromolecules* 8, 3047–3053. https://doi.org/10.1021/bm0703741

Parandi, E., Mousavi, M., Kiani, H., Rashidi Nodeh, H., Cho, J., Rezania, S., 2023a. Optimization of microreactor-intensified transesterification reaction of sesame cake oil (sesame waste) for biodiesel production using magnetically immobilized lipase nano-biocatalyst. *Energy Conversion and Management* 295. https://doi.org/10.1016/j.enconman.2023.117616

Parandi, E., Safaripour, M., Abdellattif, M.H., Saidi, M., Bozorgian, A., Rashidi Nodeh, H., Rezania, S., 2022. Biodiesel production from waste cooking oil using a novel biocatalyst of lipase enzyme immobilized magnetic nanocomposite. *Fuel* 313. https://doi.org/10.1016/j.fuel.2021.123057

Parandi, E., Safaripour, M., Mosleh, N., Saidi, M., Rashidi Nodeh, H., Oryani, B., Rezania, S., 2023b. Lipase enzyme immobilized over magnetic titanium graphene oxide as catalyst for biodiesel synthesis from waste cooking oil. *Biomass and Bioenergy* 173. https://doi.org/10.1016/j.biombioe.2023.106794

Paraskar, P.M., Major, I., Ladole, M.R., Doke, R.B., Patil, N.R., Kulkarni, R.D., 2023. Dimer fatty acid – A renewable building block for high-performance polymeric materials. *Industrial Crops and Products*. https://doi.org/10.1016/j.indcrop.2023.116817

Patil, S.S., Rathod, V.K., 2022. Combined effect of enzyme co-immobilized magnetic nanoparticles (MNPs) and ultrasound for effective extraction and purification of curcuminoids from Curcuma longa. *Industrial Crops and Products* 177. https://doi.org/10.1016/j.indcrop.2021.114385

Polizzi, K.M., Bommarius, A.S., Broering, J.M., Chaparro-Riggers, J.F., 2007. Stability of biocatalysts. *Current Opinion in Chemical Biology*. https://doi.org/10.1016/j.cbpa.2007.01.685

Prakasham, R.S., Devi, G.S., Laxmi, K.R., Rao, C.S., 2007. Novel synthesis of ferric impregnated silica nanoparticles and their evaluation as a matrix for enzyme immobilization. *Journal of Physical Chemistry C* 111, 3842–3847. https://doi.org/10.1021/jp0670182

Pugh, S., McKenna, R., Moolick, R., Nielsen, D.R., 2011. Advances and opportunities at the interface between microbial bioenergy and nanotechnology. *Canadian Journal of Chemical Engineering* 89, 2–12. https://doi.org/10.1002/cjce.20434

Qi, B., Luo, J., Wan, Y., 2018. Immobilization of cellulase on a core-shell structured metal-organic framework composites: Better inhibitors tolerance and easier recycling. *Bioresource Technology* 268, 577–582. https://doi.org/10.1016/j.biortech.2018.07.115

Rajnish, K.N., Samuel, M.S., John, J. A., Datta, S., Chandrasekar, N., Balaji, R., Jose, S., Selvarajan, E., 2021. Immobilization of cellulase enzymes on nano and micro-materials for breakdown of cellulose for biofuel production—A narrative review. *International Journal of Biological Macromolecules*. https://doi.org/10.1016/j.ijbiomac.2021.05.176

Rao, S. V., Anderson, K.W., Bachas, L.G., 1998. Oriented immobilization of proteins. *Microchimica Acta*, 128, 127–143.

Reshmy, R., Narisetty, V., Tarafdar, A., Bachan, N., Madhavan, A., Tiwari, A., Chaturvedi, P., Varjani, S., Sirohi, R., Kumar, V., Awasthi, M.K., Binod, P., Nagoth, J.A., Sindhu, R., 2023. An overview of cellulase immobilization strategies for biofuel production. *Bioenergy Research* 16, 4–15. https://doi.org/10.1007/s12155-022-10431-3

Rodrigues, J., Canet, A., Rivera, I., Osório, N.M., Sandoval, G., Valero, F., Ferreira-Dias, S., 2016. Biodiesel production from crude Jatropha oil catalyzed by non-commercial immobilized heterologous Rhizopus oryzae and Carica papaya lipases. *Bioresource Technology* 213, 88–95. https://doi.org/10.1016/j.biortech.2016.03.011

Roth, H.C., Schwaminger, S.P., Peng, F., Berensmeier, S., 2016. Immobilization of cellulase on magnetic nanocarriers. *ChemistryOpen* 5, 183–187. https://doi.org/10.1002/open.201600028

Sahu, A., Badhe, P.S., Adivarekar, R., Ladole, M.R., Pandit, A.B., 2016. Synthesis of glycinamides using protease immobilized magnetic nanoparticles. *Biotechnology Reports* 12, 13–25. https://doi.org/10.1016/j.btre.2016.07.002

Salwan, R., Sharma, A., Sharma, V., 2020. *Nanomaterial-immobilized biocatalysts for biofuel production from lignocellulose biomass* (pp. 213–250). https://doi.org/10.1007/978-981-13-9333-4_9

Sampaio, C.S., Angelotti, J.A.F., Fernandez-Lafuente, R., Hirata, D.B., 2022. Lipase immobilization via cross-linked enzyme aggregates: Problems and prospects – A review. *International Journal of Biological Macromolecules*, 215, 434–449. https://doi.org/10.1016/j.ijbiomac.2022.06.139

Santos, S., Puna, J., Gomes, J., 2020. A review on bio-based catalysts (immobilized enzymes) used for biodiesel production. *Energies (Basel)* 13. https://doi.org/10.3390/en13113013

Sen, T., Bruce, I.J., Mercer, T., 2010. Fabrication of novel hierarchically ordered porous magnetic nanocomposites for bio-catalysis. *Chemical Communications* 46, 6807–6809. https://doi.org/10.1039/c0cc01747g

Shah, S., Gupta, M.N., 2007. Lipase catalyzed preparation of biodiesel from Jatropha oil in a solvent free system. *Process Biochemistry* 42, 409–414. https://doi.org/10.1016/j.procbio.2006.09.024

Shalini, P., Deepanraj, B., Vijayalakshmi, S., Ranjitha, J., 2023. Synthesis and characterisation of lipase immobilised magnetic nanoparticles and its role as a catalyst in biodiesel production. *Materials Today: Proceedings* 80, 2725–2730. https://doi.org/10.1016/j.matpr.2021.07.027

Sheikh, Z.U.D., Bajar, S., Devi, A., Rose, P.K., Suhag, M., Yadav, A., Yadav, D.K., Deswal, T., Kaur, J., Kothari, R., Pathania, D., Rani, N., Singh, A., 2023. Nanotechnology based technological development in biofuel production: Current status and future prospects. *Enzyme and Microbial Technology* 171. https://doi.org/10.1016/j.enzmictec.2023.110304

Sheldon, R.A., 2007. Cross-linked enzyme aggregates (CLEA's): Stable and recyclable biocatalysts. *Biochemical Society Transactions*, 35(6), 1583–1587. https://doi.org/10.1042/BST0351583

Shomal, R., Du, W., Al-Zuhair, S., 2022. Immobilization of lipase on metal-organic frameworks for biodiesel production. *Journal of Environmental Chemical Engineering* 10. https://doi.org/10.1016/j.jece.2022.107265

Sillu, D., Agnihotri, S., 2020. Cellulase immobilization onto magnetic halloysite nanotubes: Enhanced enzyme activity and stability with high cellulose saccharification. *ACS Sustainable Chemistry & Engineering* 8, 900–913. https://doi.org/10.1021/acssuschemeng.9b05400

Souza, R.L., Faria, E.L.P., Figueiredo, R.T., Mettedi, S., Santos, O.A.A., Lima, A.S., Soares, C.M.F., 2017. Protic ionic liquid applied to enhance the immobilization of lipase in sol–gel matrices. *Journal of Thermal Analysis and Calorimetry* 128, 833–840. https://doi.org/10.1007/s10973-016-5950-4

Matsunaga, T., Kamiya, S., 1987. Applied micwbiology biotechnology use of magnetic particles isolated from magnetotactic bacteria for enzyme immobilization. *Appl Microbiol Biotechnol* 26, 328–332.

Talekar, S., Ghodake, V., Ghotage, T., Rathod, P., Deshmukh, P., Nadar, S., Mulla, M., Ladole, M., 2012. Novel magnetic cross-linked enzyme aggregates (magnetic CLEAs) of alpha amylase. *Bioresource Technology* 123, 542–547. https://doi.org/10.1016/j.biortech.2012.07.044

Talekar, S., Joshi, A., Joshi, G., Kamat, P., Haripurkar, R., Kambale, S., 2013. Parameters in preparation and characterization of cross linked enzyme aggregates (CLEAs). *RSC Advances* https://doi.org/10.1039/c3ra40818c

Tan, H., Feng, W., Ji, P., 2012. Lipase immobilized on magnetic multi-walled carbon nanotubes. *Bioresource Technology* 115, 172–176. https://doi.org/10.1016/j.biortech.2011.10.066

Tan, L., Tan, Z., Feng, H., Qiu, J., 2018. Cellulose as a template to fabricate a cellulase-immobilized composite with high bioactivity and reusability. *New Journal of Chemistry* 42, 1665–1672. https://doi.org/10.1039/c7nj03271d

Tang, A., Zhang, Y., Wei, T., Wu, J., Li, Q., Liu, Y., 2019. Immobilization of Candida cylindracea lipase by covalent attachment on glu-modified bentonite. *Applied Biochemistry and Biotechnology* 187, 870–883. https://doi.org/10.1007/s12010-018-2838-8

Tran, D.T., Chen, C.L., Chang, J.S., 2012. Immobilization of Burkholderia sp. lipase on a ferric silica nanocomposite for biodiesel production. *Journal of Biotechnology* 158, 112–119. https://doi.org/10.1016/j.jbiotec.2012.01.018

Verma, M.L., Barrow, C.J., 2016. Recent advances in feedstocks and enzyme-immobilised technology for effective transesterification of lipids into biodiesel. *Microbial Factories: Biofuels, Waste Treatment: Volume 1* (pp. 87–103). Springer India. https://doi.org/10.1007/978-81-322-2598-0_6

Verma, M.L., Naebe, M., Barrow, C.J., Puri, M., 2013. Enzyme immobilisation on amino-functionalised multi-walled carbon nanotubes: Structural and biocatalytic characterisation. *PLoS One* 8. https://doi.org/10.1371/journal.pone.0073642

Verma, M.L., Puri, M., Barrow, C.J., 2016. Recent trends in nanomaterials immobilised enzymes for biofuel production. *Critical Reviews in Biotechnology*. https://doi.org/10.3109/07388551.2014.928811

Villeneuve, P., Muderhwa, J.M., Graille, J., Haas, M.J., 2000. Customizing lipases for biocatalysis: A survey of chemical, physical and molecular biological approaches. *Journal of Molecular Catalysis B: Enzymatic* 9(4-6), 113–148.

Wang, L., Zhi, W., Wan, J., Han, J., Li, C., Wang, Y., 2019. Recyclable β-glucosidase by one-pot encapsulation with Cu-MOFs for enhanced hydrolysis of cellulose to glucose. *ACS Sustainable Chemistry & Engineering* 7, 3339–3348. https://doi.org/10.1021/acssuschemeng.8b05489

Wang, P., 2006. Nanoscale biocatalyst systems. *Current Opinion in Biotechnology* 17(6), 574–579. https://doi.org/10.1016/j.copbio.2006.10.009

Wang, Q., Zhang, R., Liu, M., Ma, L., Zhang, W., 2023. Co-immobilization of lipases with different specificities for efficient and recyclable biodiesel production from waste oils: Optimization using response surface methodology. *International Journal of Molecular Sciences* 24. https://doi.org/10.3390/ijms24054726

Wang, S., Zheng, D., Yin, L., Wang, F., 2017. Preparation, activity and structure of cross-linked enzyme aggregates (CLEAs) with nanoparticle. *Enzyme and Microbial Technology* 107, 22–31. https://doi.org/10.1016/j.enzmictec.2017.07.008

Wei, H., Wang, Q., Zhang, R., Liu, M., Zhang, W., 2023. Efficient biodiesel production from waste cooking oil by fast co-immobilization of lipases from Aspergillus oryzae and Rhizomucor miehei in magnetic chitosan microcapsules. *Process Biochemistry* 125, 171–180. https://doi.org/10.1016/j.procbio.2022.12.025

Wong, L.S., Khan, F., Micklefield, J., 2009. Selective covalent protein immobilization: Strategies and applications. *Chemical Reviews* 109, 4025–4053. https://doi.org/10.1021/cr8004668

Wu, L.M., Zeng, Q.H., Ding, R., Tu, P.P., Xia, M.Z., 2019. Fe3O4/Acid activated montmorillonite/cellulase composites: Preparation, structure, and enzyme activity. *Applied Clay Science* 179. https://doi.org/10.1016/j.clay.2019.105129

Xie, W., Wang, J., 2012. Immobilized lipase on magnetic chitosan microspheres for transesterification of soybean oil. *Biomass and Bioenergy* 36, 373–380. https://doi.org/10.1016/j.biombioe.2011.11.006

Yagiz, F., Kazan, D., Akin, A.N., 2007. Biodiesel production from waste oils by using lipase immobilized on hydrotalcite and zeolites. *Chemical Engineering Journal* 134, 262–267. https://doi.org/10.1016/j.cej.2007.03.041

Yiu, H.H.P., Wright, P.A., 2005. Enzymes supported on ordered mesoporous solids: A special case of an inorganic-organic hybrid. *Journal of Materials Chemistry* 15, 3690–3700. https://doi.org/10.1039/b506090g

Yücel, Y., 2011. Biodiesel production from pomace oil by using lipase immobilized onto olive pomace. *Bioresource Technology* 102, 3977–3980. https://doi.org/10.1016/j.biortech.2010.12.001

Zdarta, J., Meyer, A.S., Jesionowski, T., Pinelo, M., 2018. A general overview of support materials for enzyme immobilization: Characteristics, properties, practical utility. *Catalysts*. https://doi.org/10.3390/catal8020092

Zhang, M., Jun, S.H., Wee, Y., Kim, H.S., Hwang, E.T., Shim, J., Hwang, S.Y., Lee, J., Kim, J., 2022. Activation of crosslinked lipases in mesoporous silica via lid opening for recyclable biodiesel production. *International Journal of Biological Macromolecules* 222, 2368–2374. https://doi.org/10.1016/j.ijbiomac.2022.10.023

Zhao, X., El-Zahab, B., Brosnahan, R., Perry, J., Wang, P., 2007. An organic soluble lipase for water-free synthesis of biodiesel. *Applied Biochemistry and Biotechnology* 143, 236–243. https://doi.org/10.1007/s12010-007-8043-9

Zhi, J., Wang, Y., Lu, Y., Ma, J., Luo, G., 2006. In situ preparation of magnetic chitosan/Fe3O4 composite nanoparticles in tiny pools of water-in-oil microemulsion. *Reactive and Functional Polymers* 66, 1552–1558. https://doi.org/10.1016/j.reactfunctpolym.2006.05.006

Zhou, M., Ju, X., Zhou, Z., Yan, L., Chen, J., Yao, X., Xu, X., Li, L.Z., 2019. Development of an immobilized cellulase system based on metal-organic frameworks for improving ionic liquid tolerance and in situ saccharification of bagasse. *ACS Sustainable Chemistry & Engineering* 7, 19185–19193. https://doi.org/10.1021/acssuschemeng.9b05175

Zhou, M., Yan, L., Chen, H., Ju, X., Zhou, Z., Li, L., 2021. Development of functionalized metal–organic frameworks immobilized cellulase with enhanced tolerance of aqueous-ionic liquid media for in situ saccharification of bagasse. *Fuel* 304. https://doi.org/10.1016/j.fuel.2021.121484

Zou, Z., Zhou, H., Dai, L., Liu, D., Du, W., 2023. A dual stable MOF constructed through ligand exchange for enzyme immobilization with improved performance in biodiesel production. *Renewable Energy* 208, 17–25. https://doi.org/10.1016/j.renene.2023.03.072

8 Bioreactors Designs for Microbial Biofuel Production

Ramesh Sharma, Pinku Chandra Nath, Shubhankar Debnath, Amiya Ojha, N. Sridhar, Tarun Kanti Bandyopadhyay, and Biswanath Bhunia

8.1 INTRODUCTION

A strong market for biofuels is being created by the global depletion of fossil fuels, environmental problems, rising fuel prices, and calls for oil liberty. Since most biofuels have the ability to be carbon neutral or negative and suitable for use in current agricultural transportation, they have the potential to be both domestically and internationally accessible for power security. Right now, bioethanol seems to be the most promising alternative for gasoline. Simple alkyl alcohols like bioethanol are utilized as fuel for spark-ignition vehicles. Microbes like *E. coli* and other feedstocks can be used to create bioethanol. Geothermal, solar, hydro, and biomass energy are regarded as sustainable and environmentally beneficial sources of energy. In contrast, fossil fuels, including natural gas, coal, and oil, are non-renewable resources that cannot be replenished at a sustainable rate. The majority of energy produced today comes from fossil fuels. According to reports from the Energy Information Administration (EIA) in the United States, 70% of all oil is used for transportation. According to EIA reports from 2014, 27% of the petroleum oil used in the United States is imported into other nations. The production of biofuels has been under increased scrutiny due to factors such as rising gas prices, dependence on foreign oil imports, growing concerns about national energy security, and the environmental impact of high oil convention (Das et al., 2024).

Bioreactors are fundamental components of all biological processes, and they are considered as the center of bioprocesses. The containers where a biological process or reaction occurs are called bioreactors. For the functioning of a biological system (microbe, enzyme, or cell) and to guarantee a high yield of the bioprocess, the bioreactor must fulfill the ideal operating conditions (Jiao et al., 2024; Sharma et al., 2024). To get over these restrictions, one crucial instrument to investigate is the bioreactor's design. Mass and heat transfer, mixing efficiency, and other factors

 DOI: 10.1201/9781003633709-8

that are critical to the effectiveness of enzymatic hydrolysis can all be controlled by bioreactor design optimization.

8.2 BIOREACTOR DESIGN, DEFINATION

A bioreactor is a vessel or system used in cell culture to grow substrates from plants, animals, and microbes at the commercial or required level. These apparatuses contain biological reactions and offer a regulated setting for ideal cell development or product manufacturing. Using microorganisms, fungi, and cells (derived from plants and animals), they carry out a variety of biochemical processes to produce biological products, such as (ethanol, organic acids, biodiesel, biogas, antibiotics, aromatic compounds, and biohydrogen) and biomass (which includes things such as a single-cell protein, Baker's yeast, cells from animals, and tiny algae; Ghazani et al., 2024). Biofuel production has been implemented through the use of bioreactors (Manna et al., 2024). More than US$709.4 million was projected to be the value of the bioreactor market in 2018 globally, and a compound annual growth rate of 17.6% is predicted for 2019–2025 (Ashrafizadeh & Seifollahi, 2021). By optimizing the design and operational parameters, such as the concentration of microbial species, humidity, inflow flow rate, temperature, pH, and stirring rate, this growing need drives research and development of more efficient, economical bioreactors (Kim et al., 2021). The most significant issues facing the sector are determining, putting into practice, and utilizing the ideal criteria.

8.3 ESSENTIAL FUNCTIONS AND REQUIREMENTS FOR BIOREACTOR

The primary function of the bioreactor is to supply appropriate and controlled environmental conditions for the survival of microbial cells. To reliably and economically produce higher-quality goods, a bioreactor's design and operation should satisfy output criteria and designated process variables. When constructing a bioreactor in this way, different factors are considered, including geometry, aerations, temperature, pH, lighting, sterility, and mixing systems, low energy consumption, and the appropriate size and material (Spier et al., 2011). The main considerations for scale-up criteria are aeration rate, impeller speed, and reactor geometry. The scaling up of the bioreactors is effected by oxygen supply and carbon dioxide removing efficiencies, particularly at very high cell densities, the scale-up provides the proper mass transfer rates. Thus, bioreactors need to fulfill specific specifications (Panda et al., 2018; Panda, 2011): be simple to keep an eye on and/or regulate reaction parameters (temperature, redox value, pH and agitation rate, etc.) to establish a regulated antimicrobial environment for biocatalysts (e.g., facilitating the development of cells) and, similarly, in order to sustain microbial development, to function dependably for several days or weeks. One example is the degradation of (R)-2-(2,4- dichlorophenoxy) propionate (R-2,4-DP) by a pure culture of *Delftiaacidovorans* duration of 5e11 weeks (Delavar & Wang, 2022). For this to happen, the bioreactor must maintain its cell content unaltered. While the fundamental engineering and design challenges of bioreactors are identical, maintaining desired biological activity while eliminating or restricting undesirable activities might occasionally provide greater challenges than those faced by conventional chemical reactors. To retain cells and reactants in the bioreactor and maintain aseptic conditions for extended periods of time, sterile operating protocols are employed to prevent contamination and ensure continuous, efficient bioprocessing. As previously mentioned, in these types of situations, a bioreactor must maintain appropriate aseptic conditions while preventing contamination. This is because, in contrast to chemical organisms, biological organisms are more delicate and less stable. To efficiently satisfy productivity and yield requirements and to be easily maintained and modified using different organisms for the particular reaction in the reactor, the design approach in this case should take bioreactor functionalities into account.

8.4 BIOREACTORS CLASSIFICATION

8.4.1 According to Their Operational Modes

Different types of bioreactors have identical functions and requirements. Classifications of bioreactors are based on the requirements of applications, selective microorganisms, and mode of operations, etc. Table 8.1 shows the various types of bioreactors with their advantages and disadvantages.

8.4.1.1 Batch Reactors

Microorganisms are poured into sterile culture media in a batch operation after it has been sterilized. Added feeding is not done while the bioprocess is operating. When the batch reactor is just starting its culture, all reactant are supplied (Figure. 8.1). Cells, substrates (salts, nutrients, vitamins), and concentrations of product altered over the course of this reaction period. After a set amount of time, the bioprocess promotes to finished, and the result is released. On the other hand, controlling components like bases and acids can be added.

When growing a tiny inoculum of microorganisms (such yeast or bacteria) under controlled conditions to increase their biomass before inoculating larger-scale bioreactors, batch bioreactors are frequently employed for seed culture propagation (Garduno et al., 2024). In order to guarantee the availability of a high-quality and active culture for later stages of the biofuel production process, this step is essential. Similarly, fermentation operations in the manufacture of biofuels, specifically the generation of ethanol and butanol from feedstocks comprising sugars or starch, are carried out in batch bioreactors. In a batch fermentation, the microbes that turn the substrate into alcohol are added to the bioreactor, together with the sugar or starch-containing material (Bušić et al., 2018). Until the substrate is exhausted or the required product concentration is reached, fermentation takes place. Lipid productions is one of the batch bioreactor's other key uses. Batch bioreactors are a useful tool for encouraging lipid buildup within the cells used to produce biodiesel from microalgae or oleaginous microorganisms (Zhang et al., 2021). Batch cultures can cause bacteria to store lipids as energy reserves by supplying particular growth circumstances, such as nutrition limitation or light intensity. These lipids can then be removed and transformed into biodiesel.

TABLE 8.1
Classifications of Bioreactors with Their Characterizations

Classification	Types of reactors	Advantages	Disadvantages
Modes of operation	Batch	Simple equipment and flexibility in operating conditions, short process length, produces different products in the same reactor, cost-effective and readily deployable, simple to maintain hygiene.	Excessive labor expenses per unit of production, large-scale production is challenging to keep up. Risk of explosion when reactors are being emptied, lower product quality compared to others.
	Fed-batch	The nutritional environment can be maintained, regulation of by-product production or catabolite suppression occurs when there is a restricted supply of substrates that are specifically needed for the creation of the desired product, enabling the replenishment of water lost through evaporation.	Prior examination of the microorganism, including its needs and knowing its physiology, is necessary for productivity, care should be taken when designing a cycle fed-batch culture to make sure that toxins don't build up to levels that stop the process from working and that nutrients other than those in the feed medium don't become too limited, operator expertise is required for the setup.
	Continuous	Increased efficiency in terms of output per unit of volume, control of growth rates over prolonged durations by manipulation of dilution rates, correct details about kinetic factors, maintenance energy, and growth rates, reduced cleaning and vessel handling.	Deterioration of the first product strain occurs when a more rapidly proliferating strain supersedes it, maintaining filamentous organisms is challenging due to the high viscosity and heterogeneous composition of the broth, requiring flow control and longevity.
Bacterial immobility	Stirred tank	Continuous operation, effective temperature regulation, effortlessly adjusts to dual-phase operations, optimal gas diffusion to developing cells and thorough mixing of the substances, flexible and low operating cost.	The necessity of shaft seals and bearings, restriction in size imposed by the motor's dimensions, foaming is often a problem, consumption of power is more due to the mechanical pressure pumps.
	Bubble bed	Have high volumetric productivity and excellent heat management, self-regulating, improved optimization of the plate surface area and distribution of flow.	Less efficient than other bioreactors, ineffective break the foam when it begins to form, higher catalyst consumption than the fixed-bed, higher installation cost and difficult to design.

(*Continued*)

TABLE 8.1 (Continued)
Classifications of Bioreactors with Their Characterizations

Classification	Types of reactors	Advantages	Disadvantages
	Airlift bed	The design is simple, without any components that move or agitate, resulting in reduced maintenance requirements and a lower probability of failures, easier sterilization; the energy efficiency of this system is quite high, and its productivity is equivalent to that of stirred tank bioreactors.	The agitation on the airlift bioreactor is controlled by the supply air to adjust the supply air then the higher pressure needed, no bubbles breaker; the higher pressure of air needed, the more energy consumption needed and the higher the cost.
	Packed bed	Superior catalyst efficiency compared to other catalytic reactors, simple design, efficient in extreme temperatures and pressures, uniform particle mixing.	Difficult to clean, difficult to replace catalyst, undesirable side reactions, lack of current understanding, erosion of internal components.
	Fluidized bed	Homogeneous particle distribution, levels of temperature changes, ability to operate the reactor in a continuous state.	Specifications for pumping and pressure loss, enlarged reactor vessel dimensions, lack of current understanding, deterioration of inner components.

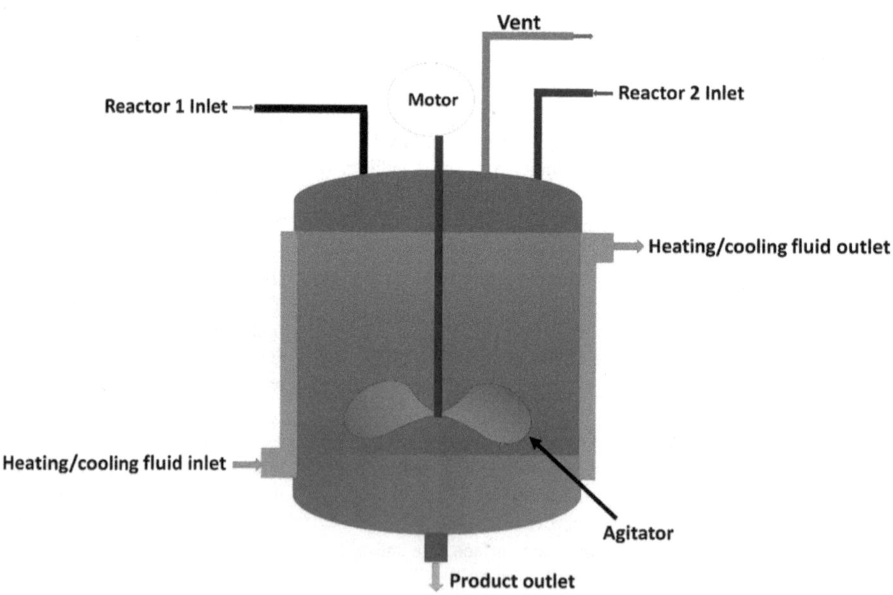

FIGURE 8.1 Schematic diagram of batch reactor.

8.4.1.2 Fed-batch Type

The fed-batch bioreactors function using operating modes of both batch and continuous processes. A fed-batch process is a semipermanent technique, in contrast to the batch process, which is characterized by intermediate operations. It is possible to modify longer processes with higher cell densities to get better yields and enhanced selectivity by modifying parameters. This is because substrates can be supplied to the bioreactor gradually using ports during the bioprocess. The products are collected from the ending sides of the production cycles, just like in a batch bioreactor. Compared to batch reactors, fed-batch reactors have higher operating safety and stability. Because of this, they are currently used in every area of the biotechnological industry's microbial biofuel production.

The fed-batch bioreactor is a popular bioreactor type for biofuel production because of its capacity to manage nutrient supply during fermentation and sustain high cell densities (Teworte et al., 2022). Better control over cell growth and metabolite production is possible with a fed-batch bioreactor because the substrate (such as sugars or other carbon sources) and other nutrients are provided incrementally throughout the fermentation process. Fed-batch bioreactors are used in the following ways to produce biofuel: Fed-batch bioreactors minimize inhibitory by-product accumulation while continually delivering nutrients, allowing microorganisms to be cultivated at high cell densities. Higher cell densities can result in higher product yields when producing biofuel from microorganisms like bacteria or yeast, therefore this is especially helpful in that regard. Substrate utilization rates have a major impact on product creation and overall process efficiency in the biofuel manufacturing process. With fed-batch bioreactors, substrate inhibition and metabolic overflow can be minimized, while substrate feeding rates are precisely controlled to maximize carbon source availability. Fermentation performance can be improved and biofuel productivity increased using this regulated feeding approach. Nutrient limitation techniques, like nitrogen or phosphorus limitation, can be applied in fed-batch bioreactors to create stress conditions that encourage the accumulation of desirable products, like lipids or ethanol (Bumbak et al., 2011). Fed-batch cultures can enhance product yields and purity by rerouting cellular metabolism toward the production of target biofuels through careful regulation of necessary nutrient supply (Puyol et al., 2017).

8.4.1.3 Continuous Reactors

Substrates are continually fed into a continuous bioreactor together with the culture, and the products are produced at the same rate. This keeps process parameters and nutrient and cell concentrations constant during the procedure. In those identical conditions, a fresh cultural medium is provided. Continuous cultures are especially helpful in situations where an excess of nutrients could impede growth because of the accumulation of toxins or improper heating. Reduced product obstruction and longer output space times are further advantages. Cells are gathered to restrict the intake and flow rates of microorganisms when the medium is removed. Additionally, there are other ways to store the cells (such as in a spin filter). The production of the products from this bioreactor can be significantly increased during continuous operation when compared to the fed-batch method. Conversely, the shorter growth time lowers the chance of contamination. When producing large volumes of reactions involving gas, liquid, or soluble solid substrates, as well as processes involving microorganisms with high mutation stability, a continuous bioreactor is often utilized. It has been common practice for the culturing of microorganisms with the production of biofuels, vinegar, baker's yeast, and treatment of waste water using this reactor.

In the manufacturing of biofuels, continuous reactors are also employed because they provide benefits over batch processes, including easier scaling up, increased productivity, and steady-state operation (Gopi et al., 2022). Here is how continuous reactors are used in the manufacturing of biofuels: For the continuous fermentation processes involved in the manufacture of biofuels, continuous reactors can be used. While the fermentation broth is continuously removed from the reactor, microorganisms like bacteria or yeast are continuously given substrate and other nutrients. Compared to batch fermentation, this enables steady-state operation and constant biofuel generation over extended periods of time, resulting in increased output. Immobilized cell technology, in which the microbial cells are immobilized on a support matrix inside the reactor, can be applied to continuous reactors. Longer retention periods and higher cell densities are made possible by this continuous operation that keeps the cells inside the reactor. When producing biofuels through processes like the conversion of sugars to ethanol or triglycerides to biodiesel, which call for frequent or extended contact to the biocatalyst, immobilized cell reactors come in handy. Enzymatic conversion methods for the manufacture of biofuels can be carried out in continuous reactors. Continuous feed of enzymes and substrate into the reactor enables continuous conversion of biomass into biofuels like ethanol and biodiesel. In comparison to chemical catalysts, continuous enzymatic reactors have benefits, including higher reaction rates, substrate selectivity, and simpler product separation, which makes them appealing for the manufacture of sustainable biofuels.

8.5 CLASSIFICATION ACCORDING TO MICROORGANISM IMMOBILITY

Efficient mixing between substrates and cells is essential for achieving high reaction rates in bioreactors. Agitation can be achieved through various modes, including pneumatic, mechanical, and static methods. Among these, mechanically agitated bioreactors are the most widely used for commercial cultivation of microbial, animal, and plant cells. In this type of bioreactor, a mechanically driven impeller stirs the medium to ensure uniform distribution of nutrients and maintain optimal environmental conditions. The impellers are used to stir the medium in mechanically agitated bioreactors, like stirred tanks. Pneumatically agitated bioreactors include airlift bioreactors and bubble columns. Airlift bioreactors use a riser section where gas is sparged. The medium flows downhill in the down-corner section of the reactor after the release of the gas at the top of the structure.

8.5.1 Stirr Tank Bioreactor

One of the most popular reactors for producing biogas is the continuous stirred tank bioreactor (CSTB), particularly when treating solid substrates. The washout of microorganisms at low hydraulic retention time (HRT), which may be linked to a high organic loading rate (OLR) is one of these reactors' drawbacks. High levels of organic materials feeding these reactors quickly cause acidification and the generation of volatile fatty acids, which impede the AD process. As a result, in a two-stage AD process, this reactor is typically utilized as the fermentative unit (Wang et al., 2020), and it is also common for biohydrogen production (Brindhadevi et al., 2021). For improved mass transfer between microorganisms and substrates and to prevent precipitation, substrates in a CSTB are agitated using a mechanical or hydraulic stirrer. The reactor's top is where the generated biogas exits the reactor. This reactor can be produced and utilized in various sizes. Using an external heat exchanger, internal tubes, or a jacket encircling the vessel, the reactor temperature can be precisely controlled at the ideal value (Middleton, 2021).

The alternative suspended-growth process is the CSTB (Ishak et al., 2012). Its foundation is a CSTB or a traditional mixed-flow reactor (MFR). The reactor must have an air supply, and the stirrer may be located at the top or bottom. The CSTB and the packed-bed reactor (PBR) are the two primary bioreactor designs that have been researched in order to determine the operating parameters and configuration that would boost the intended product yields while preventing the side reactions. Because it is simple to adjust the operational settings, CSTB is commonly used (Ri et al., 2019). However, PBR is frequently researched due to its established capacity to improve cell density and, consequently, its elevated rate of hydrogen production (Dasgupta et al., 2010). While PBR can support a higher conversion rate than CSTB, it necessitates the regeneration of the catalyst or bed. Owing to *E. aerogenes*' documented propensity for self-flocculation (Jitrwung & Yargeau, 2015), CSTB was seen to be a preferable option because of the increased chance of PBR blockage.

8.5.2 Photobioreactor

In order to maintain a clean culture operation, photobioreactors are closed systems that can operate indoors or outdoors and are inoculated with a single species. Although the quality and growth of the harvested product are higher in closed cultivation systems due to improved control over contamination and cell physiology while comparing to open systems, the manufacture cost is higher. Huge outdoor systems are made up of tubes that are exposed to the sun and can run either continuously or in batches. Numerous designs have been built or suggested at the pilot scale. Glass and plastic, like polyethylene, are the two materials used to make tubes. Overheating is an issue as a result of the tubes acting as sun light collectors. Because of this, the tubular solar receptors need a temperature-control mechanism, which is typically a swimming pool. As an alternative, using strains that are thermotolerant has been suggested as a way to get around cooling facilities. In order to provide flexibility to the system regarding culture volume and to make control and operation easier, tubes are typically organized into multiple modules. The solar receptors are equipped with CO_2 supply systems, which include carbonation towers, pumps for medium circulation, and tanks for nutrient mixing. A variety of microorganisms, including *Scenedesmus*, *Spirulina*, and *Chlorella*, have been produced in these photobioreactors (Parsy et al., 2024). Another version, built in Chile, has a cement pond lined with epoxy resin and enclosed in a polyethylene dome, covering an area of 110 m^2 of sun irradiation. Paddle wheel agitation is the method used. *Spirulina* biomass has been produced using it, and growth densities of 450–750 mg L^{-1} have been attained (Kurpan et al., 2024). Utilizing oval plastic bags floating on thermal waters is the basis of an inventive concept being used in the United States. The necessity for artificial light necessitates smaller photobioreactors when they are used indoors. Designs with interior lighting to provide maximum light incidence can be made as fermenter-like reactors made of stainless steel or as plastic tubular systems. Due to their modest throughputs, their application is somewhat restricted for the production of single-cell protein (SCP); on the other hand, they are perfectly suitable for the synthesis of high added value algae metabolites, including polysaccharides, carotenes, and other colors, polyunsaturated fatty acids, and so forth. In order to use wastewater's inorganic salts as fertilizer for microalgae growth, in order to ensure optimal light penetration and nutrient availability in microalgae culture, Chang et al. (2018) designed an annular photobioreactor using ion exchange membranes that work selectively with transfer of cations and anions from a chamber of wastewater to a chamber of microalgae cultivation while preventing the movement of suspended solids in wastewater (Chang et al., 2018). Additionally, the generation of microalgae biomass and biofuels was facilitated by the use of membrane-based tubular photobioreactor, which significantly improved the technological and financial viability of microalgae culture using membrane photobioreactors (Chang et al., 2019).

A range of biofuel and bioproduct yields from photoautotrophic algae cultivation are shown in Figure 8.2, which takes into account other conversion processes beyond transesterification. After growing, the algae are harvested, extracted, and then put through a number of conversion processes, including pyrolysis, esterification, anaerobic fermentation, gasification, and hydrothermal liquefaction.

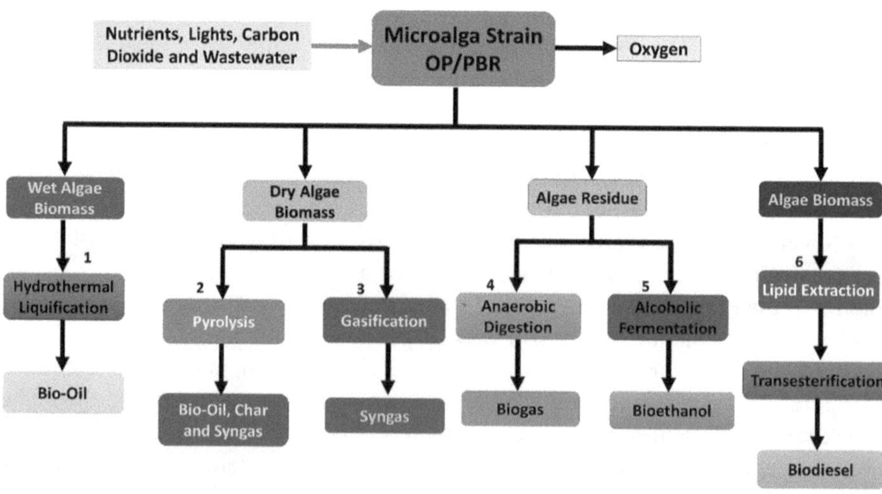

FIGURE 8.2 Algal-based process for biofuel and bioproducts productions using open pond and photobioreactor.

Specialized systems like raceway ponds, closed photobioreactors, and open ponds are used to cultivate algae. Algae use photosynthesis to develop and store biomass during cultivation. They do this by using CO_2, sunshine, and nutrients from the growth medium. The algae are taken out of the cultivation system once they have attained the required biomass concentration (Moreno-Garcia et al., 2017). After that, the biomass is processed to remove the oil, which is usually made up of lipids. Oil can be extracted using a variety of techniques, such as supercritical fluid extraction, solvent extraction, and mechanical pressing. The extracted oil contains a high concentration of triglycerides, which can be trans esterified to produce biodiesel. The extracted oil is subjected to a chemical process called transesterification, which involves the reaction of triglycerides with an alcohol (methanol or ethanol) in the presence of a catalyst (potassium hydroxide or sodium hydroxide, for example). Glycerol and biodiesel (methyl or ethyl esters) are the by-products of this process. Before being used as a renewable fuel for diesel engines, the biodiesel can next undergo additional purification and refinement to fulfill quality criteria. Pyrolysis is a potential further processing method for the algae's leftover biomass following oil extraction. Pyrolysis is a thermochemical process that breaks down materials at high temperatures without the presence of oxygen. The leftover biomass is heated during pyrolysis, which results in its breakdown into three primary products: bio-oil, syngas, and biochar. Bio-oil can be utilized as a feedstock for chemical synthesis or further processed into fuels for vehicles. Syngas, a hydrogen and carbon monoxide mixture, can be processed into fuels and chemicals or utilized to generate heat and electricity. A solid carbonaceous residue known as biochar can be added to soil as a soil supplement to increase soil fertility and sequester carbon. Anaerobic digestion is an additional option for processing the leftover biomass (Thomas et al., 2017). In the biological process known as anaerobic digestion, organic matter is broken down by microorganisms in the absence of oxygen, yielding biogas a combination of CO_2 and methane as a by-product. Biomethane, which can be converted from biogas and injected into natural gas pipelines or used as a fuel for vehicles, can also be extracted from the gas and used as a sustainable energy source for the production of heat and power. For agricultural usage, the digestate left over from anaerobic digestion can be utilized as a nutrient-rich fertilizer. The desired product will determine how many and how intense post-harvesting procedures are used, which will increase the viability of algal-based productions (Ferrell & Sarisky-Reed, 2010).

8.5.2.1 Photo Bioreactor Design Consideration

Algal lipid concentration and growth are influenced by a multitude of factors. Photosynthesis is the process that first converts sunlight into stored energy. As a result, every element involved in photosynthesis promotes growth. This chapter examines several parameters, including temperature, pH, CO_2 removal, illumination, mixing, water, and fertilizer availability. The list is not all-inclusive, and this section's last topic covers additional important topics, including genetic engineering and reactor maintenance. It is noteworthy that the ideal development conditions in each category are contingent upon the particular strain of algae used for culture.

8.5.2.1.1 Lighting

An ideal reactor increases the frequency and wavelength of light exposure for cells, as well as the strength and penetration of light (Sathinathan et al., 2023). The intensity of the light is important because algae experience light saturation at a certain point and release any extra energy as heat. The spatial dilution of light, or the dispersion of solar radiation over a larger photosynthetic surface area, can lessen light saturation. Additionally, mutual shadowing of the cells in the culture is lessened by spatial light dilution, leading to faster growth rates and less accessory pigment content (Sharma et al., 2022). Therefore, optimizing the surface area to volume ratio which may be used to compare reactor is a design principle for photobioreactors (Ramesh et al., 2021). Kunjapur and Eldridge (2010), designs with a 400 m^2/m^3 ratio value were considered state-of-the-art (Kunjapur & Eldridge, 2010). Beyond volume and surface area, a reactor's particular geometry affects how light is distributed. For instance, in a tubular reactor, the biomass density in the medium and the tube's diameter largely dictate the light gradient. The light penetration and intensity are both influenced by the biomass density. For light intensity and light penetration to continue at their ideal levels, each strain has an ideal cell density that must be maintained (Kunjapur & Eldridge, 2010). The maximum cell concentration in algal cultures without mutual shadowing, or the critical cell density, is a crucial operating parameter that Park and Lee described (Park & Lee, 2001). Because of the fact that several investigations have demonstrated that cultures grow differently when exposed to different hues of light, the wavelength of light used to produce algae is also an important design consideration. Optimizing this feature in systems with natural light illumination is, however, far more difficult than in systems with artificial light illumination, where the chosen wavelength of light may be flashed. Sadly, photosynthesis is unable to consume more than half of the incident solar radiation from natural light (Pulz & Scheibenbogen, 2006). A large-scale culture's maximum potential yield of 30–40 g/m^2d may be achieved when the light conversion efficiency reaches its upper limit when natural light is the growth-rate limiting factor (Goldman, 1979). According to Matthijs et al., (1996) red light is ideal for meeting the needs of the pigments in the first excited state of the light-harvesting antenna complexes (LHC), which are essential for photosynthesis in green algae. Additionally, he pointed out that the growth qualities remained the same when blue light was added to red LED light (Matthijs et al., 1996).

The high expense of installing and maintaining artificial light sources in conventional photobioreactors with artificial lighting systems persists as a key issue, despite the fact that many efforts have been made to produce efficient and economical photobioreactors. Moreover, Eq. (8.1) states that as the concentrations of both cells and products rise, the light intensity diminishes exponentially with distance from a reactor wall.

$$\frac{I_L}{I_O} = \exp(-\gamma L), \tag{8.1}$$

where γ is the turbidity, I_o is the original incident intensity, and I_L is the light intensity at depth L. As a result, light intensity tends to drop quickly as a result of light shading effects brought on by rises in cell and product concentrations or by the development of biofilm on reactor vessel surfaces (Chen et al., 2008).

Furthermore, because traditional light sources typically produce a significant amount of heat, they cannot be in close proximity to the microalgae growth, even though a short light path is theoretically advantageous for achieving high light efficiency. As a result, conventional photobioreactors' light conversion efficiency has been restricted. A variety of photobioreactor designs with varied lighting schemes have been created to improve the microalgae production rate and oil/lipid content due to the issues and limits associated with standard light sources (Ma & Hanna, 1999). As demonstrated earlier, cell concentration and the amount of light energy available frequently determine how well microalgae produce biomass. Therefore, several photobioreactors have been built and designed to increase light supply and microalgae biomass production performance without considering the issue of economic viability. A photobioreactor was proposed by El-Shishtawy et al. (1997), which combined a light receiving face with a reflecting sheet to transmit light sources (El-Shishtawy et al., 1997). Additionally, a photobioreactor made of three concentric glass cylinders with incandescent bulbs inserted inside was created by certain researchers (Tsygankov et al., 1994). Issarapayup et al. (2009) determined the ideal downcomer-to-riser area ratio for the culture of *Haematococcus pluvialis* NIES-144 using a flat-plate airlift photobioreactor (Issarapayup et al., 2009). While numerous photobioreactor designs have been developed and demonstrated to be successful in growing microalgae, the primary barrier to their widespread use is the high power and operating expense associated with the requirement for artificial light sources.

LEDs provide several benefits, including a longer life expectancy, lower heat generation, higher conversion efficiency, and a greater tolerance for on/off switching. They are also lightweight and tiny enough to fit into almost any photobioreactor. Furthermore, the narrow light emission spectrum of LEDs, which ranges from 20 nm to 30 nm, can be matched with photosynthetic requirements. For example, according to Yeh and Chung (2009), the adsorption wavelengths of red LED and blue LED are approximately 645–665 nm and 450–470 nm, respectively (Yeh & Chung, 2009). According to Wang et al. (2007), utilizing red LEDs in the photoautotrophic cultivation of *Spirulina platensis* produced the best specific growth rate and biomass output (Wang et al., 2007). Furthermore, a number of research revealed that different species may require different wavelength conditions. According to Katsuda et al. (2004), lighting using red LEDs is appropriate for microalgal cell growth, however switching to blue LED lighting may enhance H. pluvialis' ability to produce astaxanthin (Katsuda et al., 2004). Additionally, incoming light intensity, duty cycle, and flashing frequency were found to have definite effects on algal cultivation (Katsuda et al., 2006). Their findings demonstrated that a promising illumination technique for algal cell growth and astaxanthin production is the use of blue LED flashing light. LEDs were used by Lee and Palsson (1994) as the only light source for the green algae *Chlorella pyrenoidosa* grown indoors (Lee & Palsson, 1994). Since LEDs with a wide range of absorption wavelengths are now accessible, their study attempted to mix multiple LED with varying absorption wavelengths to increase the production of microalgae biomass and target products.

8.5.2.1.2 Mixing

Algal development in a reactor is significantly influenced by the amount of mixing occurring there. In fact, according to Suh and Lee (2003), mixing is the most important element influencing algal growth rates when environmental factors do not limit growth rates. Mixing has two main effects on growth. By enhancing mass transfer between the nutrients and the cells and the frequency at which the cells are exposed to the light and dark portions of the reactor, mixing increases productivity (Qiang & Richmond, 1996).Mixing aims to lower diffusion barriers surrounding the cells and uniformly transfer radiation to every cell in the culture (Richmond, 2004). Lighting and mixing go hand in hand since mixing is frequently what creates the light and dark cycles that are ideal for the growth of algae. Similarly, bad lighting negates the benefits of combining. Richmond confirmed the importance of this link by stating that when cultures were subjected to low light and low cell density, the rate of mixing had no effect on production (Richmond, 2004). When the reactor was scaled up by increasing the tube diameter, Ugwu et al. (2008) showed that the installation of static mixers in tubular reactors was successful in enhancing light utilization and biomass outputs. Also outlined were a few of the variables that affect a reactor's overall mass transfer coefficient (kLa), a crucial measure for comparing reactors (Ugwu et al., 2008). Temperature, surfactant/antifoam agents, sparger type, and agitation rate all affect the coefficient. Large bubbles may occur from the use of fine spargers, which reduces the contact area between the liquid and gas and results in inefficient mass transfer. The liquid flow rate determines both the bubble size and the gas bubble velocity. Static mixers or baffles can be installed to minimize the size of bubbles (Ugwu et al., 2008). According to Fu et al., (2019), a closed reactor with a dual aperture and perforated membrane sparger system that divides the air supply utilized for mixing with the supply of CO_2 (Fu et al., 2019). Compared to traditional sparging, this separation produced a fivefold increase in the amount of CO_2 transferred from the gas phase to the liquid phase. Numerous benefits of integrating two types of spargers in a dual sparging bioreactor were reported by Fu et al. (2019). Initially, the CO_2 mass transfer coefficient was raised by the membrane sparger's tiny bubbles. Second, by raising the gradient of CO_2 partial pressure between the liquid and gas phases, the independent addition of CO_2 additionally enhanced mass transfer. Third, the big air bubbles caused turbulence, which inhibited the building of the wall. A high level of dependability was also demonstrated by the dual sparging bioreactor, as there were few equipment malfunctions and little maintenance needs.

8.5.2.1.3 Water Consumptions

One notable advantage of using various strains of algae for fuel production over traditional crops is that freshwater does not need to be used for cultivation. Many regions of the world already have limited freshwater supplies, and the increased development of biofuels puts further strain on existing supplies (Berndes, 2002). Water usage is therefore a crucial factor to consider when comparing reactor designs. Algae have a far greater range of water sources in which to grow than other terrestrial crops (Demirbas & Demirbas, 2011). According to Vo et al., (2020) studies, algae can develop in brackish or salinized water, pure drinking water, and even wastewater discharge (Vo et al., 2020). Microalgae strains are typically classified into two groups according to whether they thrive best in freshwater or saltwater. Each strain of algae has different rates of lipid and carbohydrate production, and these rates are influenced by the salt level (Rao et al., 2007). To give you an idea of how salinity-optimal the growth media can get, a few strain examples are provided together with their salinity values. The ideal growing conditions for the species of *Nannochloropsis*, *Tetraselmis*, and *Isochrysis* were discovered by Abu-Rezq et al. (1999). The ideal salinity range for *Isochrysis* was 25–35 ppt, *Tetraselmis* was 20–35 ppt, and *Nannochloropsis* was 20–40 ppt (Abu-Rezq et al., 1999).

8.5.2.1.4 Carbon Dioxide Consumption

For photosynthesis to take place, CO_2 is required in addition to light and water (Nath et al., 2023b). On the other hand, too much CO_2 can also be harmful to cell development and photosynthesis. Lee and Tay (1991) discovered that the growth rates of *Chlorella* cells exposed to high CO_2 partial pressures (pCO_2) decreased (Lee & Tay, 1991). The entire culture can receive enough CO_2 by diffusion across a gas permeable membrane, which also prevents CO_2 inhibition at high gaseous pCO_2. Maximum growth is frequently achieved at CO_2 concentrations between 1% and 5% (by volume). Nevertheless, labs sometimes use 5–15% CO_2 or even pure CO_2 to aerate algal cultures (Suh & Lee, 2003). A pCO_2 value of at least 0.15 kPa is necessary to prevent kinetic CO_2 uptake limitation(Schenk et al., 2008). Furthermore, a ratio of roughly 1.7–1.8 g CO_2/g dry biomass is necessary. Because it lowers greenhouse gas emissions and the cost of producing algal biofuel, flue gas is a preferred source of CO_2 (Doucha et al., 2005). Typical coal-fired power plants' flue gas can have up to 13% CO_2 in it (Chisti, 2007). Doucha et al. (2005) compared the performance of a closed reactor using pure CO_2 to one using flue gas as a source of CO_2 (Doucha et al., 2005). Flue gas flow rate can be adjusted with an infrared analyzer that detects bypass concentration of CO_2 in the gas phase. At the downstream end of the reactor, a pCO_2 greater than 0.1 kPa was maintained to avoid CO_2 growth limitation. Productivities and photosynthetic efficiency under pure CO_2 versus flue gas circumstances were surprisingly close. The efficiency of CO_2 mass transfer for flue gas was lower than that of pure CO_2 due to the relatively low CO_2 concentration in the gas. Furthermore, the authors discovered that the proliferation of microalgae was not inhibited by the presence of CO and NOx in flue gas (Doucha et al., 2005).

8.5.2.1.5 Nutrient supply

The availability of nutrients is essential for the development and productivity of the microorganisms (such as algae or cyanobacteria) being grown in a photobioreactor used to produce biofuel (Johnson et al., 2018; Nath et al., 2021). For these microbes to perform photosynthesis and generate biomass, which can be transformed into biofuels, they need vital nutrients. In a photobioreactor, the following essential elements are usually required for the production of biofuel: The main source of carbon for photosynthesis is carbon dioxide (Nath et al., 2023a). Usually, it is added to the culture media of the photobioreactor directly: either injection or bubbling. Several industrial operations, including those in power plants and fermentation facilities, can provide CO_2. Nitrogen is necessary for photosynthesis and cellular growth; it is also a component of proteins and chlorophyll. Nitrates, nitrites, ammonium salts, and urea are examples of common sources of nitrogen. To avoid nitrogen excess or limitation, which can alter the productivity and composition of biomass, the proper concentration of nitrogen must be maintained. Phosphorus is an essential building block of ATP, phospholipids, and nucleic acids, which are required for cell structure and energy transmission (Nath et al., 2022). Usually, phosphates like sodium phosphate (Na_2HPO_4) or potassium di-hydrogen phosphate (KH_2PO_4) are used to supply phosphorus. Enzyme activity, osmoregulation, and preserving cellular turgor pressure all depend on potassium. Potassium is typically provided as potassium sulfate (K_2SO_4) or potassium chloride (KCl). As cofactors for enzymes and to sustain biological activities, microorganisms need a variety of trace elements, including iron (Fe), magnesium (Mg), calcium (Ca), manganese (Mn), zinc (Zn), copper (Cu), and others. Usually, these trace minerals are provided as soluble salts or as parts of solutions containing trace elements. Certain microbes could need particular vitamins or growth factors in order to develop and metabolize functions. These could contain vitamins that can be supplemented as needed, such as cobalamin, thiamine, or biotin.

Algae need more than just the reactants in the photosynthesis reaction to flourish. The two main nutrients, phosphorus and nitrogen, are involved in the regulation of both lipid synthesis and growth rates. Carbon, hydrogen, oxygen, sulfur, calcium, magnesium, sodium, potassium, and chlorine are additional vital nutrients. Iron, boron, manganese, copper, molybdenum, vanadium, cobalt, nickel, silicon, and selenium are among the nutrients that are required in trace amounts (Suh & Lee, 2003). At high cell densities, an examination of a common medium called N-8 showed deficiencies in iron, magnesium, sulfur, and nitrogen. Further tests revealed that balanced supplementation produced better results than the addition of each of the four elements separately, which had no effect on culture performance. Therefore, the experimenters claimed that the foundation for an effective medium design should be nutrition balancing based on the elemental makeup of the biomass (Mandalam & Palsson, 1998). Chisti did point out that certain nutrients must be present in excess. For instance, because phosphates react with metal ions, extra phosphorus must be given. Limiting nutrients, particularly nitrogen or phosphorus, can induce stress by raising the biomass's fat contents. But applying stress also slows down growth, which could result in less lipid being produced overall. The high metabolic cost of lipid production is the reason for the trade-off between productivity and lipid content. Three distinct nutrition supply scenarios were outlined by Rodolfi et al. (2009): nutrient-sufficient, nutrient-limited, and nutrient-deficient (Rodolfi et al., 2009). The distinction between the last two examples can be subtle, but the first case should be obvious. When cells are cultivated in an environment with a steady but insufficient supply of a limiting nutrient to which the cells typically adapt, nutrient limitation occurs. The culture's dependence on internal reserves due to the lack of nutrients in the surroundings is a sign of nutrient insufficiency. A few hardy strains were produced in each of the three scenarios, and microalgae that had previously been grown in a nutrient-sufficient environment were subjected to the nutrient-deficient scenario in a study by Rodolfi et al., (2009). The genus *Nannochloropsis*, the authors discovered, was an exception to the norm, having increased lipid output and content in an N-deficient environment (Rodolfi et al., 2009).

8.5.3 Bubble Column Bioreactor

The bubble column bioreactor is made up of a cylindrical tank with metal microporous sparger or pipes or plates that are perforated for gas. The schematic diagram of bubble column bioreactor is shown in Figure 8.3. Through a sparger at the column's base, air bubbles are introduced into a liquid phase or a liquid-solid suspension in bubble columns (Wilkinson et al., 1992).The structural design of these reactors is extremely basic. Features of mass transfer and hydrodynamics are dependent on gas flow rate, column diameter, sparger design, and medium qualities (e.g., viscosity). Because there isn't any back mixing in the gas phase, bubbles rise at a comparable upward flow velocity at low gas flow rates (homogeneous flow). Most often, liquid mixing occurs right after bubbling. Bubble columns often don't have any extra interior features. When the gas flow rate increases, the flow is seen as heterogeneous. Liquid flows downward toward the column wall as a result of bubbles and liquid rising mostly in the column's center. Under this regime, flow is divided (Singh et al., 2014). The mixing in bubble columns is not as strong as it is in airlift reactors and stirred vessels. As a result, bubble column reactors perform well in terms of heat and mass transfer; nevertheless, scaling up will necessitate a deeper comprehension of multiphase fluid dynamics and its implications. These benefits include compactness and low maintenance and operating expenses due to the absence of moving parts when compared to stirred tank bioreactors. These bioreactors are frequently used to create synthetic fuels like methanol, which are significantly more environmentally friendly than fuels obtained from petroleum (Singh et al., 2014). Conversely, the bubble column bioreactor is a perfect reactor for simultaneous saccharification and fermentation (SSF) because of its many advantages, including less stress, cheap energy, and operating costs, and no need

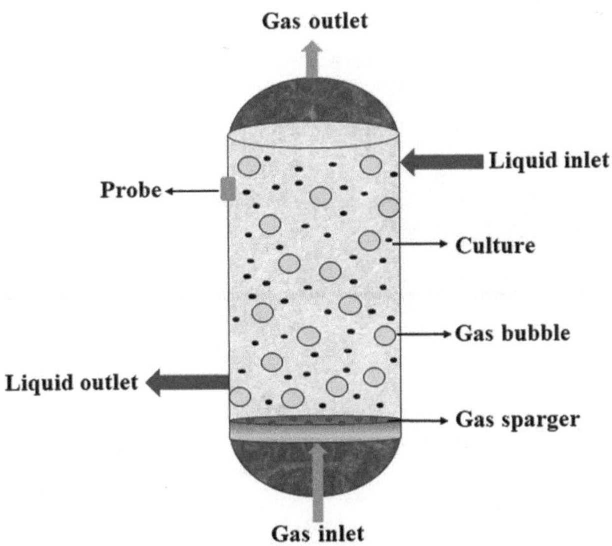

FIGURE 8.3 Schematic diagram of bubble column bioreactor.

for mechanical part movement. These reactors have been employed in the industry for biological processes as well as in gas-liquid-solid systems (Hwee, 2022). There aren't many studies on biomass conversion in a three-phase system utilizing these kinds of bioreactors, as far as we know.

8.5.4 Membrane Bioreactor

The membrane bioreactor (MBR) process is a variation on the traditional AS process, where the effluent and activated sludge are separated (solid/liquid separation) using a membrane filtration unit (such as microfiltration, UF, or ultrafiltration, MF) rather than a secondary sedimentation tank (Rahman & Al-Malack, 2006). The schematic diagram for membrane bioreactor is shown in Figure 8.4. The materials used in membrane configurations might be polymeric, metallic, or inorganic (ceramic). The most common shapes are tubular, hollow fiber, and flat sheet (plate). Some of the main features of ceramic membranes include fouling control, strong resistivity to corrosion, and backwashing. They cost more, though, than polymeric membranes, which seem to be the most extensively employed in modern applications. Examples of these include polyethylene (PE), polyvinylidene fluoride (PVDF), polyethersulfone (PES), and polysulfone (PSF) membranes. Membrane pores normally have a size of 0.01–0.45 mm (Chen & Uan, 2013). Systems using membrane bioreactors typically have two different configurations: (I) A submerged membrane bioreactor (SMBR), in which the permeate is immediately suctioned via dead-end filtration while the membrane module is submerged inside the bioreactor. Keep in mind that membranes may be immersed in the membrane tank or the

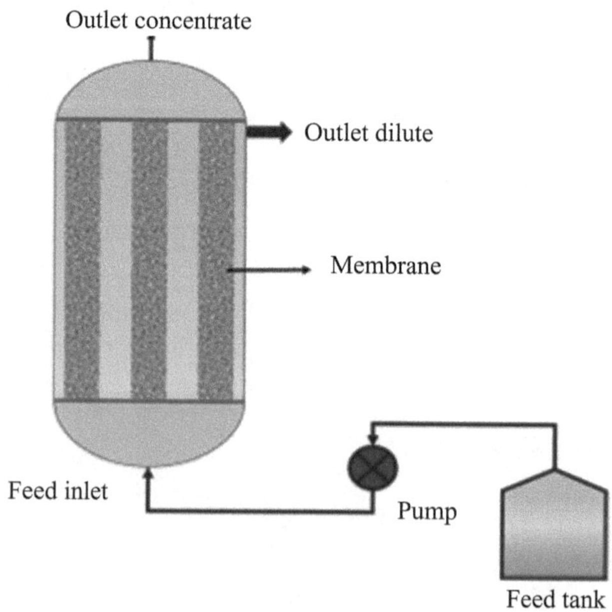

FIGURE 8.4 Schematic membrane bioreactor.

aeration tank. Air is given for biological processes and membrane scouring in submerged designs (Chen & Uan, 2013). A SMBR can apply both flat-plate (vertical) and hollow-fiber (horizontal or vertical) membranes, and it is more frequently employed to treat municipal wastewater (Pombo et al., 2011). (II) a cross-flow membrane bioreactor (CFMBR), where the aeration tank is not the location for the membrane module installation. The liquid-solids mixture known as mixed liquor is pumped to the membrane module, where it undergoes cross-flow filtration across the membrane to filter it. The excess flow, or retentate, is circulated to the aeration tank, while the permeate is released. Horizontal or vertical tubular membranes are frequently utilized in this design (Pombo et al., 2011).

The development of membrane bioreactors as a foundation for technology for manufacturing value-added goods such as chemicals, polymers, fuels, and pharmaceuticals is gaining more attention (Mazzei et al., 2017). Any bioreactor system combined with a membrane separation unit is referred to as a membrane bioreactor in general(Stephenson et al., 2000). Membranes have a high specific surface area and selectivity, which makes them useful for delivering substrates, removing products, and/or retaining biological agents. Nevertheless, the industry has not yet taken advantage of these bioreactor systems for value-added chemical production, even if they demonstrate promising results at the laboratory scale. Extensive research has been conducted to find effective fermentative organisms, inexpensive fermentation substrates, and ideal fermentation environments. Batch, continuous, fed-batch, semi-continuous, immobilized, and MBR systems are among the several fermentation systems that are employed. In addition to saving energy and water, the MBR systems have the ability to continually extract ethanol from the fermentation process, which speeds up the process and increases throughput without the need for extra tanks to be installed.

To ensure appropriate biomass accumulation, a membrane filter is used during treatment in an anaerobic membrane bioreactor (AnMBR). This reactor's general construction is comparable to that of CSTR or UAFR, with the exception that it additionally employs a membrane filter as packing (Kiani et al., 2022). Membranes for microfiltration and ultrafiltration are employed to retain sludge for this reason. Sludge retention increases solid retention time and reduces sludge output, both of which boost process efficiency (Bohdziewicz et al., 2008). Membrane filters have two possible uses: submerged mode inside the reactor and external mode outside the reactor. In the external configuration, the membrane unit is isolated from the reactor and is fed with mixed liquor under pressure from the bioreactor. The membranes operate at a tangential flow under pressure, allowing the permeate to be discharged and the retentate to be recycled back into the bioreactor. This external configuration requires the use of a mechanical stirrer inside the reactor. In the internal configuration, the membrane unit is directly submerged in the bioreactor (Zhen et al., 2019). Systems using an MBR can offer the necessary step toward treating wastewater in a sustainable manner. Since they can regulate solid concentration and HRT, produce an effluent with high-quality chemical oxygen demand (COD) and total suspended solids, and support the growth of slow-growing microbes like methanogens, they are commonly utilized in wastewater treatment. Membrane fouling, which reduces the amount of membrane area available for filtering and raises running expenses even more, is this bioreactor's primary drawback (Buntner et al., 2013). The HRT that is applied determines membrane fouling. Fouling of membranes is caused by an increase in extracellular polymeric polymers and dissolved microbial matter generation, which occurs as HRT decreases and OLR rises (de Barros et al., 2017). This reactor uses a lot of energy since it has a pumping system. Moreover, membrane blockage may have an impact on efficiency (Shoener et al., 2016).

Final microalgal biomass concentration is known to be influenced by light, nutrients, and CO_2 transfer rate (SUN et al., 2020). In order to improve the efficiency of CO_2 transmission in microalgae growing systems, hollow-fiber membranes (Mortezaeikia et al., 2016), selective CO_2 transfer membranes (Neori, 2011) and integrated alkali absorbent membrane systems (Ibrahim et al., 2018) were effectively included into their works. The outcomes showed that microalgae biomass was increased and that the carbon availability in microalgae suspensions was significantly improved.

8.5.5 Airlift Bioreactor

With the exception of the impeller, stirred tank reactors and airlift bioreactors are fairly similar. The schematic diagram for airlift bioreactor is shown in Figure 8.5. This gas liquid bioreactor operates on the basis of a draft tube. Compressed air is employed in this reactor for agitation and aeration. For optimal growth, hairy roots typically need 0.050.4 vol of air/vol of liquid/min oxygen delivery. The glass grid that serves as the reactor's aerator facilitates the passage of humidified air, which is

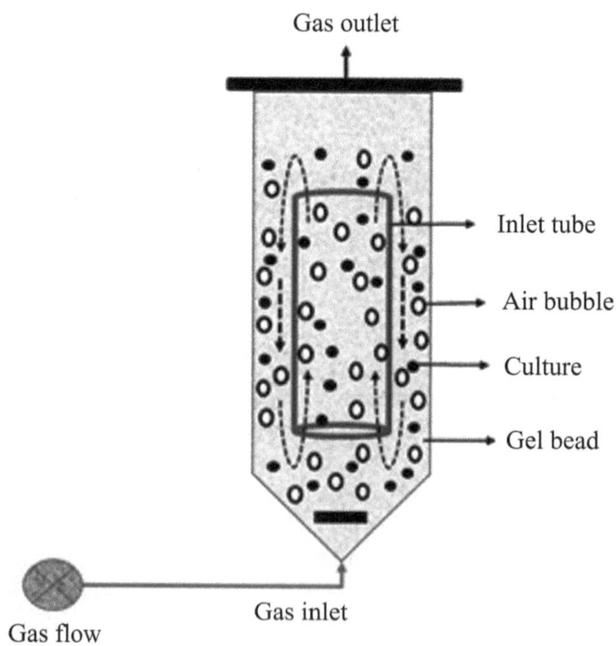

FIGURE 8.5 Schematic diagram of airlift bioreactor.

necessary for mixing and oxygenation. Although they are used to cultivate various hairy roots, airlift reactors also have disadvantages. The creation of a dead zone as a result of inadequate mixing and an uneven nutrient supply brought on by the high biomass density of the reactor is one of its main shortcomings. An airlift bioreactor uses less energy and has a straightforward design that allows for good mixing and high mass transfer (Sivasubramanian & Prasad, 2009). Air bubbles supplied through diffusers at the reactor's bottom enhance mixing in the airlift bioreactor. As a result, fungal pellet development is encouraged, and shear stresses are avoided.

8.5.6 Hybrid Bioreactor

The development of a hybrid biological reactor (HBR) entailed adding carriers to the aeration tanks of a normal suspended-growth system (activated sludge process) in order to introduce a new phase of attached biomass. The schematic diagram for hybrid bioreactor is shown in Figure 8.6. By incorporating porous materials into a standard activated sludge unit, a unique HBR was created that could handle both suspended and attached-growth biomass. An anaerobic digester that combines an upflow anaerobic sludge blanket (UASB) reactor and an anaerobic filter is called a hybrid reactor. This combination is a sophisticated version that allows for longer solid retention times. A special kind of bioreactor system that can accommodate both adherent and suspension culture is the VacciXcell Hybrid Bioreactor. For suspension culture, it applies the stir principle; for adherent culture, it applies the tide motion principle. All it takes to transition the VacciXcell Hybrid Bioreactor from suspension culture to adherent culture is to swap out the impeller for the matrix vessel. The BioNocTM II carriers, which the cells attach to and grow on, are contained in the matrix vessel. A motor oscillates the vessel up and down to submerge the carriers in the nutrient solution. The matrix vessel is half-perforated and has a cylindrical form. The non-perforated half is utilized for seeding, while the perforated half is used for medium

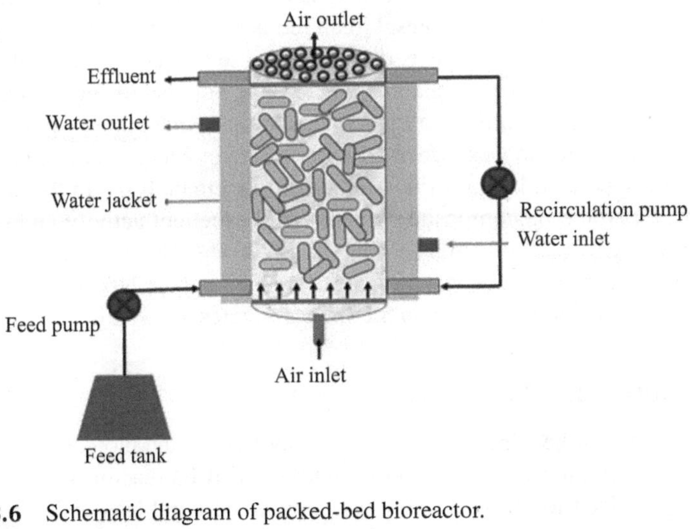

FIGURE 8.6 Schematic diagram of packed-bed bioreactor.

submersion. A magnetic stirrer that is integrated into the reactor vessel keeps the nutrient medium homogenous while it stays at the bottom. Both headspaces overlay, and the dissolved oxygen in the nutrient medium contribute to aeration. The VacciXcell Hybrid Bioreactor's multifunctionality is a benefit. Determining which kind of culture will produce the intended product most effectively is crucial in a start-up or research and development setting. The VacciXcell Hybrid Bioreactor allows users to experiment with four distinct culture types: suspension culture, fermentation, adherent culture using microcarrier in a stirred tank environment, and adherent culture using macrocarrier. Data collection and analysis would be simpler and more effective if the design of experiment (DOE) could be completed with these four cultural types in a single system. This would reduce overall costs and shorten the time needed to develop new processes.

8.5.7 Packed-bed Bioreactor

Among the fixed-bed bioreactors are packed-bed bioreactors and the schematic diagram for packed-bed bioreactor is shown in Figure 8.6. Carriers are materials that are packed or put into a vessel to aid in the immobilization of cells. It is possible to create packing materials that allow fluids to move from one end to the other using polymer, glass, ceramic, and natural materials. Immobilized cells can grow at very high densities inside macroporous carriers using packed-bed bioreactors (Meuwly et al., 2007). The materials used can be stiff, compressible, or nonporous gels. Nutrients can be added to the immobilized biocatalyst from the top or bottom of the column. Within the solid bed, a fluid that contains substrate and nutrient solution is continuously flowing and coming in touch with the immobilized biocatalyst. The concentration of the nutrients (and thus the products generated) may be increased by increasing the nutrient broth's flow velocity. The period of residence and the flow of fluid are adjusted to increase or decrease substrate contact with the bed. The packed-bed bioreactor's exit is where the outputs are extracted. Because of insufficient mixing, it is difficult to change the pH of packed-bed bioreactors by adding acid or alkali. Continuous PBRs are an excellent option for producing biodiesel in an industrial setting because they increase the rate of reaction, separate products directly, and use less energy and oil or alcohol by strengthening the mass/heat phenomenon. For the purpose of producing biodiesel through soybean oil methanolysis, a PBR system employing a lipase–Fe_3O_4 nanoparticle biocomposite catalyst was effectively constructed (Wang et al., 2011). The rate of reaction was increased by emulsification prior to methanolysis. In the single PBR, the lipase-nanoparticle biocomposite demonstrated excellent activity and stability at the ideal flow rate of 0.25 mL min^{-1} (Wang et al., 2011). The reaction continued for 240 hours, with a continuous conversion rate of 45%. Compared to the single PBR, the four PBR produced significantly greater conversion rates and stability.

8.5.8 Fluidized-bed Biorecator

With the exception of keeping solid particles suspended in the bioreactor by reducing fluid motion, the fluidized-bed bioreactor and the airlift bioreactor are similar. Solid particles are fluidized to create a two-phase fluid-solid combination in the fluidized bioreactors according to their design. A downhill or upward recirculation stream keeps the bed of solid particles suspended in the bioreactor as the fluid flows away (Özkaya et al., 2019). To guarantee ongoing contact between the biocatalysts and the reaction components, recycling the liquid is essential. High bioprocessing effectiveness is made possible by this. These reactors have constant temperature gradients and particle mixing since they operate nonstop. These reactors use immobilized microparticles called cells that move with the fluid. Additionally, it's critical to ensure that the suspended solid particles are neither unduly light nor dense. Although the smaller particle size may float, it allows for a faster rate of mass transfer, oxygen transmission, and nutrition transfer to the cells. On the other hand, too dense particles could sink to the bottom and not stay in a nice, suspended state. To resist gravity and avoid elutriation, it's also critical to make sure the fluid is flowing at an appropriate upward velocity. The fluidized-bed bioreactors can be used in reactions with fluid-suspended biocatalysts, such as immobilized enzymes, cells, and microbial flocs, that are housed in porous particles called biofilm carriers. In order to address the harvesting process where the support material was fluidizing inside the culture media, microalgae cultivation via fluidized-bed bioreactor was introduced. This allowed the microalgae to settle onto the surface of fluidized material and grow from there (Rosli et al., 2020). This mode of cultivation facilitates the harvesting process and improves *Chlorella vulgaris* growth because it makes it easier to separate the biomass from the culture media on the support material.

An anaerobic fluidized-bed reactor (AFBR) is a device that accepts biomass and has a non-fixed supporting material (Kiani et al., 2022). Recently, the technique has been employed to minimize waste's organic matter and sulfates. Wastewaters with a high level of organic debris have also been treated with it. Among the most practical AD bioreactors is this one. Because the biomass is immobilized on supporting particles, it remains in the reactor and is not washed off. Substrates exit the reactor by ascending due to drag force. The reactor's internal gas and liquid circulation system carries out the stirring process. The microbial communities that are doing the adsorption of the wastewater in this reactor transform the wastewater into tiny, fluidized particles that have changed in size, shape, density, and hydrodynamic behavior. These tiny, hefty particles are surrounded and covered by bilayers of biomass. Because the particles are heavy, the reactor has a very high precipitation rate (approximately 50 m/h), which raises the fluid velocity (10–30 m/h). Wastewater builds up and sludge activity rises as a result of this increased velocity. A tiny volume is necessary for AFBR. This reactor was used for AD in the 1970s. In 1984, the first large-scale AFBR was produced. AFBR has a height-to-diameter ratio of 14:1, and it can be either ascending or descending. In order to improve the cohesiveness of microorganisms and stop their leaching, an AFBR often contains mineral matter, gravel, sand, activated carbon, anionic and cationic exchange resins, and polystyrene particles (Kiani et al., 2022). When compared to UASB reactors, these supporting materials allow for higher operating temperatures and a higher material flow velocity. The majority of them are supported by gravel. This reactor stands out for its high mass and heat transfer rates, uniform distribution of the liquid phase, good treatment efficiency at high OLR, and stability. Additional benefits of this reactor include improved substrate digestion and the prevention of granule formation due to increased bed porosity. It is also less susceptible to external variables like pH and temperature (Saravanane & Murthy, 2000). Compared to other methods, sludge retention time is high (De Bazua et al., 1991). Numerous experiments have been performed on this reactor. Fernández et al. (2001) conducted studies with two types of supporting materials, zeolite beds and active carbon, in two experimental reactors with capacities of 750 mL and 1 L (Fernández et al., 2001). They found that both beds could remove 70% of the COD in vinasse at an OLR rate of 10 kg/m^{3}d. The production rate of methane gas was 2 L/Ld on average. While ammonium was reduced to less than 1 mg/L in the presence of zeolite, the formation of sulfide gas was negligibly reduced in the presence of active carbon. In the experiment performed (Siqueira et al., 2013), peak biogas productivity was determined to be 0.402 m^3/kg-COD and optimum COD removal efficiency (70%) was reached at an OLR value of 13.93 ± 2.18 kg-COD/m^{3}d on vinasse bio digestion at 30°C. The issues with large-scale fluidized-bed reactors, such as the stability of fluidized particles – anaerobic microorganisms – have not yet been assessed, despite their great efficiency. Furthermore, the development of microbes alters the size of the particles, which could have an impact on the system's hydrodynamics. Furthermore, this reactor requires a lot of energy to run in order to maintain the appropriate fluidization levels.

8.5.9 The Horizontal Anaerobic Immobilized Biomass (HAIB) Reactor

One type of fixed-bed horizontal flow reactor created by Damianovic et al. (2009) is the HAIB reactor (Damianovic et al., 2009). It was first applied to a paper manufacturing company's wastewater. Supporting materials like polystyrene, polyurethane foam, and sand particles are employed before the reactor outlet or along the reactor to prevent biomass washout in this reactor. When these substances are present, bacteria multiply more effectively (Damianovic & Foresti, 2007). This reactor's large length-to-diameter ratio helps to improve process stability. This reactor is similar to a tube; materials enter via one end and exit through the other when they come into touch with microorganisms and pass through the supporting material. The materials are pumped into the reactor by a pump. Along the reactor, the gas is created and exits the system. For sampling purposes, valves are included within the reactor body (Rodrigues et al., 2021). The hydrodynamic flows in the reactor architecture can be thought of as a plug flow pattern. Stirring may be done here without the need for a separate mixing device because of the reactor's axial flow and the gas's vertical rise. Increasing the concentration of biomass in this reactor increases resilience to the presence of hazardous materials and shock loads while simultaneously accelerating the rate of substrate disintegration (Mockaitis et al., 2014). The main drawback of this reactor is that it clogs the bed for wastewaters high in solids, like sugarcane vinasse, due to the excessive particles buildup and the ongoing growth of biomass. When it comes to vinasse treatment, this is an important constraint (de Aquino et al., 2017).

8.5.10 Fix-bed Reactor

There are two varieties of fixed-bed reactors (FBRs): ascending reactors and descending reactors (Sreekrishnan et al., 2004). In the descending kind, feeding enters the reactor from the top and exits from the bottom after going through the supporting materials, which are positioned inside other materials or stainless steel cylinders, and creating a biofilm layer (Zhang et al., 2012). In order to create the biofilm layer on the material surface, which separates solid particles and completes biological treatment, supporting materials such as polyethylene, gravel, granules of activated carbon (GAC), polyvinyl chloride (PVC), and plastic materials are typically utilized (Robles-González, 2011). The ascending kind involves pumping wastewater into the reactor from the bottom and allowing it to exit from the top after it has gone through the layer of biofilm. On the reactor's top, biogas and liquid effluent are separated (Liu et al., 2004). The biofilm's matrix structure needs to be mechanically stable. Both high and low total dissolved solids (TDS) values are utilized with this kind of reactor (Sreekrishnan et al., 2004). Among this reactor's qualities are its capacity to remove COD at low HRT levels and its stability in the face of OLR shocks (Robles-González, 2011). Other benefits of those reactors include the potential to increase the biomass content in order to improve degradation rates and the growing resistance to the presence of harmful compounds (Mockaitis et al., 2014). Even with the benefits of FBR that have already been discussed, processing wastewaters that are high in particles, like vinasse, may eventually cause the bed to clog due to the excessive buildup of solids. It is necessary to look into this issue in detail.

8.6 FACTOR EFFECTING THE BIOFUEL PRODUCTION IN BIOREACTOR

The right conditions should be given in bioreactors used to produce biogas for all microbial communities involved in substrate conversion, particularly methanogens. Since microbes are inhibited or inactivated in strongly acidic or basic conditions, the pH value should also be checked. Controlling the temperature is important since it affects both the living environment and the activity of microorganisms. The presence of inhibitors and the actions of microorganisms in the event of a nutrient shortage may result in subpar reactor performance. The parameters should be changed for high-strength wastewaters, such as vinasse, which are often produced at high rates, in order to maximize biogas production and remove as much COD as possible at the highest OLR in the shortest amount of time.

The majority of bacteria, with the exception of a small number, prefer a neutral pH (Hagos et al., 2017). It has been observed that the pH range of 6.8–7.2 is optimal for biogas production. Methanogens are sensitive to pH fluctuations due to their slow growth rate, and they are typically best at a pH of roughly 7. Conversely, acidogenic bacteria exhibit reduced sensitivity to pH fluctuations, with their ideal pH falling between 4 and 8.5 (Hagos et al., 2017). Vinasse's acidic nature prevents methanogen growth because it exits the distillation column at a pH of 3–5 (Boncz et al., 2012). Sufficient buffering capacity or alkalinity is needed to maintain the pH of the anaerobic digester at a minimally acceptable level (Gerardi, 2003). To create the ideal pH conditions for vinasse biodegradation, 0.004 g of NaOH per g of COD and 0.28–1 g of $NaHCO_3$ per g of COD are often utilized as buffers (Fuess et al., 2017a). Other approaches showed that combining wastewater from sugarcane processing with vinasse led to a large reduction in the amount of vinasse that needed to be treated, an increase in pH, and a reduced concentration of inhibitors (González et al., 2017). However, by mixing raw vinasse with sufficient amounts of OLR and HRT to prevent the overconsumption of chemicals throughout the process, a variety of experiences have been generated (Cabrera-Díaz et al., 2017).

Temperature is a basic factor that influences substrate conversion and determines the efficiency of biogas production. Temperature influences microbial development and metabolic pathways, which in turn influences digestate qualities as a fertilizer and biogas output and yield (Ohimain & Izah, 2017). Vinasse exiting the alcohol production facility has a temperature between 65°C and 107°C. It may be susceptible to thermophilic AD depending on the temperature – for example, at 55°C (Fuess et al., 2017b); this is more practical from an economic standpoint than carrying out the procedure at a mesophilic temperature (Barros et al., 2016). Since most hydrolytic activity is favored at higher temperatures, it is generally faster to execute the process at thermophilic settings than mesophilic conditions for the recalcitrant substrates. Mesophilic conditions (characterized by a higher microbial diversity and therefore potentially more capable of enduring operating perturbations) are associated with more robust reactors, but the higher temperatures also cause the microbial diversity to decrease as the sludge becomes more specialized.

The existence of phenols and their antioxidant properties is one of the factors that has to be taken into account in AD of vinasse. It is shown that *Clostridium* sp. plays a part in the breakdown of polyphenols (Jung et al., 2010), and during the acidogenic phase, it is one of the main species that produces hydrogen. It might be the primary cause of both the increased COD removal efficiency and the creation of hydrogen, a substitute biofuel.

When the concentration of nutrients that are available to microorganisms beyond the necessary threshold, the nutrients function as a hazardous material based on the original COD. It demonstrates that the current levels of nutrients are inadequate because of the high COD content in vinasse. Insufficient levels of macro- and micronutrients in vinasse might lead to inefficient production of biogas. Vinasse can be combined with animal dung or industrial organic wastes (cellulose) to make up for the micro and macro-nutrient deficiencies. Moraes et al. (2015) demonstrated that 15% more methane may be produced by adding cow dung to vinasse (Moraes et al., 2015). Vinasse's phosphorus content is insufficient for AD, and direct phosphorous injection cannot make up for it (Sanchez Riera et al., 1985). As an alternative, sodium phosphate and urea can be added to provide the ideal levels of phosphorus and nitrogen (Siqueira et al., 2013). According to Boncz et al. (2012), 0.215 g/g-COD is the ideal urea concentration for maximizing biogas production. There are differences in the ideal C/N ratio for producing biogas from different sources. For easy digestion of the carbon sources, a C/N ratio of 20–30 is recommended (Hagos et al., 2017). Agricultural goods have been shown to have a high carbon-to-nitrogen ratio because of their low residual nitrogen content. Addits with a high carbon content, like lignocelluloses and agricultural leftovers, should be added to the digester when the C/N ratio is low. Few research have been conducted on the co-digestion of sugarcane processing industry residues thus far, but this is an important alternative to consider in order to improve the perception and application of biorefineries. A recent example showed that co-digesting sugarcane vinasse and glycerol, a C-rich feedstock, is a feasible alternative for managing the primary residual streams from two significant biofuel industries: biodiesel and bioethanol plants (do Vale Borges et al., 2021).

As previously stated, the amount of vinasse varies from nation to nation and is determined by industrial regulation. Cell growth is inhibited when the water content is below the necessary threshold, while the digester's acidity is enhanced when the solid concentration is higher (Gerardi, 2003). Maximum biogas yield has been reported to be achievable at a vinasse-to-water ratio of 1:3 (Gerardi, 2003).

The OLR, with respect to the availability of the effective substrate (organic matter convertible into biogas), is the quantity of organic material added to the reactor (per unit of volume) in a specific amount of time (Ohimain & Izah, 2017). In order for the reactor to remove significant amounts of pollutants and produce biogas at the highest potential output, the value of OLR should be set (de Barros et al., 2017). HRT is a function of OLR; that is, (while keeping a fixed COD) the longer the HRT, the lower the OLR. HRT influences the proliferation and metabolic activity of microorganisms, which has a major impact on the effectiveness of wastewater treatment (Sreekrishnan et al., 2004).

Because the requirements of methanogenic and acidogenic bacteria varies, as do the properties of the substrates, there is no one rule that can be used to determine the optimal HRT in anaerobic reactors (de Barros et al., 2017). The nature and structure of the reactor determine optimal OLR and HRT (Ohimain & Izah, 2017). While the washing of slow-growing bacteria is impeded, hard, degradable chemical digestion is given enough time at high HRT (de Barros et al., 2017). In this instance, a bigger reactor volume is needed, which could raise the plant's implementation costs (Sreekrishnan et al., 2004). The combined APBR (acidogenic)-ASTBR (methanogenic) configuration was subjected to a maximum OLR of 30 kg-COD/m^{3}d (Fuess et al., 2017b), ultimately creating 0.329 Nm^3/kCOD methane while eliminating 84.6% of the COD in the vinasse. During the 3.6-day HRT, an MBR at an OLR of 2.5 kg-COD/m^{3}d produced the greatest COD removal of 97.2% (Mota et al., 2013). Using an AFBR operating at an OLR of 25.32 m^3/kg-COD over the course of a day, a maximum methane gas production rate of 0.386 m^3/kg-COD was achieved (Siqueira et al., 2013). The combo of an APBR with a UASB working at an OLR of 25 kg-COD/m^{3}d, where 59.4% of COD was eliminated and methane gas generated at 0.306 m^3/kg-COD, produced the shortest HRT of 10.2 h (Júnior et al., 2016).

The development of economical, energy-efficient, and environmentally sustainable methods is crucial for the commercial production of bioethanol. Two key techniques used for the industrial manufacture of second-generation bioethanol are pre-saccharification and simultaneous fermentation (PSSF) and SSF (Hu & Rodrigues Reis, 2017). In the first, the PSSF is split into two stages: the saccharification, which occurs between 8 and 24 hours at 50°C (the enzyme's optimal temperature), and the fermentation process, which takes place between 30°C and 37°C. In the second, the SSF, enzymatic hydrolysis and fermentation are done at a higher temperature (30°C–37°C; Moraes et al., 2014). Yeast strains need to adjust to higher temperatures in order to produce more ethanol, which is a significant alternative and approach to improve the operating conditions of enzymes and microorganisms. Microorganisms can ferment a wide range of inexpensive carbon-rich sources to make biodiesel without interfering with food production. *Oleaginous* microorganisms are defined as microorganisms that have a microbial lipid content of more than 20%. The schematic diagram for the production of biodiesel from *Oleaginous* microorganisms is shown in Figure 8.7. They are regarded as an alternative feedstock for the synthesis of oil and fat (Ma et al., 2018). Global interest has been piqued in the manufacture of

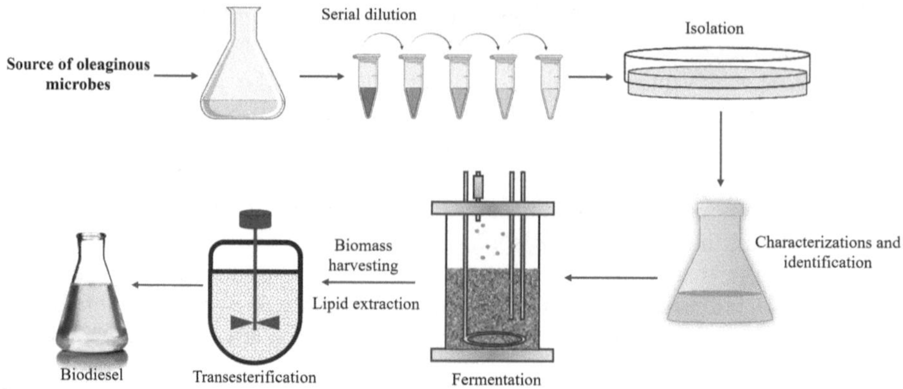

FIGURE 8.7 Flow diagram for biodiesel production from *Oleaginous* bacteria.

biodiesel from microbial lipids known as single-cell oils Mhlongo et al., (2021). According to Patel et al. (2020), *Oleaginous* bacteria primarily synthesize lipids with lengths ranging from 4 to 28 unbranched carbon chains (Patel et al., 2020). These lipids can be either saturated or unsaturated fatty acids, contingent on the kind of carbonated hydro chain and the quantity of double bonds present.

8.7 CONCLUSIONS

Biofuels are acknowledged as a viable solution to the present energy crisis, particularly for nations like China that heavily rely on crude oil for social and economic development and were affected by the oil shortage. The primary goal of biofuels is to reduce costs because, despite significant advancements in co-fermentation, biomass pretreatment, and cellulose production in recent decades, biofuels are still not as economically viable as fuels derived from petroleum. A thorough understanding of the properties of lignocellulosic biomass, modernizations for generating more skilled microbial strains and cellulose for higher yields, process integration, reaction rates, and optimization for the reduction of energy utilization that necessitate persistent effort are all included in the multidisciplinary study of the entire process. Moreover, upgrading a feedstock-based biorefinery to use biomass more widely and produce more beneficial by-products during the interim, such as bio-based materials derived from the lignin component, would level the playing field and increase the viability of bioethanol production.

REFERENCES

Abu-Rezq, T.S., Al-Musallam, L., Al-Shimmari, J., Dias, P. 1999. Optimum production conditions for different high-quality marine algae. *Hydrobiologia*, 403, 97–107.

Ashrafizadeh, S.N., Seifollahi, Z. 2021. Trends in biotechnology and ties with chemical engineering. *Journal of Biotechnology and Biomedicine*, 4(4), 169–186.

Barros, V.G.D., Duda, R.M., Oliveira, R.A.D. 2016. Biomethane production from vinasse in upflow anaerobic sludge blanket reactors inoculated with granular sludge. *Brazilian Journal of Microbiology*, 47, 628–639.

Berndes, G. 2002. Bioenergy and water—The implications of large-scale bioenergy production for water use and supply. *Global Environmental Change*, 12(4), 253–271.

Bohdziewicz, J., Neczaj, E., Kwarciak, A. 2008. Landfill leachate treatment by means of anaerobic membrane bioreactor. *Desalination*, 221(1–3), 559–565.

Boncz, M., Formagini, E., Santos, L.d.S., Marques, R., Paulo, P. 2012. Application of urea dosing for alkalinity supply during anaerobic digestion of vinasse. *Water Science and Technology*, 66(11), 2453–2460.

Brindhadevi, K., Shanmuganathan, R., Pugazhendhi, A., Gunasekar, P., Manigandan, S. 2021. Biohydrogen production using horizontal and vertical continuous stirred tank reactor-a numerical optimization. *International Journal of Hydrogen Energy*, 46(20), 11305–11312.

Bumbak, F., Cook, S., Zachleder, V., Hauser, S., Kovar, K.J. 2011. Best practices in heterotrophic high-cell-density microalgal processes: Achievements, potential and possible limitations. *Applied Microbiology and Biotechnology*, 91, 31–46.

Buntner, D., Sánchez, A., Garrido, J. 2013. Feasibility of combined UASB and MBR system in dairy wastewater treatment at ambient temperatures. *Chemical Engineering Journal*, 230, 475–481.

Bušić, A., Marđetko, N., Kundas, S., Morzak, G., Belskaya, H., Ivančić Šantek, M., Komes, D., Novak, S., Šantek, B.J. 2018. Bioethanol production from renewable raw materials and its separation and purification: A review. *Food Technology and Biotechnology*, 56(3), 289–311.

Cabrera-Díaz, A., Pereda-Reyes, I., Oliva-Merencio, D., Lebrero, R., Zaiat, M. 2017. Anaerobic digestion of sugarcane vinasse through a methanogenic UASB reactor followed by a packed bed reactor. *Applied Biochemistry and Biotechnology*, 183, 1127–1145.

Chang, H., Fu, Q., Zhong, N., Yang, X., Quan, X., Li, S., Fu, J., Xiao, C. 2019. Microalgal lipids production and nutrients recovery from landfill leachate using membrane photobioreactor. *Bioresource Technology*, 277, 18–26.

Chang, H., Quan, X., Zhong, N., Zhang, Z., Lu, C., Li, G., Cheng, Z., Yang, L. 2018. High-efficiency nutrients reclamation from landfill leachate by microalgae Chlorella vulgaris in membrane photobioreactor for bio-lipid production. *Bioresource Technology*, 266, 374–381.

Chen, C.-Y., Saratale, G.D., Lee, C.-M., Chen, P.-C., Chang, J.-S. 2008. Phototrophic hydrogen production in photobioreactors coupled with solar-energy-excited optical fibers. *International Journal of Hydrogen Energy*, 33(23), 6886–6895.

Chen, J., Uan, D.K. 2013. Low dissolved oxygen membrane bioreactor processes (LDO-MBRs): A review. *International Journal of Environmental Engineering*, 5(2), 129–149.

Chisti, Y. 2007. Biodiesel from microalgae. *Biotechnology Advances*, 25(3), 294–306.

Damianovic, M.H.R.Z., Foresti, E. 2007. Anaerobic degradation of synthetic wastewaters at different levels of sulfate and COD/sulfate ratios in horizontal-flow anaerobic reactors (HAIB). *Environmental Engineering Science*, 24(3), 383–393.

Damianovic, M.H.R.Z., Moraes, E.M., Zaiat, M., Foresti, E. 2009. Pentachlorophenol (PCP) dechlorination in horizontal-flow anaerobic immobilized biomass (HAIB) reactors. *Bioresource Technology*, 100(19), 4361–4367.

Das, P., Jha, C.K., Saxena, S., Ghosh, R.K. 2024. Can biofuels help achieve sustainable development goals in India? A systematic review. *Renewable and Sustainable Energy Reviews*, 192, 114246.

Dasgupta, C.N., Gilbert, J.J., Lindblad, P., Heidorn, T., Borgvang, S.A., Skjanes, K., Das, D. 2010. Recent trends on the development of photobiological processes and photobioreactors for the improvement of hydrogen production. *International Journal of Hydrogen Energy*, 35(19), 10218–10238.

de Aquino, S., Fuess, L.T., Pires, E.C. 2017. Media arrangement impacts cell growth in anaerobic fixed-bed reactors treating sugarcane vinasse: Structured vs. randomic biomass immobilization. *Bioresource Technology*, 235, 219–228.

de Barros, V.G., Duda, R.M., da Silva Vantini, J., Omori, W.P., Ferro, M.I.T., de Oliveira, R.A. 2017. Improved methane production from sugarcane vinasse with filter cake in thermophilic UASB reactors, with predominance of Methanothermobacter and Methanosarcina archaea and Thermotogae bacteria. *Bioresource Technology*, 244, 371–381.

De Bazua, C., Cabrero, M., Poggi, H. 1991. Vinasses biological treatment by anaerobic and aerobic processes: Laboratory and pilot-plant tests. *Bioresource Technology*, 35(1), 87–93.

Delavar, M.A., Wang, J. 2022. Modeling microbial growth of dynamic membrane in a biohydrogen production bioreactor. *International Journal of Hydrogen Energy*, 47(12), 7666–7681.

Demirbas, A., Demirbas, M.F. 2011. Importance of algae oil as a source of biodiesel. *Energy Conversion and Management*, 52(1), 163–170.

do Vale Borges, A., Fuess, L.T., Alves, I., Takeda, P.Y., Damianovic, M.H.R.Z. 2021. Co-digesting sugarcane vinasse and distilled glycerol to enhance bioenergy generation in biofuel-producing plants. *Energy Conversion and Management*, 250, 114897.

Doucha, J., Straka, F., Lívanský, K. 2005. Utilization of flue gas for cultivation of microalgae (Chlorella sp.) in an outdoor open thin-layer photobioreactor. *Journal of Applied Phycology*, 17, 403–412.

El-Shishtawy, R.M., Kawasaki, S., Morimoto, M. 1997. Biological H 2 production using a novel light-induced and diffused photoreactor. *Biotechnology Techniques*, 11, 403–407.

Fernández, N., Fdz-Polanco, F., Montalvo, S., Toledano, D. 2001. Use of activated carbon and natural zeolite as support materials, in an anaerobic fluidised bed reactor, for vinasse treatment. *Water Science and Technology*, 44(4), 1–6.

Ferrell, J., Sarisky-Reed, V. 2010. *National algal biofuels technology roadmap*. EERE Publication and Product Library, Washington, DC (United States).

Fu, J., Huang, Y., Liao, Q., Xia, A., Fu, Q., Zhu, X. 2019. Photo-bioreactor design for microalgae: A review from the aspect of CO_2 transfer and conversion. *Bioresource technology*, 292, 121947.

Fuess, L.T., de Araújo Júnior, M.M., Garcia, M.L., Zaiat, M. 2017a. Designing full-scale biodigestion plants for the treatment of vinasse in sugarcane biorefineries: how phase separation and alkalinization impact biogas and electricity production costs? *Chemical engineering research and design*, 119, 209–220.

Fuess, L.T., Kiyuna, L.S.M., Júnior, A.D.N.F., Persinoti, G.F., Squina, F.M., Garcia, M.L., Zaiat, M. 2017b. Thermophilic two-phase anaerobic digestion using an innovative fixed-bed reactor for enhanced organic matter removal and bioenergy recovery from sugarcane vinasse. *Applied Energy*, 189, 480–491.

Garduno, R., Smith, M., Baranova, E., Kinsley, C. 2024. Hydrogen production in dark fermentation of a brewery sludge pre-treated with white-rot fungi in submerged culture. *Bioresource Technology Reports*, 25, 101773.

Gerardi, M.H. 2003. *The microbiology of anaerobic digesters*. John Wiley & Sons.

Ghazani, S.M., Boroujeni, Y.S., Shaw, N., Marangoni, A.G. 2024. Scientific issues and challenges with production and refining edible oils and fats. *Cellular Agriculture*, 215–232.

Goldman, J.C. 1979. Outdoor algal mass cultures—II Photosynthetic yield limitations. *Water Research*, 13(2), 119–136.

González, L.M.L., Reyes, I.P., Romero, O.R. 2017. Anaerobic co-digestion of sugarcane press mud with vinasse on methane yield. *Waste Management*, 68, 139–145.

Gopi, R., Thangarasu, V., Ramanathan, A. 2022. A critical review of recent advancements in continuous flow reactors and prominent integrated microreactors for biodiesel production. *Renewable and Sustainable Energy Reviews* 154, 111869.

Hagos, K., Zong, J., Li, D., Liu, C., Lu, X. 2017. Anaerobic co-digestion process for biogas production: Progress, challenges and perspectives. *Renewable and Sustainable Energy Reviews*, 76, 1485–1496.

Hu, B., Rodrigues Reis, C. 2017. Vinasse from sugarcane ethanol production: better treatment or better utilization? *Frontiers in Energy Research*, 5, 7.

Hwee, N. 2022. Methane remediation using biocatalysts in batch gas-solid mass transfer system challenges and prospects. Technical Report. Lawrence Livermore National Laboratory.

Ibrahim, M., Johari, M.A.M., Maslehuddin, M., Rahman, M.K. 2018. Influence of nano-SiO2 on the strength and microstructure of natural pozzolan based alkali activated concrete. *Construction and Building Materials*, 173, 573–585.

Ishak, S., Malakahmad, A., Isa, M.H. 2012. Refinery wastewater biological treatment: A short review. *Journal of Scientific & Industrial Research*, 71(4). 251–256.

Issarapayup, K., Powtongsook, S., Pavasant, P. 2009. Flat panel airlift photobioreactors for cultivation of vegetative cells of microalga Haematococcus pluvialis. *Journal of Biotechnology*, 142(3–4), 227–232.

Jiao, H., Tsigkou, K., Elsamahy, T., Pispas, K., Sun, J., Manthos, G., Schagerl, M., Sventzouri, E., Al-Tohamy, R., Kornaros, M. 2024. Recent advances in sustainable hydrogen production from microalgae: Mechanisms, challenges, and future perspectives. *Ecotoxicology and Environmental Safety*, 270, 115908.

Jitrwung, R., Yargeau, V. 2015. Biohydrogen and bioethanol production from biodiesel-based glycerol by Enterobacter aerogenes in a continuous stir tank reactor. *International Journal of Molecular Sciences*, 16(5), 10650–10664.

Johnson, T.J., Katuwal, S., Anderson, G.A., Gu, L., Zhou, R., Gibbons, W.R. 2018. Photobioreactor cultivation strategies for microalgae and cyanobacteria. *Biotechnology Progress*, 34(4), 811–827.

Jung, K.-W., Kim, D.-H., Shin, H.-S. 2010. Continuous fermentative hydrogen production from coffee drink manufacturing wastewater by applying UASB reactor. *International Journal of Hydrogen Energy*, 35(24), 13370–13378.

Júnior, A.D.N.F., Koyama, M.H., de Araújo Júnior, M.M., Zaiat, M. 2016. Thermophilic anaerobic digestion of raw sugarcane vinasse. *Renewable Energy*, 89, 245–252.

Katsuda, T., Lababpour, A., Shimahara, K., Katoh, S. 2004. Astaxanthin production by Haematococcus pluvialis under illumination with LEDs. *Enzyme and Microbial Technology*, 35(1), 81–86.

Katsuda, T., Shimahara, K., Shiraishi, H., Yamagami, K., Ranjbar, R., Katoh, S. 2006. Effect of flashing light from blue light emitting diodes on cell growth and astaxanthin production of Haematococcus pluvialis. *Journal of Bioscience and Bioengineering*, 102(5), 442–446.

Kiani, M.K.D., Parsaee, M., Ardebili, S.M.S., Reyes, I.P., Fuess, L.T., Karimi, K. 2022. Different bioreactor configurations for biogas production from sugarcane vinasse: A comprehensive review. *Biomass and Bioenergy*, 161, 106446.

Kim, T., Hite, M., Rogacki, L., Sealock, A.W., Sprouse, G., Novak, P.J., LaPara, T.M. 2021. Dissolved oxygen concentrations affect the function but not the relative abundance of nitrifying bacterial populations in full-scale municipal wastewater treatment bioreactors during cold weather. *Science of the Total Environment*, 781, 146719.

Kunjapur, A.M., Eldridge, R.B. 2010. Photobioreactor design for commercial biofuel production from microalgae. *Industrial & engineering chemistry research*, 49(8), 3516–3526.

Kurpan, D., Idà, A., Körner, F.G., Bombelli, P., da Silva Aguiar, J.P., Marinho, L.M., do Valle, A.F., Acién, F.G., Trasatti, S.P., Schievano, A. 2024. Long-term evaluation of productivity and harvesting efficiency of an industrial Spirulina (Arthrospira platensis) production facility. *Bioresource Technology Reports* 25, 101741.

Lee, C.G., Palsson, B.Ø. 1994. High-density algal photobioreactors using light-emitting diodes. *Biotechnology and bioengineering*, 44(10), 1161–1167.

Lee, Y.-K., Tay, H.-S. 1991. High CO 2 partial pressure depresses productivity and bioenergetic growth yield of Chlorella pyrenoidosa culture. *Journal of Applied Phycology*, 3, 95–101.

Liu, J., Olsson, G., Mattiasson, B. 2004. Monitoring and control of an anaerobic upflow fixed-bed reactor for high-loading-rate operation and rejection of disturbances. *Biotechnology and Bioengineering*, 87(1), 43–53.

Ma, F., Hanna, M.A. 1999. Biodiesel production: A review. *Bioresource Technology*, 70(1), 1–15.

Ma, Y., Gao, Z., Wang, Q., Liu, Y. 2018. Biodiesels from microbial oils: Opportunity and challenges. *Bioresource Technology*, 263, 631–641.

Mandalam, R.K., Palsson, B. 1998. Elemental balancing of biomass and medium composition enhances growth capacity in high-density Chlorella vulgaris cultures. *Biotechnology and Bioengineering*, 59(5), 605–611.

Manna, D., Chowdhury, R., Calay, R.K., Mustafa, M.Y. 2024. Experimental insights into the fermentation of pyro-syngas to ethanol in a semi-batch and continuous stirred bioreactor with mathematical modelling and optimization. *Energies*, 17(3), 562.

Matthijs, H.C., Balke, H., Van Hes, U.M., Kroon, B.M., Mur, L.R., Binot, R.A. 1996. Application of light-emitting diodes in bioreactors: Flashing light effects and energy economy in algal culture (Chlorella pyrenoidosa). *Biotechnology and Bioengineering*, 50(1), 98–107.

Mazzei, R., Piacentini, E., Yihdego Gebreyohannes, A., Giorno, L. 2017. Membrane bioreactors in food, pharmaceutical and biofuel applications: State of the art, progresses and perspectives. *Current Organic Chemistry*, 21(17), 1671–1701.

Meuwly, F., Ruffieux, P.-A., Kadouri, A., Von Stockar, U. 2007. Packed-bed bioreactors for mammalian cell culture: Bioprocess and biomedical applications. *Biotechnology Advances* 25(1), 45–56.

Mhlongo, S.I., Ezeokoli, O.T., Roopnarain, A., Ndaba, B., Sekoai, P.T., Habimana, O., Pohl, C.H. 2021. The potential of single-cell oils derived from *Filamentous Fungi* as alternative feedstock sources for biodiesel production. *Frontiers in Microbiology*, 12, 637381.

Middleton, B. 2021. Design of an undergraduate laboratory experiment utilizing a stirred-tank, jacketed bioreactor.

Mockaitis, G., Pantoja, J., Rodrigues, J.A.D., Foresti, E., Zaiat, M. 2014. Continuous anaerobic bioreactor with a fixed-structure bed (ABFSB) for wastewater treatment with low solids and low applied organic loading content. *Bioprocess and Biosystems Engineering*, 37, 1361–1368.

Moraes, B., Triolo, J.M., Lecona, V.P., Zaiat, M., Sommer, S.G. 2015. Biogas production within the bioethanol production chain: Use of co-substrates for anaerobic digestion of sugar beet vinasse. *Bioresource Technology*, 190, 227–234.

Moraes, B.S., Junqueira, T.L., Pavanello, L.G., Cavalett, O., Mantelatto, P.E., Bonomi, A., Zaiat, M. 2014. Anaerobic digestion of vinasse from sugarcane biorefineries in Brazil from energy, environmental, and economic perspectives: Profit or expense? *Applied Energy*, 113, 825–835.

Moreno-Garcia, L., Adjallé, K., Barnabé, S., Raghavan, G.2017. Microalgae biomass production for a biorefinery system: Recent advances and the way towards sustainability. *Renewable and Sustainable Energy Reviews* 76, 493–506.

Mortezaeikia, V., Yegani, R., Hejazi, M., Chegini, S. 2016. CO_2 biofixation by Dunaliella Salina in batch and semi-continuous cultivations, using hydrophobic and hydrophilic poly ethylene (PE) hollow fiber membrane photobioreactors. *Iranian Journal of Chemical Engineering (IJChE)*, 13(1), 47–59.

Mota, V.T., Santos, F.S., Amaral, M.C. 2013. Two-stage anaerobic membrane bioreactor for the treatment of sugarcane vinasse: Assessment on biological activity and filtration performance. *Bioresource Technology*, 146, 494–503.

Nath, P.C., Bandyopadhyay, T.K., Mahata, N., Pabbi, S., Tiwari, O.N., Indira, M., Bhunia, B., 2022. Optimization and kinetic study of ultrasonic-mediated phycoerythrin extraction from Anabaena sp. *Biomass Conversion and Biorefinery* 14(21), 1–12.

Nath, P.C., Bandyopadhyay, T.K., Mahata, N., Tiwari, O.N., Bobby, M.N., Indira, M., Bhunia, B., 2023a.C-phycoerythrin production from Anabaena sp. BTA 903: Optimization, production kinetics, thermodynamic, and stability analysis 14(16), *Biomass Conversion and Biorefinery* 1–13.

Nath, P.C., Bhunia, B., Bandyopadhyay, T.K. 2023b. Carbon dioxide capture and its enhanced utilization using microalgae. In *Green Sustainable Process for Chemical and Environmental Engineering and Science* (pp. 531–546). Elsevier.

Nath, P.C., Tiwari, O.N., Devi, I., Bandyopadhyay, T.K., Bhunia, B., 2021. Biochemical and morphological fingerprints of isolated Anabaena sp.: A precious feedstock for food additives. *Biomass Conversion and Biorefinery* 11, 2723–2733.

Neori, A. 2011. "Green water" microalgae: The leading sector in world aquaculture. *Journal of Applied Phycology*, 23, 143–149.

Ohimain, E.I., Izah, S.C. 2017. A review of biogas production from palm oil mill effluents using different configurations of bioreactors. *Renewable and Sustainable Energy Reviews*, 70, 242–253.

Özkaya, B., Kaksonen, A.H., Sahinkaya, E., Puhakka, J.A. 2019. Fluidized bed bioreactor for multiple environmental engineering solutions. *Water Research*, 150, 452–465.

Panda, S., Akcil, A., Mishra, S., Erust, C. 2018. A novel bioreactor system for simultaneous mutli-metal leaching from industrial pyrite ash: Effect of agitation and sulphur dosage. *Journal of Hazardous Materials*, 342, 454–463.

Panda, T. 2011. *Bioreactors: Analysis and design*. McGraw-Hill Education.

Park, K.-H., Lee, C.-G. 2001. Effectiveness of flashing light for increasing photosynthetic efficiency of microalgal cultures over a critical cell density. *Biotechnology and Bioprocess Engineering*, 6, 189–193.

Parsy, A., Monlau, F., Guyoneaud, R., Sambusiti, C. 2024. Nutrient recovery in effluents from the energy sectors for microalgae and cyanobacteria biomass production: A review. *Renewable and Sustainable Energy Reviews*. 191, 114207.

Patel, A., Karageorgou, D., Rova, E., Katapodis, P., Rova, U., Christakopoulos, P., Matsakas, L. 2020. An overview of potential oleaginous microorganisms and their role in biodiesel and omega–3 fatty acid-based industries. *Microorganisms*, 8(3), 434.

Pombo, F., Magrini, A., Szklo, A. 2011. Technology roadmap fo r wastewater reuse in petroleum refineries in Brazil in *Environmental ManagementPractice*, 425.

Pulz, O., Scheibenbogen, K. 2006. Photobioreactors: design and performance with respect to light energy input. *Bioprocess and Algae Reactor Technology, Apoptosis*, 123–152.

Puyol, D., Batstone, D.J., Hülsen, T., Astals, S., Peces, M., Krömer, J.O. 2017. Resource recovery from wastewater by biological technologies: *Opportunities, Challenges, and Prospects*. 7, 2106.

Qiang, H., Richmond, A. 1996. Productivity and photosynthetic efficiency of Spirulina platensis as affected by light intensity, algal density and rate of mixing in a flat plate photobioreactor. *Journal of Applied Phycology*, 8, 139–145.

Rahman, M.M., Al-Malack, M.H. 2006. Performance of a crossflow membrane bioreactor (CF–MBR) when treating refinery wastewater. *Desalination*, 191(1–3), 16–26.

Ramesh, S., Nath, P., Vanitha, K., Tiwari, O., Bandyopadhyay, T., Bhunia, B. 2021. Effects of different monosaccharides on thermal stability of phycobiliproteins from Oscillatoria sp. (BTA-170): Analysis of kinetics, thermodynamics, colour and antioxidant properties. *Food Bioscience*, 44, 101354.

Rao, A.R., Dayananda, C., Sarada, R., Shamala, T., Ravishankar, G. 2007. Effect of salinity on growth of green alga Botryococcus braunii and its constituents. *Bioresource Technology*, 98(3), 560–564.

Ri, P.-C., Kim, J.-S., Kim, T.-R., Pang, C.-H., Mun, H.-G., Pak, G.-C., Ren, N.-Q. 2019. Effect of hydraulic retention time on the hydrogen production in a horizontal and vertical continuous stirred-tank reactor. *International Journal of Hydrogen Energy*, 44(33), 17742–17749.

Richmond, A. 2004. Principles for attaining maximal microalgal productivity in photobioreactors: An overview. *Asian Pacific Phycology in the 21st Century: Prospects and Challenges: Proceeding of the Second Asian Pacific Phycological Forum*, held in Hong Kong, China, 21–25 June 1999 (pp. 33–37). Springer.

Robles-González, V. 2011. Integrated treatment of mezcal vinasses for depuration and discharge, Doctoral Thesis. ENCB del IPN. México DF.

Rodolfi, L., Chini Zittelli, G., Bassi, N., Padovani, G., Biondi, N., Bonini, G., Tredici, M.R. 2009. Microalgae for oil: Strain selection, induction of lipid synthesis and outdoor mass cultivation in a low-cost photobioreactor. *Biotechnology and Bioengineering*, 102(1), 100–112.

Rodrigues, B.C.G., de Mello, B.S., da Silva, B.F., Pozzi, E., de Lima Gomes, P.C.F., Sarti, A. 2021. Limonene removal using a horizontal-flow anaerobic immobilized biomass bioreactor. *Journal of Water Process Engineering*, 43, 102225.

Rosli, S., Lim, J., Lam, M., Ho, Y., Yeong, Y., Zaid, H.M., Chew, T., Merican, Z.A., Mohamad, M. 2020. Cultivation of microalgae in fluidized bed bioreactor: impacts of light intensity and CO_2 concentration. *IOP Conference Series: Materials Science and Engineering* (p. 022018). IOP Publishing.

Sanchez Riera, F., Cordoba, P., Sineriz, F. 1985. Use of the UASB reactor for the anaerobic treatment of stillage from sugar cane molasses. *Biotechnology and Bioengineering*, 27(12), 1710–1716.

Saravanane, R., Murthy, D. 2000. Application of anaerobic fluidized bed reactors in wastewater treatment: a review. *Environmental Management and Health*, 11(2), 97–117.

Sathinathan, P., Parab, H., Yusoff, R., Ibrahim, S., Vello, V., Ngoh, G. 2023. Photobioreactor design and parameters essential for algal cultivation using industrial wastewater: A review. *Renewable and Sustainable Energy Reviews*, 173, 113096.

Schenk, P.M., Thomas-Hall, S.R., Stephens, E., Marx, U.C., Mussgnug, J.H., Posten, C., Kruse, O., Hankamer, B. 2008. Second generation biofuels: High-efficiency microalgae for biodiesel production. *Bioenergy Research*, 1, 20–43.

Sharma, A., Rana, A., Sharma, N., Sharma, S. 2024. Application of fermentation techniques in the production of genetically engineered microorganisms (GMOs). In *Genetically Engineered Organisms in Bioremediation* (pp. 63–73). CRC Press.

Sharma, R., Nath, P.C., Pabbi, S., Bandyopadhyay, T.K., Vanitha, K., Mahata, N., Bhunia, B., Tiwari, O.N. 2022. Production of Oscillatoria sp. BTA-170 biomass in photobioreactor: Analysis of composition, drying behavior, sorption isotherm, and powder flow characteristics. *Journal of Food Process Engineering* 45(10), e14044.

Shoener, B.D., Zhong, C., Greiner, A.D., Khunjar, W.O., Hong, P.-Y., Guest, J.S. 2016. Design of anaerobic membrane bioreactors for the valorization of dilute organic carbon waste streams. *Energy & Environmental Science*, 9(3), 1102–1112.

Singh, J., Kaushik, N., Biswas, S. 2014. Bioreactors–technology & design analysis. *The Scitech Journal*, 1(6), 28–36.

Siqueira, L.M., Damiano, E.S., Silva, E.L. 2013. Influence of organic loading rate on the anaerobic treatment of sugarcane vinasse and biogas production in fluidized bed reactor. *Journal of Environmental Science and Health, Part A*, 48(13), 1707–1716.

Sivasubramanian, V., Prasad, B.N. 2009. Effects of superficial gas velocity and fluid property on the hydrodynamic performance of an airlift column with alcohol solution. *International Journal of Engineering, Science and Technology*, 1(1), 245–253.

Spier, M.R., Vandenberghe, L., Medeiros, A.B.P., Soccol, C.R. 2011. Application of different types of bioreactors in bioprocesses. *Bioreactors: Design, Properties and Applications*, 53–87.

Sreekrishnan, T., Kohli, S., Rana, V. 2004. Enhancement of biogas production from solid substrates using different techniques–A review. *Bioresource Technology*, 95(1), 1–10.

Stephenson, T., Brindle, K., Judd, S., Jefferson, B. 2000. *Membrane Bioreactors for Wastewater Treatment*. IWA publishing.

Suh, I.S., Lee, C.-G. 2003. Photobioreactor engineering: design and performance. *Biotechnology and Bioprocess Engineering*, 8, 313–321.

Sun, Z.-C., Li, G.-T., Zhang, C.-L., Wang, Z.-M., Lin, Q.-M., Zhao, X.-R. 2020. Contrasting resilience of soil microbial biomass, microbial diversity and ammonification enzymes under three applied soil fumigants. *Journal of Integrative Agriculture*, 19(10), 2561–2570.

Teworte, S., Malcı, K., Walls, L.E., Halim, M., Rios-Solis, L. 2022. Recent advances in fed-batch microscale bioreactor design. *Biotechnology Advances* 55, 107888.

Thomas, P., Soren, N., Rumjit, N.P., James, J.G., Saravanakumar, M., 2017. Biomass resources and potential of anaerobic digestion in Indian scenario. *Renewable and Sustainable Energy Reviews* 77, 718–730.

Tsygankov, A.A., Laurinavichene, T.V., Gogotov, I.N. 1994. Laboratory scale photobioreactor. *Biotechnology Techniques*, 8, 575–578.

Ugwu, C., Aoyagi, H., Uchiyama, H. 2008. Photobioreactors for mass cultivation of algae. *Bioresource Technology*, 99(10), 4021–4028.

Vo, H.N.P., Ngo, H.H., Guo, W., Chang, S.W., Nguyen, D.D., Chen, Z., Wang, X.C., Chen, R., Zhang, X. 2020. Microalgae for saline wastewater treatment: A critical review. *Critical Reviews in Environmental Science and Technology* 50(12), 1224–1265.

Wang, C.-Y., Fu, C.-C., Liu, Y.-C. 2007. Effects of using light-emitting diodes on the cultivation of Spirulina platensis. *Biochemical Engineering Journal*, 37(1), 21–25.

Wang, X., Liu, X., Zhao, C., Ding, Y., Xu, P. 2011. Biodiesel production in packed-bed reactors using lipase–nanoparticle biocomposite. *Bioresource Technology*, 102(10), 6352–6355.

Wang, Z., Ma, T., Xing, L. 2020. Process performance and microbial interaction in two-stage continuously stirred tank reactors for sludge anaerobic digestion operated at different temperatures. *Biochemical Engineering Journal*, 161, 107682.

Wilkinson, P.M., Spek, A.P., van Dierendonck, L.L. 1992. Design parameters estimation for scale-up of high-pressure bubble columns. *AIChE Journal*, 38(4), 544–554.

Yeh, N., Chung, J.-P. 2009. High-brightness LEDs—Energy efficient lighting sources and their potential in indoor plant cultivation. *Renewable and Sustainable Energy Reviews*, 13(8), 2175–2180.

Zhang, D., Zhu, W., Tang, C., Suo, Y., Gao, L., Yuan, X., Wang, X., Cui, Z. 2012. Bioreactor performance and methanogenic population dynamics in a low-temperature (5–18 C) anaerobic fixed-bed reactor. *Bioresource Technology*, 104, 136–143.

Zhang, L., Loh, K.-C., Kuroki, A., Dai, Y., Tong, Y.W. 2021. Microbial biodiesel production from industrial organic wastes by oleaginous microorganisms: current status and prospects. *Journal of Hazardous Materials* 402, 123543.

Zhen, G., Pan, Y., Lu, X., Li, Y.-Y., Zhang, Z., Niu, C., Kumar, G., Kobayashi, T., Zhao, Y., Xu, K. 2019. Anaerobic membrane bioreactor towards biowaste biorefinery and chemical energy harvest: Recent progress, membrane fouling and future perspectives. *Renewable and Sustainable Energy Reviews*, 115, 109392.

9 Renewable Energy Policies and Their Impact on Biofuel Development

Shubhnam and Diksha Garg

9.1 INTRODUCTION

Energy and related services are needed to meet basic human needs like health, lighting, cooking, space comfort, mobility, and communication, as well as to serve as generative processes. As a result, the United Nations (UN) General Assembly proposed a set of global Sustainable Development Goals (SDGs) that included 17 goals and 169 targets at the UN in New York by the Open Working Group (Owusu and Asumadu-Sarkodie, 2016). In order to address climate change, renewable energy, food, health, and water provision requires a coordinated worldwide monitoring and modeling of many socially significant earth system variables, socioeconomic indicators, and environmental stressors to inform policy, guide interventions, and ensure sustainable development (Hák et al., 2016).

Policies pertaining to renewable energy are crucial in determining the course of sustainable energy development worldwide. Governments and international organizations have made the switch from fossil fuels to renewable energy sources – such as bioenergy, direct solar energy, geothermal energy, hydropower, wind, and ocean energy (tide and wave) – a top priority due to the pressing need to combat climate change, lower greenhouse gas emissions, and improve energy security. When fossil fuels are substituted with renewable energy sources, greenhouse gas emissions are greatly reduced. Renewable energy sources should be sustainable since they are derived organically from continuous energy flows in our environment. Renewable energy must be infinite and offer environmentally friendly products and services in order to be considered sustainable. A sustainable biofuel, for example, shouldn't raise net CO_2 emissions, negatively impact food security, or endanger biodiversity (Owusu and Asumadu-Sarkodie, 2016). Since energy use accounts for the bulk of global greenhouse gas (GHG) emissions, a transition of the global energy system is extremely important. A shift to greater proportions of renewable energy will make it easier to provide everyone with inexpensive, clean energy, lower GHG emissions, and lessen water scarcity by doing away with the need for freshwater in thermal power plants. The majority of studies on the global energy transition show routes that, even in 2050, produce CO_2 emissions that are incompatible with the objectives of the Paris Agreement. In order to meet the growing energy demands of the global civilization of the future, the majority of the studies outline strategies for gradually replacing

DOI: 10.1201/9781003633709-9

nonsustainable technology with sustainable renewable energy sources. According to a number of international studies using various models and hypotheses, such a shift is feasible by 2050. To achieve complete sustainability, biofuels should only be used for inevitable residues, and synthetic fuels must be used more frequently to prevent fuel production from competing with food crops (Bogdanov et al., 2021).

Strict environmental regulations are in place in several European nations, such as Germany, France, and the Netherlands, to deter polluting industries and promote renewable energy sources and environmentally friendly logistics techniques. The relationship between renewable energy consumption and economic growth demonstrates that both economic and environmental performance are positively impacted by renewable energy consumption, and transportation-related activities, such as lengthy lead times and traffic congestion, use more energy to complete their logistics tasks. Waste biomass is unavoidably regarded as an appropriate way to reduce reliance on energy from fossil fuels (Khan et al., 2020). The decrease in air pollution from motor vehicles is a major advantage of switching to biofuels from fossil fuels. Biofuels typically burn more cleanly. Compared to conventional fuels, the growing use of biodiesel and ethanol blends in automobile engines results in lower emissions of PM, CO, and unburned volatile organic compounds. The primary explanation for these lower emissions in the literature is that because ethanol and biodiesel have higher oxygen contents than gasoline and diesel, they burn more completely, which lowers particulate matter and exhaust gas. Although biofuels are still producing CO_2, the carbon footprint is smaller, as biomass feedstocks act as a carbon sink by absorbing CO_2 for photosynthesis during crop growth (Sahu et al., 2024). In addition, biofuels also have the potential to be "carbon negative" through strategic choices of producers and the use of applications such as carbon capture and storage. However, this is assuming carbon neutrality is problematic, as biofuels produced in areas such as North America and Europe are often termed "carbon positive" due to the large input of fossil fuel activities and fertilizer in their production (Khan et al., 2020).

Policies around grid connection and energy distribution can significantly impact the feasibility of renewable energy projects for start-ups. An example of how grid connection policies can impact renewable energy projects is evident in Germany's Renewable Energy Sources Act (Pata et al., 2023). This legislation guarantees renewable energy producers priority access to the grid and fixed feed-in tariffs for a specified period. Such policies have been instrumental in enabling small- and medium-sized enterprises (SMEs) and start-ups to enter the renewable energy market by providing predictable revenue streams and reducing financial risks associated with grid connection. Conversely, in some regions, grid connection policies may impose high costs or bureaucratic hurdles, delaying or preventing the integration of renewable energy projects. This discrepancy might discourage investment and innovation by impeding the scalability and profitability of start-ups in the renewable energy industry. Therefore, regulations that support the viability and success of renewable energy projects led by start-ups are essential (Bańka et al., 2022). These policies should facilitate grid connection, assure equitable access to the grid, and create transparent and stable frameworks. Availability and accessibility of subsidies, tax incentives, and grants are vital for start-ups, and legislative frameworks oversee these processes. As evidence, we may point to the US Investment Tax Credit as a prime illustration of how regulatory regimes affect renewable energy start-ups (Dobrovolska et al., 2024). Significant investments in solar, wind, and other clean energy technologies have been encouraged by the Investment Tax Credit (ITC), which provides a federal tax credit of up to 26% for eligible renewable energy projects. This policy has played a crucial role in lowering the upfront costs for start-ups, thereby enhancing their financial viability and stimulating growth in the renewable energy sector. There is increasing public support for sustainable energy solutions. Regulatory frameworks need to reflect societal demands and environmental imperatives (Trevisan et al., 2023). Ensuring that regulatory policies promote inclusive access to renewable energy solutions is vital for equitable development and broader acceptance. Ensuring that regulatory policies promote inclusive access to renewable energy solutions is vital for equitable development and broader acceptance. Examining regulatory approaches in countries that are leaders in renewable energy can provide insights and best practices that can be adopted or adapted. Harmonizing regulations and fostering international cooperation can help overcome cross-border challenges and promote global entrepreneurship in renewable energy (Al-Sarihi and Mansouri, 2022).

Some limitations in the biofuel futures studies are the overall shortcoming of model results to some of the mutual futures drawbacks determinism with the novel technologies. Moreover, many studies, which have a shortcoming in the hypothetical foundation, design the impacts of technology policies in their representation of a biofuel shift, posing a hypothesis on the impacts of biofuel dissemination strategies (Mignogna et al., 2024). The number of experts in the process of futures studies for the biofuel future was often low, which also imposes concerns on the transparency of the process. The roadmaps developed are not distinct and clear about the assumptions to create a vision. Top-down approaches focusing on international and regional driving forces were often employed, overlooking local problems and specific regions' opportunities. Several studies analyzed the bioeconomy's more extensive sustainability effects. Most of the studies treated the potential growth in biofuels in an isolated context – e.g., overlooking the wider systems transition needed for the envisioned biofuel futures. Several future scenarios appeared biased, treating the drawbacks of biofuels primarily as technological challenges (Cao et al., 2022).

Three key policy perspectives of the bioeconomy can be identified based on the development process. Environmental sustainability is the primary driver for the future bioeconomy emphasizing sustainability-based policy routes for the bioeconomy (Ramcilovic-Suominen and Pülzl, 2018). In this context, some studies suggest creating global environmental awareness based on academic projections, some others suggest more specific measures such as (i) civic training programs, (ii) enhanced research and development, (iii) tax tariffs for cars and incentives for biofuels, and (iv) improving the infrastructure (Aguilar et al., 2019). The policy recommendations also include mandates or goals for zero-emission cars, standards and codes, support to encourage confidence in biofuel investments, development of biofuel industry and governmental support, funding for sustainable and renewable energies, sound transition plans to increase confidence and decrease uncertainty, and enhance skills base in biofuels. These programs would be necessary for adding the request share within the environmental sustainability perspective (Halonen et al., 2022). The alternate perspective is the request dominance of biofuels, where the main motorist would be technological improvements to make bioeconomy cost-effective in the energy and transportation requests. The policy recommendations include an essential dilemma between cinch- heft and winner- selecting. Specifically, in a winner-picking policy, selecting the leading technology at an early stage is highly risky and uncertain, potentially resulting in policy failure. On the other hand, an incremental policy approach in which winners are not chosen by creating an ideal-acquainted policy structure (e.g., incentives for zero-carbon buses) could lead to a lock-in to existing technological pathways precluding further disruptive pathways for a bioeconomy. The third perspective is to protect the investments in biofuel development driven by economic and supply-side energy security.

The purpose of this chapter is to explain the role of various energy policies for the growth of biofuel as a sustainable energy source and the effect of various policies formed by government of many nations to analyze their effect on biofuel innovation and discusses the challenges with these policies and their benefits. By discussing global case studies and future strategies, the chapter provides valuable insights for policymakers, researchers and industries and the importance of well-designed policies in driving a sustainable and inclusive energy transition cannot be overstated.

9.2 OVERVIEW OF BIOFUELS

Today, the production of bioethanol and biodiesel has once again emerged as a viable and sustainable alternative to petroleum-based fuels, thanks to recent developments in crop engineering and fermentation techniques, as well as environmental regulations that aim to reduce GHG emissions. Corn and sugar cane, which together make up 84% of the world's total production, are the sources of bioethanol in the US and Brazil, respectively (Liu et al., 2021). Fossil fuels are seen as possible sources of energy. However, concerns about rising oil prices and GHG emissions from burning fossil fuels made the development of renewable energy sources necessary. Biofuels, such as advanced biofuels, bioethanol, and biodiesel, are seen as an essential energy source to address the issue of GHG emissions. The biodegradability and less toxicity of biofuels make biofuels an appropriate replacement of fossil reserves. The demand for biofuels, particularly in the form of fuel blends, has risen in recent years. The quantity of bioethanol blended with gasoline consumed for transportation in the United States increased from 1.4 billion gallons in 1995 to 14.4 billion gallons in 2016 (Kaur and K N, 2019).

9.3 TYPES OF BIOFUELS

9.3.1 Liquid Biofuels

9.3.1.1 Bioethanol

There are several feedstocks from which bioethanol can be made. Bioethanol can be classified into three groups based on the type of feedstocks: first-, second-, and third-generation biofuels. The disadvantage of first-generation biofuels is that they are made from food crops, such as corn and sugar cane, which directly compete with food production because they need soil, water, and fertilizer (Abdul Kareem Joyia et al., 2024). Their employment in environmentally friendly transportation is further restricted by strict laws governing the use of pesticides and genetically modified crops. To produce second-generation biofuels, nonedible lignocellulose remains of plants or agricultural leftovers are utilized as the primary substrate. The net carbon footprint (emitted carbon minus consumed carbon) of these biofuels can be neutral or even negative, and they are more cost-effective than crops because they are produced from white wood chips or agricultural and forest leftovers (Jain and Kumar, 2024). However, the extraction of fermentable sugars through thermal, chemical, and/or biochemical pretreatment is necessary for the energetic and economical cost of employing lignocellulose for the manufacture of biofuel. In the upcoming years, using algae to produce bioethanol is thought to be a sustainable way to address the drawbacks of first- and second-generation biofuels (Abdul Kareem Joyia et al., 2024). Due to the exceptional ability of algae to grow on land unsuitable for food production using wastewater or brackish water, the interest in developing algae as a resource of bioethanol has exploded in the present era. Additionally, the cultivation of algae for bioethanol production is more beneficial in terms of yield in contrast to the cultivation of other feedstocks. Compared to land plants, the energy requirement for the production of supporting tissue is less in algae. Also, algae take up the nutrients over their entire surface, which results in increased bioethanol yield by algae (Kaur and K N, 2019).

9.3.1.2 Biodiesel

At the 1900 World's Fair in Paris, Rudolf gave a demonstration of the first diesel engine that ran on peanut oil (Frankelius and Lindahl, 2021). Then, in 1977, Brazilian scientist Expedito Parente created the first industrial method for producing biodiesel, and in 1989, the first biodiesel plant that used rapeseed as a feedstock was operational in Austria (Matos Fernandes et al., 2022). Because it helps benefit the environment by reducing the amount of direct and indirect GHGs such as CO_2, CO, SO_2, and HC, biodiesel is considered a clean energy source. Therefore, compared to using fossil fuels, using biodiesel can aid in preserving ecological equilibrium (Mathew et al., 2021). Four primary methods for producing biodiesel have been documented: pyrolysis (thermal cracking), microemulsion, transesterification, and direct use/blending of oils. Pure biodiesel i.e. 100% FAME (fatty acid methyl ester) is called B100, whereas biodiesel blends have lower concentrations often blended with petroleum diesel in different ratios like B2 (2% concentration), B5 (concentration of biodiesel up to 5%) and B20 (6%–20% biodiesel; Fardilah et al., 2023).

9.3.1.3 Advanced Biofuels

Petroleum-based fuels are a blend of linear, branched, and aromatic hydrocarbons. A number of pathways in living systems that produce hydrocarbons can create molecules that are similar to those found in gasoline, diesel, and jet fuel, such as fatty acids, polyketides, and isoprenoids. In addition to these hydrocarbons, higher alcohols (hydrocarbon chains longer than ethanol) can be used to make advanced biofuels and are suitable substitutes for gasoline (Keasling et al., 2021).

9.3.2 Gaseous biofuels

9.3.2.1 Biogas

Biogas is also known as biomethane, which is mainly produced from anaerobic digestion or biological fermentation of organic substances – for example, liquid manure and other digestible biomass feedstocks. Biogas can be used as a substitute for natural gas. However, there are substantial discrepancies between the characteristics of biogas and natural gas (Kabeyi and Olanrewaju, 2022). It is noted that natural gas is mostly constituted of methane (95%) and ethane (5%) whereas biogas mainly comprises methane (45%–65%) and carbon dioxide (30%–40%). In comparison to natural gas, biogas has much less energy due to its greater carbon dioxide component (30%–40%). However, biogas has a few advantages over natural gas: (a) the raw feedstock is renewable; (b) it is cost-effective; (c) it doesn't release GHGs into the atmosphere (the CO_2 that is produced and included can be fixed to form raw feedstock through photosynthesis); (d) it is produced locally without relying on foreign natural gas or oil supplies; (e) it helps reduce the pollution caused by organic wastes; (f) it helps alleviate waste management issues (Ruan et al., 2019).

9.3.2.2 Biohydrogen

In general, there are two main groups of methods that can be used to produce biohydrogen: (a) thermochemical conversion and (b) biological conversion. Biohydrogen is defined as hydrogen produced from renewable sources, which can include energy crops, crop residues, agricultural residues, forest residues, livestock residues, algae biomass, municipal wastes, industrial wastes, waste oil, wastewater, etc. (Ananthi et al., 2024). Thermochemically converted hydrogen is also known as biohydrogen because biomass is the initial raw material. Water-gas shift reaction of (gasification or pyrolysis) syngas, steam reforming of natural gas (or methane), steam reforming of pyrolysis biooil, and thermochemical breakdown of water and biomass are the primary examples of thermochemical conversion (Li et al., 2022). Because the reactions are mostly endothermic, external energy is typically needed even when thermochemical conversion procedures are economical. Conversely, the production of hydrogen using biological technologies typically has the following benefits: (a) it can be carried out at room temperature and pressure (usually in an aqueous environment with the use of a catalyst); (b) it uses little to no energy; (c) it may be reasonably priced (Xu et al., 2022).

9.4 HISTORICAL DEVELOPMENT OF BIOFUELS

Historical changes in the usage of biofuel are strongly associated with other measures of human activity, such as land use patterns, agricultural production, and the need for food (Figure 9.1). However, biofuel usage was either disregarded in earlier historical inventories of atmospheric emissions or was only taken into account when extrapolating current per capita consumption rates using demographic trends. The following timeline shows the historical development of biofuels (Sajid et al., 2021).

9.4.1 Key Drivers for Biofuel Adoption

Climate change mitigation: In order to meet the demand for fuel while reducing GHG emissions, biofuel technology advancements must concentrate on three main areas: (1) optimizing current biofuel-production technology for higher productivity and efficiency of lignocellulosic biomass conversion, (2) diversifying feedstocks to

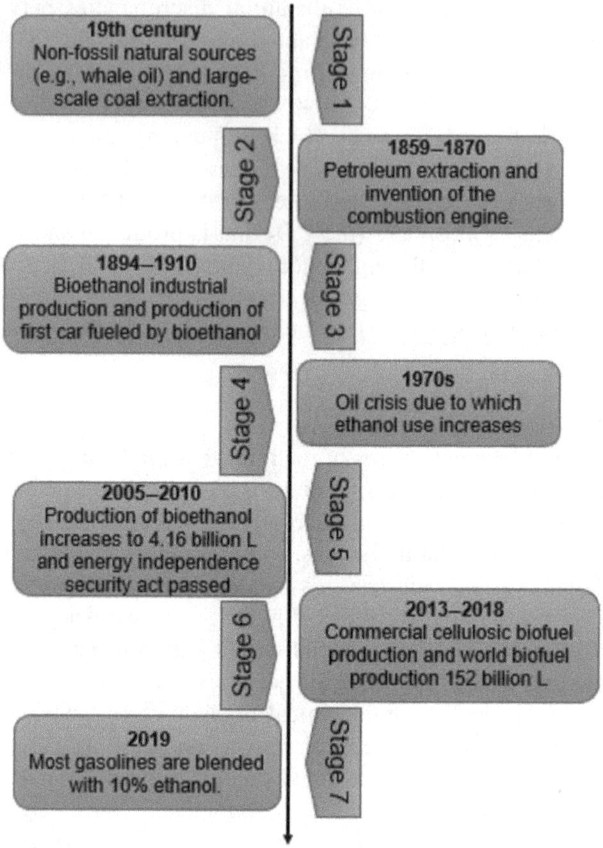

FIGURE 9.1 Historical development of biofuels.

ensure the viability of biofuel production within current ecological and economic constraints (e.g., carbon fixation through photosynthetic and electrochemical means as well as conversion of biowaste into value-added products), and (3) expanding the chemical space toward designer molecules that improve fuel economy and performance while lowering carbon emissions (Liu et al., 2021).

9.5 RENEWABLE ENERGY POLICY LANDSCAPE

9.5.1 Global Renewable Energy Policy Frameworks – The Paris Agreement and Sustainable Development Goals

Two of the most important international initiatives for resolving the conflict between environmental stewardship and economic growth are the Paris Agreement under the United Nations Framework Convention on Climate Change (UNFCCC) and the SDGs in the 2030 Agenda for Sustainable Development of the United Nations Development Program (UNDP; Olabi et al., 2022). The former seeks to abolish hunger and poverty, advance justice, peace, and human rights, as well as achieve sustainable, ecologically friendly economic growth and prosperity for all (Iacobuţă et al., 2022). The latter advocates for financing low-carbon, climate-resilient global development and limiting the increase in global temperatures in the 21st century to 2°C (over preindustrial levels; Rosegrant et al., 2022). SDGs pertaining to infrastructure and industrialization (SDG 9), access to next-generation energy (SDG 7), clean water and sanitization (SDG 6), and other objectives are in line with the Nationally Determined Contributions (NDCs) to the Paris Agreement. Still, both frameworks entail substantial investments (Tolliver et al., 2019).

The objectives of the Paris Agreement are as follows: (1) Maintaining the rise in the average global temperature well below 2°C over preindustrial levels and working to keep it to 1.5°C, as this would greatly lessen the risks and effects of climate change. (2) Improving climate resilience and promoting the development of low GHG emissions while enhancing the capacity to adjust to the negative effects of climate change without endangering food production. (3) Aligning financial flows with a course toward development that is climate resilient and low in GHG emissions (Mitchell et al., 2018).

The greatest chance yet to fortify the connections between climate and development is presented by the remarkable global accomplishments of the SDGs and the Paris Agreement. The proof is very obvious. Without considering the implications for development, it is difficult, if not impossible, to meet the ambitious goals of the Paris Agreement (Olabi et al., 2022). Furthermore, as the effects of climate change worsen, it would be challenging, if not impossible, to meet the SDG objectives and targets. The process of evaluating these connections is becoming easier thanks to new techniques and strategies (Rosegrant et al., 2022). This includes the development of a new family of scenarios that aims to integrate qualitative analysis with biophysical, economic, and social change study. Policymakers will also be able to identify which policies are better suited to their circumstances thanks to the methodology being developed for evaluating the connections between and among SDGs. Although it is a good beginning step, nations' pledges to take action through their NDCs are far from sufficient (Iacobuţă et al., 2022). Based on the periodic assessment and review, these will need to be more ambitious. The size and scope of the necessary action, as well as the limited window of opportunity, are clearly demonstrated by the scientific data. Focusing on regions that have a significant impact and making a return in terms of emissions reductions is one strategy and a crucial first step. A few of these were displayed here. Take the large, anticipated increase in spending on urban infrastructure, for instance. One crucial strategy to significantly contribute to decarbonization would be to "do it right" from the beginning using low-carbon technology, taking use of the massive urban infrastructure that is anticipated to be built over the next few decades, especially in Africa and Asia (Gomez-Echeverri, 2018).

9.6 REGIONAL AND NATIONAL POLICIES

9.6.1 European Union Renewable Energy Directive (RED II)

Although it has been implemented very differently, the European Union's (EU) approach to renewable energy legislation has been just as ambitious. A comprehensive framework for promoting renewable energy across all industries was established in 2018 with the implementation of the RED II (Mai-Moulin et al., 2021). The directive contains precise measures for sustainable biofuel production and usage, as well as binding targets for member states. Advanced biofuels made from non-food crop feedstocks are given special attention in RED II, which reflects the EU's commitment to tackling issues of food security while advancing the development of renewable energy (Krug et al., 2022).

The policy approaches between the United States and EU exhibit notable differences in their structure and implementation mechanisms. The US system, primarily through the Renewable Fuel Standard (RFS), focuses on volume-based mandates and employs a market-based credit trading system known as Renewable Identification Numbers (RINs; Afkhami and Ghoddusi, 2022). This approach has effectively created a stable market for biofuels while providing flexibility for fuel producers and importers to meet their obligations.

The policy framework of the EU, on the other hand, places more emphasis on sustainability standards and targets that are based on percentages. With particular sub-targets for the transportation sector, RED II requires member states to guarantee that at least 32% of their overall energy consumption originates from renewable sources by 2030. More rigorous sustainability standards are also part of the EU's strategy, especially with relation to food crop-based biofuels and indirect land use change (ILUC; Abreu et al., 2022).

Certification programs have been put in place in both areas to guarantee adherence to their own regulations. The EU has created extensive sustainability certification programs that monitor the production of biofuel from feedstock to final usage, whereas the US system mostly depends on Environmental Protection Agency (EPA) oversight and third-party verification (Mai-Moulin et al., 2021). As the global trade in biofuels has grown, so too have these certification criteria.

In order to ensure a sustainable biomass and bioenergy supply in the EU in the near future, the RED II establishes binding sustainability criteria for the whole bioenergy sector. However, the RED II also introduces additional challenges for the implementation of sustainability standards at the level of the EU Member States (MS) (Mai-Moulin et al., 2021). These challenges include expanding the scope to encompass advanced biofuels, new and updated sustainability requirements for agriculture and forest biomass, and new end-use industries like heat and power. This study also demonstrates that the RED II sustainability standards lack socioeconomic criteria that are pertinent for biomass feedstocks imported into the EU; lack strict protection of air, soil, and water resources; and are inadequate in preventing some dangers of unsustainable forest management.

9.6.2 US Renewable Fuel Standard (RFS)

The RFS requires refiners and importers to blend a minimum amount of biofuel into the country's fuel supply, with the amount increasing yearly until 2022. The EPA has the authority to establish volume amounts after 2022 (Taheripour et al., 2022). The RFS has four interconnected mandates – total advanced biofuel, cellulosic biofuel, biomass-based diesel, and total renewable fuel – all of which have volume criteria outlined in the statute. Advanced biofuel comprises a category for additional advanced biofuels without any particular requirements, and cellulosic and biomass-based diesel are both subcategories of this category (Lark et al., 2022). The distinction between the advanced biofuel and total renewable fuel regulations also suggests a need for conventional biofuels, such as corn ethanol. In order to be eligible as a biofuel under the RFS, the Act also specifies a number of environmental and production conditions. The RFS's designers aimed to assist achieve energy independence and lessen pollution from the use of fossil fuels. The RFS has had negative effects on refiners, customers, and the environment in its 15 years of operation, all while failing to achieve its declared goals (Younes et al., 2022).

Ethanol producers, environmentalists, refiners, oil firms, and consumer activists are engaged in continuous political disputes over the RFS (Stokes and Breetz, 2018). E15, gasoline containing up to 15% ethanol, was permitted for year-round sale by President Donald Trump's EPA in 2019. In 2011, the Obama administration banned it during the summer months to reduce smog. In order to alleviate some of the financial burden on small refineries, EPA has historically used its statutory authority to give more small refiner exemptions, which judges have determined had been mistakenly refused during the Obama administration. The exemptions avoid damaging the major farming constituency because gasoline retailers and wholesalers continue to blend ethanol with the fuel produced by small refiners, and the EPA has pledged to offset waivers and maintain fuel targets. Since there is uncertainty about future exclusions of this kind, EPA's commitment to combating market manipulation of bogus RINs is even more important. However, the ethanol sector continues to demand concessions at the expense of consumers, the environment, and overall economic efficiency, even though they stand to gain from these changes. President Trump promised to permit the sale of E15 blends through the current infrastructure, which was previously restricted to the sale of E10 blends, prior to the 2020 election (Tang et al., 2023). He also rejected numerous pending waiver petitions from small refiners in response to pressure from lawmakers in farm states.

9.6.3 Policies in Emerging Economies

9.6.3.1 Biofuel Policy in India

The Indian government has implemented a number of programs and legislative initiatives to support biofuels throughout the nation during the past 20 years (Figure 9.2) (Ebadian et al., 2020). The National Biodiesel Mission, the Ethanol Blending Program, and the National Biofuel Policy of 2008 are the most relevant of them. In order to set new goals that are in line with the advancements in other renewable energy sectors, the government recently adopted the National Policy on Biofuel, 2018 after reviewing these programs and learning from past experiences (Saravanan et al., 2020).

9.6.3.2 Ethanol Blending Program

The Power Alcohol Act of 1948, the nation's first biofuel law, was passed in order to meet future energy demands, make use of molasses surpluses, and encourage the growth of the power alcohol sector in India (Dey et al., 2023). Because there is a limited supply of ethanol, Section 6 of the Act, which requires a 20% (E20) blending of power alcohol with gasoline, has never been put into practice. This statute was

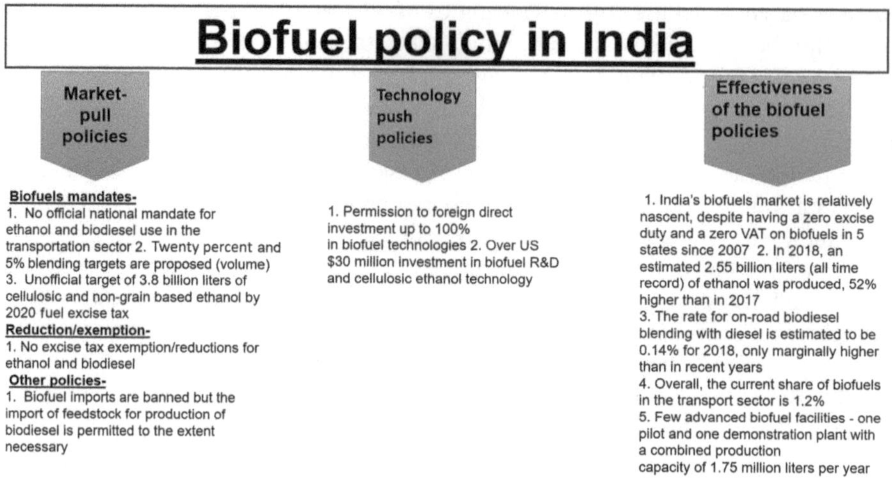

FIGURE 9.2 Biofuel policies and their effectiveness on the biofuels markets in India.

repealed by the Parliament in 2000. The Petroleum Ministry then released a notice regarding the Ethanol Blending Program (EBP) in 2002. In states that produce a lot of sugar, it required 5% (E5) ethanol to be blended with gasoline starting in January 2003. Because of the severe lack of blended biofuel and the rising expense of producing ethanol, the government made the E5 optional in 2004. With the exception of the states of Jammu and Kashmir, the northeastern states, and other island areas, the government reinstated the ethanol blending standard in October 2007 and made E5 mandatory nationwide. Due to feedstock shortages, nonadoption of the ethanol price formula, and procurement delays, the EBP has remained unprofitable (Saravanan et al., 2020).

9.6.3.3 National Biodiesel Mission

The Indian government made the decision to successfully implement a biodiesel program after taking into account the advantages of clean energy sources and rural jobs. Under the direction of the Indian Planning Commission at the time, a Committee on the Development of Biofuel was established in this regard (Mittal and Lim, 2024). Nonedible oil seeds were recommended by the Committee as a suitable feedstock for biofuel production. As a result, in 2003, the Indian government started the National Biodiesel Mission (NBM). The Mission proposed to blend 5% (B5) of biodiesel with diesel in 2007 and progressively rise to 20% (B20) in 2011, with a concentration on the Jatropha plantation on wastelands. The Mission was carried out in two stages: the first was experimentation and demonstration, and the second was a proposal to extend the Jatropha crop and develop the infrastructure. The government launched the Biodiesel Purchase Policy in 2006, which allowed oil marketing corporations to purchase biodiesel starting in January 2006 at a price of Rs. 26.50 per liter (Suhara et al., 2024). However, it was unlikely to be implemented because the cost of producing biodiesel was 50% more than the price specified in the purchase policy. Lack of feedstock availability, inadequate Jatropha seed collection and extraction, and a lack of stakeholder confidence all hampered the NBM's growth. Because of these concerns, the government reviewed the biofuel program and, in September 2008, approved the National Policy on Biofuels.

9.6.3.4 Analysis of the Indian Biofuel Policy (2018)

Resilient biomass selection, other biomass co-products, the type of biofuel generated, blending requirements, and strict biofuel policies are some of the elements that affect the sustainable production of biofuel (Datta, 2022). The policy's aims to support both domestic feedstock production and the development of advanced technologies using novel feedstocks. The 2018 policy is probably going to lower manufacturing costs and increase user affordability (Narwane et al., 2021). The technology framework, financial incentives, pricing, and institutional interventions enable the biofuels development vision, goals, and strategy. The goal of this program is to blend 5% biodiesel into diesel and 20% ethanol into gasoline (E20) by 2030 (Datta, 2022). But as of right now, the nation has only blended 2% (E2) of ethanol and less than 0.1% (B0.1) of biodiesel. As a result, it is still unclear how the nation would meet a challenging goal by 2030. According to Clause 2 of the Biofuel Policy (2018), the policy restricts the use of biofuels to bioethanol; biodiesel; advanced biofuels, such as lignocellulosic feedstocks; and drop-in fuels made from plastic waste, industrial waste, municipal solid waste, and agricultural residues (Usmani et al., 2023).

9.6.3.5 Biofuel Policy in China

To encourage the production of biofuel, China has implemented a number of policies and programs, including tax breaks, financial incentives, national plans, and technical standards (Figure 9.3). For example, the Ministry of Finance offered renewable energy generation subsidies of up to 4,000 Yuan per kW annually in 2012 (Han et al., 2022). Other noteworthy initiatives to support research and development

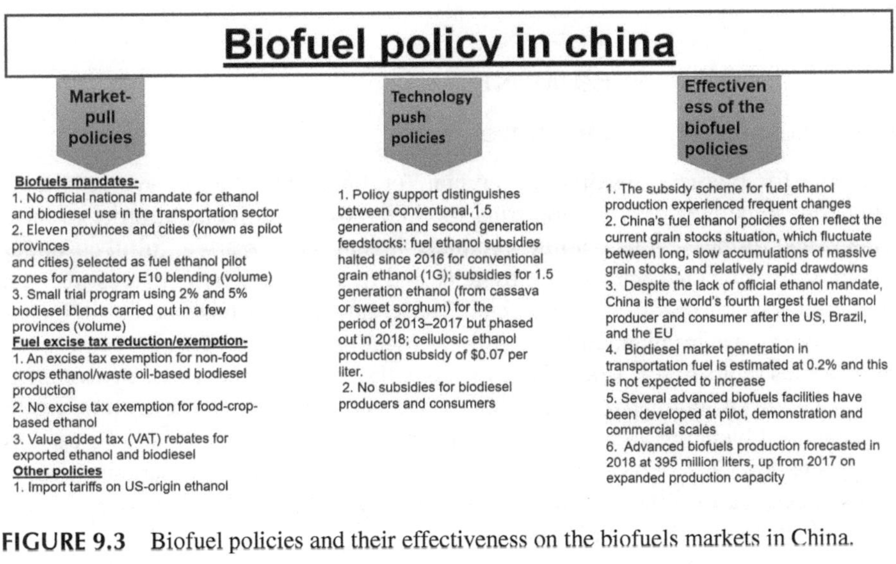

FIGURE 9.3 Biofuel policies and their effectiveness on the biofuels markets in China.

in the biomass energy sector during the 11th Five-Year Plan (2006–2010) include the S&T Support Program, the 863 Program, and the 973 Program. The 973 Program provided funding for large-scale projects involving lignocellulose materials, forestry biomass, and microalgae energy production (Foong et al., 2022). The industrialization of biomass liquid fuel and the development of sweet potato-based ethanol fuel in the southwest were primarily supported by the 863 Program. It is important to remember that China initially set the same goals and objectives for the development of biofuels, including biodiesel and ethanol (Han et al., 2022). However, because of the lack of feedstock and technology, implementation strategies, planning, policy, and research and development have been conducted in different stages (Saravanan et al., 2020).

Bioethanol policy initiatives – Chinese bioethanol development has undergone two phases: the pilot phase (1986–2005) and the implementation phase (2006–current).

National demonstration (pilot) phase (1986–2005): During this stage, pilot testing, research, and development of pertinent technologies for the production of biofuels were the main priorities (Usmani et al., 2023). The National High Technology Research and Development Initiative, also referred to as Plan 863, was implemented by the Chinese government in 1986 in order to finance and incentivize biofuel research. In order to adopt ethanol standards and establish quality control standards for ethanol composition, packaging, testing, storage, and technical requirements for gasoline ethanol, China passed legislation in 2001 (Osiolo et al., 2023). The Bioethanol Utilization Plan was legally incorporated into the government's tenth Five-Year Plan (2001–2005). Experimenting with ethanol production, marketing, and other complementary measures was the plan's primary goal. Stale wheat and maize were regarded as the main feedstocks for the synthesis of bioethanol (Osiolo et al., 2023).

Implementation phase (2006–present): As a part of this phase, China has adopted different national standards (GB 18350 and GB 18351) related to production and use of biofuels. The standard GB 18351 permitted 10% (E10) blending of ethanol with petrol (Saravanan et al., 2020).

9.7 KEY POLICY INSTRUMENTS

9.7.1 Subsidies and Tax Incentives

Programs for biofuel subsidies are crucial for lowering reliance on fossil fuels for energy since they assist in lowering the high cost of biofuels. In order to raise demand and further regulate biofuel products, it is also necessary to fulfill customer demands. Finding environmentally friendly and renewable energy sources that can be combined with fossil fuels is crucial. These new sources must be feasible economically and environmentally, and able to meet energy demands. Biofuels made from various biomass feedstocks have the potential to reduce the use of environmentally hazardous reactionary energies, but their development was aided by duty credit programs and blending authorizations that permitted many feedstocks to be used as energy sources. Current biodiesel production, limited by various restrictions on natural resources, has experienced costs that are 20%–50% higher than the price of fossil fuel–derived energy (Kuo et al., 2019).

Government subsidies and biofuel policies apply to the production of biofuel. The following elements need to be taken into account when assessing these different generations: (1) the social costs, like land use and CO_2 emissions from agricultural production; (2) manufacturing costs, like land cost, production cost, facility investment, tax, and by-product revenue; (3) government subsidies that encourage carbon reduction by making it more profitable; (4) the potential revenue of the by-product (e.g., lignin that could be used to power boiler systems); and (5) the domestic gross requirement, which includes figuring out how many vehicles are using cellulosic ethanol (Leong et al., 2018).

Fuel efficiency determines the ratios used to combine biofuels with fossil fuels. (2) E5: a 5% ethanol (E100) blended with 95% gasoline, (3) B5: a 5% biodiesel (B100) blended with 95% diesel, (4) E10 and B10: 10% biofuels blended with 90% fossil fuels in volume, and (1) E3: a 3% ethanol (E100) blended with 97% gasoline. The study evaluated the social costs of fossil fuels and biofuels per functional transportation unit, considering both external and private costs. The findings indicated that, at the given discount rate, reducing the use of B5 and B10 biofuels would be more cost-effective than continuing to use diesel (Malode et al., 2021).

Through a variety of subsidies and legislative incentives, policies now in place encourage the development of biofuels and the production objectives. Prior research has been done on the possibility of commercializing biofuels in many nations. In Brazil, incentives have been put in place to make biofuels more competitive with fossil fuels, and government promotion is spurring innovation in sugar cane species and growing techniques. The price of biofuel is still comparatively greater than that of traditional fuels, nevertheless (Hasan et al., 2023). Prior research has also demonstrated that the price of gasoline, GHG prices, and emission reductions have a major positive effect on the well-being of society. As can be seen from the aforementioned, when developing a model to assess the growth of biofuels, several factors need to be taken into account, such as the (1) social cost, (2) manufacturing cost, (3) subsidy by government, and (4) the domestic gross requirement (Kuo et al., 2019).

9.7.2 Renewable Portfolio Standards (RPS)

The RPS is essentially a quantitative policy tool designed to raise the market share of renewable electricity. Theoretically, if policy tools are created with the size and scope of externality-driven market inefficiencies in mind, they can address market failures with the same result whether they are based on prices or quantities. It is anticipated that the RPS will only partially counteract the negative externalities related to the energy sector. This is due to two factors: (i) The RPS solely targets the electricity sector of a larger energy economy, and (ii) the RPS frequently requires a specific percentage (i.e., not 100%) of electricity to come from renewable sources. States that were early adopters have been updating RPS designs frequently by enacting more mandates, although the RPS still spreads widely in terms of both its coverage and its goals (Yin and Powers, 2010). As a result, the twofold characteristics lead to the twofold implications about the partial RPS contributions in mitigating the negative externalities of fossil fuels. First, it takes time for renewable electricity to mature, which should reduce the risks to energy security and environmental externalities. As a result, the full transformation is regarded as long-term feasible. Second, with the ensuing attempts to increase the market share of renewable electricity, it helps to reduce the short-term externalities that already exist in the energy sector (Joshi, 2021).

9.7.3 Carbon Pricing and Emissions Trading Systems

Governments everywhere concur that the most economical and efficient way to achieve their emission reduction goals is to price carbon emissions. Emitters are given the option to either stop their activities, change their activities, or pay for the emissions when carbon pricing is implemented (Fageda and Teixidó, 2022). Governments use the money raised by carbon pricing to fund green projects. Three primary carbon pricing schemes are currently in use: (1) carbon tax, (2) carbon cap and trade, and (3) policies pertaining to carbon offsets. Emitters are assessed a fee for each unit of carbon they release under the carbon tax program. Since the emitter would be liable for additional emission fines if they do not switch to a cleaner technology, it is thought that a high carbon tax rate encourages the substitution of cleaner alternatives for fossil fuels (Ji et al., 2021). The carbon tax policy may encourage the production of greener energy, depending on the price of carbon emissions and the cost of switching to cleaner technologies. The economic and environmental effects of various carbon pricing schemes on supply chains vary. A number of earlier studies examined how carbon pricing regulations affected the best choices, costs, and emissions of models for optimizing the biomass supply chain. These studies fall into three categories: those that addressed the levels of strategic, tactical, and operational planning. Studies with a strategic focus created optimization models to pinpoint the best locations for bio-conversion plants that turn biomass into energy (Lovcha et al., 2022).

9.8 IMPACT OF RENEWABLE ENERGY POLICIES ON BIOFUEL DEVELOPMENT

9.8.1 Economic Impacts

The relationship between the consumption of renewable energy and economic growth has drawn the attention of environmental economists and policymakers around the world. To investigate this relationship, three different types of datasets were used: time-series, panel, and cross-country analyses (Gradziuk et al., 2021). The first hypothesis states that increasing the consumption of renewable energy promotes positive output, and that if consumption of renewable energy decreases, energy conservation policies will have a significant and negative impact on economic growth. The second hypothesis, known as the growth hypothesis, suggests that the consumption of renewable energy causes economic growth and that a unidirectional causality relationship connects economic growth to the consumption of renewable energy exits, meaning that a decrease or increase in energy consumption does not affect economic growth (Usmani et al., 2023). Third, the feedback hypothesis proposes a reciprocal causal link between economic growth and the use of renewable energy.

Any increase in the usage of renewable energy would have the opposite impact and be crucial in promoting economic growth. Fourth, these two variables are independent, as demonstrated by the neutrality theory. The majority of the energy research currently in publication examined the connections between the usage of renewable energy sources and economic expansion, but the empirical findings were not entirely consistent (Subramaniam and Masron, 2021).

Many countries throughout the world have promoted the growth of renewable energy initiatives by offering various incentives, such as tax breaks and subsidies. Renewable energy sources are now more affordable since these nations have been providing additional incentives to promote the use of renewable energy (Usmani et al., 2023). In many nations, renewable energy sources may compete with conventional energy sources. Renewable energy initiatives have surfaced in Africa, Latin America, and Asia in addition to the United States and Europe. Additionally, it has supplied top manufacturers outside of the United States and Europe, as well as cutting-edge renewable energy technologies (Shahbaz et al., 2020).

By improving technology, raising the productivity of energy sources, and lowering energy costs, economic globalization helps nations meet the need for renewable energy. In this case, the demand for renewable energy is impacted by economic globalization through trade openness, technology, and foreign direct investment inflows (Hussein and Abdullahi, 2024). Trade influences the use of renewable energy through the effects of technique, content, and size. While the composition and technique have a favorable environmental quality, the scale effect has a negative one. According to the scale effect, greater trade leads to larger production scales, and economic growth may result in lower environmental quality because of higher output and energy use. The rise in emissions will then cause the quality of the environment to deteriorate (Wang et al., 2024). According to the composition impact, growing trade openness has led to structural changes in the composition, which have made the natural resource sector more significant and seem to be good for the planet's future. According to the technique effect, trade-induced economic expansion lowers pollution emissions by boosting the spread of renewable energy technologies to developing nations. As a result, the effects of trade openness on the production of biofuels tend to differ based on the composition, technology, and scale effects (Subramaniam and Masron, 2021).

A sizable amount of employment has already or will be created by biofuels. Three hundred thousand jobs, or one job per hectare, would be created if Tanzania switched from importing gasoline for transportation to producing its own biofuels (Subramaniam and Masron, 2021). The majority of these positions will be in rural regions, which is advantageous for rural economic development; nonetheless, there are several restrictions: (1) The more lucrative biofuels systems need more work and create fewer jobs. Employment is declining in nations with more developed industrial systems. For instance, the overall number of workers in Brazil's sugar cane industry decreased by one-third between 1992 and 2003 (Hussein and Abdullahi, 2024). (2) Production is very seasonal, and the proportion of temporary to permanent employees is rising in Brazil. (3) Jobs in on-farm production are likely to be unskilled and poorly paid because the feedstock prices will be forced down to reduce costs (Wang et al., 2024).

9.8.2 Environmental Impacts

A common method for estimating the global warming potential (GWP) and other environmental effects of biofuels is life cycle assessment, or LCA. However, due to a variety of uncertainties and a wide range of methodological decisions in LCA, the estimates differ significantly among the research. In spite of this, the body of current evidence is informative. First, it demonstrates that first-generation biofuels can, on average, have fewer GHG emissions than fossil fuels when no LUC is involved (Rocha et al., 2014). However, the GHG savings for the majority of the feedstocks are insufficient to meet the RED's requirements. Second, if there is no LUC, second-generation biofuels generally have a higher potential to lower GHG emissions than first-generation biofuels. However, it will take time to produce second-generation biofuels, and the industry's confidence to spend will probably depend on first-generation fuels' sustained backing (Jeswani et al., 2020). Third, it is evident that third-generation biofuels derived from algae are unlikely to contribute to the transportation sector at this stage of development because their GHG emissions are larger than those of fossil fuels. Furthermore, the algal feedstock will remain limited to high-value applications like cosmetics and dietary supplements due to its unproven nature and high production costs. When evaluating the sustainability of biofuels, numerous other sustainability concerns must be taken into account in addition to the effects on the environment (Jeswani et al., 2020). These include production costs and competitiveness with fossil fuels; the security of food, energy, and water; the creation of jobs; rural development; and the effects on human health (Table 9.1).

To prevent shifting the burdens from one stage of the life cycle or supply chain to another, it is imperative that the sustainability elements of biofuels be assessed on a life cycle basis throughout whole supply chains. It's also critical to remember that if the results are unreliable, LCA and more general sustainability assessments are useless. Strong auditing of biofuel supply chains is therefore essential to preventing adverse socioeconomic effects, ensuring fuel traceability, and reducing the danger of fraud. Furthermore, if LCA is to be reliable and helpful for policy, it is imperative to enhance transparency, data availability, and sharing. Similar to how national inventories were created for GHG reporting under the Kyoto Protocol, this might be accomplished by creating open national and international databases (He et al., 2022). To prevent abuse and misunderstanding, it's also critical to make sure that the data and models from other disciplines that are utilized in LCA maintain appropriate levels of transparency, rigor, and robustness (Jeswani et al., 2020).

9.9 FUTURE DIRECTIONS

Energy security, oil prices, resource depletion, and climate change are just a few of the many issues facing the modern world, all of which have an adverse effect on the environment. Research and development of fuels and energy derived from biomass has advanced significantly as a result of all these difficulties. Biofuels are therefore anticipated to be the most beneficial in this regard in order to address such issues in a very sustainable manner. Biofuels are seen to be the most practical solutions for

TABLE 9.1
Major Country Emits CO_2, a Carbon Tax, and Indirect Implications on the Renewable Scheme

Country	Amount (Billion Tons CO_2)	Carbon Tax	Implications on Renewable	References
China	9.8	No carbon tax. However, the country has announced an allocation for renewable infrastructure improvement	Incorporation of 20% of renewable energy into primary energy.	Wang et al. (2021
United States	5.3	Suggested to be $25 per metric ton/emission	The United States provide a mechanism to provide dividend and incentives to companies to invest in innovation and alternative energy sources.	Suali and Suali (2023)
India	2.5	Carbon cess since 2010 implemented in place of a carbon tax, on coal at Rs. 400/ton	About 60% of revenue from carbon cess was used or planned to invest in renewable energy.	Suali and Suali (2023)
Japan	1.2	JPY 289/t – CO_2	The government provides subsidies for renewable energy. Japan has developed an RE model for 80% emission reduction in 2050.	Suali and Suali (2023)

lowering carbon dioxide emissions in the transportation industry. Furthermore, using locally accessible indigenous resources, biofuels can be manufactured with ease. Algal biofuels have received a lot of attention lately and are thought to be the most promising solution to the world's energy issue. The primary benefits of using algae include the possibility of a high yield and the absence of competition for freshwater and land resources from food crops. Around the world, a lot of research is being conducted to increase the production of biofuels. Although there is no denying that biofuel is a rapidly expanding field of study and a rapidly evolving industry, significant technological advancements in biofuel production have been made, and a thorough understanding of biofuel production processes has also been gained. However, biofuels cannot yet fully replace fossil fuels, and several integrated engineering and biological approaches are still needed to optimize biofuel production at the commercial scale. Furthermore, so that a sustainable biofuels economy may be realized, it is critical to comprehend how future climate changes will impact biofuel production. Since biofuels may improve supply security and lower car emissions, they will undoubtedly be a major sustainable energy source in the future as an alternative to fossil fuels.

9.10 CONCLUSION

The transition to renewable energy is essential for sustainable development, addressing climate change, and ensuring energy security. The UN SDGs emphasize renewable energy's role in reducing GHG emissions and providing clean, accessible energy. Policies worldwide promote renewable sources like bioenergy, solar, wind, and hydropower while discouraging fossil fuels. Biofuels, although cleaner, can pose environmental and food security concerns. Effective regulations, such as Germany's Renewable Energy Sources Act and the US ITC, support renewable energy start-ups by ensuring grid access and financial incentives. However, biofuel policies face challenges, including limited expert input, unclear roadmaps, and inadequate integration with broader sustainability goals. Future policies must balance technological advancements, environmental sustainability, and energy security. By examining global case studies and policy recommendations, this chapter highlights the importance of well-designed policies for a successful and inclusive energy transition, enabling growth and innovation in the renewable energy sector.

REFERENCES

Abdul Kareem Joyia, M., Ahmad, M., Chen, Y.-F., Mustaqeem, M., Ali, A., Abbas, A., Ashraf Gondal, M., 2024. Trends and advances in sustainable bioethanol production technologies from first to fourth generation: A critical review. *Energy Convers Manag* 321, 119037. https://doi.org/10.1016/j.enconman.2024.119037

Abreu, M., Silva, L., Ribeiro, B., Ferreira, A., Alves, L., Paixão, S.M., Gouveia, L., Moura, P., Carvalheiro, F., Duarte, L.C., Fernando, A.L., Reis, A., Gírio, F., 2022. Low Indirect Land Use Change (ILUC) energy crops to bioenergy and biofuels—A review. *Energies (Basel)* 15, 4348. https://doi.org/10.3390/en15124348

Afkhami, M., Ghoddusi, H., 2022. Pricing renewable identification numbers under uncertainty. *Quant Finance* 22, 725–742. https://doi.org/10.1080/14697688.2021.1996625

Aguilar, A., Twardowski, T., Wohlgemuth, R., 2019. Bioeconomy for sustainable development. *Biotechnol J* 14. https://doi.org/10.1002/biot.201800638

Al-Sarihi, A., Mansouri, N., 2022. Renewable energy development in the Gulf cooperation council countries: Status, barriers, and policy options. *Energies (Basel)* 15, 1923. https://doi.org/10.3390/en15051923

Ananthi, V., Ramesh, U., Balaji, P., Kumar, P., Govarthanan, M., Arun, A., 2024. A review on the impact of various factors on biohydrogen production. *Int J Hydrogen Energy* 52, 33–45. https://doi.org/10.1016/j.ijhydene.2022.08.046

Bańka, M., Salwin, M., Kukurba, M., Rychlik, S., Kłos, J., Sychowicz, M., 2022. Start-up accelerators and their impact on sustainability: Literature analysis and case studies from the energy sector. *Sustainability* 14, 13397. https://doi.org/10.3390/su142013397

Bogdanov, D., Ram, M., Aghahosseini, A., Gulagi, A., Oyewo, A.S., Child, M., Caldera, U., Sadovskaia, K., Farfan, J., De Souza Noel Simas Barbosa L., Fasihi, M., Khalili, S., Traber, T., Breyer, C., 2021. Low-cost renewable electricity as the key driver of the global energy transition towards sustainability. *Energy* 227, 120467. https://doi.org/10.1016/j.energy.2021.120467

Cao, B., Bai, C., Zhang, M., Lu, Y., Gao, P., Yang, J., Xue, Y., Li, G., 2022. Future landscape of renewable fuel resources: Current and future conservation and utilization of main biofuel crops in China. *Sci Total Environ* 806, 150946. https://doi.org/10.1016/j.scitotenv.2021.150946

Datta, B., 2022. An economic analysis of biofuels: policies, trade, and employment opportunities. In *Handbook of biofuels* (pp. 3–29). Elsevier. https://doi.org/10.1016/B978-0-12-822810-4.00001-4

Dey, B., Roy, B., Datta, S., Singh, K.G., 2023. Comprehensive overview and proposal of strategies for the ethanol sector in India. *Biomass Convers* Biorefin 13, 4587–4618. https://doi.org/10.1007/s13399-021-01546-2

Dobrovolska, O., Ortmanns, W., Podosynnikov, S., Halynskyi, D., Miniailo, A., 2024. Start-ups and entrepreneurship in renewable energy: investments and risks. *FMIR* 8, 213–240. https://doi.org/10.61093/fmir.8(2).213-240.2024

Ebadian, M., van Dyk, S., McMillan, J.D., Saddler, J., 2020. Biofuels policies that have encouraged their production and use: An international perspective. *Energy Policy* 147, 111906. https://doi.org/10.1016/j.enpol.2020.111906

Fageda, X., Teixidó, J.J., 2022. Pricing carbon in the aviation sector: Evidence from the European emissions trading system. *J Environ Econ Manage* 111, 102591. https://doi.org/10.1016/j.jeem.2021.102591

Fardilah, V., Pusparizkita, Y.M., Tauviqirrahman, M., Bayuseno, A.P., 2023. An overview of biodiesel, a fuel blend made from palm oil. *IOP Conf Ser Earth Environ Sci* 1268, 012057. https://doi.org/10.1088/1755-1315/1268/1/012057

Foong, S.Y., Chan, Y.H., Loy, A.C.M., How, B.S., Tamothran, A.M., Yip, A.J.K., Liew, R.K., Peng, W., Alstrup, A.K.O., Lam, S.S., Sonne, C., 2022. The nexus between biofuels and pesticides in agroforestry: Pathways toward United Nations sustainable development goals. *Environ Res* 214, 113751. https://doi.org/10.1016/j.envres.2022.113751

Frankelius, P., Lindahl, M., 2021. Energy solutions for agricultural machinery: from the oil era towards a sustainable bioeconomy. In *Bio#Futures* (pp. 319–348). Springer International Publishing. https://doi.org/10.1007/978-3-030-64969-2_15

Gomez-Echeverri, L., 2018. Climate and development: Enhancing impact through stronger linkages in the implementation of the Paris Agreement and the Sustainable Development Goals (SDGs). *Philos Tran A Math Phys Eng Sci* 376, 20160444. https://doi.org/10.1098/rsta.2016.0444

Gradziuk, P., Jończyk, K., Gradziuk, B., Wojciechowska, A., Trocewicz, A., Wysokiński, M., 2021. An economic assessment of the impact on agriculture of the proposed changes in EU biofuel policy mechanisms. *Energies (Basel)* 14, 6982. https://doi.org/10.3390/en14216982

Hák, T., Janoušková, S., Moldan, B., 2016. Sustainable development goals: A need for relevant indicators. *Ecol Indic* 60, 565–573. https://doi.org/10.1016/j.ecolind.2015.08.003

Halonen, M., Näyhä, A., Kuhmonen, I., 2022. Regional sustainability transition through forest-based bioeconomy? Development actors' perspectives on related policies, power, and justice. *For Policy Econ* 142, 102775. https://doi.org/10.1016/j.forpol.2022.102775

Han, X., Chen, Y., Wang, X., 2022. Impacts of China's bioethanol policy on the global maize market: A partial equilibrium analysis to 2030. *Food Secur* 14, 147–163. https://doi.org/10.1007/s12571-021-01212-5

Hasan, M., Abedin, M.Z., Amin, M. Bin, Nekmahmud, M., Oláh, J., 2023. Sustainable biofuel economy: A mapping through bibliometric research. *J Environ Manage* 336, 117644. https://doi.org/10.1016/j.jenvman.2023.117644

He, R., Luo, L., Shamsuddin, A., Tang, Q., 2022. Corporate carbon accounting: a literature review of carbon accounting research from the Kyoto Protocol to the Paris Agreement. *Accounting & Finance* 62, 261–298. https://doi.org/10.1111/acfi.12789

Hussein, O.A., Abdullahi, A.M., 2024. Dynamic impacts of globalization and environmental performance on renewable energy. *Discover Sustainability* 5, 365. https://doi.org/10.1007/s43621-024-00604-9

Iacobuţă, G.I., Brandi, C., Dzebo, A., Elizalde Duron, S.D., 2022. Aligning climate and sustainable development finance through an SDG lens. The role of development assistance in implementing the Paris Agreement. *Global Environmental Change* 74, 102509. https://doi.org/10.1016/j.gloenvcha.2022.102509

Jain, S., Kumar, S., 2024. A comprehensive review of bioethanol production from diverse feedstocks: Current advancements and economic perspectives. *Energy* 296, 131130. https://doi.org/10.1016/j.energy.2024.131130

Jeswani, H.K., Chilvers, A., Azapagic, A., 2020. Environmental sustainability of biofuels: A review. *Proceedings of the Royal Society A: Mathematical, Physical and Engineering Sciences* 476. https://doi.org/10.1098/rspa.2020.0351

Ji, C.-J., Hu, Y.-J., Tang, B.-J., Qu, S., 2021. Price drivers in the carbon emissions trading scheme: Evidence from Chinese emissions trading scheme pilots. *J Clean Prod* 278, 123469. https://doi.org/10.1016/j.jclepro.2020.123469

Joshi, J., 2021. Do renewable portfolio standards increase renewable energy capacity? Evidence from the United States. *J Environ Manage* 287, 112261. https://doi.org/10.1016/j.jenvman.2021.112261

Kabeyi, M.J.B., Olanrewaju, O.A., 2022. Biogas production and applications in the sustainable energy transition. *J Energy* 2022, 1–43. https://doi.org/10.1155/2022/8750221

Kaur, J. K. N. Y., 2019. Exploring the potential of carbohydrate rich algal biomass as feedstock for bioethanol production. In *Liquid Biofuel Production* (pp. 167–195). Wiley. https://doi.org/10.1002/9781119459866.ch5

Keasling, J., Garcia Martin, H., Lee, T.S., Mukhopadhyay, A., Singer, S.W., Sundstrom, E., 2021. Microbial production of advanced biofuels. *Nat Rev Microbiol* 19, 701–715. https://doi.org/10.1038/s41579-021-00577-w

Khan, S.A.R., Zhang, Y., Kumar, A., Zavadskas, E., Streimikiene, D., 2020. Measuring the impact of renewable energy, public health expenditure, logistics, and environmental performance on sustainable economic growth. *Sustainable Development* 28, 833–843. https://doi.org/10.1002/sd.2034

Krug, M., Di Nucci, M.R., Caldera, M., De Luca, E., 2022. Mainstreaming community energy: Is the renewable energy directive a driver for renewable energy communities in Germany and Italy? *Sustainability* 14, 7181. https://doi.org/10.3390/su14127181

Kuo, T.C., Lin, S.-H., Tseng, M.-L., Chiu, A.S.F., Hsu, C.-W., 2019. Biofuels for vehicles in Taiwan: Using system dynamics modeling to evaluate government subsidy policies. *Resour Conserv Recycl* 145, 31–39. https://doi.org/10.1016/j.resconrec.2019.02.005

Lark, T.J., Hendricks, N.P., Smith, A., Pates, N., Spawn-Lee, S.A., Bougie, M., Booth, E.G., Kucharik, C.J., Gibbs, H.K., 2022. Environmental outcomes of the US renewable fuel standard. *Proceedings of the National Academy of Sciences* 119. https://doi.org/10.1073/pnas.2101084119

Leong, W.-H., Lim, J.-W., Lam, M.-K., Uemura, Y., Ho, Y.-C., 2018. Third generation biofuels: A nutritional perspective in enhancing microbial lipid production. *Renewable and Sustainable Energy Reviews* 91, 950–961. https://doi.org/10.1016/j.rser.2018.04.066

Li, S., Li, F., Zhu, X., Liao, Q., Chang, J.-S., Ho, S.-H., 2022. Biohydrogen production from microalgae for environmental sustainability. *Chemosphere* 291, 132717. https://doi.org/10.1016/j.chemosphere.2021.132717

Liu, Y., Cruz-Morales, P., Zargar, A., Belcher, M.S., Pang, B., Englund, E., Dan, Q., Yin, K., Keasling, J.D., 2021. Biofuels for a sustainable future. *Cell* 184, 1636–1647. https://doi.org/10.1016/j.cell.2021.01.052

Lovcha, Y., Perez-Laborda, A., Sikora, I., 2022. The determinants of CO_2 prices in the EU emission trading system. *Appl Energy* 305, 117903. https://doi.org/10.1016/j.apenergy.2021.117903

Mai-Moulin, T., Hoefnagels, R., Grundmann, P., Junginger, M., 2021. Effective sustainability criteria for bioenergy: Towards the implementation of the european renewable directive II. *Renew Sustain Energy Rev* 138, 110645. https://doi.org/10.1016/j.rser.2020.110645

Malode, S.J., Prabhu, K.K., Mascarenhas, R.J., Shetti, N.P., Aminabhavi, T.M., 2021. Recent advances and viability in biofuel production. *Energy Conversion and Management: X* 10, 100070. https://doi.org/10.1016/j.ecmx.2020.100070

Mathew, G.M., Raina, D., Narisetty, V., Kumar, V., Saran, S., Pugazhendi, A., Sindhu, R., Pandey, A., Binod, P., 2021. Recent advances in biodiesel production: Challenges and solutions. *Sci Total Environ* 794, 148751. https://doi.org/10.1016/j.scitotenv.2021.148751

Matos Fernandes, F., Silva Martins, L.O., Santos Marques, R., Santana Silva, M., Gaudêncio Mendonça Freires, F., 2022. Second-generation biodiesel in Brazil: An analysis of research on animal fats through social and complex networks. *J Technol Manag Innov* 17, 28–39. https://doi.org/10.4067/S0718-27242022000400028

Mignogna, D., Szabó, M., Ceci, P., Avino, P., 2024. Biomass energy and biofuels: Perspective, potentials, and challenges in the energy transition. *Sustainability* 16, 7036. https://doi.org/10.3390/su16167036

Mitchell, D., Allen, M.R., Hall, J.W., Muller, B., Rajamani, L., Le Quéré, C., 2018. The myriad challenges of the Paris Agreement. *Philos Trans A Math Phys Eng Sci* 376, 20180066. https://doi.org/10.1098/rsta.2018.0066

Mittal, V., Lim, E., 2024. Aligning advances in biodiesel technology with the needs of the defense community. *Eng* 5, 2709–2727. https://doi.org/10.3390/eng5040142

Narwane, V.S., Yadav, V.S., Raut, R.D., Narkhede, B.E., Gardas, B.B., 2021. Sustainable development challenges of the biofuel industry in India based on integrated MCDM approach. *Renew Energy* 164, 298–309. https://doi.org/10.1016/j.renene.2020.09.077

Olabi, A.G., Obaideen, K., Elsaid, K., Wilberforce, T., Sayed, E.T., Maghrabie, H.M., Abdelkareem, M.A., 2022. Assessment of the pre-combustion carbon capture contribution into sustainable development goals SDGs using novel indicators. *Renew Sustain Energy Rev* 153, 111710. https://doi.org/10.1016/j.rser.2021.111710

Osiolo, H.H., Marwah, H., Leach, M., 2023. The emergence of large-scale bioethanol utilities: Accelerating energy transitions for cooking. *Energies (Basel)* 16, 6242. https://doi.org/10.3390/en16176242

Owusu, P.A., Asumadu-Sarkodie, S., 2016. A review of renewable energy sources, sustainability issues and climate change mitigation. *Cogent Eng* 3, 1167990. https://doi.org/10.1080/23311916.2016.1167990

Pata, U.K., Kartal, M.T., Erdogan, S., Sarkodie, S.A., 2023. The role of renewable and nuclear energy R&D expenditures and income on environmental quality in Germany: Scrutinizing the EKC and LCC hypotheses with smooth structural changes. *Appl Energy* 342, 121138. https://doi.org/10.1016/j.apenergy.2023.121138

Ramcilovic-Suominen, S., Pülzl, H., 2018. Sustainable development – A 'selling point' of the emerging EU bioeconomy policy framework? *J Clean Prod* 172, 4170–4180. https://doi.org/10.1016/j.jclepro.2016.12.157

Rocha, M.H., Capaz, R.S., Lora, E.E.S., Nogueira, L.A.H., Leme, M.M.V., Renó, M.L.G., Olmo, O.A. Del, 2014. Life cycle assessment (LCA) for biofuels in Brazilian conditions: A meta-analysis. *Renew Sustain Energy Rev* 37, 435–459. https://doi.org/10.1016/j.rser.2014.05.036

Rosegrant, M.W., Sulser, T.B., Wiebe, K., 2022. Global investment gap in agricultural research and innovation to meet Sustainable Development Goals for hunger and Paris Agreement climate change mitigation. *Front Sustain Food Syst* 6. https://doi.org/10.3389/fsufs.2022.965767

Ruan, R., Zhang, Y., Chen, P., Liu, S., Fan, L., Zhou, N., Ding, K., Peng, P., Addy, M., Cheng, Y., Anderson, E., Wang, Y., Liu, Y., Lei, H., Li, B., 2019. Biofuels: Introduction. In *Biofuels: Alternative feedstocks and conversion processes for the production of liquid and gaseous biofuels* (pp. 3–43). Elsevier. https://doi.org/10.1016/B978-0-12-816856-1.00001-4

Sahu, S., Kunj, P., Kaur, A., Khatri, M., Singh, G., Arya, S.K., 2024. Catalytic strategies for algal-based carbon capture and renewable energy: A review on a sustainable approach. *Energy Convers Manag* 310, 118467. https://doi.org/10.1016/j.enconman.2024.118467

Sajid, Z., da Silva, M., Danial, S., 2021. Historical analysis of the role of governance systems in the sustainable development of biofuels in Brazil and the United States of America (USA). *Sustainability* 13, 6881. https://doi.org/10.3390/su13126881

Saravanan, A.P., Pugazhendhi, A., Mathimani, T., 2020. A comprehensive assessment of biofuel policies in the BRICS nations: Implementation, blending target and gaps. *Fuel* 272, 117635. https://doi.org/10.1016/j.fuel.2020.117635

Shahbaz, M., Raghutla, C., Chittedi, K.R., Jiao, Z., Vo, X.V., 2020. The effect of renewable energy consumption on economic growth: Evidence from the renewable energy country attractive index. *Energy* 207, 118162. https://doi.org/10.1016/j.energy.2020.118162

Stokes, L.C., Breetz, H.L., 2018. Politics in the U.S. energy transition: Case studies of solar, wind, biofuels and electric vehicles policy. *Energy Policy* 113, 76–86. https://doi.org/10.1016/j.enpol.2017.10.057

Suali, E., Suali, L., 2023. Impact assessment of global biofuel regulations and policies on biodiversity. In *Environmental sustainability of biofuels* (pp. 137–161). Elsevier. https://doi.org/10.1016/B978-0-323-91159-7.00012-6

Subramaniam, Y., Masron, T.A., 2021. The impact of economic globalization on biofuel in developing countries. *Energy Convers Manag: X* 10, 100064. https://doi.org/10.1016/j.ecmx.2020.100064

Suhara, A., Karyadi Herawan, S.G., Tirta, A., Idris, M., Roslan, M.F., Putra, N.R., Hananto, A.L., Veza, I., 2024. Biodiesel sustainability: Review of progress and challenges of biodiesel as sustainable biofuel. *Clean Technologies* 6, 886–906. https://doi.org/10.3390/cleantechnol6030045

Taheripour, F., Baumes, H., Tyner, W.E., 2022. Economic impacts of the U.S. renewable fuel standard: An ex-post evaluation. *Front Energy Res* 10. https://doi.org/10.3389/fenrg.2022.749738

Tang, T., McCaffery, C., Ma, T., Hao, P., Durbin, T.D., Johnson, K.C., Karavalakis, G., 2023. Expanding the ethanol blend wall in California: Emissions comparison between E10 and E15. *Fuel* 350, 128836. https://doi.org/10.1016/j.fuel.2023.128836

Tolliver, C., Keeley, A.R., Managi, S., 2019. Green bonds for the Paris agreement and sustainable development goals. *Environ Res Lett* 14, 064009. https://doi.org/10.1088/1748-9326/ab1118

Trevisan, R., Ghiani, E., Pilo, F., 2023. Renewable energy communities in positive energy districts: A governance and realisation framework in compliance with the italian regulation. *Smart Cities* 6, 563–585. https://doi.org/10.3390/smartcities6010026

Usmani, R.A., Mohammad, A.S., Ansari, S.S., 2023. Comprehensive biofuel policy analysis framework: A novel approach evaluating the policy influences. *Renew Sustain Energy Rev* 183, 113403. https://doi.org/10.1016/j.rser.2023.113403

Wang, Y., Guo, C., Du, C., Chen, X., Jia, L., Guo, X., Chen, R., Zhang, M., Chen, Z., Wang, H., 2021. Carbon peak and carbon neutrality in China: Goals, implementation path, and prospects. *China Geology* 4, 1–27. https://doi.org/10.31035/cg2021083

Wang, Y., Shinwari, R., Naeem, M.A., 2024. The effects of green technology and globalization on energy demand in emerging economies. *Renew Energy* 237, 121706. https://doi.org/10.1016/j.renene.2024.121706

Xu, X., Zhou, Q., Yu, D., 2022. The future of hydrogen energy: Bio-hydrogen production technology. *Int J Hydrogen Energy* 47, 33677–33698. https://doi.org/10.1016/j.ijhydene.2022.07.261

Yin, H., Powers, N., 2010. Do state renewable portfolio standards promote in-state renewable generation?. *Energy Policy* 38, 1140–1149. https://doi.org/10.1016/j.enpol.2009.10.067

Younes, A., Fingerman, K.R., Barrientos, C., Carman, J., Johnson, K., Wallach, E.S., 2022. How the U.S. renewable fuel standard could use garbage to pay for electric vehicles. *Energy Policy* 166, 112916. https://doi.org/10.1016/j.enpol.2022.112916

10 Commercialization of Microbial Biofuels around the World

Navneet Kaur, Simran Jot Kaur, Parmjit Singh Panesar, and Shashi Kant Bhatia

10.1 INTRODUCTION

Before the millennium, people relied on natural fuels such as wood and straw to fulfill their energy requirements. Later on, with the emergence of the Industrial Revolution, natural fuels were ousted by fossil fuels like coal, natural gas, and petroleum (Priya et al., 2021). In the 21st century, with a tremendous increase in population and urbanization, the demand for sustainable energy has become a major challenge. Most of the energy used worldwide is supplied by fossil fuels, which account for 84% of total energy, out of which 96% of this is utilized in transportation. However, fossil fuels are a limited reserve and not sustainable (Shiru and Shiru 2021). Additionally, they are also associated with several threatening conditions like the release of toxic gases into the environment, water and soil pollution, urban smog, acid rain, emission of greenhouse gases (GHGs), and subsequently climate change (Awogbemi et al., 2021). GHG emissions by burning of fossil fuels have led to an average increase in earth's temperature by 1.9°C. It has also been predicted that the amount of GHG emissions will elevate by 50% in the coming years and will become a major concern to global warming and climate change by 2050. This scenario has prompted government and international organizations to outline potent energy policies to curb GHG emissions into the environment and to switch toward renewable energy sources to stabilize climate change (Subramaniam et al., 2020).

Renewable energy sources are natural sources such as biomass, biofuels, geothermal, hydropower, tidal power, solar, and wind energy. Among these energy sources, biofuels have gained considerable recognition throughout the world due to several benefits offered by them – e.g., being renewable they are also a sustainable source of energy, are environment friendly as they fix carbon dioxide and lower air pollution by preventing rotting and burning of biomass in the fields, create energy security in countries with low fossil fuel reserves and give rise to new jobs, thereby leading to economic development (Balan 2014). Owing to the aforementioned advantages, interest in biofuels is growing, and theoretically, they are believed to solve global challenges related to energy scarcity and climate change (Solomon 2010). However, despite having numerous benefits, several questions have been raised against the

DOI: 10.1201/9781003633709-10

successful production and commercialization of biofuels; historically, sustainability has been mostly disregarded in the case of biofuels; the production is energy inefficient and uneconomical, has costly infrastructure requirements, etc. (Priya et al., 2021; Amerit et al., 2023). Thus, the chapter is aimed at highlighting the recent commercial status of some of the major microbial biofuels like bioethanol and biobutanol with emphasis on the potential difficulties in production and technological advancements needed for commercialization of these microbial biofuels along with the current commercial status of these fuels around the world.

10.2 BIOFUELS AND THEIR CLASSIFICATION

The term "biofuel" was first used in the early 1970s, and since then, it has been defined by various authors in various contexts. Precisely, biofuel may be defined as fuel produced using natural sources such as plant and plant based products, organic waste, or manure in any form – i.e., solid, liquid, or gas – having the ability to be transformed into another type of biofuel. Similar to the definition of biofuel, various authors have categorized biofuels into different classes based on different criteria. Generally, the classification of biofuels is made based on the following criteria – i.e., physical state/ nature of biofuel, technology maturity or commercialization status of biofuel, type of feedstock, and conversion routes used (Awogbemi et al., 2021; Osman et al., 2021). However, from the commercialization point of view, classification based on technology maturity or commercialization status of biofuels (Figure 10.1), and based on the

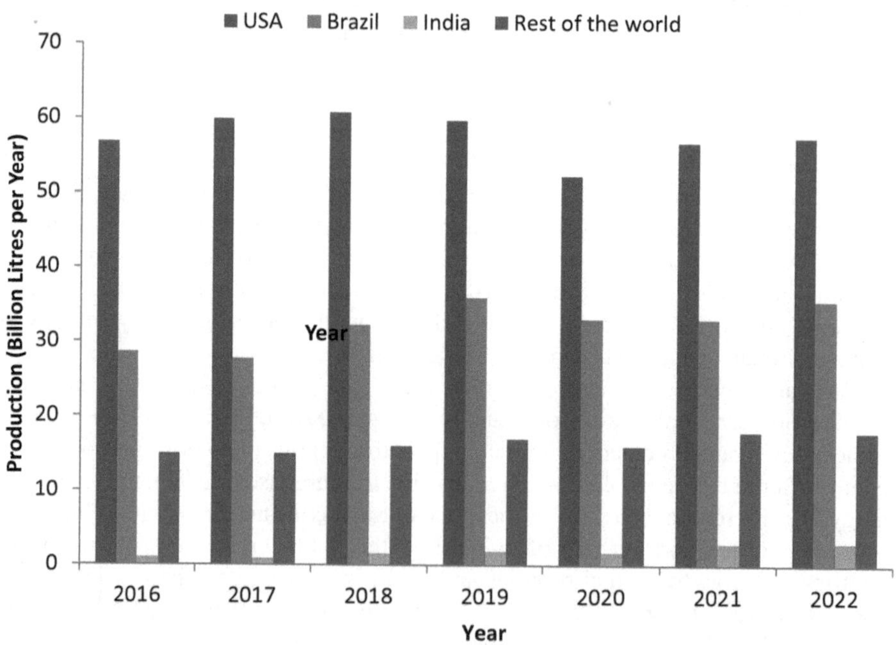

FIGURE 10.1 Global bioethanol production from year 2016–2022 in the USA, Brazil, India, and the rest of the world.

type of feedstock is crucial for a better understanding of the constraints confronted in production and worldwide distribution and commercialization of biofuels.

10.2.1 On the Basis of Technological Maturity and Commercialization Status

a. *Conventional Biofuels*: Biofuels obtained using existing commercial technology such as anaerobic digestion, fermentation, and dark fermentation are termed as conventional biofuels. The technology for these biofuels has already been evolved and commercialized, and these biofuels are even used at small-scale and housing facilities (Figure 10.1). The examples of feedstock used to produce these biofuels include animal fat, corn, sugar beet, sugar cane, wheat, waste oil, and fat, etc. (Ruan et al., 2019; Awogbemi et al., 2021). The examples of commercial biofuels include bioethanol produced from sugar and starch via fermentation, biomethane produced via anaerobic digestion, and biodiesel produced via transesterification (Ruan et al., 2019)
b. *Advanced Biofuels*: The technology to manufacture these biofuels is either at the research, demonstration, or early commercial stage. They are generally produced using inedible feedstock such as hydrotreated vegetable oil (HVO), lignocellulose, microbial lipids, microalgae, or macroalgae (Awogbemi et al., 2021). Among advanced biofuels, HVO represents the early commercial stage while bioethanol produced using lignocellulose, biomass-to-liquid (BtL) biodiesel, microalgae biodiesel, and biohydrogen represent demonstration and research stages (Figure 10.1; Ruan et al., 2019).

10.2.2 On the Basis of Feedstock

a. *First-generation feedstock (1GF)*: They mainly comprise of edible or food crops, and these crops are readily available sources of starch and sugar such as corn, maize, soybean, sugarcane, sugar beet, rapeseed, potato, oil crops (coconut, canola), and wheat. They undergo limited mechanical and thermal pre-treatments followed by fermentation to obtain biofuels. Bioalcohols are generally produced using first-generation feedstock. However, commercial use of this feedstock is associated with several drawbacks such as high cost of production, unsustainable nature, and food insecurity issues due to the requirement of large arable land, high water and fertilizer requirements, and deforestation. Some of the crops like soybean even resulted in an overall negative energy balance. Hence, due to the aforementioned limitations and the presence of better substitutes in the form of second-generation feedstock, their value is slowly declining (Nanda et al., 2018; Callegari et al., 2020). The biofuels produced using 1GF are called first-generation biofuels (1GB).
b. *Second-generation feedstock (2GF)*: They comprises of nonedible feedstock or feedstock with no or minimal food value such as agricultural residues, nonfood crops, and other biomass waste, rich in cellulose, hemicellulose,

and lignin fractions. If grown, these feedstocks must have minimal land requirements and a nominal effect on the environment in terms of water and fertilizer demand. The examples of 2GF include agro-industrial residues (barley straw, flax straw, paddy straw, wheat straw, corncob, corn stover, cotton stalk, mustard stalk, canola meal, canola hull, flax fiber, jute bast, coconut coir, coconut shell, palm seeds, rice husk, walnut shell, almond shell, cashew nut shell, hazelnut shell, peanut shell, apricot pits, peach pits, plum pits, olive pits, sugarcane bagasse, etc.), forest (stems, barks, twigs, cones, and needles) and wood residues (sawdust, wood chips, lumps, bales, pellets, and briquettes), energy crops (switchgrass, *miscanthus*, Indian grass, Timothy grass, elephant grass, reed canary grass, ryegrass, alfalfa, bamboo, hybrid polar, *jatropha*, *camelina*, oil palm, and rapeseed) waste vegetable oil and municipal solid waste (namely, landfill gas, sludge, grass and yard clippings). These feedstocks offer greater advantages over 1GF such as low carbon footprints, high net energy yields, rapid growth, and minimal requirements in terms of nutrients and land. However, they require several pretreatment steps to obtain fermentable sugars which escalate operational costs for biofuels, thus making the process commercially uneconomical and unsustainable (Nanda et al., 2018; Callegari et al., 2020). The biofuels produced using 2GF are called second-generation biofuels (2GB's).

c. *Third-generation feedstocks (3GFs)*: They comprise of micro and macroalgae and have recently been put into their separate class due to excellent yield and numerous benefits provided by algae in comparison to other feedstock. Algae, being autotrophic organisms, convert biomass into carbon dioxide (CO_2) and in the process release oxygen (O_2) into the atmosphere. They produce biofuels by direct and indirect approach. In the direct approach, lipids or oils secreted by algae are utilized to generate biodiesel or gasoline fractions, while in the indirect approach, algal biomass is acted upon by special fermentative bacteria or oleaginous microorganisms to produce biobutanol (Callegari et al., 2020). The potential benefits offered by algae in comparison to other feedstock include ease of cultivation, as they show rapid growth; do not require arable land or fertilizers; can be grown naturally or under artificial conditions; and can be grown in wastewater and thus are helpful in bioremediation of wastewater and help in CO_2 mitigation; biofuels produced via algae show compatibility with traditional gasoline engines and last, but not least, give high energy yield. Despite a myriad of advantages, the high cost of production, complexities of conversion methods, and technical difficulties with merging of existing technology with problematic infrastructure are some of the limitations in sustainable commercial implementation of 3GFs (Awogbemi et al., 2021). The biofuels produced by these processes are called third-generation biofuels (3GB's).

d. *Fourth generation feedstocks (4GFs)*: The group consists of genetically engineered algae, though they are still at the research stage and mostly limited to laboratories. Metabolic engineering is employed to enhance oil or lipid content produced by algae, improvement in carbon capture pathways, and intensification of growth and betterment of harvesting and fermentation methods. Even nanofarming has been employed recently to enhance oil production in algae. The biofuels obtained by 4GF are called fourth generation biofuels (4GBs) and include green biodiesel, biogasoline, and green aviation fuels (Awogbemi et al., 2021).

10.3 MICROBIAL BIOFUELS – CHALLENGES AND OPPORTUNITIES IN THEIR COMMERCIALIZATION

As discussed earlier, fossil fuels are nonrenewable sources of energy that pose a substantial threat to the environment by contributing to GHG emissions and are a major force behind climate change. Thus, the world needs to shift toward more inexpensive, practical, and environment friendly renewable energy sources that can be produced from biomass or biological waste (Keasling et al., 2021). When these biofuels are produced utilizing microorganisms, they are called microbial biofuels. The fascination toward microbial biofuels is constantly escalating due to metabolic diversity shown by various microbial strains that permit the production of biofuels using diverse strategies (Keasling et al., 2021) which include the following:

- Fermentation of sugars obtained from biomass (plant residue or agro-industrial residues or biological waste) to obtain biofuels such as bioethanol or biobutanol
- Cultivation of photosynthetic microorganisms such as algae using photobioreactors in the presence of sunlight and CO_2 for direct synthesis of biofuel such as, bioethanol and biohydrogen
- CO_2 fixation by microbial cells
- Metabolic engineering of microbes

Major microbial biofuels that have been widely used, researched, and commercialized include bioethanol. On the other hand, biobutanol has superior qualities than bioethanol but its commercialization faces critical challenges. Moreover, though the production of bioethanol has been achieved at large or pilot scale, still it suffers from various hurdles in successful commercialization, such as low-yield substrate limitations, including high cost, availability, pretreatment cost, technological constraints, high infrastructure cost, downstream problems, fuel performance, policy, and regulations. The authors have tried to compile difficulties encountered in the production and commercialization of microbial biofuels with special emphasis on bioethanol, and biobutanol, along with possible solutions to overcome the problems.

10.3.1 BIOETHANOL

Low molecular weight alcohols such as ethanol, butanol, methanol, and propanol have marked capability to substitute fossil fuels. Among these, ethanol has generated a lot of interest in its production and utilization as a biofuel, as it can be produced by employing diverse feedstock and conversion technologies. However, mass scale production of bioethanol is achieved mainly through fermentation of starch- and sugar-based raw materials, also called 1GF, and from lignocellulose provided by 2GF mainly using yeast *Saccharomyces cerevisae*. The bioethanol produced via fermentation is generally used in the food and beverage industry, while for utilization in the fuel industry, it is mostly mixed with small amounts of methanol and gasoline, which makes it unfit for human consumption. Ethanol offers various advantages as an alternative fuel. Some of them include high octane number, high heat of vaporization, clean combustion, and hence low GHG emission and sustainability. On account of the aforementioned benefits, ethanol has already captured the biofuel market worldwide with the United States and Brazil being constant leaders by together adding to 90% of the world's total bioethanol production using corn and sugarcane as feedstocks (Nanda et al., 2018). It is important to recognize here that the worldwide production of bioethanol is based on 1GFs, which suffer several limitations, as discussed earlier. Not only this but also the production process and distribution witness surplus hardships, as discussed in detail in the following section.

10.3.1.1 Feedstock Barriers

10.3.1.1.1 Type of Feedstock Generations and Pretreatment

The feedstock for the production of bioethanol on a mass scale includes directly fermentable sugars such as pentoses and hexoses (1GF) and lignocellulosic fractions from agricultural and forest residues (2GF). Among 1GF, corn is widely employed in the United States, wheat and corn in Canada, sugarcane in Brazil while wheat, potatoes, and sugar beet in Europe for the generation of ethanol. On a commercial scale, 1GF provides high yields by producing high quality sugars and requires little or no pretreatment. Nevertheless, utilization of these feedstocks is condemned due to high-price, low-energy balance, resulting in the food versus fuel debate – i.e., when these feedstocks are used at large scale for the production of biofuels, they affect the world food supply, and thus are not regarded as sustainable (Alam and Tanveer 2020; Tse et al., 2021). Moreover, the supply of these crops on the world scale is not enough to fulfill the requirement of ethanol as transport fuel (Lahtchev 2012).

The 2GF on the other hand is composed of nonedible fractions such as agricultural, industrial, and forest residues. These raw materials do not act as competition to food crops. They are relatively cheap and available in surplus quantities. However, they require pretreatment of lignocellulosic biomass. Lignocellulose comprises cellulose, hemicellulose, and lignin. These molecules being recalcitrant in nature are hard to break by chemical and biological methods. Therefore, pretreatment becomes necessary to remove lignin and release fermentable sugars prior to fermentation. Pretreatment of 2GF has an eloquent effect on the overall process of ethanol production both in terms of yield and production cost. Another important issue with 2GF is variability – i.e., different types of 2GF require different pretreatment steps, making it the most expensive step in ethanol production due to the complex technologies and infrastructure involved (Azhar et al., 2017; Tse et al., 2021). Another important feature of pretreatment is to reduce the development of degradation products, as they pose an inhibitory effect on subsequent stages of hydrolysis and fermentation. The generation of these products severely interferes with the production of ethanol besides leading to escalated costs of the final product owing to supplementary steps needed in the removal of these inhibitors. As a result, the adoption of lignocellulose feedstock at a commercial scale is much slower than expected (Lahtchev 2012).

3GF for bioethanol production includes algal biomass (macro and microalgae). Algae offer diverse benefits in comparison to first and 2GF, as it readily absorbs CO_2, contains a huge reserve of carbohydrates needed for ethanol production, and requires less space for propagation and ease of cultivation (Tse et al., 2021).

For mass production of algae, two methods are mainly employed – namely, open and closed photo-bioreactors, along with modifications in these methods. An open pond system is cost-effective method due to the ample availability of light in the form of natural sunlight for cultivation. The method also acts as a superior carbon sink by capturing CO_2 from flue gases. However, the method is not flawless, the limitations include easy contamination by unwanted bacteria, fungi, and protozoa and tremendous loss of water due to high evaporation. Moreover, the method is not suitable in temperate regions with low sunlight (Lam et al., 2019). To solve the limitations of the open pond system, closed bioreactors came into being. The advantages associated with these are high biomass productivity, low chances of contamination, re-circulation of water, and strictly controlled process parameters like pH, temperature, mixing, CO_2 concentration etc. Despite numerous advantages, these bioreactors are confederated with negative energy balance, as they consume huge amounts of electricity in mixing and pumping operations (Lam et al., 2019).

In addition to cultivation problems, algae also require pretreatment in the form of cell disruption methods to release carbohydrates. Moreover, in the absence of sufficient pretreatment, side products such as formic acid, acetic acid, and furanic compounds can be formed, which results in lower yield. Feedstock variability is another major problem, as algal species vary in the composition of neutral sugars, amino sugars, and uronic acids. This can result in different cultivation methods and pretreatment approaches leading to process complications and additional costs (Tse et al., 2021).

The theoretical yields and commercial constraints in the utilization of different generations of feedstock for bioethanol production at the industrial scale are compiled in Table 10.1.

TABLE 10.1
Classification of Biofuels According to Technology Maturity and Commercialization Status

→	*Conventional*	*Advanced*			Example of Biofuels
→	*Commercial*	*Early Commercial*	*Demonstration*	*Research*	
1	Sugar and Starch	Cellulosic			Bioethanol
2	Transesterification	Hydrogenated Vegetable oil	BtL		Biodiesel
3	Biogas		Syngas		Biomethane
4			Biohydrogen		Biohydrogen

Source: Adapted from Ruan et al. (2019).

10.3.1.1.2 Hydrolysis

After selection of feedstock and its pretreatment, the next step is to carry out hydrolysis. The aim of hydrolysis is to release fermentable sugars in the form of pentoses and hexoses to act upon by yeast and produce bioethanol. Generally, hydrolysis is carried out in the presence of acid or enzymes (Azhar et al., 2017).

Among the acid and enzymatic methods, the former is widely used as it's cheap and rapid. However, it liberates harmful environmental toxicants in the form of furfural and hydroxymethyl furfural, which also act as inhibitors in the fermentation process and thereby affect the yield of bioethanol. Other major commercial constraints to deal in acid hydrolysis are difficulties in the retrieval and reuse of acid, which makes the production process uneconomical and costly (Lahtchev 2012; Azhar et al., 2017).

On the other side, the enzymatic process is more environmentally friendly and effective. The enzymes operate at mild conditions, and the process can be merged with fermentation in the form of simultaneous saccharification and fermentation (SSF), thus reducing the operational and production costs. Moreover, the method is profitable in terms of energy and chemical utilization. In terms of economic influence on the process, the major limitation encountered in enzymatic hydrolysis is a high price associated with enzyme consortia and the production of low sugar hydrolysate, which ultimately shoots production costs (Azhar et al., 2017).

Several scientists and research groups have been working extensively toward lowering the cost of enzymes used for hydrolysis. For this purpose, some workers have proposed the addition of various additives like nonionic surfactants and non-catalytic proteins and salts to boost the ability of enzymes in hydrolysis. Nevertheless, the studies are limited to 1GF such as corn straw, sugarcane bagasse, and wheat straw only and those too pretreated mainly by physico-chemical methods (Amandio et al., 2023).

To enhance the economic feasibility of enzymatic hydrolysis, another important aspect to be considered is the improvement in the sugar content of hydrolysate. High solid content (negligible quantity of water in slurry before the start of enzymatic hydrolysis) of more than 15% w/v generally solves the problem. The financial benefits of this strategy include low capital and operational cost, increased yield, low purification and recovery costs, and low waste disposal costs. The method is beneficial in terms of energy also (da Silva et al., 2020).

10.3.1.2 Fermentation Obstacles

a. *Ethanol producing microorganisms*: The widely used microorganism for the production of bioethanol from 1GF or sugar containing feedstock is *Saccharomyces cerevisiae* because of its ability to efficiently convert sucrose into hexose sugars – i.e., glucose and fructose. Another important property of *S. cerevisae* which makes it an attractive candidate for ethanol fermentation is its ability to thrive at a wide range of pH, which makes the fermentation less prone to contamination. Earlier, baker's yeast was employed to carry out bioethanol fermentation due to its cost-effective and readily available nature, but its use was discontinued due to contamination by wild-type yeasts at industrial scale. Other major challenges associated with the use of *S. cerevisiae* are its intolerance to higher sugar and salt concentrations in the medium, high ethanol concentration (above 20%), higher temperatures (35°C–45°C), and bacterial contamination. Cane molasses media, with high osmolarity due to the presence of elevated sugar and salt concentration, negatively impact ethanol synthesis in *S. cerevisiae*. The other major issue demanding attention is the inability of *S. cerevisiae* to ferment pentose sugars while using lignocellulosic feedstocks (Azhar et al., 2017; Busic et al., 2018).

To overcome the aforementioned hindrances and increase the efficiency of ethanol production at an industrial scale, ethanol-tolerant, as well as thermo-tolerant yeasts, which can efficiently ferment hexose as well as pentose sugars, need to be employed. One such example is *Kluyveromyces marxianus*, the strain that can tolerate temperatures in the range 42°C–45°C and can co-ferment hexose and pentose sugars as well. Moreover, genetic engineering techniques, hybrid strains, and co-culture can be used to solve the problem of hexose and pentose simultaneous utilization. Hybrid strains such as *Pachysolen tannophilus* and *Pichia stipites* have been produced via protoplast fusion, which can efficiently perform co-fermentation. Similarly, genetic engineering techniques have been utilized in which engineered *S. cerevisae* strains have been developed by introducing xylose reductase and xylitol dehydrogenase genes – for example, *S. cerevisiae* ZU-10. In the co-culture process, two different yeasts are simultaneously cultured and grown in the same fermenter vessel – for example, *Pichia fermentans* and *P. stipitis* (xylose fermenting yeasts) – are often used along with *S. cerevisae* (Lahtchev 2012; Azhar et al., 2017).

b. *Fermentation strategies*: For the production of bioethanol at a commercial scale, major strategies employed include, submerged or liquid state fermentation/separate hydrolysis and fermentation (SHF), solid state fermentation (SSF), simultaneous saccharification and co-fermentation (SSCF), consolidated bioprocessing (CBP) and very high gravity (VHG) fermentation (Tse et al., 2021).

   1) *Submerged or liquid state fermentation or SHF* is generally employed with 1GF containing a high amount of sugar and starch. Hence, the raw material is first mixed with water to get sugars in a liquid medium or in case of starchy raw-materials cooking or enzymatic hydrolysis is employed. The approach, however, suffers several limitations at the commercial scale as it requires high energy input and large quantities of water, along with the generation of huge quantities of waste, which eventually leads to high operational costs (Tse et al., 2021).
   2) *SSF*: The method involves the growth of desired microorganisms on solid or insoluble support without the addition of water. The strategy is mainly utilized with 2GF. In this approach, enzymatic hydrolysis is carried out continuously to release reducing sugars in fermentation media in comparison to submerged fermentation, which requires a high concentration of enzymes initially to sustain saccharification. Reducing sugars produced in SSF are converted to ethanol via simultaneous

saccharification and fermentation, thus leading to high ethanol yields. The method offers superiority in the form of low product inhibition, high ethanol yield, and low cost of equipment, as saccharification and fermentation are both carried out in a single vessel. The only disadvantage includes close process monitoring and longer incubation time (Tse et al., 2021).

3) *SSCF*: An alternative strategy is an SSCF that involves the use of a single unit to carry out both hydrolysis and fermentation. It is more suitable for 2GF as the strains employed result in co-fermentation of hexoses and pentoses simultaneously thereby leading to high ethanol yield. The sole drawback associated with the method is longer fermentation time due to the successive use of substrates by the culture involved (Baroda et al., 2022).
4) *Consolidated Bioprocessing (CBP)*: In this process single unit is used to perform enzyme production, substrate hydrolysis, and fermentation. It involves the use of genetically engineered strains that can both produce enzymes to cause biomass degradation and at the same time produce ethanol by fermenting sugars. The method can be of significant value at a commercial scale due to low infrastructure and operational cost and moreover being environment friendly. The only factor limiting its use at industrial scale is low conversion efficiency (Baroda et al., 2022)
5) *VHG Fermentation*: Another advancing technique that rely on the high sugar content of biomass to produce a significant amount of ethanol. The advantages include low waste generation and low energy and water input leading to low production and operational cost. The method is also environmentally friendly. However, the method can only be applied to1GF and 3GF only (Tse et al., 2021)

c. *Modes of Fermentation*: The fermentation of ethanol can be conducted in three modes –namely, batch, continuous, and fed-batch mode. Among all the methods, fed-batch mode is mostly employed at the commercial scale, as it offers maximum viable cell concentration, the extended life span of cells, high ethanol concentration, low by-product accumulation, and better control of factors such as pH, temperature and dissolved oxygen (DO) (Tse et al., 2021; Baroda et al., 2022).

10.3.1.3 Downstream Processing Constraints

The recovery of bioethanol produced via fermentation undergoes two costs intensive and energy consuming steps namely distillation and dehydration. The steps are important for the conversion of hygroscopic ethanol to fuel grade ethanol. However, they also lead to a notable increase in GHG emissions. Recovery of an insignificant amount of ethanol also significantly leads to increased production and operational costs, along with a negative energy balance. High ethanol concentration is achieved by utilizing 1GF (5%–12% v/v) while meager concentration is obtained by using 2GF and 3GF – i.e., below 40g/l – making the process of distillation quite expensive in comparison to its price (Beluhan et al., 2023).

Various distillation methods for the separation of ethanol include azeotropic distillation, extractive distillation, pressure swing distillation, and simple vacuum distillation. Though the aforementioned methods are used at large scale, they are not economically feasible and involve enormous amounts of energy. Simple vacuum distillation can yield anhydrous ethanol, but its industrial application is limited by high operational costs (Beluhan et al., 2023).

An alternative strategy to distillation methods is adsorption, which is quite an economic and eco-friendly approach used at a commercial scale to recover ethanol. However, the same physic-chemical characteristics of ethanol and water like high polarity and strong binding with the same adsorbants make this method a little challenging. But this is still preferred at the industrial scale, as regeneration of adsorbant is not required, as it is directly used as feedstock to produce bioethanol (Beluhan et al., 2023).

The development of new innovative methods to overcome the high operational costs of separation methods is ongoing. Some of the techniques involving the unification of separation and fermentation involve combining fermentation with adsorption, solvent extraction, flash fermentation, gas stripping, membrane perforation, and vacuum fermentation (Beluhan et al., 2023). Thus, the techno-economic analysis of all three generations of feedstock and fermentation approaches available for ethanol manufacture suggests that 1GF is presently the ideal choice for large-scale production of bioethanol. Although 2GF has the ability to fulfill the world's demand for energy in a renewable fashion, major progress is required to lower production cost and to amplify co-product usage. Moreover, the unification of both 1GF and 2GF can result in increased ethanol yield and decreased capital investment at the industrial scale. Lastly, though 3GF is revolutionary and the best replacement for 1GF and 2GF for bioethanol production, it needs breakthrough technology in order to be inexpensive and economically sustainable (Tse et al., 2021).

10.3.1.4 Engine Compatibility Issues

Nowadays, alternative biofuels such as biodiesel and bioethanol are blended with diesel and gasoline in quite high concentrations in an effort to reduce the use of fossil fuels (Khuong et al., 2017). Different concentrations of bioethanol are blended with gasoline from 10% to 85%, such as E15 and E85, with the former containing 15% bioethanol and 85% gasoline, while the latter contains 85% bioethanol and 15% gasoline (Callegari et al., 2020). Even some biofuel utilizing engines have been designed to work on pure hydrated bioethanol containing 93% bioethanol and 7% water. Moreover, use of bioethanol in countries like Argentina, Brazil, the United States, and some parts of America and Europe has become an energy policy. Blended bioethanol and gasoline mixes are also served at gas stations due to high GHG emissions by using fuel oil (Saifudin et al., 2018). Despite a long list of advantages bioethanol as fuel contains some components that pose serious engine issues such as chloride ions present in bioethanol reacting with oxide film of engine metal and resulting in corrosion. Similarly, aluminum hydroxide present in bioethanol as a soluble element results in clogging of the fuel system (Saifudin et al., 2018). Bioethanol being hygroscopic in nature can absorb water from the environment and result in the corrosion of the fuel injector and electric fuel pump, which can ultimately decrease the combustion energy and result in vehicle start-up problems in cold weather. It can also impact the performance and properties of the lubricating oils, as it has a higher tendency to enter the oil sump of the engine because of its high heat of evaporation in comparison to gasoline. Mixing of bioethanol with lubricant can significantly deteriorate the performance of engine oil as well. Emulsion of ethanol, water, and oil formed due to ethanol-water mixture considerably damages engine properties in the form of high wear and tear along with engine failure. However, the problem can be overcome by draining the engine oil on a regular basis. In a recent study conducted to evaluate the effect of various gasoline and bioethanol blends (E0, E10, E20, E30, and E85) on properties of commercial engine oil, it was clearly stated that higher concentrations of bioethanol in gasoline greatly reduce oil viscosity while increasing acid, which remarkably intensified oxidation and degradation of oil. It even knocked down the lubricating efficiency of engine oil. Enhanced friction and wear losses were seen with all blended samples. Nevertheless, only E10 was recommended by the workers to be used as transport fuel with existing gasoline engines, as it showed only a few deleterious effects in comparison to blends with higher concentrations of bioethanol (Khuong et al., 2017).

10.3.2 Biobutanol

Buty alcohol, also called biobutanol, is now considered a potential biofuel. It is a promising energy source of biological origin. It is a nonhygroscopic, colorless, flammable alcohol. It is noncorrosive, has a high calorific value, and lower affinity to water. Butanol has lower heat of vaporization, which enables the easier starting of a motor in cold weather. Owing to its low vapor pressure and low flashpoint, it is easier to handle. Moreover, it is easily biodegradable in the environment. Conventionally, the production of butanol is made by a process called acetone-butanol- ethanol fermentation, which is carried out mostly by the bacteria *Clostridium acetobutylicum* (Antil, 2019). Butanol offers high combustion heat, boiling point, and capacity to mix with gasoline in higher proportion without requiring any change in prevailing Otto Cycle Engines (Mishra and Dubey, 2017). Currently, most of the butanol produced worldwide is obtained from petroleum derived chemicals. Butanol market has its most dominance over North America and the United States; Canada, and Mexico have significantly contributed to the growth of the bio-butanol market in the region. But due to the introduction of legislation in various nations regarding the rise of clean fuels and reduction of greenhouse emissions, moderate growth is anticipated in the European region, which strongly advocates the need for biofuels for a safer environment.

However, Asia-Pacific is the fastest growing region (Figure 10.2) for biobutanol market due to the rising demand of the solvent in a variety of applications (www.custommarketinsights.com). The expansion of the global biobutanol market is estimated to be 22.3 billion by 2030, which marks a boost as compared to present years (Figure 10.3; www.custommarketinsights.com). Although biobutanol has shown tremendous opportunities to be safeguarded as an alternative biofuel, it has shown certain constraints regarding the production process and other major limitations, as mentioned in detail in the following section.

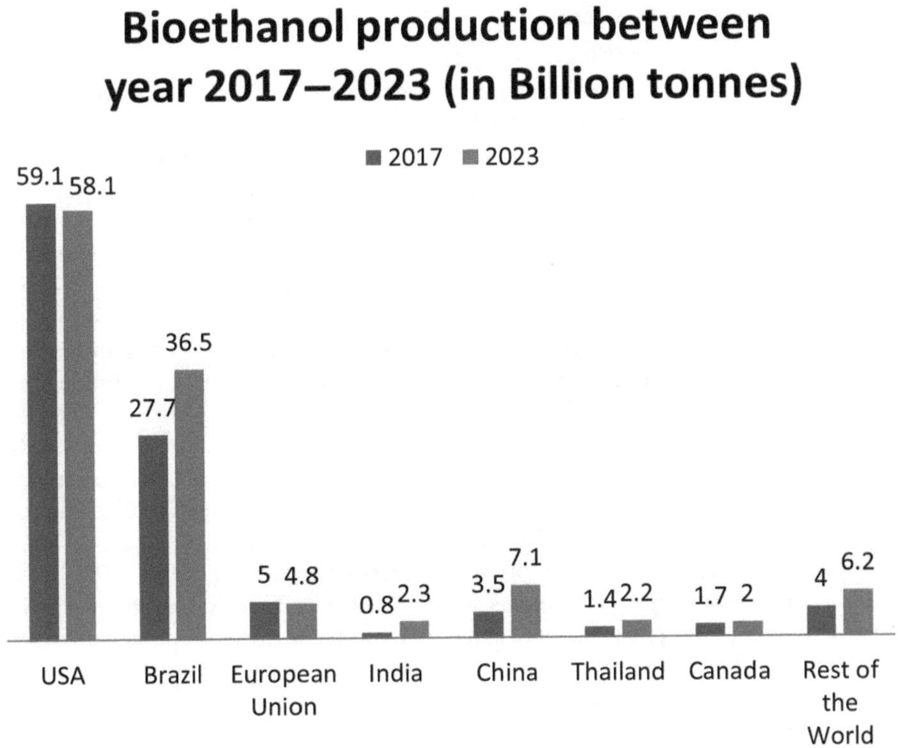

FIGURE 10.2 Bioethanol production between year 2017–2023.

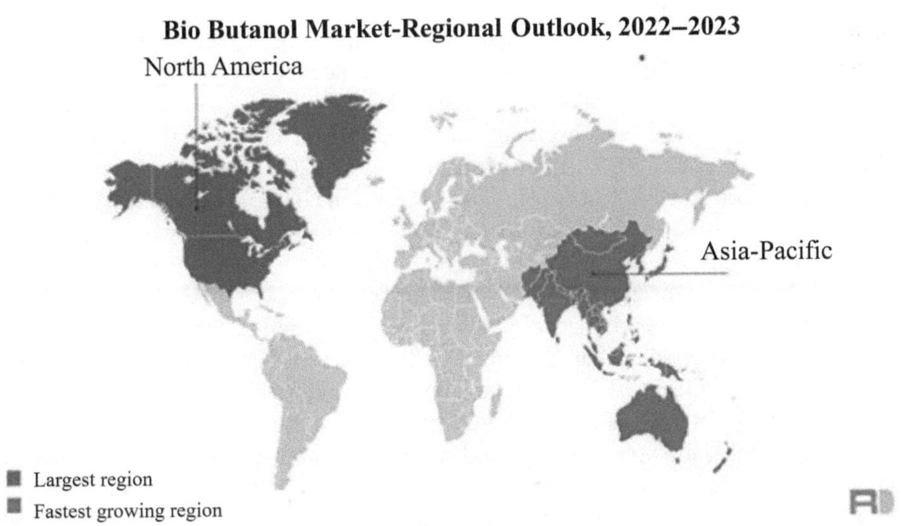

FIGURE 10.3 Regional outlook of biobutanol market from the year 2022–2027.

10.3.2.1 Feedstock Barriers

10.3.2.1.1 Type of Feedstock Generations and Pretreatment

The selection of feasible substrates plays a remarkable role in biobutanol production. Currently, four generations of feedstock have been utilized to produce butanol. Starch and sucrose feedstocks constitute the first-generation feedstocks and are most used for generating butanol. These largely include food crops, such as sugarcane, maize, and cereal grains, and starchy materials such as cassava, potatoes, and sweet potatoes. Although they provide sufficient fermentable substrates through simplified pretreatment processes with higher butanol yields and productivity, their utility is discouraged as they occupy large lands for cultivation, and compete with food supply, when these materials are made available for the production of biofuels at a global scale, significantly enhancing the overall cost of fermentation (Gottumukkala et al., 2017).

Lignocellulose biomass, which is basically a nonedible biomass, comprises the second-generation biobutanol feedstock, and it is predominantly obtained from agricultural, forestry, and municipal residues. Huge carbon resources are available in this category. Owing to its cost-effective and minimal impact on the food supply, it is a preferred substitute for biobutanol production. However, the major limitation lies in attaining sufficient fermentable sugars from complex and recalcitrant biomass. So, these require complex physicochemical processes to break down the structures of this biomass for delignification and release of fermentable sugars, making pretreatment a requisite step before fermentation. But this substantially multiplies the production cost, making the process more expensive (Rastogi and Shrivastava, 2017).

Moreover, different substrates utilize diverse pretreatment approaches and technologies, as lignocellulosic biomass is a complex feedstock whose physicochemical properties and chemical compositions vary with different biomass. Thus, making the process nonuniform and aiding a burden on the overall cost of biobutanol production. Another important aspect of pretreatment is to remove inhibitors that usually produce and can completely halt microbial activity, impacting further steps of fermentation. So, the necessary detoxification steps are needed to facilitate subsequent biological processes, but it markedly increases the costs and reduces the productivity of biobutanol (Karthick and Nanthagopal, 2021). Fruit pomace (Jin et al., 2019), wheat straw (Qi et al., 2019), rice (Mohapatra et al., 2020), potato peels (Abedini et al., 2020), corn stover (Desta et al., 2021), sugarcane bagasse (Chacon et al., 2021) and barley straws (Kuittinen et al., 2022) are some of the most commonly reported 1GF in the literature.

Algal biomass (macroalgae and microalgae) represents the third-generation feedstock for biobutanol production and is marked as a potential candidate for producing biobutanol. Microalgae can grow faster than terrestrial crops due to their higher photosynthetic efficiency. It can even be cultivated using ponds, seawater, and wastewater, and does not need lands like conventional agriculture. Moreover, it offers no competition with cultivable land and food supply. It can accumulate considerable carbohydrates that can be used for fermentation, they fix CO_2, and grow faster and at a higher rate than terrestrial plants (Bradley et al., 2022). Moreover, it contains no lignin and only a small amount of hemicellulose to block the decomposition of the glucose-based carbohydrates (Demuez et al., 2015). Thus, the conversion of microalgal biomass to glucose is more effective than that of lignocellulosic material. However, it comes with its own set of limitations. One major drawback is significantly increasing the cost of downstream processing by encompassing highly sophisticated technologies due to lower production and productivity of butanol. Another complication is to obtain sufficient fermentable sugars from the algal biomass (Wang et al., 2017).

Fourth generation feedstocks include syngas. This feedstock caters to increased CO_2 capture ability. Moreover, no pretreatment steps are needed to obtain fermentable sugars. It can directly utilize *Clostridia* with a high production rate of biobutanol. However, this requires an extensive study, as it is still at its infancy as a technology. There are several unknown factors that limit the production of biobutanol. A poor mass transfer from gas to liquid has been encountered in this feedstock (Dutta et al., 2014).

10.3.2.1.2 Hydrolysis

Following pretreatment of the feedstocks, hydrolysis plays a pivotal role as it is involved in the release of fermentable sugars that can be utilized by microbes to produce biobutanol in the process of fermentation. Typically, hydrolysis is performed using enzymatic or acid reactions (Serrano-Echeverry et al., 2023). Although, acid hydrolysis can easily dissolve lignin without prior chemical pretreatment. However, the by-products of this reaction (e.g., acetic acid, formic acid, hydroxymethylfurfural, phenolic compounds) have been found to be toxic to microorganisms, which can inhibit fermentation. Moreover, this process is not cost-effective, as many hurdles are encountered in the reusability of acid (Serrano-Echeverry et al., 2023).

Enzymatic hydrolysis, on the other hand, avoids the aforementioned drawbacks. It is more eco- friendly and thus sustainable in nature. However, it requires different enzymes and specific conditions (pH, temperature, etc.) for each substrate due to the high specificity of enzymes. Nowadays, the SHF process is being explored for biobutanol production. In this, after pretreatment, delignified cellulose is hydrolyzed to glucose by cellulases under optimum conditions, particularly around 50°C, and then resulting glucose is transferred to a fermentation reactor, which is then acted upon by butanol-producing microorganisms at temperatures around 35°C (Wang et al., 2016). However, the accumulation of glucose during enzymatic hydrolysis may inhibit cellulases activity in the hydrolysis process. Moreover, two separated enzymatic hydrolysis and fermentation processes inevitably increase the cost of equipment and the risk of microbial contamination, which makes the process uneconomically feasible (Cao et al., 2020).

10.3.2.2 Fermentation Obstacles

a. *ABE Fermentation*: In industrial ABE fermentation processes, two distinct phases, acidogenesis, and solventogenesis are involved. During acidogenesis, the bacterial strains use sugar substrates for biomass generation as well as to boost their rapid growth and produce acetic and butyric acid as by-products. The synthesis of these acids is essential for ATP generation, which is necessary for cellular metabolism (Sukumaran et al., 2011). Subsequently, in the solventogenesis phase, in response to the high acid concentration in the environment, acetate and butyrate are reassimilated as substrates for solvent biosynthesis, leading to a cessation of bacterial growth (Kolesinska et al., 2019). The primary product of this stage is butanol, along with a mixture of acetone and ethanol, with a typical molar ratio of 6:3:1, respectively.

b. *Microbes Producing Biobutanol*: The most well-known ABE fermenters are *Clostridia*, which are obligate anaerobes and spore forming bacteria. *C. acetobutylicum* (Ibrahim et al., 2015), *C. beijerinckii* (Plaza et al., 2017), *C. saccharoperbutylacetonicum* (Zetty-Arenas et al., 2019), and *C. pasteurianum* (Lipovsky et al., 2016) are the most important industrial strains for biobutanol production. These bacteria produce biobutanol by fermenting sugars through acetone, butanol and ethanol fermentation (Zetty-Arenas et al., 2019; Ashani et al., 2020). The main advantage of these anaerobic bacteria is that they possess similar metabolic pathways capable of fermenting a broad range of carbon substrates, such as disaccharides (sucrose, cellobiose, lactose, etc.), pentoses (xylose and arabinose), hexoses (glucose, galactose, and fructose), and starch fuels (Guo et al., 2022). However, the biggest challenge for the industrialization of butanol fermentation is low butanol yield due to the formation of by-products acetone and ethanol and the low butanol tolerance capability of these bacteria. These by-products affect the bacterial cell membrane, and once the concentration of butanol and other products reaches a certain level, bacterial metabolism is inhibited. In addition to this, bacterial strains are exposed to inappropriate pH, temperature, dissolved oxygen, and osmotic pressure due to the inhomogeneity of large-scale fermentation systems (Guo et al., 2022). Moreover, current butanol-producing microorganisms either *C. acetobutylicum* or *C. beijerinckii* cannot ferment the pentose sugars into butanol efficiently decreasing the titer of biobutanol production.

Taking into consideration the aforementioned challenges to enhancing the productivity of biobutanol, developing robust strains with improved butanol yield and tolerance is needed of an hour for butanol fermentation. Different researchers have utilized various strategies for modifying strains to enhance biobutanol yield and titer through evolutionary engineering, metabolic engineering, genetic engineering, and mutagenesis (Jang et al., 2013; Schwarz et al., 2017; Gao et al., 2020). In a study conducted by Xu et al. (2015), hyperbutanol-producing strain *C. acetobutylicum* JB200 acquired from *C. acetobutylicum* ATCC 55025 by long-term adaptation with intermittently butanol challenges could tolerate and produce butanol as high as about 20 g/L. *C. cellulovorans* was metabolically engineered, resulting in 4.96 g/L butanol titer from alkali extracted corn cob xylose (Wen et al., 2022). A synthetic small regulatory RNA (sRNA)–based system, consisting of a target recognition site, MiC sRNA scaf fold, and an RNA chaperone Hfq, was developed to control or knock down gene expression for *C. acetobutylicum* with stability and efficiency (Cho et al., 2019). The thiolase of *C. acetobutylicum* was specifically engineered to reduce sensitivity toward coenzyme A (CoASH), significantly alleviating feedback inhibition through three amino acid substitutions (R133G, H156N, G222V) and, increasing the production of butanol by 18% (Mann and Lütke-Eversloh, 2013).

c. *Fermentation Strategies*: Advanced fermentation processes utilized for biobutanol production at the industrial level comprise SHF, SSF, consolidated bioprocesses (CBP), SSCF, and two-stage and multistage fermentation (Guo et al., 2022).

1) *SHF*: This process was discussed earlier in regard to hydrolysis of the feedstock.
2) *SSF*: This is the process in which cellulose hydrolysis and fermentation take place in one step. It can be successfully applied for 1GF and delignated 2GF. This process has relatively few processing steps, lower operating costs, less risk of contamination, and lower feedback inhibition of sugars on enzymes, thus improving the yield of butanol. Although this method has a lot of merits, it still suffers some challenges, such as matching the optimum temperatures between enzymolysis (45°C–50°C) and the ABE fermentation (35°C–37°C) to achieve high cellulase activity, sugar yield, and butanol production (Li et al., 2019). Consequently, the efficiency of the enzymatic hydrolysis is inevitably compromised, and a much longer time is needed to complete the hydrolysis. In addition to this, enzymatic hydrolysis and sugar consumption by *Clostridia* is another challenge (Qureshi et al., 2008a, 2008b).
3) *SSCF*: This process is applied in butanol production in which chemically prepared water soluble oligomers and pretreated cellulose from lignocellulosic biomass were simultaneously subjected to enzymatic hydrolysis and ABE fermentation. This process takes into account the unique properties of the hydrolysate, which includes both C5 and C6 sugars (Cai et al., 2016). This process results in a higher yield of butanol as compared to other methods.
4) *Consolidated Bioprocesses (CBP)*: It is an integrated approach that combines cellulose production, lignocellulose hydrolysis, and fermentation of sugars completed in a single step with the help of a single microbe, which can be genetically engineered or a microbial consortium (Guo et al., 2022). In recent times, this process has been proposed as the most economically attractive configuration for lowering the cost of hydrolysis and fermentation of cellulosic biomass. CBP shows numerous merits, including reduction of contamination risk, increased overall production yield, and ability to reduce the reactor and enzyme costs (Vivek et al., 2019; Olguin-Maciel et al., 2020; Tri and Kamei, 2020). However, there is no perfect CBP microorganism that can degrade lignocellulosic biomass efficiently and at the same time produce butanol at a desired yield. So, Co-culturing biofuel producing strains with saccharolytic strains has been widely studied to address the limitations in substrate utilization by individual strains for the production of various bioproducts.

   In one study, it was reported that the co-culture of *Bacillus* sp. SGP1 and *C. tyrobutyricum* ATCC 25755T enhanced butyric acid production from sucrose (Dwidar et al., 2013). However, such consolidated bioprocessing implementation strongly depends on pH regulation (Wen et al., 2020).

Solventogenic Clostridium usually produces solvents after acidogenesis, when the culture pH decreases to 4.5 from 5.5 (Wen et al., 2020), which, however, would inhibit growth and hamper cellulose degradation by cellulolytic Clostridium (Raita et al., 2021).

5) *Two-Stage and Multistage Fermentation*: This process involves a diversified approach of collaborating many stages of fermentation together. These processes can completely utilize substrates and successfully assimilate acids to solvent concentration on long-term fermentation (Guo et al., 2022).

d. *Modes of Fermentation*: Different fermentation processes such as immobilized cell, batch, fed batch, and continuous have been utilized for butanol production. Each process has its own set of merits and limitations. However, continuous processes (employing free cell, immobilized cells, and cell recycling) provide more lucrative and productive alternatives for commercial production of butanol. This method caters to reduction in sterilization and reinoculation time, appreciable productivity, and fewer substrate and biobutanol inhibition of the strains by removing the no-production time from the bioreactor sterilization and inoculum preparation (Ezeji et al., 2007; Li et al., 2020).

10.3.2.3 Hardships Occurring During Recovery of Biobutanol

It has been found that the separation and purification of biobutanol present greater complexity and higher costs in the downstream stages of biobutanol production due to various reasons. As the concentration of butanol in the fermentation broth is significantly lower, typically around 2 w/w%, it is somewhat tedious to recover. Another reason is the boiling point of the butanol/water azeotrope, which is very close to that of water (93°C vs. 100°C) at atmospheric pressure, making distillation-based separation more difficult and complex. Then, the final concentration of butanol in the distilled aqueous azeotrope obtained is only 55.5 w/w%, which is notably low. However, it is quite vital to develop efficient, cost-effective, and sophisticated methods for the recovery of biobutanol to increase its economic feasibility as a fuel (Chadni et al., 2023). Conventionally, different methods such as vapor, liquid, adsorbent, and membrane-based techniques are exploited for the downstream processing of biobutanol. These methods have their own array of advantages and limitations.

Distillation is a preferred method for biobutanol recovery. In this technique, solvents are fractionated based on their varying volatility levels. Its industrial-scale operation is simple, yielding dehydrated solvents of high purity and greater recovery rates. However, the process is energy-intensive and requires high temperatures to operate. While in membrane distillation, there is a separation of ABE via a microporous hydrophobic membrane at different temperatures. This causes no damage to the culture. But it is limited by vapor liquid equilibrium and has small selectivity. Moreover, it is a complicated process (Cai et al., 2022).

Researchers are exploring the innovative and novel techniques to maximize the butanol production and to minimize the costs that occur during the recovery process. In situ product recovery (ISPR) techniques have been utilized to further alleviate butanol toxicity during the fermentation process and to enhance the yield of butanol.

These processes amalgamated pervaporation, adsorption, liquid-liquid extraction, and gas stripping, with the fermentation process for higher butanol productivity (Li et al., 2020). Lee et al. (2008) combined fed-batch fermentation and gas stripping methods, and the productivity of biobutanol significantly increased to 1.16 g/L/h. So, the implementation of such hybrid techniques would not only address the complexities and higher costs associated with biobutanol recovery but also offer competitiveness of biobutanol as a sustainable biofuel (Serrano-Echeverry et al., 2023).

10.3.2.4 Butanol as a Fuel or Additive

In recent decades, various countries, including the United States, Canada, Sweden, India, Australia, Thailand, China, Peru, Paraguay, and Brazil, have enacted legislation mandating the blending of alcohols with gasoline and diesel to reduce the burden on fossil fuels and for a sustainable solution for the growing energy demands (Balat and Balat, 2008). Butanol can be blended with gasoline at higher concentrations, reaching up to 85% due to its lower vapor pressure. Furthermore, it is possible to directly supply and store butanol using existing gasoline pipelines, providing additional logistical convenience (Pachapur et al., 2016). Blending of butanol with gasoline has been shown to have positive effects on motor performance, combustion conditions, and pollutant emissions and exhibited numerous advantages in key parameters such as brake thermal efficiency (BTE), brake specific fuel consumption (BSFC), and carbon monoxide emissions, as compared to other bioalcohols (Göktaş et al., 2021).

It has been noted that butanol has a higher energy density with an LHV (lower heating value) of 99,800 btu/gal, in contrast to ethanol at 76,000 btu/gal. Even Ramey found in his test of a 1992 Buick driven across the country on B100 (100% butanol) that the gas mileage was 8% better (based on the volume of butanol used) than on gasoline (Szulczyk, 2010). The blending of butanol may require engine modifications due to the lower vapor pressure of butanol. The effects of mixing n-butanol at 5%, 10%, and 15% v/v with 20% palm oil methyl ester blended with ultralow sulfur diesel fuel mixed with (B20) indicated an increase in BTE at medium and high engine loads, while BSFC increased across all scenarios. Moreover, the inclusion of n-butanol led to decreased emissions of PM2.5 and polycyclic aromatic hydrocarbons (PAHs), which are known for their carcinogenic and cytotoxic properties (Zhang and Balasubramanian, 2014).

In spite of all these advantages, along with numerous investigations carried out in this domain with promising results, the main reason that butanol has not become a major biofuel is that it costs much more to produce. So, the inclusion of biobutanol as fuel is still in the early stages.

10.4 CURRENT COMMERCIALIZATION STATUS OF MICROBIAL BIOFUELS

The world production of biofuels in the years 2019–2024 was expected to increase by CAGR of 8% (Shahid et al., 2021). The major microbial biofuels include bioethanol and biobutanol. The full business potential of the commercialization of biofuels is still limited by sentiments of society attached toward fossil fuels. The livelihood of developing countries mainly revolves around fossil fuel usage in the form of electricity, transport fuel, etc. Thus, to get rid of fossil fuels is quite a difficult task (Emetere and Joel 2020).

However, the major drivers behind the commercialization of biofuels include energy security, rural development, and employment generation. Nowadays, even policies focusing on combating climate change, such as low carbon footprints in transportation, have greatly impacted biofuel policy development. Along with this, increasing population and energy demand have a profound effect on the depletion of fossil fuels, which have necessitated the use of low-cost alternatives for transportation and industrial operations in the form of biofuels. The price and market of biofuels depend on the availability of feedstock, the rate of crude oil in the world market, the structure of biorefinery, and most importantly the demand of biofuel in that particular place. In 1990s, in an effort to build rural economies and to generate jobs in the energy sector, the production of biofuels was greatly enhanced. This also resulted in less trade of crude or processed oil from Middle East countries (Shahid et al., 2021).

Presently, the leading producers and consumers of biofuels include the United States and Brazil. Moreover, the overall share of biofuels in the global market is expected to get a boost of around 28% for Europe by 2025. At present, bioethanol is leading the global biofuel market but biobutanol is steadily gaining pace because of its better fuel properties than bioethanol (Callegari et al., 2020). The only limitation in the commercialization of biobutanol is its low production and hence it is yet to be commercialized at a large scale (Anandharaj et al., 2020).

Currently, the global ethanol production has increased tremendously from 105.5 billion liters to 119 billion liters between the years 2017 to 2023. The United States and Brazil are among the topmost producers, producing 58.1 and 36.5 billion liters, respectively. The global bioethanol production of leading countries is compared in Figure 10.3 (IEA, 2024). Out of the total bioethanol produced in the United States, 90% is utilized within the country itself to fulfill its energy requirements, while the remaining is exported to other countries. Most of the ethanol in the United States is mainly produced using corn and grains. However, this trend is slowly diminishing due to the popularity of 2GF. Brazil on the other hand mainly utilizes sugarcane for bioethanol production. The other major ethanol producers include China (7.1 billion liters), India (2.3 billion liters), Thailand (2.2 billion liters), and Canada (2.0 billion liters). Bioethanol production in China is mainly achieved from maize, cassava, and molasses, while it also imports a notable quantity from the United States as well (Halder et al., 2019). India is also emerging as a major bioethanol-producing country with production of 2.3 billion liters in 2023 in comparison to 0.8 billion liters in 2017. The main reason behind this enhanced production is encouragement in the form of national biofuel policies adopted by various countries, which are mainly established and executed to enhance the production and consumption of biofuels in the form of blending mandates, tax exemptions, making standards in the form of low carbon footprints, monetary help and public financing (Ebadian et al., 2020).

The United States, being a leader in bioethanol production, had more than 200 ethanol production units with an average production of 260,000 m^3/year (Figure 10.4). However, most of these plants are no longer in operation because they

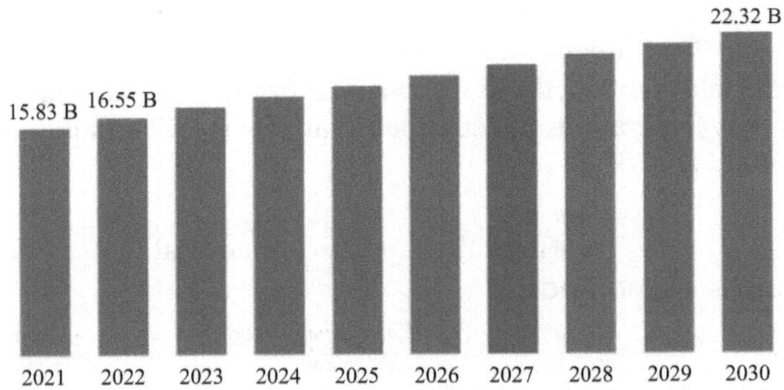

FIGURE 10.4 Global biobutanol market from year 2021–2030.

were mainly using 1GF, such as corn and sorghum, which are facing severe criticism worldwide due to the food versus fuel debate. Only a few facilities with 1GF are in a working state while most of the plants operational are now using 2GF or lignocellulose for the commercial production of bioethanol (Halder et al., 2019). In 2020, Brazil had almost 360 ethanol producing plants with a production capacity of 42,800 million liters. The plants were mainly run on sugarcane and corn. However, new facilities are under construction that will utilize 2GF. Europe comprises 57 1GF-based units for ethanol production as per 2020 with an average capacity of 8.15 billion liters. However, only 58% capacity was utilized (Hoang & Nghiem, 2021). The information on some of the commercial scale, pilot, and demonstration phase facilities worldwide is given in Tables 10.2 and 10.3.

TABLE 10.2
Theoretical Yields Obtained and Commercial Constraints Associated with Different Generations of Feedstock Available for Bioethanol Production at Large Scale

Feedstock	Theoretical Yield (L/t)	Commercial Constraints	References
First-Generation Feedstock (1GF)			
Sugar beet	110	High raw material cost, competence with food, high requirement of land, fertilizers and water	Marzo et al. (2019), Edeh (2020), Tse et al. (2021)
Sugarcane	70–75		
Cassava	137–180		
Maize	400		
Rice	430		
Wheat	340		

(*Continued*)

TABLE 10.2 (Continued)
Theoretical Yields Obtained and Commercial Constraints Associated with Different Generations of Feedstock Available for Bioethanol Production at Large Scale

Feedstock	Theoretical Yield (L/t)	Commercial Constraints	References
Second-Generation Feedstock (2GF)			
Corn stover	362–571	High operational cost, (due to additional cost of transport, storage, and pretreatment approaches) and high variability	Alia et al. (2019), Callegari et al. (2020), Tse et al. (2021)
Wheat straw	406–507		
Barley straw	540		
Sugarcane bagasse	318–532		
Switchgrass	392–457		
Sorghum	268–380		
Poplar	419–635		
Agave	347		
Agave Americana	347		
Agave tequilana	401		
Corn grain	470		
Rice straw	416–570		
Hardwood sawdust	381		
Softwood	543		
Willow	515		
Eucalyptus	578		
Pine	571		
Third-Generation Feedstock (3GF)			
Microalgae	167–501	Geographic and technical issues, high capital and operating cost involved in algal cultivation, water consumption cost, transport, storage, pretreatment cost	Alia et al. (2019), Lam et al. (2019), Tse et al. (2021)
Brown seaweeds (macroalgae)	12–1128		
Seagrass (macroalgae)	747		
Green seaweeds (macroalgae)	72–608		
Red seaweeds (macroalgae)	12–595		

TABLE 10.3
Bioethanol Production Facilities (Based on Fermentation) Around the World and Their Scale of Operation

Name of Facility	State/Country	Feedstock Used
Commercial Scale Facilities		
Sarina-Mackay	Austria	Sugarcane molasses
AustroCel Biorefinery	Austria	Lignocellulose/cellulose waste
Agrana Biorefinery	Pischlsdorf, Austria	Wheat and maize
Zeitz	Germany	Sugar beet, grains
Alco Energy Rotterdam Biorefinery	Netherland	Cereals/grains

TABLE 10.3 (Continued)
Bioethanol Production Facilities (Based on Fermentation) Around the World and Their Scale of Operation

Name of Facility	State/Country	Feedstock Used
Domsjo Fabriker	Sweden	Forestry raw material (wood)
Lantmannen Agroethanol	Sweden	Grains, waste residues from the food industry, agriculture, and forestry
ST1 Gothenburg, Ethanol (Etanolix) Plant (NEOT)	Sweden	Residues, biowaste, and process residue from local bakeries and bread from shops
Borregaard (Industries Chemcell Ethanol)	Norway	Spent sulfite liquor
Beijing Shougang LanzaTech New Energy Technology Co. Ltd	China	Waste gases
Cane Technology Center	Brazil	Bagasse
Gevo	Luverne, United States	Forest residues
GranBio	Sao Miguel, Brazil	Sugarcane and bagasse straw
Henan Tianguan Group	Zhenping, China Nanyang, China	Wheat and corn stover Lignocellulose
LongLive Biotechnology Co. Ltd.	Yucheng, China	Corn cob
POET-DSM Advanced Biofuels	Emmetsburg, United States of America	Agricultural residues
Quad Country Corn Processors	Galva, United States	Corn kernel fiber
Raizen Energia	Costa Pinto, Brazil	Bagasse
Bajaj Hindustan Ltd.	Mumbai, India Uttar Pradesh (UP), India	
Balarampur Chinni Mills	Kolkata, India	Molasses
Dalmia Bharat Sugar and Industries Ltd.	Ramgarh, UP, India	Molasses
Shree Renuka Sugars Ltd.	Karnataka, India Maharashtra, India	Molasses
Triveni Engineering and Sugar Industries	UP, India	Molasses
Pilot Scale Facilities		
Ethtec, Hunter Pilot Biorefinery (HPB)	Muswellbrook, New South Wales, Australia	Lignocellulosic material (sugarcane bagasse, crop stubbles and forest material)
Versalis Biorefinery	Crescentino, Italy	Hardwood (poplar), Agricultural residues
Chempolis Ltd.	Finland	Wheat straw, corn stover, bagasse
IFP, Futurol	France	Wheat straw, corn stover, bagasse
Aemetis	Butte, United States	Switchgrass, grass seed, grass straw and corn stalks
Energy and Chemical Department of East China University of Science and Technology	Shanghai, China	Crop and forestry residues

(Continued)

TABLE 10.3 (Continued)
Bioethanol Production Facilities (Based on Fermentation) Around the World and Their Scale of Operation

Name of Facility	State/Country	Feedstock Used
Greenfield Ethanol	Chatham, Canada	Lignocellulosics
Iowa State University	Boone, United States	Lignocellulosics
Lignol	Vancouver, Canada	Sugarcane bagasse
Lignol Innovations Ltd.	Bernaby, Canada	Hardwood and softwood
Lignol	Grand Juction, United States	Woody biomass
NREL	Golden, United States	Lignocellulosics
Petrobras	Rio de Janeiro, Brazil	Sugarcane bagasse
POET	Scotland, United States	Corn fiber, corn cob and corn stalks
PROCETHOL 2G	Pomacle, France	Woody and agricultural by-products, residues, energy crops
Woodland biofuels	Sarnia, Canada	Wood waste
American Process	Thomaston, United States	Forest residues
Anhui BBCA Biochemical	Bengbu, China	Corncob/corn stover
Borregaard AS	Sarpsborg, Norway	Sugarcane bagasse, straw, wood, energy crops, other lignocellulosics
Chempolis Ltd.	Oulu, Finland	Straw, reed, empty fruit bunch, bagasse,corn stalks, wood residues
COFCO Zhaodong Co.	Zhaodong, China	Corn stover
DuPont	Vonore, United States	Corn stover, cobs and fiber; switchgrass
Iogen Corporation	Ottawa, Canada	Wheat, barley and oat straw, corn stover, sugarcane
Jilin Fuel Alcohol	Jilin, China	Straw
LanzaTech	Soperton, United States	Woody biomass syngas
LanzaTech	Beijing, China	Industrial off gas
North European Oil Trade Oy	Gothenburg, Sweden	Organic residues and waste streams
North European Oil Trade Oy	Kajaani, Finland	Sawdust
Renmatix Rome	United States	Wood chips, switchgrass, and other raw materials
Shandong Zesheng Biotech Co.	Dongping, China	Straw
SP/EPAP	Ornskoldsvik, Sweden	Wood chips, sugarcane, bagasse, wheat, corn stover, energy grass, recycled waste
Tembec Chemical Group	Temiscaming, Canada	Spent sulfite liquor feedstock
Woodland Biofuels	Sarnia, Canada	Wood waste
ZeaChem	Boardman, United States	Poplar trees, wheat straw

Source: Halder et al. (2019), ETIP Bioenergy Report (2020)

10.5 CONCLUSION AND FUTURE PROSPECTS

In a nutshell, the chapter established a detailed account of challenges, including pre-treatment, hydrolysis, microbial strains, fermentation strategies, downstream issues, engine compatibility, etc., faced by industries to commercialize microbial biofuels.

Producing biofuels from food-based feedstock won't be a viable option in the near future; hence, researchers need to focus on utilizing alternative feedstock and technological interventions, such as microbial-based conversion of 2GF into biofuels. The technological limitations like low-yield, inefficient pretreatment methods, high cost of enzymes, high operational cost, purification cost, etc., need to be looked at with a deeper perspective. Not only this, but microbial biofuel commercialization also suffers from other hindrances in the form of low funds or subsidies, lack of blending mandates and market standards regarding biofuels, and constant competition from inexpensive gasoline and diesel. Thus thorough research and investigation are mandatory to enhance microbial biofuel production while simultaneously reducing cost. Even, government policies and regulations should incorporate funding programs, tax rebates, and awareness initiatives regarding the benefits of biofuels in combating climate change so as to increase their acceptance worldwide. Reforms in biofuel related policies are required to encourage industrial production of microbial biofuels. Thorough knowledge of logistics, operational parameters, process gaps, and assessment of risk on each parameter must be done to diminish any operational and monetary setbacks. Production costs of biofuels generally exceed their market price, which is a major knock-back in the scale up process of biofuels. Thus, to overcome this, high value co-product utilization must be followed. Hence, for the successful commercialization of biofuels, cooperation from industry, academicians, and governments is crucial.

REFERENCES

Abedini, A., Amiri, H., Karimi, K., 2020. Efficient biobutanol production from potato peel wastes by separate and simultaneous inhibitors removal and pretreatment. *Renew. Energy*, 160, 269–277.

Alam, M.S., Tanveer, M.S., 2020. Conversion of biomass into biofuel: a cutting-edge technology. *Bioreactors*, 55–74, Elsevier.

Alia, K.B., Rasul, I., Azeem, F., Hussain, S., Siddique, M.H., Muzammil, S., Riaz, M., Bari, A., Liaqat, S., Nadeem, H., 2019. Microbial production of ethanol. In: Inamuddin, Ahmer, A.F., Asiri, A.M. (Eds.), *Microbial Fuel Cells – Materials and Applications* (Vol. 46, pp. 307–334). Materials Research Foundations. http://doi.org/10.21741/9781644900116-12

Amandio, M.S.T., Rocha, J.M.S., Xavier, A.M.R.B., 2023. Enzymatic hydrolysis strategies for cellulosic sugars production to obtain bioethanol from *Eucalyptus globulus* bark. *Ferment.*, 9, 241. https://doi.org/10.3390/fermentation9030241

Amerit, B., Ntayi, J.M., Ngoma, M., Bashir, H., Echegu, S., Nantongo, M. 2023. Commercialization of biofuel products: A systematic literature review. *Renew. Energy Focus*, 44, 223–236. https://doi.org/10.1016/j.ref.2022.12.008

Anandharaj, S.J., Gunasekaran, J., Udayakumar, G.P., Meganathan, Y., Sivarajasekar, N. 2020. Biobutanol: Insight, production and challenges. In V. Sivasubramanian, et al. (Eds.), *Sustainable Development in Energy and Environment. Springer Proceedings in Energy* (pp. 26–37). Springer Nature.

Antil, S., 2019. Biobutanol: Production, scope and challenges. *Int. J. Curr. Microbiol. App. Sci.* 8(11), 580–584.

Ashani, P.N., Shafiei, M., Karimi, K., 2020. Biobutanol production from municipal solid waste: technical and economic analysis. *Biores. Technol.* 308, 123267.

Awogbemi, O., Kallon, D.V.V., Onuh, E.I., Aigbodion, V.S., 2021. An overview of the classification, production and utilization of biofuels for internal combustion engine applications. *Energies*, 14, 5687. https://doi.org/10.3390/en14185687

Azhar, S.H.M., Abdulla, R., Jambo, S.A., Marbawi, H., Gansau, J.A., Faik, A.A.M., Rodrigue, K.F., 2017. Yeasts in sustainable bioethanol production: A review. *Biochem. Biophys. Rep*. 10, 52–61. https://doi.org/10.1016/j.bbrep.2017.03.003

Balan, V., 2014. Current challenges in commercially producing biofuels from lignocellulosic biomass. *ISRN Biotech*, 2014. https://doi.org/10.1155/2014/463074

Balat, M., Balat, H., 2008. A critical review of bio-diesel as a vehicular fuel. *Energy conver. Managem*. 49(10), 2727–2741.

Beluhan, S., Mihajlovski, K., Santek, B., Ivancic Santek, M., 2023. The production of bioethanol from lignocellulosic biomass: Pretreatment methods, fermentation, and downstream processing. *Energies*, 16, 7003. https://doi.org/10.3390/en16197003

Bradley, T., Ling-Chin, J., Maga, D., Speranza, L.G. and Roskilly, A.P., 2022. 5.18-Life cycle assessment (LCA) of algae biofuels, *Molecules*, 5, 387–404.

Baroda, M., Yelle, D.J., Serwanska, K., 2022. Bioethanol Production from Lignocellulosic Biomass-Challenges and Solutions. *Molecules*, 27, 8717. https://doi.org/10.3390/molecules27248717

Busic, A., Mardetko, N., Kundas, S., Morzak, G., Belskaya, H., Šantek, M.I., Komes, D., Novak, S., and Santek, B., 2018. Bioethanol production from renewable raw materials and its separation and purification: A review. *Food Technol. Biotechnol*, 56, 289–311. https://doi.org/10.17113/ftb.56.03.18.5546

Cai, D., Li, P., Luo, Z., Qin, P., Chen, C., Wang, Y., Wang, Z. and Tan, T., 2016. Effect of dilute alkaline pretreatment on the conversion of different parts of corn stalk to fermentable sugars and its application in acetone–butanol–ethanol fermentation. *Biores. Technol*., 211, 117–124.

Cai, D., Wen, J., Zhuang, Y., Huang, T., Si, Z., Qin, P. and Chen, H., 2022. Review of alternative technologies for acetone-butanol-ethanol separation: Principles, state-of-the-art, and development trends. *Separ. Purific. Technol*. 298, 121244.

Callegari, A., Bolognesi, S., Cecconet, D., Capodaglio, A.G., 2020. Production technologies, current role, and future prospects of biofuels feedstocks: A state-of-the-art review. *Crit. Rev. Environ. Sci. Technol*. 50, 384–436. https://doi.org/10.1080/10643389.2019.1629801

Cao, X., Chen, Z., Liang, L., Guo, L., Jiang, Z., Tang, F., Yun, Y., Wang, Y., 2020. Co-valorization of paper mill sludge and corn steep liquor for enhanced n-butanol production with *Clostridium tyrobutyricum* Δcat1:: adhE2. *Biores. Technol*. 296, 122347.

Chacon, S.J., Matias, G., Ezeji, T.C., Maciel Filho, R. and Mariano, A.P., 2021. Three-stage repeated-batch immobilized cell fermentation to produce butanol from non-detoxified sugarcane bagasse hemicellulose hydrolysates. *Biores. Technol*. 321, 124504.

Chadni, M., Moussa, M., Athès, V., Allais, F., Ioannou, I., 2023. Membrane contactors-assisted liquid-liquid extraction of biomolecules from biorefinery liquid streams: A case study on organic acids. *Separ. Purific. Technol*. 317: 123927.

Cho, C., Hong, S., Moon, H.G., Jang, Y.S., Kim, D. Lee, S.Y., 2019. Engineering *clostridial* aldehyde/alcohol dehydrogenase for selective butanol production. *MBio*, 10(1), 10–1128.

da Silva, A.S., Espinheira, R.P., Teixeira, R.S.S. et al. 2020. Constraints and advances in high-solids enzymatic hydrolysis of lignocellulosic biomass: a critical review. *Biotechnol. Biofuels*, 13, 58. https://doi.org/10.1186/s13068-020-01697-w

Demuez, M., Mahdy, A., Tomás-Pejó, E., González-Fernández, C. and Ballesteros, M., 2015. Enzymatic cell disruption of microalgae biomass in biorefinery processes. *Biotechnol. Bioeng*. 112(10), 1955–1966.

Desta, M., Lee, T., Wu, H., 2021. Well-to-wheel analysis of energy use and greenhouse gas emissions of acetone-butanol-ethanol from corn and corn stover. *Renew. Energy*, 170, 72–80.

Dutta, K., Daverey A., Lin, J.G., 2014. Evolution retrospective foe alternative fuels: First to Fourt generation. *Renew. Energy*, 69, 114–122.

Dwidar, M., Kim, S., Jeon, B.S., Um, Y., Mitchell, R.J., Sang, B.I., 2013. Co-culturing a novel *Bacillus* strain with *Clostridium tyrobutyricum* ATCC 25755 to produce butyric acid from sucrose. *Biotechnol. Biofuels*, 6, 1–10.

Ebadian, M., Dyk, S.V., McMillan, J.D., Saddler, J., 2020. Biofuels policies that have encouraged their production and use: An international perspective. *Energy Policy*, 147, 111906. https://doi.org/10.1016/j.enpol.2020.111906

Edeh, I., 2020. Bioethanol Production: An Overview, In: *Bioethanol Technologies* (pp. 1–22). IntechOpen. https://doi.org/10.5772/intechopen.94895

Emetere, M.E., Joel, W.O., 2020. Biofuel commercialization in developing countries: Readiness and prospects. *JREE*, 10(4), 119–130. https://doi.org/10.30501/jree.2023.379263.1533

ETIP Bioenergy Report (2020) Current status of advanced biofuels demonstration in Europe. European commission. https://www.etipbioenergy.eu/wpcontent/uploads/2025/05/ETIP%20Bioenergy_Current_Status_of_Adv_Biofuels_Demonstrations_in_Europe.pdf

Ezeji, T., Qureshi, N., Blaschek, H.P., 2007. Butanol production from agricultural residues: impact of degradation products on *Clostridium beijerinckii* growth and butanol fermentation. *Biotechnol. Bioeng.*, 97(6), 1460–1469.

Gao, Y., Zhang, M., Zhou, X., Guo, X., Lei, C., Li, W., Lu, D., 2020. Effects of carbon ion beam irradiation on butanol tolerance and production of *Clostridium acetobutylicum. Front. Microbiol.*, 11, 602774.

Göktaş, M., Balki, M.K., Sayin, C., Canakci, M., 2021. An evaluation of the use of alcohol fuels in SI engines in terms of performance, emission and combustion characteristics: A review. *Fuel*, 286, 119425.

Gottumukkala, L.D., Haigh, K., Gorgens, J., 2017. Trends and advances in conversion of lignocellulosic biomass to biobtanol: Microbes, bioprocesses and industrial viability. *Renewable Sustainable Energy Rev.*, 76, 963–973.

Guo, Y., Liu Y., Guan M., Tang H., Wang Z., Li, L., Pang H., 2022. Production of butanol from lignocellulosic biomass: Recent advances, challenges and prospects. *RSC Adv.*, 12, 18848–18863.

Halder, P. Azad, K., Shah, S., Sarker, E., 2019. Prospects and technological advancement of cellulosic bioethanol ecofuel production, in: Azad, A.K., (Eds.), *Advances in ecofuels for a sustainable environment* (pp. 211–236). Elsevier. https://doi.org/10.1016/B978-0-08-102728-8.00008-5

Hoang, T.D., Nghiem, N. 2021. Recent developments and current status of commercial production of fuel ethanol. *Ferment.*, 7, 314. https://doi.org/10.3390/fermentation7040314

Ibrahim, M.F., Linggang, S., Jenol, M.A., Yee, P.L., Abd-Aziz, S., 2015. Effect of buffering system on acetone-butanol-ethanol fermentation by *Clostridium acetobutylicum* ATCC 824 using pretreated oil palm empty fruit bunch. *Biores*, 10(3), 3890–3907.

IEA (2024), *World Energy Outlook 2024*, IEA. https://www.iea.org/reports/world-energy-outlook-2024, Licence: CC BY 4.0 (report); CC BY NC SA 4.0 (Annex A)

IEA Biofuel production by country/region and fuel type, 2016–2022, IEA. https://www.iea.org/data-and-statistics/charts/biofuel-production-by-country-region-and-fuel-type-2016-2022, IEA. Licence: CC BY 4.0

IEA Share of global ethanol output by country between 2017 and 2023, IEA. https://www.iea.org/data-and-statistics/charts/share-of-global-ethanol-output-by-countrybetween-2017-and-2023, IEA. Licence: CC BY 4.0

Jang, Y.S., Malaviya, A., Lee, S.Y., 2013. Acetone–butanol–ethanol production with high productivity using *Clostridium acetobutylicum* BKM19. *Biotechnol. Bioeng*, 110(6), 646–1653.

Jin, Q., Qureshi, N., Wang, H., Huang, H., 2019. Acetone-butanol-ethanol (ABE) fermentation of soluble and hydrolyzed sugars in apple pomace by *Clostridium beijerinckii* P260. *Fuel*, 244, 536–544.

Karthick, C., Nanthagopal, K., 2021. A comprehensive review on ecological approaches of waste to wealth strategies for production of sustainable biobutanol and its suitability in automotive applications. *Energy Conver. Managem*, 239, 114219.

Keasling, J., Garcia Martin, H., Lee, T.S. et al., 2021. Microbial production of advanced biofuels. *Nat. Rev. Microbiol.*, 19, 701–715. https://doi.org/10.1038/s41579-021-00577-w

Khuong, L.S., Masjuki, H.H., Zulkifli, N.W.M., Niza Mohamad, E., Kalam, M.A., Abdullah Alabdulkarem, Arslan, A., Mosarof, M.H., Syahira, A.Z., Jamshaida, M., 2017. Effect of gasoline–bioethanol blends on the properties and lubrication characteristics of commercial engine oil. *RSC Adv.*, 7, 15005. https://doi.org/10.1039/c7ra00357a

Kolesinska, B., Fraczyk, J., Binczarski, M., Modelska, M., Berlowska, J., Dziugan, P., Antolak, H., Kaminski, Z.J., Witonska, I.A., Kregiel, D., 2019. Butanol synthesis routes for biofuel production: Trends and perspectives. *Materials*, 12(3), 350.

Kuittinen, S., Hietaharju, J., Bhattarai, I., Hassan, M.K., Kupiainen, L., Kangas, J., Tanskanen, J., Pappinen, A., 2022. Technoeconomic analysis and environmental sustainability estimation of bioalcohol production from barley straw. *Biocat. Agric. Biotechnol*, 43, 102427.

Lahtchev, K., 2012. The contribution of second generation bioethanol to sustainable energy independence. *Ecol. Eng. Environ*, 3, 46–57.

Lam, M., Khoo, C.G., Lee, K.T., 2019. Scale-up and commercialization of algal cultivation and biofuels production, in: Pandey, A., Mohan, S.V., Chang, J., Hallenbeck, P.C., Larroche, C. (Eds.), *Biomass, Biofuels and Biochemicals* (pp. 475–505). Elsevier.

Lee, S.Y., Park, J.H., Jang, S.H., Nielsen, L.K., Kim, J. and Jung, K.S., 2008. Fermentative butanol production by *Clostridia. Biotechnol. Bioeng*. 101(2): 209–228.

Li, S., Huang, L., Ke, C., Pang, Z., Liu, L., 2020. Pathway dissection, regulation, engineering and application: lessons learned from biobutanol production by solventogenic *clostridia. Biotechnol. Biofuels*, 13(1), 1–25.

Li, Y., Tang, W., Chen, Y., Liu, J., Chia-Fon, F.L., 2019. Potential of acetone-butanol-ethanol (ABE) as a biofuel. *Fuel*, 242, 673–686.

Lipovsky, J., Patakova, P., Paulova, L., Pokorny, T., Rychtera, M., Melzoch, K., 2016. Butanol production by *Clostridium pasteurianum* NRRL B-598 in continuous culture compared to batch and fed-batch systems. *Fuel Process. Technol.*, 144, 139–144.

Mann, M.S., Lütke-Eversloh, T., 2013. Thiolase engineering for enhanced butanol production in *Clostridium acetobutylicum. Biotechnol. Bioeng.*, 110(3), 887–897.

Marzo, C., Diaz, A.B., Caro, I., Blandino. A., 2019. Status and perspectives in bioethanol production from sugar beet, in: Ray, R.C., Ramachandran, S. (Eds.), *Bioethanol production from food crops: Sustainable sources, interventions and challenges* (pp. 61–79). Academic Press. https://doi.org/10.1016/B978-0-12-813766-6.00004-7

Mishra, N., Dubey, A., 2017. Biobutanol: an alternative biofuel. *Adv. Biofeedstocks and Biofuels*, 2, 155–175.

Mohapatra, S., Mishra, R.R., Nayak, B., Behera, B., Mohapatra, P.K.D., 2020. Development of co-culture yeast fermentation for efficient production of biobutanol from rice straw: A useful insight in valorization of agro - industrial residues. *Biores. Technol.* 318, 124070.

Nanda, S., Rana, R., Sarangi, P.K., Dalai, A.K., Kozinski, J.A., 2018. A broad introduction to first-, second- and third-generation biofuels, in: Sarangi, P.K., et al. (Eds.), *Recent advancements in biofuels and bioenergy utilization* (pp. 1–25). Springer Nature. https://doi.org/10.1007/978-981-13-1307-3_1

Olguin-Maciel, E., Singh, A., Chable-Villacis, R., Tapia-Tussell, R., Ruiz, H.A., 2020. Consolidated bioprocessing, an innovative strategy towards sustainability for biofuels production from crop residues: An overview. *Agronomy*, 10(11), 1834.

Osman, A., Mehta, N., Elgarahy, A.M., Al-Hinai, A., Al-Muhtaseb, A.H., Rooney, D.W., 2021. Conversion of biomass to biofuels and life cycle assessment: a review. *Environ. Chem. Lett*, 19, 4075–4118. https://doi.org/10.1007/s10311-021-01273-0

Pachapur, V.L., Sarma, S.J., Brar, S.K., Chaabouni, E., 2016. Platform chemicals: Significance and need. *Platform Chemical Biorefinery*, 1–20.

Plaza, P.E., Gallego-Morales, L.J., Peñuela-Vásquez, M., Lucas, S., García-Cubero, M.T., Coca, M., 2017. Biobutanol production from brewer's spent grain hydrolysates by *Clostridium beijerinckii*. *Biores. Technol.*, 244, 166–174.

Priya, Deora, P.S., Verma, Y., Muhal, R.A., Goswami, C., Singh, T., 2021. Biofuels: An alternative to conventional fuel and energy source. *Mater. Today. Proc.*, 48, 1178–1184. https://doi.org/10.1016/j.matpr.2021.08.227

Qi, G., Huang, D., Wang, J., Shen, Y., Gao, X., 2019. Enhanced butanol production from ammonium sulfite pretreated wheat straw by separate hydrolysis and fermentation and simultaneous saccharification and fermentation. *Sust. Energy Technol. Assess.*, 36, 100549.

Qureshi, N., Saha, B.C., Cotta, M.A., 2008b. Butanol production from wheat straw by simultaneous saccharification and fermentation using *Clostridium beijerinckii*: part-II -Fed-batch fermentation. *Biomass Bioenergy*, 32(2), 176–183.

Qureshi, N., Saha, B.C., Hector, R.E., Hughes, S.R., Cotta, M.A., 2008a. Butanol production from wheat straw by simultaneous saccharification and fermentation using *Clostridium beijerinckii*: Part I—Batch fermentation. *Biomass. Bioenergy*, 32(2), 168–175.

Raita, S., Spalvins, K., Blumberga, D., 2021. Prospect on agro-industrial residues usage for biobutanol production. *Agron. Res.*, 19(S1), 877–895.

Rastogi, M., Shrivastava, S., 2017. Recent advances in second generation bioethanol production: An insight to pretreatment, saccharification and fermentation processes. *Renew. Sustain. Energy Rev*, 80, 330–340.

Ruan, R., Zhang, Y., Chen, P., Liu, S., Fan, L., Zhou, N., Ding, K., Peng, P., Addy, M., Cheng, Y., Anderson, E., Wang, Y., Liu, Y., Lei, H., Li, B., 2019. Biofuels: Introduction, in: Pandey, A., Larroche, C., et al., (Eds.), *Biofuels: Alternative feedstocks and conversion processes for the production of liquid and gaseous biofuels* (pp. 3–43). Academic Press, Elsevier. https://doi.org/10.1016/B978-0-12-816856-1.00001-4

Saifudin, M.A., Jamina, A.W., Nugrohoa, R., Karmiadji, D.W., 2018. The effect of ethanol on fuel tank corrosion rate. *Jurnal. Teknologi.*, 80, 19–25. https://doi.org/10.11113/jt.v80.12324

Schwarz, K.M., Grosse-Honebrink, A., Derecka, K., Rotta, C., Zhang, Y., Minton, N.P., 2017. Towards improved butanol production through targeted genetic modification of *Clostridium pasteurianum*. *Metabolic Eng*, 40, 124–137.

Serrano-Echeverry, V.A., Mendieta-Reyes, N.E., Guerrero, C.A., 2023. Biobutanol an eco-friendly fuel of the future, history, current advances, and trends: A review. *Heliyon*, 1–68, https://doi.org/10.2139/ssrn.4498670

Shahid, M.K., Batool, A., Kashif, A., Nawaz, M.H., Aslam, M., Iqbal, N., Young Gyun Choi G, 2021 Biofuels and biorefineries: Development, application and future perspectives emphasizing the environmental and economic aspects. *J. Environ. Manage*, 297, 113268. https://doi.org/10.1016/j.jenvman.2021.113268

Shiru, S., Shiru, M.S., 2021. Towards commercialization of third-generation biofuel industry for sustainable energy production in Nigeria. *Chem. Bio. Eng. Rev*. 8, 1–20. https://doi.org/10.1002/cben.202100015

Solomon, B.D., 2010. Biofuels and sustainability. *Ann. N.Y Acad. Sci.* 1185, 119–134.

Subramaniam, Y., Masron, T.A., Azman, N.H.N., 2020. Biofuels, environmental sustainability, and food security: A review of 51 countries. *Energy Res. Soc. Sci.*, 68, 101549. https://doi.org/10.1016/j.erss.2020.101549

Sukumaran, R.K., Gottumukkala, L.D., Rajasree, K., Alex, D., Pandey, A., 2011. Butanol fuel from biomass: Revisiting ABE fermentation. In *Biofuels* (pp. 571–586). Academic Press.

Szulczyk, K.R., 2010. Which is a better transportation fuel--butanol or ethanol? *Internat. J. Energy Environ.*, 1(3), 501–512.

Tri, C.L., Kamei, I., 2020. Butanol production from cellulosic material by anaerobic co-culture of white-rot fungus *Phlebia* and bacterium *Clostridium* in consolidated bioprocessing. *Bioresour. Technol.* 305, 123065.

Tse, T.J., Wiens, D.J., Reaney, M.J.T., 2021. Production of bioethanol—A review of factors affecting ethanol yield. *Ferment.*, 7, 268. https://doi.org/10.3390/fermentation7040268

Vivek, N., Nair, L.M., Mohan, B., Nair, S.C., Sindhu, R., Pandey, A., Shurpali, N., Binod, P., 2019. Bio-butanol production from rice straw–recent trends, possibilities, and challenges. *Biores. Technol. Rep.* 7, 100224.

Wang, Y., Ho, S.H., Yen, H.W., Nagarajan, D., Ren, N.Q., Li, S., Hu, Z., Lee, D.J., Kondo, A. and Chang, J.S., 2017. Current advances on fermentative biobutanol production using third generation feedstock. *Biotechnol. Adv.*, *35*(8), 1049–1059.

Wang, Y.F., Tian, J., Ji, Z.H., Song, M.Y., Li, H., 2016. Intracellular metabolic changes of *Clostridium acetobutylicum* and promotion to butanol tolerance during biobutanol fermentation. *The Internat. J. Biochem. Cell Biol.*, 78, 297–306.

Wen, Z., Li, Q., Liu, J., Jin, M., Yang, S., 2020. Consolidated bioprocessing for butanol production of cellulolytic Clostridia: development and optimization. *Microb. Biotechnol*, 13(2), 410–422.

Wen, Z., Tu, Z., Al Makishah, N., Fang, D., Ledesma-Amaro, R., Zhu, J., Li, Q., Li, Y., Jin, M., 2022. Genetic Manipulation of a twin *Clostridia* consortium for co-production of n-butanol and isobutanol by consolidated bioprocessing. *ACS Sustain. Chem. Eng.* 10(45), 14671–14684.

Xu, M., Zhao, J., Yu, L., Tang, I.C., Xue, C., Yang, S.T., 2015. Engineering *Clostridium acetobutylicum* with a histidine kinase knockout for enhanced n-butanol tolerance and production. *Applied Microbiol. Biotechnol.*, 99, 1011–1022.

Zetty-Arenas, A.M., Alves, R.F., Portela, C.A.F., Mariano, A.P., Basso, T.O., Tovar, L.P., Maciel Filho, R., Freitas, S., 2019. Towards enhanced n-butanol production from sugarcane bagasse hemicellulosic hydrolysate: Strain screening, and the effects of sugar concentration and butanol tolerance. *Biomass and Bioenergy*, *126*, 190–198.

Zhang, Z.H., Balasubramanian, R., 2014. Influence of butanol addition to diesel–biodiesel blend on engine performance and particulate emissions of a stationary diesel engine. *Appl. Energy*, 119, 530–536.

11 Life Cycle Assessment of Microbial Biofuel
Environmental and Economic Impacts

Ruchi Pathania, Prem Pritam, and Richa Guleria

11.1 INTRODUCTION

Rapid industrialization and population growth are driving increased energy demand, causing resource shortages and environmental challenges like rising CO_2 levels and global warming. Fossil fuels contribute 98% of carbon emissions, making it urgent to find alternative solutions that address both the energy crisis and environmental concerns (Kazemi et al., 2023; Shah et al., 2024). Exploring solutions to meet growing energy needs while mitigating environmental impacts is crucial (Osman et al., 2021). Renewable energy resources are being used to address the energy issue and reduce greenhouse gas emissions (GHGe). According to the US Energy Information Administration (EIA), global energy demand will increase by nearly 50% by 2050 compared to current levels (EIA projects nearly 50% increase in world energy use by 2050, led by growth in renewables; US EIA (n.d.).

Biomass, categorized as either non-lignocellulosic or lignocellulosic, includes woody, herbaceous, aquatic debris, agricultural manure, and by-products. Technologies like thermochemical processes, enzymatic hydrolysis, and fermentation can convert biomass into fuels and chemicals, including 1,3-propanediol, acetone, n-butanol, itaconic acid, and xylitol (Kaloudas et al., 2021). Biofuels are crucial for reducing GHGe in the transportation sector, alongside strategies such as electric vehicles and fuel efficiency. In 2018, biobased substitutes for gasoline and diesel accounted for 7.7% of global fuel consumption, which highlights their growing importance. However, further technological advancements and sustainable practices are needed to increase their contribution to the global transition toward cleaner, low-carbon energy systems ("Biofuels Dashboard 2020 | IFPEN," n.d.).

The development of long-term energy strategies focuses on sustainability, economic growth, and environmental protection, including climate change mitigation. Microbes are already known for the production of the majority of commercial biofuels, mainly from simpler feedstocks like sugars and starches, with ethanol and biodiesel reducing fossil fuel dependence ("Biofuels and the Environment | US EPA," n.d.). However, there is an urgent need for advanced biofuels from complex

DOI: 10.1201/9781003633709-11

feedstocks like lignocellulosic biomass, algae, or waste, which can replace petroleum-based fuels, such as gasoline and diesel. These advanced biofuels offer a sustainable, scalable solution to decarbonize the transport sector, a major contributor to global carbon emissions. To support the transition to a low-carbon economy, further research and technological progress are essential for large-scale commercial production of these biofuels.

Various international and national organizations have made mid- and long-term projections for global biofuel production, offering a range of estimates for the growth of liquid biofuels in transportation. The International Energy Agency (IEA) anticipates that biofuels could supply up to one-third of global transportation fuel by 2050. In contrast, the Organization for Economic Cooperation and Development (OECD) and British Petroleum (BP) project that biofuels will make up around 7% of the transportation fuel mix by 2030 (Jeswani et al. 2020). The Circular Bio-Society 2050 Vision outlines a future where food, renewable products, and energy are sustainably provided through the principles of a circular bioeconomy (Biobased Industries Consortium (BIC), n.d.).

The Life Cycle Assessment (LCA) technique is widely used to evaluate the environmental impact of microbial biofuel production across its entire life cycle, from raw material extraction to production, distribution, use, and disposal. LCA provides insights into the sustainability of microbial biofuels, identifying areas for improvement and comparing them with fossil fuels and other biofuels. There is several software tools commonly used for LCA studies in microbial biofuels, each offering unique features, capabilities, and approaches for analyzing data. The concept of LCA was introduced by the Society of Environmental Toxicology and Chemistry (SETAC) in 1990 and later integrated into the International Organization for Standardization (ISO) 14000 Environmental Management Standards in 1993 (Liu et al. 2024). The environmental impact assessed through LCA includes energy use, water consumption, GHGe, air pollution, waste generation, and more (Teillard et al., 2016).

LCA, combined with techno-economic analysis (TEA), provides valuable insights into the economic and environmental feasibility of biofuel production at a commercial scale. It employs various methodologies, system boundaries, impact categories, functional units, allocation approaches, and assumptions about by-products and co-products, along with different reference systems for comparison. Key factors like GHGe and global warming potential (GWP) are crucial in evaluating biofuel production's environmental sustainability and its potential role in mitigating climate change (Czyrnek-Delêtre et al., 2017). With the rapid growth of biomass-based biofuel production, LCA has gained significant importance, aligning with policies focused on reducing GHGe. To support the sustainable development of the industry, updated LCAs are essential to assess the carbon intensity of biofuels. In this chapter, we discuss the types of biofuels produced from different organisms, environmental impacts, biomass-to-biofuel conversion approaches, and various relevant case studies.

11.2 MICROBIAL BIOFUELS

Bioproducts are categorized into three main groups: bioenergy, biomaterials (such as bioplastics, biofoams, and biocomposites), and biochemicals (including succinic acid (SA), ethylene, lactic acid, etc.; Scarlat et al., 2015). Biofuels/bioenergy can be in solid, liquid, or gaseous forms, such as bioethanol, biodiesel, bio-oils, hydrogen, biogas, biomethane, and syngas (Singhal and Prashant, 2020). These biofuels are used in transportation (e.g., blending with gasoline for ethanol), renewable thermal energy, and renewable electricity generation. The development of bioproducts is primarily going in Europe, while biofuels development is mainly focused on North America. The United States is the world's largest producer of bioethanol, with Brazil and the European Union ranking second and third, respectively. In the United States, most bioethanol is produced from corn, while Brazil primarily uses sugarcane, and corn is also the main crop for bioethanol production in Europe ("Alternative Fuels Data Center: Maps and Data," n.d.). By 2025, the United States is expected to have over 500 ethanol production facilities, each with an average capacity of 1.06 million barrels per day ("EIA increases forecasts for 2025 and 2026 ethanol production | Ethanol Producer Magazine," n.d.).

11.2.1 Biofuels Classification Based on Biomass

Biomass-derived fuels are considered attractive and feasible energy sources. Converting biomass to biofuels not only reduces the release of toxic compounds but also significantly lowers the net CO_2 emissions into the atmosphere (Voloshin et al., 2016). The chemical composition of biomass is crucial in the production of biobased products, as different biomass types influence both yield and solubilization efficiency. Depending on the type of feedstock, biofuels are categorized into four generations (Alalwan et al., 2019; Mat Aron et al., 2020), as illustrated in Figure 11.1.

First-generation biofuels are primarily derived from food crops rich in sugars or starches (such as sugarcane, potatoes, and corn) and oilseeds (like soybean and sunflower). Sugar-rich feedstocks are used to produce bioethanol through fermentation by yeast (*Saccharomyces cerevisiae*), while oil-rich feedstocks are converted into biodiesel through transesterification (Alalwan et al., 2019; Lee and Lavoie, 2013). The main advantage of first-generation biofuels is their low-cost and simple production process. However, they face disadvantages, such as competition for arable land (food vs. fuel controversy) and significant freshwater consumption (Naik et al., 2010).

Second-generation biofuels are produced from lignocellulosic feedstocks, such as wood residues, forest dregs, organic waste, agricultural residues,

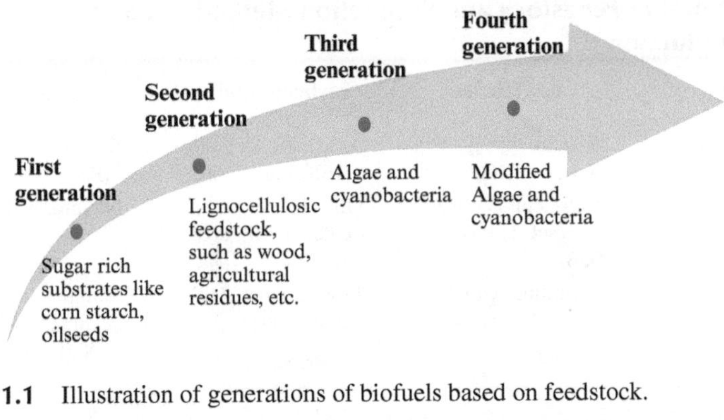

FIGURE 11.1 Illustration of generations of biofuels based on feedstock.

and municipal or industrial wastes. These biofuels have the advantage of utilizing agricultural and forest waste, reducing the impact on the food sector. However, their production requires energy-intensive and time-consuming conversion processes (Lee and Lavoie, 2013; Shrestha et al., 2017). Additionally, commercial viability is still a challenge. Given that lignocellulosic biomass is renewable and widely available, various methods are being explored to convert it into bioenergy (Hassan et al., 2019).

Third-generation biofuels are derived from photosynthetic microorganisms, such as algae and cyanobacteria (Hannon et al., 2010). The biomass produced from photosynthetic microorganisms do not require complex treatments compared to lignocellulosic enriched biomass to generate fermentable sugars or lipids to produce fuels. In cyanobacteria, carbohydrates (primarily glycogen) are the primary storage compound, while algae store lipids. Carbohydrates can be used to produce bioethanol and higher alcohols through fermentation, while lipids can be used to make biodiesel and other lipid-based compounds. While this generation overcomes some of the challenges of first- and second-generation biofuels, commercial-scale production still faces challenges due to low product yields. Therefore, new strains from the local environment with desirable properties must be explored.

Fourth-generation biofuels are produced from genetically modified algae and cyanobacteria (Abdullah et al., 2019). A combination of genetic or metabolic engineering strategies is employed to enhance bioproduct formation. Modified cyanobacterial and algal strains have been successfully engineered to produce industrially relevant chemicals, including 2,3-butanediol, acetone, 1-butanol, ethylene, ethanol, fatty acids, SA, and isoprene (Hendry et al., 2020). Cyanobacteria, in particular, have become a popular research host for fourth-generation biofuels due to their advantages. However, significant efforts are still required to improve the commercial viability of these processes.

The environmental benefits of biofuel production are largely dependent on feedstock and production methods used, as shown in Table 11.1.

TABLE 11.1
Environmental of Feedstock and Production Method Used for Biofuel Production

Metric	Corn Ethanol	Soybean Biodiesel	Advanced Biofuels
GHG Reduction	12% more than gasoline (Hill et al., 2006)	41% more than diesel (Hill et al., 2006)	60% more than petroleum (Fore et al., 2011)
Nitrogen Emission	Eightfold higher than biodiesel (Hill et al. 2006)	Only 1% of corn ethanol (Hill et al., 2006)	No information
Land Usage	Competition with food crops (Manuel, 2007)	Less land use (Hill et al., 2006)	Marginal land use (Fore et al., 2011; Manuel, 2007)

11.2.2 Comparison Between Biofuels and Fossil Fuels

Biofuels are renewable, derived from organic materials like plants, algae, and waste, and are part of a sustainable cycle, while fossil fuels are nonrenewable, formed over millions of years. Biofuels generally have a lower carbon footprint as plants absorb CO_2 during growth, although they still emit GHG during production and use. Biofuels have complex environmental impacts, involving trade-offs between GHG reductions, resource consumption, and ecosystem health compared to fossil fuels. Fossil fuels, on the other hand, are major contributors to GHGe and climate change. Biofuels typically have lower energy density than fossil fuels, requiring larger volumes to produce the same energy. While biofuels can be more expensive to produce, advances in technology are making them more cost-effective, unlike fossil fuels, which are cheaper to extract but subject to volatile prices. Biofuels can enhance energy security and support rural economies, whereas fossil fuels rely on imports and can lead to geopolitical tensions. Biofuels, especially first-generation types, may compete with food production, while fossil fuels do not, but their extraction causes environmental harm. Biofuels are more sustainable in the long term as long as biomass is available, while fossil fuels are finite and will eventually deplete, making alternatives necessary. The environmental effects of biofuels depend on feedstock type, land-use policies, and production processes, and some are listed in Table 11.2.

TABLE 11.2
Environmental Impacts Across Fuel Types

Environmental Impact	Fossil Fuels	Conventional Biofuels (First Generation)	Advanced Biofuels (Second/Third Generation)	References
GHGe	Three- to elevenfold lower land-use emissions per MJ than biofuels	12%–40% reduction vs. fossil fuels; LUC risks negate benefits	Up to 95% reduction (waste biomass); algae still has high emissions	Fargione et al. (2008), Hertel et al. (2010), Searchinger et al. (2008)
Air Pollution	Higher SOx, PM2.5; lower agrichemical pollutants	Higher VOCs, N_2O and PM10 from agrichemicals	Lower SOx and PM10 than fossil refineries	Chisti (2007), Hill et al. (2009)
Water Pollution	Oil spills; fracking contaminants	100 times more nitrogen runoff than biodiesel; eutrophication	Minimal runoff when using marginal lands/waste	Davis et al. (2012), Donner and Kucharik (2008), Vidic et al. (2013)
Land-Use Efficiency	Localized extraction; smaller footprint per MJ	17% cropland expansion (2008–2022); habitat fragmentation	Potential for marginal lands; less competition with food crops	Gelfand et al. (2013), Lark et al. (2022)

(Continued)

TABLE 11.2 (Continued)
Environmental Impacts Across Fuel Types

Environmental Impact	Fossil Fuels	Conventional Biofuels (First Generation)	Advanced Biofuels (Second/ Third Generation)	References
Net Energy Balance (NEB)	High energy density but nonrenewable	1.25:1 (corn ethanol);5.4:1 (soy biodiesel)	10–15:1 (cellulosic feedstocks)	Pimentel and Patzek (2005), Schmer et al. (2008)
Policy Challenges	Subsidies favor extraction over renewables	RFS increased fertilizer use by 8%	RED II mandates 65% GHG cuts; scalability issues	("Directives directive (EU) 2018/2001 of the European parliament and of the council of 11 December 2018 on the promotion of the use of energy from renewable sources (recast) (Text with EEA relevance)," n.d.

First-generation biofuels provide modest GHG reductions (e.g., 12% for ethanol, 41% for biodiesel) compared to fossil fuels, but often fail to meet regulatory criteria, like the European Union's 60%–65% savings target (Hill et al., 2006). LUC, such as converting grasslands or forests to agriculture, can offset these benefits by releasing stored carbon (Jeswani et al. 2020; Lark et al. 2022). Second-generation biofuels, such as cellulosic ethanol, show more promise, offering almost-neutral or negative emissions when using waste biomass ("Biofuels and the Environment I US EPA" n.d.; Jeswani et al. 2020). However, third-generation biofuels are currently more carbon-intensive than fossil fuels due to energy-intensive processing (Jeswani et al. 2020). Fossil fuels generally have lower land-use emissions per unit of energy generated compared to biofuels. For instance, petroleum extraction results in 3–11 times lower carbon emissions per MJ of energy than biofuel feedstock farming ("ITS Comparing Land Use Impacts of Petroleum and Biofuels – ITS," n.d.).

Biofuel production contributes to air and water pollution throughout its agricultural and refining stages. Corn ethanol, for example, uses significantly more nitrogen and pesticides than soybean biodiesel, fueling eutrophication and harmful algal blooms like the dead zone in the Gulf of Mexico (Hill et al. 2006; Manuel 2007). Ethanol refineries also emit pollutants such as PM10, VOCs, and ozone precursors, while biodiesel refining releases hazardous air pollutants like formaldehyde, surpassing fossil fuel refineries in some areas ("Biofuels and the Environment – U.S. Energy Information Administration (EIA)," n.d.; "Environmental Integrity Despite Green Image, Biofuels Manufacturing Releases Large Amounts of Hazardous Air Pollutants," n.d.). Although soybean biodiesel has 59% lower emissions than diesel, fertilizer-derived nitrous oxide can offset 40%–50% of the GHG benefits of maize ethanol (Hill et al. 2006; Jeswani et al. 2020).While fossil fuels have higher sulfur oxide and particulate emissions, biofuels present unique water quality issues due to agricultural inputs ("Biofuels and the environment – U.S. Energy Information Administration (EIA)," n.d.; Fargione et al., 2008). Biofuel crop expansion leads to deforestation and habitat loss. A rise of 17% in the corn and soy farming in the US Midwest from 2008 to 2022 replaced 4.2 million acres of natural land close to biorefineries, thus reducing biodiversity, disrupting ecosystems, and accelerating soil erosion (Lark et al. 2022). In contrast, fossil fuel extraction causes land disturbances, but these are generally smaller per energy unit ("ITS Comparing Land Use Impacts of Petroleum and Biofuels – ITS," n.d.).

Regulatory authorities like the US Renewable Fuel Standard (RFS) and EU Renewable Energy Directive (RED) struggle to prevent unintended consequences. The RFS, for example, has driven a 26% increase in farmland and an 8% rise in fertilizer use, leading to higher emissions ("Biofuels and the Environment | US EPA" n.d.; Lark et al. 2022). The US Department of Agriculture's (USDA) 2025 Climate-Smart Agriculture recommendations aim to standardize GHG accounting for biofuel feedstocks, promoting practices like cover crops and no-till farming to store carbon ("USDA Publishes Interim Rule on Technical Guidelines for Climate-Smart Agriculture Crops Used as Biofuel Feedstocks | Home," n.d.). In conclusion, while biofuels offer a pathway to decarbonize transportation, strict controls on LUC and pollution are essential. Policies like the USDA's 2025 guidelines support second-generation technologies and climate-smart agriculture, improving sustainability ("Biofuels and the Environment | US EPA" n.d.). To minimize environmental trade-offs, net-zero targets must integrate biofuels with solar, wind, and electrification strategies.

11.2.3 Technical Challenges in Scaling Up Production of Biofuels

To scale up microbial biofuel production, technical challenges must be addressed across feedstock logistics, strain optimization, bioprocess engineering, and economic feasibility. While microbes such as bacteria, algae, and yeast offer potential, their industrial adoption faces hurdles in infrastructure, efficiency, and integration with existing energy systems. Some of challenges for large-scale production are discussed next.

Feedstock Logistics and Preprocessing: Feedstock logistics and preprocessing are major bottlenecks in scaling microbial biofuels. Lignocellulosic biomass, with low bulk densities, requires large-scale transportation to meet energy needs. For instance, meeting the IEA's 2015 bioenergy target would require transporting 15 billion metric tons of biomass annually, surpassing current supply chain capacities (Richard, 2010). Preprocessing methods can help improve bulk density but add costs and energy demands. Seasonal availability and storage challenges also require advanced preservation techniques (Gallego-García et al., 2023; Richard, 2010). For algal biofuels, water and nutrient control are crucial. Open pond systems, while affordable, face evaporation losses and unsustainable nutrient demands. Substituting algal biofuels for US petroleum would require phosphate amounts exceeding current mining output. Early-stage solutions like anaerobic digestion and wastewater integration are key for long-term sustainability (Hannon et al., 2010).

Bioreactor Design and Scale-Up: Scaling microbial biofuels from lab-scale to industrial bioreactors presents challenges such as heat dissipation, gas exchange inefficiency, and mixing issues that affect light penetration and nutrient distribution (Ambaye et al., 2021; Hannon et al., 2010). Aeration control in high-density cultures is energy-intensive, and contamination risks remain high. Open ponds are vulnerable to invasive species, and sterilization costs in closed systems are significant (Tan et al., 2021).

Microbial Strain Limitations: Producing high lipid content in microbes strains growth, creating a trade-off between biomass output and oil accumulation. While genetic engineering shows promise, societal concerns and legal barriers hinder its deployment. Lignocellulosic inhibitors also affect microbial growth, requiring robust microbes or detoxification methods (Gallego-García et al., 2023; Tan et al., 2021).

Downstream Processing and Energy Costs: Harvesting and product extraction are energy-intensive, accounting for 20%–30% of production costs (Ambaye et al., 2021; Hannon et al., 2010). Fuel compatibility, especially with ethanol's corrosive properties, adds complexity to infrastructure integration (Hannon et al., 2010; Richard, 2010).

Economic Viability and Biorefinery Integration: Microbial biofuels are currently expensive, with costs two to ten times higher than conventional fuels. A biorefinery model that maximizes co-products like proteins and carotenoids is crucial for economic sustainability (Gallego-García et al., 2023). Lignocellulosic biorefineries and waste valorization can reduce costs, but scaling these technologies requires advancements in genetic engineering, policy frameworks, and global collaboration (Ambaye et al., 2021).

11.2.4 Environmental Impacts of Microbial Biofuel Production

The environmental impacts of microbial biofuel production encompass a range of factors that can influence both the natural ecosystem and human health. Microbial biofuels, produced using microorganisms (such as bacteria, yeast, or algae) to convert organic matter into biofuels, offer potential advantages over traditional fossil fuels. Understanding these impacts is necessary for developing sustainable methods and regulations that have the potential of maximizing the benefits of microbial biofuels while minimizing their environmental footprint. Here are several key aspects to consider when evaluating the environmental impacts of microbial biofuel production and some are discussed in ("What are the environmental impacts of biofuel production?" n.d.; Chowdhury et al. 2025; Jeswani et al. 2020; Wu et al. 2018).

GHGe: Microbial biofuels generally produce lower GHGe compared to fossil fuels. The CO_2 released during their combustion is partially offset by the carbon absorbed during the cultivation of feedstocks like agricultural residues or algae. However, depending on the production process, energy used, and feedstock choice, microbial biofuels can still generate emissions. Additionally, the lifecycle emissions (including cultivation, conversion, and transportation) must be considered to determine their net impact on the environment. For example, some microbial biofuel production systems could require significant energy inputs, which could be sourced from nonrenewable electricity, reducing the overall emissions benefits.

Land Use and Resource Allocation: While microbial biofuels often rely on waste materials or nonfood feedstocks like agricultural residues or algae, there can still be LUC that affect biodiversity. Large-scale production of biofuel feedstocks might lead to land conversion, which could disrupt natural habitats and biodiversity. Furthermore, competition for land between biofuel crops and food production could strain food security, particularly if food prices rise due to biofuel demand. Therefore, sustainable sourcing of raw materials is critical to mitigating the impact on land use.

Water Usage and Quality: Microbial biofuel production often requires significant water resources, particularly for growing feedstocks like algae or managing fermentation processes. Depending on feedstock type, geographical location, and manufacturing techniques, the water needed for biofuel generation differs greatly, leading to water scarcity. Additionally, the production of biofuels could contribute to water quality issues through nutrient runoff (e.g., nitrogen and phosphorus) from fertilizer use in feedstock cultivation, or from wastewater produced during fermentation processes. Proper wastewater management and resource-efficient cultivation techniques are essential to minimizing the impact on water systems (D'Odorico et al., 2018).

Energy Use and Efficiency: The energy required to produce microbial biofuels can significantly impact the overall sustainability of these biofuels. If the energy input comes from nonrenewable sources, the carbon footprint of microbial biofuels can be higher. Optimizing energy efficiency in microbial biofuel production is essential to improve their environmental benefits. For instance, integrating renewable energy sources like solar or wind to power the production processes could significantly reduce their environmental footprint (Farghali et al., 2023).

Toxicity and Eco-toxicity: Some biofuel production processes may lead to the release of toxic substances into the environment such as like formaldehyde, acetaldehyde, hexane, and acrolein, posing risks to public health and the environment ("Biofuels and the Environment | US EPA"). The impact of these substances on ecosystems and human health is a concern. The use of biocatalysts, enzymes, and other sustainable technologies can help mitigate these concerns.

Waste Generation: Biofuel production generates waste, including residual biomass and wastewater that must be properly managed to prevent environmental contamination. However, many microbial biofuel processes aim to reduce waste by utilizing waste products from other industries. The management of waste and the potential for waste valorization – where waste is turned into valuable by-products – can help reduce the environmental impact of microbial biofuel production (Jamil et al., 2024).

Ecological Footprint of Microbial Strains: The use of genetically engineered microrganisms (GMOs) can also raise concerns. While these strains could improve efficiency, there is the potential for unintended ecological consequences if these modified microbes escape into the environment. The risks associated with GMOs need to be carefully evaluated to prevent disruption of local ecosystems or contamination of wild microbial populations (Prakash et al., 2011).

11.2.5 Technological Innovations and Economic Viability

Recent advances in biofuel technology focus on improving efficiency and scalability. The US Department of Energy's Co-Optima Initiative has identified 37 biofuel candidates from domestic biomass, with 10 pathways reducing lifecycle GHGe by over 60% and seven matching petroleum fuel costs ($2.50–$3.00 /gallon) by optimizing feedstock-refinery integration (Benavides et al. 2022; "Co-Optima Transportation Ingenuity Capitalizes on Strengths of US Energy Resources and Existing Infrastructure | News | NREL" n.d.). For example, engineered *Lactobacillus* strains are used to minimize acetate inhibition and improve ethanol yields from cellulosic biomass (Nanda et al., 2023). At the same time, HTL (hydrothermal liquefaction) process transforms sewage sludge into biological crude, reducing energy use by 35% compared to pyrolysis and lowering production costs to $1.75–$2.25/gallon (Grande et al., 2021). From 2005 to 2025, the production costs of corn ethanol increased from $0.46 to $0.63/energy-equivalent liter (EEL), while biodiesel rose to $0.82/EEL (Carriquiry et al., 2014).

Despite all this, Policy incentives like the USDA's Higher Blends Infrastructure Incentive Program (HBIIP) are justified by biodiesel's superior NEB (93%) and 60% lower particulate emissions, with $500 million allocated to upgrading fueling infrastructure ("Biden-Harris Administration Announces Funding for Homegrown Biofuels as Part of Investing in America Agenda | Home" n.d.; "USDA Invests in Additional Domestic Biofuels and Clean Energy Projects to Lower Costs and Increase American Energy Security Across the Nation | Home" n.d.). Despite subsidies, biodiesel offers 4.5 times more GHG savings per dollar than corn ethanol, highlighting the need for feedstock-specific policies ("Biofuels," 2008; Rafa et al., 2021). Multicomponent learning curve models predict CAPEX reductions of 10%–25% by 2050 for biofuels but achieving a 40% reduction would require significant capacity growth (Karka et al., 2021). Balancing conversion technologies with financial and environmental strategies is essential. While microbial biofuels provide renewable energy, they risk land-use conflicts and biodiversity loss. Combined strategies – new technologies, strict regulations, and ecosystem restoration – are vital to meeting climate goals and protecting biodiversity. These advancements highlight the dual need to improve conversion technologies while ensuring that financial strategies align with environmental goals.

11.3 LIFE CYCLE ASSESSMENT (LCA)

LCA is a critical method used to evaluate the environmental impact of various processes, materials, and biofuel production. It offers a comprehensive framework for understanding the environmental consequences of a product system throughout its entire life cycle. By analyzing the entire life cycle of a product, LCA provides insights into the resource consumption, emissions, and potential environmental harm associated with each phase of production, including manufacturing, transportation, and use. This approach helps identify areas where improvements can be made to minimize adverse environmental effects, such as the consumption of nonrenewable fossil fuels, the release of GHG, and other toxic emissions (Pacañot, 2022). One of LCA's main goals is to optimize product design and processes to minimize their environmental footprint. By focusing on resource efficiency, it identifies key stages where energy use, material inputs, and emissions can be reduced. This may involve enhancing energy efficiency, reducing waste, or shifting to cleaner alternatives like renewable energy, ultimately lowering carbon emissions across the product's life cycle (Brusseau, 2019).

It is important to define whether the system is cradle-to-gate (covering processes up to the point of biofuel production) or cradle-to-grave (covering the entire life cycle, including end-of-life treatment of the biofuel). For example, cradle-to-gate analyses often focus on biofuel production, while cradle-to-grave includes the use phase and final disposal. LCA is often referred to as "cradle-to-grave" analysis, as it assesses environmental impacts from raw material extraction ("cradle") to the final disposal or recycling of a product ("grave"). This approach considers the entire lifespan of a product, allowing for a more comprehensive assessment of its environmental footprint. Alternatively, "cradle-to-gate" analysis focuses on the stages leading up to production, including raw material extraction, manufacturing, and delivery, excluding the final disposal phase (Muthu, 2020; Yung et al., 2018). Through a detailed examination of a product's life cycle, LCA provides decision-makers with a well-rounded understanding of the environmental trade-offs and benefits associated with various options. This is essential for selecting the most sustainable materials, production processes, and technologies. By recognizing both the advantages and disadvantages of different choices, LCA ensures that stakeholders can make informed decisions aimed at minimizing environmental harm (Bergerson and Lave, 2004).

A key feature of LCA is the identification and assessment of key inputs, such as energy, water, and raw materials, which are required at various stages of the product life cycle. The potential environmental impacts from these inputs, including emissions to air, water, and soil, are carefully evaluated to understand how they contribute to environmental degradation or resource depletion. LCA findings provide essential information that can be used to guide decisions aimed at enhancing the sustainability of products and processes, whether through improved material choices, energy optimization, or waste reduction strategies (Klöpffer and Grahl, 2014). In essence, the term "life cycle" refers to a thorough evaluation process that includes not just the production and raw material generation stages, but also the subsequent stages of product marketing, use, and interim transport. This holistic approach ensures that the long-term environmental consequences of a product are fully understood and considered in decision-making, supporting the development of more sustainable practices across industries (Vance et al., 2022).

11.3.1 Key Stages of LCA

The International Organization for Standardization (ISO) is a worldwide federation of ISO member bodies. ISO technical committees produce international standards on a variety of topics. The LCA research has been carried out in accordance with the guidelines provided by ISO 14044:2006 standard. PAS (Publicly Available Specification for assessing the life cycle GHGe of goods and services) follows ISO 14044 in terms of general guidelines but focuses specifically on GHGe. The first step in conducting LCA is to clearly define the **purpose and scope of the study**. This involves outlining the objectives of the assessment – whether it's for product development, comparison of alternatives, or regulatory compliance. The next step is to create a **life cycle inventory (LCI)**, the most data-intensive phase of an LCA. This phase involves collecting and compiling data on the inputs and outputs of all processes involved in the life cycle. Inputs include materials, energy, and water, while outputs refer to emissions, waste, and other by-products. The **life cycle impact assessment (LCIA)** is the phase where the environmental impacts of the product or service are evaluated. The goal is to understand how the inputs and outputs identified in the LCI affect key environmental categories, such as GWP, GHGe, acidification, eutrophication, resource depletion, carbon footprints, and ecosystem damage. This step involves applying scientific methods to quantify the potential environmental impacts and determine their significance. Different impact categories can be used depending on the goal of the study, and methods may vary by the software or methodology being applied. The final step is the **interpretation of the data** collected and analyzed in the previous phases. This involves drawing conclusions based on the LCI and LCIA results and identifying areas for improvement. The interpretation phase also includes sensitivity analysis to assess the robustness of the results, considering potential uncertainties in the data or assumptions made. The findings from this phase are crucial for decision-making and can guide the development of more sustainable practices or policies. The description of these phases in detail is already discussed in Moutik et al. (2023) and Swathe Sriee et al. (2024). Different forms of LCA are discussed in Liu et al. (2024).

LCA evaluates the environmental impacts of raw material sourcing, including agricultural waste, algae, and industrial by-products, focusing on cultivation, harvesting, and transportation. For microbial biofuel production, it assesses energy use and emissions from processes like fermentation and anaerobic digestion, emphasizing microbial conversion efficiency. Energy consumption and GHGe (CO_2, CH_4, N_2O) are analyzed, comparing microbial biofuels with fossil fuels. LCA also examines water use and pollution, considering consumption and wastewater runoff. It assesses waste generation and by-products, such as spent biomass, and their reuse to minimize impact. End-use impacts focus on the energy efficiency and emissions of microbial biofuels in applications like transportation. Lastly, sustainability considerations address land-use change, food security, and social factors like job creation, offering a holistic view of biofuel production's long-term viability. The challenges in setting guidelines for LCA of biofuel systems are discussed in detailed in (Czyrnek-Delêtre et al., 2017; Singh and Olsen, 2012).

11.3.2 Functional Units, System Boundaries and Allocations in LCA

The functional units (FUs) in an LCA are a crucial element because it defines the basis for comparison and ensures consistency in the analysis (Arzoumanidis et al., 2020). It represents the unit of measurement that reflects the function of the product or service being assessed and a quantified description of the performance requirements fulfilled by a product system, focusing on the product's functions rather than its physical form. There are a number of ways to define FUs such as input basis, output basis, areas basis, based on the year, fuel volume or carbon content. The choice of FUs should align with the goals and scope of the study, as well as the specific characteristics of the biofuel being evaluated. In microbial biofuels, the FUs serves as a basis to evaluate and compare biofuel production efficiency, environmental impact, and sustainability across different processes or systems. By using appropriate FUs such as per hectare for land use or per km for transportation, the LCA can offer more accurate and context-specific results. It was observed that approximately 32% of studies used units of bioenergy in J or kwh as FU, while around 22% reported LCA results based on the amount of biofuels produced, typically measured in kg (Osman et al., 2021). Additionally, breaking down the results by each life cycle stage, with suitable FU for each stage, ensures a comprehensive understanding of the environmental impacts associated with biofuel production and use (Gaffey et al., 2024). The IEA in its BIOMITRE manual (BIOmass-based Climate Change MITigation through Renewable Energy) recommends allocation based on market value or physical relationships, such as mass, volume, or calorific value. The BIOMITRE manual further provides more detailed recommendations, suggesting that the FU should be based on mass, volume, or energy content, and offers guidelines and examples for defining biofuels' system boundaries.

System boundaries define the limits of LCA, determining which stages of production are included and assessing their environmental impacts ("Understanding System Boundaries in Life Cycle Assessment," n.d.). These boundaries are essential for assessing the sustainability of microbial biofuel production and understanding the full ecological footprint. The system boundaries are predefined, but there is the option to create new entries. The overview of system boundaries is shown in Figure 11.2. Establishing the right system boundaries for microbial biofuels is challenging due to variability in feedstock sources (e.g., waste products, algae, etc.), microbial strains, and cultivation techniques. Additionally, different biofuel production methods (e.g., fermentation, transesterification) may involve different system boundaries.

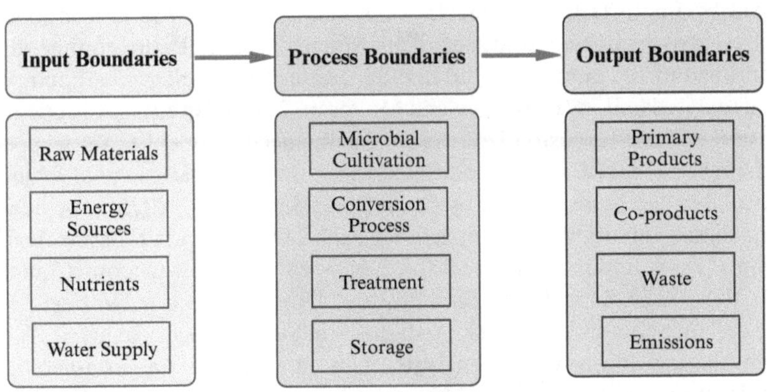

FIGURE 11.2 The overview of system boundaries for microbial biofuels.

Allocation refers to the process of assigning environmental impacts (such as energy use, water use, and GHGe) or resources to different products or outputs that arise from the same process or system (Azapagic and Clift, 1999). In microbial biofuel production, allocation can be challenging due to the simultaneous production of biofuels and valuable by-products (e.g., proteins, enzymes, or other chemicals). The chosen allocation method can significantly affect the results of both environmental and economic assessments (Swathe Sriee et al., 2024). For instance, if biofuels are co-produced with high-value compounds, the allocation method should reflect the use and economic importance of these by-products. Although selecting an appropriate allocation methodology is a critical step in LCA, various organizations recommend different approaches. Studies have shown that the choice of allocation method can lead to significantly different results.

11.3.3 Software Tools Used for LCA Studies

There are various software tools available to perform LCA for microbial biofuels studies. Notably, these tools integrate extensive databases related to LCI and impact assessment methodologies (Linares et al., 2024). Selecting the appropriate software depends on the specific needs of the study, such as the complexity of the microbial biofuel production system and the level of detail required for analysis. Many LCIA tools are part of larger LCA software, such as SimaPro, GaBi, and OpenLCA. These tools are designed to help users quantify environmental impacts, but they do so in different ways, depending on the software's underlying methodology, databases, and user interface, functionality, data quality management and modeling principles for building product systems (Lopes Silva et al., 2019). These software packages allow researchers to model, simulate, and assess the environmental impacts of biofuel production processes. Some of the most widely used LCA software include:

SimaPro is one of the most widely used LCA software tool, offering a comprehensive platform to assess the entire lifecycle of microbial biofuels, including stages such as raw material extraction, production, transportation, and disposal. Released in 1990 and developed by PRé Consultants, based in the Netherlands, SimaPro is distributed globally ("SimaPro Craft - PRé Sustainability," n.d.). **GaBi (now Sphera)**, developed by PE INTERNATIONAL, a German company, was first introduced in 1992 and is also available worldwide. Both SimaPro and GaBi are commercial software tools and most commonly used in the literature and market. Their comparisons are discussed in detail in (Herrmann and Moltesen, 2015). **OpenLCA** is open-source (free to use) and can be customized to suit specific needs ("openLCA.org | openLCA is a free, professional LCA and footprint software with a broad range of features and many available databases, created by GreenDelta since 2006," 2025). The qualitative comparison and Environmental impacts for each LCA software tool are clearly discussed in ("How Important Is the LCA Software Tool You Choose Comparative Results from GaBi, OpenLCA, SimaPro and Umberto," n.d.). **Umberto** enables users to conduct comprehensive LCA studies, including process analysis, life cycle cost analysis, and material flow analysis, with an emphasis on sustainability. It is commonly used to evaluate the resource efficiency and environmental performance of biofuel production systems, including microbial biofuels. Notably, Umberto is the only LCA software that integrates cost analysis, such as Life Cycle Costing, allowing users to create different scenarios based on technological, legal, market, price, and demand trends, all in relation to both cost and environmental criteria ("Umberto Life Cycle Assessment Software (LCA)," n.d.).

The European Commission's Joint Research Centre offers an LCA tool through the European Life Cycle Database (ELCD), designed for sustainability assessments. This tool is particularly useful for evaluating biofuels, providing access to a wide range of data sets and impact assessment methods focused on the European market. It is frequently used in European biofuel research, including microbial biofuels, to assess the environmental implications across different stages of production and use ("European Platform on LCA | EPLCA," n.d.). NREL and its partners created the USLCI Database to help LCA practitioners answer questions about environmental impact ("US Life Cycle Inventory Database | Energy Analysis | NREL," n.d.). CCLUB (Carbon Calculator for Land Use and Land Management Change from Biofuels Production) has been developed as an integral part of Argonne's GREET (GHG, Regulated Emissions, and Energy use in Technologies) model to analyze GHGe from LUC and land management change (LMC) in the context of overall biofuel LCA. It relies on biofuel production scenarios, LUC and LMC scenarios, and emission factors to generate GHGe of LUC and LMC for biofuel production in gCO_2e/MJ of fuel produced. Aspen Plus is a process simulation software that can be used for the modeling of chemical processes, including fermentation, to estimate energy usage, carbon footprints, and other environmental impacts. It is often used in the industrial scale-up of microbial biofuel processes to evaluate sustainability ("Aspen Plus Sustainability Applications," n.d.).

11.3.4 Regulatory and Policy Frameworks for LCA of Microbial Biofuels

Regulatory and policy frameworks for the LCA of microbial biofuels are vital for evaluating and minimizing environmental impacts. These frameworks guide the development, implementation, and evaluation of microbial biofuels, aiming to promote sustainability, mitigate environmental harm, and encourage innovation. These frameworks include international standards like ISO 14044 (LCA – Requirements and Guidelines) and ISO 14040 (Environmental Management – LCA – Principles and Framework), national regulations such as RFS and RED, and financial incentives for biofuels that meet sustainability criteria (Blanco-Davis et al., 2024). LCA helps ensure biofuels meet environmental standards by assessing their entire life cycle, promoting low GHGe, and preventing negative effects like land degradation or deforestation. These policies guide the development and implementation of sustainable microbial biofuels. Voluntary certification programs like the Roundtable on Sustainable Biomaterials (RSB) ("Certification – RSB," n.d.) and the International Sustainability and Carbon Certification (ISCC) ("ISCC System – Solutions for sustainable and deforestation free supply chains," n.d.) use LCA to assess the sustainability of microbial biofuels, ensuring they meet environmental, social, and economic criteria. Global harmonization efforts, led by organizations like the IEA ("IEA – International Energy Agency," n.d.) and the Global Bioenergy Partnership

(GBEP) ("Global Bioenergy Partnership (GBEP) | Department of Economic and Social Affairs," n.d.), aim to standardize LCA methodologies to support international trade and sustainability goals. Additionally, regulatory frameworks require environmental impact reporting to track progress toward sustainability and promote transparency. These frameworks, incorporating LCA, help governments and industries to ensure the responsible production and consumption of microbial biofuels, supporting a cleaner, more sustainable future.

11.4 LCA STUDIES FROM EXISTING LITERATURE

There are several studies in the literature that offer in-depth analyses/comparisons of LCA studies and some are discussed here. Angili et al. evaluated 48 papers published between 2002 and 2021, focusing on bioethanol production studies that covered a complete set of environmental impact categories. The review found that approximately 80% of the papers indicated a reduction in GWP, while some studies showed increased impacts on acidification, eutrophication, and photochemical oxidant formation due to feedstock provision. Technically, most studies emphasized pretreatment as a crucial step in bioconversion. However, despite the numerous LCA studies on bioethanol production, there remains limited attention given to uncertainty analysis (Soleymani Angili et al., 2021). Bouter et al. reviewed over 580 LCA articles on biofuels published between 2012 and 2020, with 172 usable articles providing 566 carbon footprints. The study focused on optimizing installations, assessing biomass types, comparing pathways, and evaluating methodological impacts in LCA. It revealed significant variation in biofuel carbon footprints, with feedstock type being a key factor. Some biofuel pathways showed clear climate change mitigation potential, while methodological choices, such as co-product allocation and LUC, contributed to differing LCA outcomes. Carbon neutrality also influenced results (Bouter et al., 2024).

Gaffey et al. reviewed studies published between January 2016 and March 2022, analyzing contemporary LCA biorefinery research across different feedstock categories. The review focuses on key methodological decisions, such as system boundaries, allocation, LUC, and uncertainty analysis. Of the initial 81 studies, 59 were included: 22 on dedicated feedstocks, 34 on residue feedstocks, and 3 on both. Many studies lacked comprehensive LCA assessments, omitting details on key decisions, value-chain segments, uncertainty analyses, and impact categories. Only 28% included primary data, 39% didn't conduct uncertainty or sensitivity analysis, and 8% addressed direct or indirect LUC. Just 14% considered product end-of-life. The authors emphasize the need for more transparency, a full value-chain approach, and the inclusion of renewable energy and circularity considerations in biobased value chains (Gaffey et al., 2024). Patel et al. reviewed 53 peer-reviewed studies on biofuels derived from over 25 feedstocks, focusing on publications from 2012 to 2022. The review revealed that second-generation biofuels have a greater potential to reduce GHGe compared to first-generation biofuels. However, third-generation biofuels exhibited higher GHGe than conventional fuels, ranging from 10.2 to 1910 g CO_2 eq./MJ. The net energy ratios for biofuels varied: first-generation (1.23–12.49), second-generation (0.003–15.04), and third-generation (3–18.5). Additionally, the water footprint of biofuels was found to be 50–240 times larger than that of fossil fuels. The study also identified key methodological challenges and uncertainties in LCA and proposed solutions. It underscores the need for a comprehensive approach to evaluating the environmental impact of biofuels and highlights the importance of continued research for sustainable biofuel development (Patel and Singh, 2023).

11.4.1 LCA Studies in Bacteria, Yeast, and Fungi

LCA studies on bacteria, yeast, and fungi have gained significance in biofuel production, food manufacturing, pharmaceuticals, and waste treatment. LCA evaluates the environmental impacts of these processes, from raw material procurement to product disposal. Microbial oils are lipid-rich compounds produced by microorganisms such as bacteria, yeast, and algae. Single-cell oils (SCOs) are intracellular storage lipids primarily composed of triacylglycerols (TAGs). Produced by oleaginous microorganisms, SCOs can accumulate between 20% and 80% lipids/dcw during the stationary growth phase, typically under nutrient limited conditions (e.g., nitrogen or phosphorus) coupled with an excess of carbon source (Ochsenreither et al., 2016). The main challenge in producing SCOs is the high cost of production, primarily due to expensive fermentation processes. One potential solution is combining SCO production with other biological processes, such as the biotreatment of agricultural and industrial waste by-products. Recycling organic waste from agriculture and agro-industrial activities is not only cost-effective but also beneficial for the environment. Oleaginous microorganisms like algae and fungi have recently drawn attention for their ability to produce high-purity polyunsaturated fatty acids (PUFAs) in SCOs from a wide variety of substrates. When PUFAs enter microbial cells, they are often esterified with glycerol to form structural lipids such as phospholipids and glycolipids, which are essential for maintaining membrane fluidity and function (Kothri et al., 2020).

Bacterial cellulose (BC), produced through fermentation, is a versatile material with various potential applications. An attributional LCA was conducted on BC production via static culture. The FU was defined as 1 kg of dry BC, produced with 138.8 kg of water. Water consumption was highest in the raw material phase (17.8 tons/kg), mainly for carton packaging, culture medium, and sodium hydroxide for washing. Fermentation and downstream processes accounted for the remainder. Among the environmental impacts, raw material production contributed most, particularly in climate change, fossil depletion, human toxicity (noncancer), and terrestrial toxicity. Sodium dihydrogen phosphate production had the highest impacts on human toxicity and terrestrial eco-toxicity, followed by corn syrup and carton production. Static culture fermentation and downstream processes had significant impacts on climate change and fossil depletion, while the BC production itself contributed minimally to overall resource consumption and environmental impacts (Forte et al., 2021).

A major challenge in biodiesel production is the high cost of raw materials, which account for 75% of production costs (Subramaniam et al., 2010). Barla et al. explored bacterial biomass cultivation and biodiesel production using a biorefinery approach.

Their process achieved removal efficiencies of 96%, 93%, and 98% for CO_2, SO_2, and NO, respectively, while producing 274 g of biomass in a 20 L bioreactor. The biomass yielded 58% w/w lipids, which were converted into biodiesel with a 91% w/w yield. The metabolic pathways for flue gas reduction and lipid production were analyzed to optimize biodiesel production. LCA revealed that using safer, alternative chemicals could reduce environmental impacts and GWP100. A TEA showed that the bacterial biorefinery had a net present value of $193/L of biodiesel, indicating its potential for a circular economy by promoting sustainable practices, optimizing resource use, and reducing waste (Barla et al., 2025).

Oleaginous yeast can be a potential feedstock for biodiesel production, with the de-oiled biomass used for energy-dense bio-crude production via HTL, a precursor for advanced biofuels. A detailed LCA of biodiesel and bio-crude production in a yeast biorefinery model was conducted and revealed that HTL and transesterification had higher impacts than other sub-processes. The HTL step emitted 4.02E3 kg CO_2 eq., and transesterification showed a resource depletion cost of $773, mainly due to high energy/electricity use. Solvents like methanol and glycerol contributed 60%–70% to the impact categories. Sensitivity analysis indicated that higher lipid productivity could reduce CO_2 emissions by 16% and resource depletion by 6.5%. Environmental performance could improve by replacing fossil fuel electricity with renewable energy sources, such as solar or hydroelectric power, and recycling solvents and waste heat (Chopra et al., 2020). Another study aims to develop a full-scale plant for producing yeast biodiesel from cardoon stalks using the biorefinery concept and assess its environmental impact through a cradle-to-gate, scaled-up LCA approach. The results identified key environmental hotspots across the entire process, including enzyme production for biomass hydrolysis, and the heat and organic solvents needed for lipid extraction and distillation. LCA normalization revealed that human carcinogenic toxicity accounted for the highest environmental burden (69.2%). Interestingly, GHG emissions for yeast biodiesel were negative (−1.5 g CO_2 eq/MJ), due to environmental credits from co-products. A comparison with other biodiesel sources indicated that yeast biodiesel from cardoon stalks could become a competitive and environmentally sustainable alternative (Barbanera et al., 2021).

A study by Brancoli et al. found that integrating *Neurospora intermedia* filamentous fungi into first-generation ethanol production significantly reduces GHGe. The most substantial climate benefits occurred when both dried distillers grains with solubles (DDGS) and fungal biomass were used as cattle feed, yielding an environmental abatement of −50.9 g CO_2eq/MJ, compared to −39.1 g CO_2eq/MJ in the base scenario. This resulted in a 6.5% GHG emission reduction, with both scenarios meeting the EU's 50%–60% GHG reduction target for biofuels. However, using fungal biomass as fish feed worsened the impact. Sensitivity analysis showed that GHG benefits are sensitive to DDGS and fungal biomass yields, but even with a 30% yield reduction, emissions remain significantly lower than the fossil fuel baseline. The study concludes that *Neurospora intermedia* cultivation could enhance ethanol production sustainability and mitigate climate change, especially when waste products are used effectively (Brancoli et al., 2018).

11.4.2 LCA Studies in Cyanobacteria

LCA studies in cyanobacteria and microalgae focus on assessing the environmental impacts of using these microorganisms for various applications. Few applications/case studies and importance of LCA analysis for biofuel production are discussed here. LCA and TEA analysis were conducted to assess the sustainability of this process, analyzing different nutrient sources (wastewater vs. BG-11) and PHB contents (15%–50% dcw). Increasing PHB content reduced environmental impacts by 67%–75%, mainly due to construction materials and chemicals like chloroform. The economic analysis showed a minimum selling price of 135€ per kg PHB (50% dcw, 12.5 g PHB $m^{-3}d^{-1}$). To reach a market price of 4€ per kg, PHB productivity must increase to 810 mg $L^{-1}d^{-1}$, requiring improved reactor design and genetic modification of strains (Rueda et al., 2023). An LCA was conducted for biobutanol production using genetically engineered cyanobacteria *Synechocystis* PCC 6803. The study applied a prospective, upscaled LCA to assess the environmental impacts of large-scale (20 ha) biobutanol production, providing a better comparison with conventional butanol production. The results showed that engineered cyanobacteria improved environmental performance due to higher productivity, but the high electricity demand for cultivation and harvesting reduced this benefit. However, using renewable electricity lowered GHGe to 3.1 kg CO_2 eq/kg biobutanol, 20% lower than fossil-based butanol (2.45 kg CO_2 eq/kg). The study emphasizes the importance of genetic engineering and transitioning to renewable electricity to maximize the environmental potential of microorganisms for sustainable biofuel production, as well as the need for upscaled LCA scenarios for fair comparisons with conventional technologies (Villacreses-Freire et al., 2022). Another study evaluates the environmental impacts and cumulative energy demand (CED) of cyanobacteria-produced n-butanol using a cradle-to-grave consequential LCA. The assessment considered a hypothetical plant in northern Sweden (1 ha, producing 5–85 m^3 n-butanol/year) with various cultivation and productivity scenarios. GHGe ranged from 16.9 to 58.6 gCO_2eq/MJBuOH, and CED varied from 3.8 to 13 MJ/MJBuOH. Sustainability goals (60% GHG reduction) were only met when a nearby paper mill supplied heat and CO_2, while northern Sweden's cold climate reduced reactor cooling needs. The study highlights the high CED across scenarios, emphasizing the need for significant metabolic engineering, such as >90% carbon partitioning to n-butanol and better light utilization, to make n-butanol competitive with fossil fuels and bioethanol (Nilsson et al., 2020). One study presents a concurrent TEA and LCA of a facility producing fuels and methyl laurate, an oleochemical from *Synechocystis* PCC 6803. The biorefinery model covers cultivation, methyl laurate separation, biomass harvesting, and fuel processing via HTL. Monte Carlo analysis is used to account for uncertainty in key input parameters. For the biorefinery producing both fuels and methyl laurate, the average minimum fuel selling price (MFSP) is \$2.47/dm^3 (\$9.34 per gallon of gasoline equivalent), with a GWP of 118 g CO_2.eq/MJ. For the fuel-only scenario, the MFSP is \$2.01/dm^3 (\$7.60 per gallon), with an environmental impact of 100 g CO_2.eq/MJ. These results highlight the limitations of non-stochastic methods and suggest improvements in pond productivity for both biomass and methyl laurate production (Beattie et al., 2021). Roja et al performed TEA and LCA for cyanobacterial SA production. Experimental data for cyanobacteria cultivation producing disodium succinate were scaled up to 100 tons/year of SA, with high-temperature (HT) and low-temperature (LT) crystallization (Process I) and methanol purification (Process II) considered for downstream separation. Using the ReCiPe 2016 impact assessment method and data from ecoinvent® 3.9, the production costs were INR 1,654/kg ($21.04/kg) for Process I and INR 1,501/kg ($19.09/kg) for Process II. GWP was 7.95 kg CO_2eq for Process I and 6.32 kg CO_2eq for Process II per kg of SA. A 36% reduction in production cost and 52%–56% reduction in GWP was achieved by increasing succinate titer from 2 g/L to 5 g/L (Raja and Shastri, 2025). One study evaluates the NERs and GHGe of a Photosynthetic Cyanobacterial Biorefinery (PSBR) using LCA. It focuses on engineered cyanobacterium cultures that secrete biosynthetic biofuels, including bisabolane, heptadecane, and ethanol. Results show that cyanobacterial ethanol has higher yields, lower NER, and lower GHGe compared to bisabolane and heptadecane in *Synechocystis sp*. PCC 6803. Sensitivity analysis highlights that biofuel productivity, energy consumption in vapor compression, distillation, and culture mixing are key factors influencing environmental impacts. This LCA provides a baseline for stakeholders to guide improvements in cyanobacterial biofuel production and sustainable bioprocesses (Quiroz-Arita et al., 2017).

11.4.3 LCA Studies in Microalgae

Microalgae have been regarded as a potential source of low-carbon transport fuel for the future. Few examples of LCA studies related to biomass cultivation, conversion method, and biofuel generation are discussed here. Microalgae are valuable for mitigating elevated CO_2 levels and treating wastewater, but LCA evaluations have mostly focused on microalgal biofuels and heterotrophic cultivation. Comparative LCA for various algal products, such as algal powder, bio-oil, and fatty acids, is limited, particularly for autotrophic cultivation. One study compares the LCA of several autotrophic microalgae species, focusing on *Chlorella* sp. The results show that production costs and environmental impacts can be reduced by selecting the right algal species, optimizing cultivation, biomass separation, and drying methods, and using renewable energy and nearby thermal power plants for CO_2 or flue gas. This provides insights into microalgal consumption under current global carbon sequestration challenges (Zhang et al. 2022). One study analyzed the impact of cultivating microalgae in two media (Bold's Basal (BB) media and greywater) for biofuel production. Replacing BB media with greywater reduced energy requirements by 16.3%, increasing net energy recovery by 79.3%. LCA showed CO_2 mitigation increased from 1.7 to 2.4 kg/kg biofuel respectively. Freely available greywater showed lower energy compared to synthetically prepared BB media. Notably, greywater cultivation lowered GWP by 23.1% and acidification potential by 86.2% compared to BB media. Out of all the impacts in seven environmental categories, eutrophication potential showed the biggest reduction –9287%, which was primarily due to the removal of nitrogen and phosphorus from the greywater (Mohit et al., 2024).

Another LCA study evaluated two systems for treating (i) urban wastewater and (ii) industrial wastewater (from the food industry), while recovering bioproducts (natural pigments, biofertilizer) and bioenergy (biogas). The results showed that the unialgal culture system for treating industrial wastewater had a smaller environmental impact than the mixed culture system for urban wastewater. Recovering bioproducts from microalgae wastewater treatment reduced environmental impacts by up to five times compared to conventional systems, mainly due to lower chemical consumption for microalgae cultivation. Food industry effluent was found to be the most promising for bioproduct recovery due to its higher quality compared to urban wastewater, which supports single-species microalgae cultivation (Arashiro et al., 2022). A comparative LCA and TEA were used to assess the sustainability and economic performance of two nutrient strategies. Scenario 1 uses source-separated nutrient delivery method with microalgae in tertiary wastewater treatment, while Scenario 2 employs non-separated point nutrient delivery approach with microalgae in secondary wastewater treatment. The results show that the source-separated nutrient strategy is more effective in reducing environmental impacts and improving the commercial potential of microalgae biofuel (Li et al., 2022).

In another study, LCA, TEA and data envelopment analysis (DEA) was done to optimize energy and environmental impact. Open space and greenhouse cultivation methods for microalgae were analyzed for energy requirements, productivity, net return, and environmental impact. The open space and greenhouse methods require an average input energy of 1,5920.40 and 17,691.60 MJ/kg, with energy ratios of 0.89 and 0.80 and productivity energy indexes of 0.06 and 0.02 kg/MJ, respectively. The analysis concluded that the open space production method is preferred due to its lower energy consumption (Kazemi et al. 2023). Hossain et al. conducted an LCA of microalgae-based bioethanol production in Brunei Darussalam, focusing on CO_2 emissions and energy balance. The study used carbohydrate-rich microalgae, like *Chlorella vulgaris* (55%), and cultivated 220 tons of biomass on 6 ha of offshore land. This process absorbed 366.66 tons of CO_2 annually, resulting in a net CO_2 balance of 218.86 tons from the bioethanol plant. The project also required low water and land footprints. Photobioreactor cultivation had water and land footprints of 2 m^3/GJ and 2 m^2/GJ, respectively, while open pond cultivation had 87 m^3/GJ and 13 m^2/GJ. Thus, the LCA of microalgae-bioethanol proved itself as an energy-saving approach for large-scale applications (Hossain et al., 2019).

Microalgae biomass can be transformed into liquid biofuel using various bioenergy conversion technologies. In one study, LCA evaluated two bioenergy conversion systems for microalgae-based biofuel production: (1) transesterification, HTL and pyrolysis for renewable diesel, and (2) anaerobic digestion with/without hydrothermal pretreatment for biogas. The results showed that anaerobic digestion with hydrothermal pretreatment is more industrially feasible and environmentally friendly, with a low NER of 0.71 and GHGe of −60.84 g CO_2eq/kg of biogas (Sun et al., 2019). Biofilm cultivation systems offer high biomass concentration, making them ideal for microalgae liquid biofuel conversion, yet research is limited. One study showed LCA of microalgae liquid biofuel production using three conversion technologies: transesterification, HTL, and pyrolysis. The results show HTL as the most feasible technology, achieving NER of 0.87 and GHGe of −39.91 g CO_2eq/kg. In contrast, transesterification and pyrolysis had NERs of 2.18 and 2.06, but GHGe were positive, indicating negative energy and environmental impacts due to chemical use and high energy consumption in drying. Sensitivity analysis showed that bio-crude mass yield significantly impacted the energy and environmental outcomes for HTL (Wei et al., 2024).

Bradley et al. conducted LCA studies using real-world data from an industrial facility that uses a combination of photobioreactor and fermenter systems and found that the facility's performance falls short compared to petroleum-derived diesel unless productivity is increased. Improvements in microalgae strains, photobioreactor designs, and cultivation systems are crucial for microalgae to become a sustainable fuel feedstock. Reducing electricity use and addressing the high impacts of yeast production – either through co-location or sourcing from lower-impact providers – is essential. Lipid extraction may waste valuable products, while the waste will contain unextracted lipids, polysaccharides, proteins, and minerals. Harmonizing assessments is needed for future studies to determine whether microalgae should be used for fuel or other high-value products like feed (Bradley et al., 2023). LCA of biodiesel production from microalgae and valorization of value-added products was conducted, comparing three feedstock cultivation scenarios: (i) no nutrient input, (ii) wastewater input, and (iii) fertilizer input. Two transesterification techniques were analyzed: acid catalyst and biocatalyst. The results showed higher eutrophication and photochemical oxidation emissions with the fertilizer-based scenario. However, wastewater use and biocatalyst transesterification significantly reduced environmental impacts and GHGe. The fossil energy requirement varied from 3.6 to 5.7 MJ/kg, with GHGe between 0.85 and 1.46 kg CO_2eq/kg of biodiesel. The wastewater-biocatalyst scenario had the highest NER of 18.8 and offered the added benefit of low-cost wastewater remediation (Saranya and Ramachandra, 2020). LCA was conducted for the production and combustion of 1 MJ of algal methylester using open raceways under greenhouses. The study found that improving energy balance is crucial for sustainable microalgal cultivation and reducing GHGe. While technological breakthroughs in production can reduce GHGe, a significant portion of impacts stems from electricity production for cultivation, harvesting, and transformation. A scenario with 45% local renewable electricity was compared to improvements from higher productivity. The results showed that local renewable electricity significantly reduced climate change impacts and increasing biomass productivity (10–15 g/m^2/day) before considering electricity mix changes is a more effective strategy. Renewable electricity also improved energetic efficiency, resulting in a positive cumulative energy balance (Collet et al., 2014). A comparative LCA study of a notional biodiesel production system in Australia compared algae biodiesel (with three CO_2 supplementation scenarios and two production rates) to canola and ULS diesel. Algae GHGe (−27.6 to 18.2 g CO_2.e/tkm) were significantly lower than canola (35.9) and ULS diesel (81.2), but algae costs (2.2–4.8 ¢/tkm) were higher than canola (4.2) and ULS diesel (3.8), emphasizing the need for higher production rates to improve algae biodiesel's economic viability (Campbell et al., 2011)

Ketzer et al. conducted a comparative analysis of 16 recent studies on bioenergy production from microalgae, evaluating the energetic performance based on assumptions related to biomass productivity, lipid content, required energy, and system output (Ketzer et al., 2018). Microalgal biorefineries offer a potential sustainable alternative to fossil-based refineries. Ubando et al. examines studies on microalgal biorefineries, identifies strategic technologies for sustainable biofuel production, introduces metrics to support LCA studies and discussed challenges in comparing LCA results (Ubando et al., 2022). A review discusses algae-based biofuel production, highlighting the need for industrial symbiosis to reduce environmental impacts. The use of waste heat, CO_2, renewable energy, and co-location with industries like renewable energy and dairy is beneficial for biofuel production. It also emphasizes the role of Dynamic LCA in improving climate change indicators such as GWP, GTP, and CTP, with waste-based bioenergy production showing lower GHGe compared to other bioenergy pathways (Chowdhury et al., 2023).

11.5 CHALLENGES AND FUTURE DIRECTIONS

The environmental performance of biomass-based energy, including GHG reductions and energy balances, depends on factors such as feedstock type, conversion technology, LUC, co-product management, data quality, and impact assessment methods. Choosing the right processes is essential for sustainable energy production (Prasad et al., 2020). While microbial biofuel technologies offer potential, challenges remain in process efficiency, scalability, and policy frameworks. Additionally, issues like resource competition, feedstock limitations, and high infrastructure costs persist. For example, using US corn and soybean yields for biofuels could only meet a small portion of gasoline and diesel demand, and upgrading biorefineries to process cellulosic feedstocks might cost over $1 billion per facility (Benavides et al. 2022). Future bioenergy systems will require integrated strategies to reduce environmental impacts and maximize resource efficiency. New biorefinery models using agricultural residues (e.g., corn stover, forestry by-products) can sustainably supply 30% of transportation fuel demand without affecting food security. Advances in fermentation and gasification technologies enable the conversion of diverse waste streams into drop-in biofuels, with lifecycle emissions 50%–70% lower than petroleum fuels (Jafri et al., 2022). Combining biofuel production with carbon capture and storage (BECCS) in renewable-powered refineries can achieve net-negative emissions of less than -50 gCO_2eq/MJ. These strategies highlight the need for interdisciplinary research in crop genetics, waste valorization, and hybrid energy systems to expand sustainable bioenergy solutions (He et al., 2022).

LCA results can vary significantly due to differences in microbial strains, reactor designs, and conversion processes, making it difficult to generalize LCA results across different biofuel production methods. Low ethanol yields from lignocellulosic waste require strain engineering (Pillai et al., 2024) and maintaining microbial consortia stability in large reactors necessitates AI-driven monitoring (Narisetty et al., 2022). Therefore, technological advancements and variations in operational scale must be considered when assessing the environmental impact of microbial biofuels. Furthermore, innovations in bioreactor design, biocatalysts, nanotechnology, omics-technologies, and database systems are also improving biofuel production (Chaudhary et al., 2025). While LCA is widely used for environmental decision-making, it has both theoretical and practical limitations. One primary goal of LCA in microbial biofuels is to compare their environmental impacts with conventional fossil fuels, accounting for the entire life cycle of both. In addition to GHGe, LCA of biofuels considers other environmental impacts, like acidification, eutrophication, photochemical smog, human toxicity, and eco-toxicity. These categories provide a broader view of the environmental consequences of biofuel production and use, helping to identify the most sustainable alternatives to fossil fuels. However, LCA studies must account for factors beyond carbon emissions. Liska et al. discussed eight uncertainty principles in LCA of biofuel systems, including system complexity, method variability, data gaps, spatial/temporal factors, impact categories, indirect impacts like LUC, and transparency. These factors contribute to varying results depending on feedstock choice, conversion technologies, end-use, land use, and data quality (Liska, 2015). The variable quality of data inputs significantly influences the reliability of the outcome. McKone et al. also highlight seven key challenges, including understanding feedstock options, predicting biofuel technologies, characterizing emissions, accounting for spatial variability, and uncertainty. Addressing these challenges is crucial for improving LCA reliability and informing biofuel policy (McKone et al. 2011). Accurate LCA of biofuels requires comprehensive data from all production stages, which is often difficult to obtain, especially for emerging microbial biofuel technologies. Gaps in data related to feedstock cultivation, microbial conversion efficiencies, and process-scale variations introduce uncertainties in results. Other challenges include limited data for computational modeling and deep learning, high processing costs, and integrating biofuels with other renewable energy sources. Addressing these issues is essential to fully unlocking the potential of biomass, biofuels, bioenergy, and hybrid systems.

11.6 CONCLUSION

Biofuel production from microorganisms offers several advantages over traditional fossil fuel production, including lower GHGe, reduced reliance on depleting oil reserves, and the promotion of local economies using regionally grown feedstocks. While microbial biofuels hold promise as a renewable and lower carbon alternative to fossil fuels, a thorough understanding of their potential environmental consequences is necessary. By improving efficiency, minimizing waste, and using sustainable practices, microbial biofuels could play a key role in the transition to a more sustainable energy future. Thermochemical and biochemical approaches are the primary methods used to convert biomass into biofuels. Recent progress in biomass conversion technologies has greatly enhanced the efficiency and environmental advantages of renewable energy production. LCA is a sophisticated tool that bridges the fields of science, engineering, and policy. LCA of microbial biofuels is crucial for assessing the environmental sustainability of biofuel production, highlighting areas for improvement in terms of energy efficiency, emissions reduction, and resource utilization. However, translation from laboratory to industry remains a challenge, including the requirement for versatile, readily engineered, and industrially compatible microbial biocatalysts with the necessary molecular tools and regulatory sequences, further optimization of engineered pathways to increase productivity, and ongoing TEA and LCA of these opportunities. Since energy from agricultural residues can be produced through various methods, LCA can be used to assess the most suitable type and route for sustainable energy generation based on feedstock and geography.

By providing a comprehensive view of the entire life cycle, LCA helps guide the development of more sustainable microbial biofuel technologies and informs policy decisions related to biofuel adoption and climate change mitigation.

REFERENCES

Abdullah, B., Syed Muhammad, S.A.F., Aziz, M.M.A., 2019. Fourth generation biofuel: A review on risks and mitigation strategies. *Renewable and Sustainable Energy Reviews* 107, 37–50. https://doi.org/10.1016/j.rser.2019.02.018

Alalwan, H.A., Alminshid, A.H., Aljaafari, H.A.S., 2019. Promising evolution of biofuel generations. Subject review. *Renewable Energy Focus*, 28, 127–139. https://doi.org/10.1016/j.ref.2018.12.006

Alternative Fuels Data Center: Maps and Data. (n.d.) https://afdc.energy.gov/data

Ambaye, T.G., Vaccari, M., Bonilla-Petriciolet, A., Prasad, S., van Hullebusch, E.D., Rtimi, S., 2021. Emerging technologies for biofuel production: A critical review on recent progress, challenges and perspectives. *Journal of Environmental Management*, 290, 112627. https://doi.org/10.1016/J.JENVMAN.2021.112627

Arashiro, L.T., Josa, I., Ferrer, I., Van Hulle, S.W.H., Rousseau, D.P.L., Garfí, M., 2022. Life cycle assessment of microalgae systems for wastewater treatment and bioproducts recovery: Natural pigments, biofertilizer and biogas. *Science of the Total Environment*, 847, 157615. https://doi.org/10.1016/J.SCITOTENV.2022.157615

Arzoumanidis, Ioannis, D'Eusanio, M., Raggi, A., Petti, L., Arzoumanidis, I., Eusanio, Raggi, A., Petti, L., 2020. Functional unit definition criteria in life cycle assessment and social life cycle assessment: A discussion. 1–10. https://doi.org/10.1007/978-3-030-01508-4_1

Aspen Plus Sustainability Applications (n.d.) https://www.aspentech.com/en/insights/models/applications-by-products/aspen-plus-sustainability-applications

Azapagic, A., Clift, R., 1999. Allocation of environmental burdens in multiple-function systems. *Journal of Cleaner Production*, 7, 101–119.

Barbanera, M., Castellini, M., Tasselli, G., Turchetti, B., Cotana, F., Buzzini, P., 2021. Prediction of the environmental impacts of yeast biodiesel production from cardoon stalks at industrial scale. *Fuel*, 283, 118967. https://doi.org/10.1016/J.FUEL.2020.118967

Barla, R.J., Gupta, S., Raghuvanshi, S., 2025. Optimization of bacterial biorefineries for sustainable biodiesel production and flue gas reduction: A holistic approach to climate change mitigation and circular economy. *Sustainable Energy & Fuels* https://doi.org/10.1039/D4SE01516A

Beattie, A., Vermaas, W., Darzins, A., Holland, S.C., Li, S., McGowen, J., Nielsen, D., Quinn, J.C., 2021. A probabilistic economic and environmental impact assessment of a cyanobacteria-based biorefinery. *Algal Research*, 59, 102454. https://doi.org/10.1016/J.ALGAL.2021.102454

Benavides, P.T., Bartling, A.W., Phillips, S.D., Hawkins, T.R., Singh, A., Zaimes, G.G., Wiatrowski, M., Harris, K., Burli, P.H., Hartley, D., Alleman, T., Fioroni, G., Gaspar, D., 2022. Identification of key drivers of cost and environmental impact for biomass-derived fuel for advanced multimode engines based on techno-economic and life cycle analysis. *ACS Sustainable Chemistry & Engineering*, 10, 10465–10475. https://doi.org/10.1021/ACSSUSCHEMENG.2C00944

Bergerson, J., Lave, L., 2004. Life cycle analysis of power generation systems. *Encyclopedia of Energy*, 635–645. https://doi.org/10.1016/B0-12-176480-X/00234-5

Biofuels, 2008. *ITF round tables*. https://doi.org/10.1787/9789282101803-EN

Biofuels and the environment – U.S. Energy Information Administration (EIA) (n.d.) https://www.eia.gov/energyexplained/biofuels/biofuels-and-the-environment.php.

Biofuels and the Environment | US EPA (n.d.) https://www.epa.gov/risk/biofuels-and-environment

Biofuels Dashboard 2020 | IFPEN (n.d.) https://www.ifpenergiesnouvelles.com/article/biofuels-dashboard-2020

Blanco-Davis, E., Yang, F., Li, H., Roux, M., Ekvall, T., Schmidt, J., Vega, G.C., Møller-Maersk, A.P., 2024. Which rules to follow? how differences in renewable fuel standards obscure the potential climate impact of transportation fuels. *Frontiers in Energy Research*, 12, 1490137. https://doi.org/10.3389/FENRG.2024.1490137

Bouter, A., Duval-Dachary, S., Besseau, R., 2024. Life cycle assessment of liquid biofuels: What does the scientific literature tell us? A statistical environmental review on climate change. *Biomass and Bioenergy*, 190, 107418. https://doi.org/10.1016/J.BIOMBIOE.2024.107418

Bradley, T., Rajaeifar, M.A., Kenny, A., Hainsworth, C., del Pino, V., del Valle Inclán, Y., Povoa, I., Mendonça, P., Brown, L., Smallbone, A., Roskilly, A.P., Joyce, S., Heidrich, O., 2023. Life cycle assessment of microalgae-derived biodiesel. *International Journal of Life Cycle Assessment*, 28, 590–609. https://link.springer.com/article/10.1007/s11367-023-02140-6

Brancoli, P., Ferreira, J.A., Bolton, K., Taherzadeh, M.J., 2018. Changes in carbon footprint when integrating production of filamentous fungi in 1st generation ethanol plants. *Bioresource Technology*, 249, 1069–1073. https://doi.org/10.1016/J.BIORTECH.2017.10.085

Brusseau, M.L., 2019. Sustainable development and other solutions to pollution and global change. *Environmental and Pollution Science*, 585–603. https://doi.org/10.1016/B978-0-12-814719-1.00032-X

Campbell, P.K., Beer, T., Batten, D., 2011. Life cycle assessment of biodiesel production from microalgae in ponds. *Bioresource Technology*, 102, 50–56. https://doi.org/10.1016/J.BIORTECH.2010.06.048

Carriquiry, M.A., Du, X., Timilsina, G.R., 2014. Production costs of biofuels, Natural Resource Management and Policy. 33–46. https://doi.org/10.1007/978-1-4939-0518-8_3

Certification – RSB (n.d.) https://rsb.org/certification/

Chaudhary, A., Rathour, R.K., Solanki, P., Mehta Kakkar, P., Pathania, S., Walia, A., Baadhe, R.R., Bhatia, R.K., 2025. Recent technological advancements in biomass conversion to biofuels and bioenergy for circular economy roadmap. *Renewable Energy*, 244, 122714. https://doi.org/10.1016/J.RENENE.2025.122714

Chisti, Y., 2007. Biodiesel from microalgae. *Biotechnology Advances*, 25, 294–306. https://doi.org/10.1016/J.BIOTECHADV.2007.02.001

Chopra, J., Tiwari, B.R., Dubey, B.K., Sen, R., 2020. Environmental impact analysis of oleaginous yeast based biodiesel and bio-crude production by life cycle assessment. *Journal of Cleaner Production*, 271, 122349. https://doi.org/10.1016/J.JCLEPRO.2020.122349

Chowdhury, P., Mahi, N.A., Yeassin, R., Chowdhury, N.U.R., Farrok, O., 2025. Biomass to biofuel: Impacts and mitigation of environmental, health, and socioeconomic challenges. *Energy Conversion and Management: X*, 25, 100889. https://doi.org/10.1016/J.ECMX.2025.100889

Chowdhury, R., Caetano, N., Franchetti, M.J., Hariprasad, K., 2023. Life cycle based GHG emissions from algae based bioenergy with a special emphasis on climate change indicators and their uses in dynamic LCA: A review. *Sustainability*, 15, 1767. https://doi.org/10.3390/SU15031767

Collet, P., Lardon, L., Hélias, A., Bricout, S., Lombaert-Valot, I., Perrier, B., Lépine, O., Steyer, J.P., Bernard, O., 2014. Biodiesel from microalgae – Life cycle assessment and recommendations for potential improvements. *Renew Energy*, 71, 525–533. https://doi.org/10.1016/J.RENENE.2014.06.009

Co-Optima Transportation Ingenuity Capitalizes on Strengths of U.S. Energy Resources and Existing Infrastructure | News | NREL (n.d.) https://www.nrel.gov/news/program/2018/co-optima-transportation-ingenuity-capitalizes-on-strengths-of-us-energy-resources-and-existing-infrastructure.html

Czyrnek-Delêtre, M.M., Smyth, B.M., Murphy, J.D., 2017. Beyond carbon and energy: The challenge in setting guidelines for life cycle assessment of biofuel systems. *Renewable Energy*, 105, 436–448. https://doi.org/10.1016/J.RENENE.2016.11.043

D'Odorico, P., Davis, K.F., Rosa, L., Carr, J.A., Chiarelli, D., Dell'Angelo, J., Gephart, J., MacDonald, G.K., Seekell, D.A., Suweis, S., Rulli, M.C., 2018. The global food-energy—Water nexus. *Reviews of Geophysics*, 56, 456–531. https://doi.org/10.1029/2017RG000591

Davis, S.C., Parton, W.J., Del Grosso, S.J., Keough, C., Marx, E., Adler, P.R., Delucia, E.H., 2012. Impact of second-generation biofuel agriculture on greenhouse-gas emissions in the corn-growing regions of the US. *Frontiers in Ecology and the Environment*, 10, 69–74. https://doi.org/10.1890/110003

Directives Directive (EU) 2018/2001 of the European Parliament and of the Council of 11 December 2018 on the promotion of the use of energy from renewable sources. (n.d.)

Donner, S.D., Kucharik, C.J., 2008. Corn-based ethanol production compromises goal of reducing nitrogen export by the Mississippi River. *Proceedings of the National Academy of Sciences of the United States of America*, 105, 4513–4518. https://doi.org/10.1073/PNAS.0708300105/ASSET/34B918FB-DED2-4B9E-8C52-E67FF008C971/ASSETS/GRAPHIC/ZPQ0080896520005.JPEG

EIA increases forecasts for 2025 and 2026 ethanol production | Ethanol Producer Magazine (n.d.) https://ethanolproducer.com/articles/eia-increases-forecasts-for-2025-and-2026-ethanol-production

EIA projects nearly 50% increase in world energy use by 2050, led by growth in renewables – U.S. Energy Information Administration (EIA) (n.d.) https://www.eia.gov/todayinenergy/detail.php?id=49876

Environmental Integrity Despite Green Image, Biofuels Manufacturing Releases Large Amounts of Hazardous Air Pollutants (n.d.) https://environmentalintegrity.org/news/biofuels-manufacturing-releases-large-amounts-of-hazardous-air-pollutants/

European Platform on LCA | EPLCA (n.d.) https://eplca.jrc.ec.europa.eu/ELCD3/

Farghali, M., Osman, A.I., Chen, Z., Abdelhaleem, A., Ihara, I., Mohamed, I.M.A., Yap, P.S., Rooney, D.W., 2023. Social, environmental, and economic consequences of integrating renewable energies in the electricity sector: A review. *Environmental Chemistry Letters*, 21, 1381–1418. https://doi.org/10.1007/S10311-023-01587-1

Fargione, J., Hill, J., Tilman, D., Polasky, S., Hawthorne, P., 2008. Land clearing and the biofuel carbon debt. *Science* (1979), 319, 1235–1238. https://doi.org/10.1126/SCIENCE.1152747

Fore, S.R., Porter, P., Lazarus, W., 2011. Net energy balance of small-scale on-farm biodiesel production from canola and soybean. *Biomass Bioenergy*, 35, 2234–2244. https://doi.org/10.1016/J.BIOMBIOE.2011.02.037

Forte, A., Dourado, F., Mota, A., Neto, B., Gama, M., Ferreira, E.C., 2021. Life cycle assessment of bacterial cellulose production. *International Journal of Life Cycle Assessment*, 26, 864–878.

Gaffey, J., Collins, M.N., Styles, D., 2024. Review of methodological decisions in life cycle assessment (LCA) of biorefinery systems across feedstock categories. *Journal of Environmental Management*, 358, 120813. https://doi.org/10.1016/J.JENVMAN.2024.120813

Gallego-García, M., Susmozas, A., Negro, M.J., Moreno, A.D., 2023. Challenges and prospects of yeast-based microbial oil production within a biorefinery concept. *Microbial Cell Factories*, 22, 1–15. https://doi.org/10.1186/S12934-023-02254-4

Gelfand, I., Sahajpal, R., Zhang, X., Izaurralde, R.C., Gross, K.L., Robertson, G.P., 2013. Sustainable bioenergy production from marginal lands in the US Midwest. *Nature*, 493, 514–517. https://doi.org/10.1038/nature11811

Global Bioenergy Partnership (GBEP) | Department of Economic and Social Affairs. (n.d.) https://sdgs.un.org/partnerships/global-bioenergy-partnership-gbep

Grande, L., Pedroarena, I., Korili, S.A., Gil, A., 2021. Hydrothermal liquefaction of biomass as one of the most promising alternatives for the synthesis of advanced liquid biofuels: A review. *Materials*, 14, 5286. https://doi.org/10.3390/MA14185286

Hannon, M., Gimpel, J., Tran, M., Rasala, B., Mayfield, S., 2010. Biofuels from algae: Challenges and potential importance & challenges of algal biofuels. *Biofuels* 1, 763–784.

Hassan, S.S., Williams, G.A., Jaiswal, A.K., 2019. Moving towards the second generation of lignocellulosic biorefineries in the EU: Drivers, challenges, and opportunities. *Renewable and Sustainable Energy Reviews*, 101, 590–599. https://doi.org/10.1016/J.RSER.2018.11.041

He, Y., Jaiswal, D., Liang, X.Z., Sun, C., Long, S.P., 2022. Perennial biomass crops on marginal land improve both regional climate and agricultural productivity. *GCB Bioenergy*, 14, 558–571. https://doi.org/10.1111/GCBB.12937

Hendry, J.I., Bandyopadhyay, A., Srinivasan, S., Pakrasi, H.B., Maranas, C.D., 2020. Metabolic model guided strain design of cyanobacteria. *Current Opinion in Biotechnology*, 64, 17–23. https://doi.org/10.1016/j.copbio.2019.08.011

Herrmann, I.T., Moltesen, A., 2015. Does it matter which life cycle assessment (LCA) tool you choose? – A comparative assessment of SimaPro and GaBi. *Journal of Cleaner Production*, 86, 163–169. https://doi.org/10.1016/J.JCLEPRO.2014.08.004

Hertel, T.W., Golub, A.A., Jones, A.D., O'Hare, M., Plevin, R.J., Kammen, D.M., 2010. Effects of US maize ethanol on global land use and greenhouse gas emissions: Estimating market-mediated responses. *Bioscience*, 60, 223–231. https://doi.org/10.1525/BIO.2010.60.3.8

Hill, J., Nelson, E., Tilman, D., Polasky, S., Tiffany, D., 2006. Environmental, economic, and energetic costs and benefits of biodiesel and ethanol biofuels. *Proceedings of the National Academy of Sciences of the United States of America*, 103, 11206–11210. https://doi.org/10.1073/PNAS.0604600103/SUPPL_FILE/04600TABLE9NEW.PDF

Hill, J., Polasky, S., Nelson, E., Tilman, D., Huo, H., Ludwig, L., Neumann, J., Zheng, H., Bonta, D., 2009. Climate change and health costs of air emissions from biofuels and gasoline. *Proceedings of the National Academy of Sciences of the United States of America*, 106, 2077–2082. https://doi.org/10.1073/PNAS.0812835106/SUPPL_FILE/0812835106SI.PDF

Home Page | Bio-based Industries Consortium (BIC) (n.d.) https://biconsortium.eu/

Hossain, N., Zaini, J., Indra Mahlia, T.M., 2019. Life cycle assessment, energy balance and sensitivity analysis of bioethanol production from microalgae in a tropical country. *Renewable and Sustainable Energy Reviews*, 115, 109371. https://doi.org/10.1016/J.RSER.2019.109371

How important is the LCA software tool you choose comparative results from GaBi, openLCA, SimaPro and Umberto. (n.d.) https://www.researchgate.net/publication/318217178_How_important_is_the_LCA_software_tool_you_choose_Comparative_results_from_GaBi_openLCA_SimaPro_and_Umberto

International Energy Agency [IEA]. (n.d.) https://www.iea.org/

ISCC System. Solutions for sustainable and deforestation free supply chains. (n.d.) https://www.iscc-system.org/

ITS comparing land use impacts of petroleum and biofuels – ITS (n.d.) https://its.ucdavis.edu/research/research-findings/comparing-land-use-impacts-of-petroleum-and-biofuels/

Jafri, Y., Ahlström, J.M., Furusjö, E., Harvey, S., Pettersson, K., Svensson, E., Wetterlund, E., 2022. Double yields and negative emissions? Resource, climate and cost efficiencies in biofuels with carbon capture, storage and utilization. *Frontiers in Energy Research*, 10, 797529. https://doi.org/10.3389/FENRG.2022.797529

Jamil, F., Inayat, A., Hussain, M., Akhter, P., Abideen, Z., Ghenai, C., Shanableh, A., Abdellatief, T.M.M., 2024. Valorization of waste biomass to biofuels for power production and transportation in optimized way: A comprehensive review. *Advanced Energy and Sustainability Research*, 5, 2400104. https://doi.org/10.1002/AESR.202400104

Jeswani, H.K., Chilvers, A., Azapagic, A., 2020. Environmental sustainability of biofuels: A review. *Proceedings of the Royal Society A*, 476. https://doi.org/10.1098/RSPA.2020.0351

Kaloudas, D., Pavlova, N., Penchovsky, R., 2021. Lignocellulose, algal biomass, biofuels and biohydrogen: A review. *Environmental Chemistry Letters*, 19, 2809–2824. https://doi.org/10.1007/S10311-021-01213-Y

Karka, P., Johnsson, F., Papadokonstantakis, S., 2021. Production costs of advanced biofuels using a multi-component learning curve model. *Computer Aided Chemical Engineering*, 50, 1937–1942. https://doi.org/10.1016/B978-0-323-88506-5.50300-4

Kazemi, N., Gholami Parashkoohi, M., Mohammadi, A., Mohammad Zamani, D., 2023. Environmental life cycle assessment and energy-economic analysis in different cultivation of microalgae-based optimization method. *Results in Engineering*, 19, 101240. https://doi.org/10.1016/J.RINENG.2023.101240

Ketzer, F., Skarka, J., Rösch, C., 2018. Critical review of microalgae LCA studies for bioenergy production. *Bioenergy Research*, 11, 95–105. https://link.springer.com/article/10.1007/s12155-017-9880-1

Klöpffer, W., Grahl, B., 2014. *Life cycle assessment (LCA): A guide to best practice*. 1–396. https://doi.org/10.1002/9783527655625

Kothri, M., Mavrommati, M., Elazzazy, A.M., Baeshen, M.N., Moussa, T.A.A., Aggelis, G., 2020. Microbial sources of polyunsaturated fatty acids (PUFAs) and the prospect of organic residues and wastes as growth media for PUFA-producing microorganisms. *FEMS Microbiology Letters*, 367. https://doi.org/10.1093/FEMSLE/FNAA028

Lark, T.J., Hendricks, N.P., Smith, A., Pates, N., Spawn-Lee, S.A., Bougie, M., Booth, E.G., Kucharik, C.J., Gibbs, H.K., 2022. Environmental outcomes of the US renewable fuel standard. *Proceedings of the National Academy of Sciences USA*, 119, e2101084119. https://doi.org/10.1073/PNAS.2101084119/SUPPL_FILE/PNAS.2101084119.SAPP.PDF

Lee, R.A., Lavoie, J.M., 2013. From first- to third-generation biofuels: Challenges of producing a commodity from a biomass of increasing complexity. *Animal Frontiers*, 3, 6–11. https://doi.org/10.2527/af.2013-0010

Li, P., Luo, Y., Yuan, X., 2022. Life cycle and techno-economic assessment of source-separated wastewater-integrated microalgae biofuel production plant: A nutrient organization approach. *Bioresource Technology*, 344, 126230. https://doi.org/10.1016/J.BIORTECH.2021.126230

Linares, R., López-Uceda, A., Piccinali, A., Martínez-Ruedas, C., Galvín, A.P., 2024. LCA applied to comparative environmental evaluation of aggregate production from recycled waste materials and virgin sources. *Environmental Science and Pollution Research*, 31, 44023–44035. https://doi.org/10.1007/S11356-024-33868-9

Liska, A.J., 2015. Chapter 10. Eight principles of uncertainty for life cycle assessment of biofuel systems. *Sustainable Biofuels* (pp. 243–268). https://doi.org/10.1515/9783110275896-014

Liu, M., Zhu, G., Tian, Y., 2024. The historical evolution and research trends of life cycle assessment. *Green Carbon*, 2, 425–437. https://doi.org/10.1016/J.GREENCA.2024.08.003

Lopes Silva, D.A., Nunes, A.O., Piekarski, C.M., da Silva Moris, V.A., de Souza, L.S.M., Rodrigues, T.O., 2019. Why using different life cycle assessment software tools can generate different results for the same product system? A cause–effect analysis of the problem. *Sustainable production and consumption*, 20, 304–315. https://doi.org/10.1016/J.SPC.2019.07.005

Manuel, J., 2007. Battle of the biofuels. *Environmental Health Perspectives*, 115, A92. https://doi.org/10.1289/EHP.115-A92

Mat Aron, N.S., Khoo, K.S., Chew, K.W., Show, P.L., Chen, W.H., Nguyen, T.H.P., 2020. Sustainability of the four generations of biofuels – A review. *International Journal of Energy Research*, 44, 9266–9282. https://doi.org/10.1002/er.5557

McKone, T.E., Nazaroff, W.W., Berck, P., Auffhammer, M., Lipman, T., Torn, M.S., Masanet, E., Lobscheid, A., Santero, N., Mishra, U., Barrett, A., Bomberg, M., Fingerman, K., Scown, C., Strogen, B., Horvath, A., 2011. Grand challenges for life-cycle assessment of biofuels. *Environmental Science & Technology*, 45, 1751–1756. https://doi.org/10.1021/ES103579C/ASSET/IMAGES/LARGE/ES-2010-03579C_0002.JPEG

Mohit, A., Mishel, V.K., Remya, N., 2024. Life cycle assessment and technoeconomic analysis of biofuels produced from polyculture microalgae cultivated in greywater. *Journal of Environmental Management*, 356, 120711. https://doi.org/10.1016/J.JENVMAN.2024.120711

Moutik, B., Summerscales, J., Graham-Jones, J., Pemberton, R., 2023. Life cycle assessment research trends and implications: A bibliometric analysis. *Sustainability*, 15, 13408. https://doi.org/10.3390/SU151813408

Muthu, S.S., 2020. Ways of measuring the environmental impact of textile processing: An overview. *Assessing the environmental impact of textiles and the clothing supply chain* (pp. 33–56). https://doi.org/10.1016/B978-0-12-819783-7.00002-8

Naik, S.N., Goud, V. V., Rout, P.K., Dalai, A.K., 2010. Production of first and second generation biofuels: A comprehensive review. *Renewable and Sustainable Energy Reviews*, 14, 578–597. https://doi.org/10.1016/j.rser.2009.10.003

Nanda, S., Pattnaik, F., Patra, B.R., Kang, K., Dalai, A.K., 2023. A review of liquid and gaseous biofuels from advanced microbial fermentation processes. *Fermentation*, 9, 813. https://doi.org/10.3390/FERMENTATION9090813

Narisetty, V., Reshmy, R., Maitra, S., Tarafdar, A., Alphy, M.P., Kumar, A.N., Madhavan, A., Sirohi, R., Awasthi, M.K., Sindhu, R., Varjani, S., Binod, P., 2022. Waste-derived fuels and renewable chemicals for bioeconomy promotion: A sustainable approach. *Bioenergy Research*, 16, 16–32. https://doi.org/10.1007/S12155-022-10428-Y

Nilsson, A., Shabestary, K., Brandão, M., Hudson, E.P., 2020. Environmental impacts and limitations of third-generation biobutanol: Life cycle assessment of n-butanol produced by genetically engineered cyanobacteria. *Journal of Industrial Ecology*, 24, 205–216. https://doi.org/10.1111/JIEC.12843

Ochsenreither, K., Glück, C., Stressler, T., Fischer, L., Syldatk, C., 2016. Production strategies and applications of microbial single cell oils. *Frontiers in Microbiology*, 7, 209534. https://doi.org/10.3389/FMICB.2016.01539

openLCA.org | openLCA is a free, professional life cycle assessment (LCA) and footprint software with a broad range of features and many available databases, created by GreenDelta since 2006 (2025) https://www.openlca.org/

Osman, A.I., Mehta, N., Elgarahy, A.M., Al-Hinai, A., Al-Muhtaseb, A.H., Rooney, D.W., 2021. Conversion of biomass to biofuels and life cycle assessment: A review. *Environmental Chemistry Letters*, 19, 4075–4118. https://doi.org/10.1007/S10311-021-01273-0

Pacañot, V.D.J., 2022. Evaluating environmental impacts with life cycle assessment. *Nature Reviews Earth & Environment*, 3, 224–224. https://doi.org/10.1038/s43017-022-00288-7

Patel, K., Singh, S.K., 2023. Environmental sustainability analysis of biofuels: A critical review of LCA studies. *Clean Technologies and Environmental Policy*, 25, 2489–2510. https://doi.org/10.1007/S10098-023-02596-Y

Pillai, S.M., Srivastava, R.K., Singh, S.P., 2024. Microbial induced biotechnological processes for biofuel production from waste organics conversion. *Journal of Pure and Applied Microbiology*, 18, 1593–1600. https://doi.org/10.22207/JPAM.18.3.08

Pimentel, D., Patzek, T.W., 2005. Ethanol production using corn, switchgrass, and wood; Biodiesel production using soybean and sunflower. *Natural Resources Research*, 14, 65–76. https://doi.org/10.1007/S11053-005-4679-8

Prakash, D., Verma, S., Bhatia, R., Tiwary, B.N., 2011. Risks and precautions of genetically modified organisms. *International Scholarly Research Notices*, 2011, 369573. https://doi.org/10.5402/2011/369573

Prasad, S., Singh, A., Korres, N.E., Rathore, D., Sevda, S., Pant, D., 2020. Sustainable utilization of crop residues for energy generation: A life cycle assessment (LCA) perspective. *Bioresource Technology*, 303, 122964. https://doi.org/10.1016/J.BIORTECH.2020.122964

Quiroz-Arita, C., Sheehan, J.J., Bradley, T.H., 2017. Life cycle net energy and greenhouse gas emissions of photosynthetic cyanobacterial biorefineries: Challenges for industrial production of biofuels. *Algal Research*, 26, 445–452. https://doi.org/10.1016/J.ALGAL.2017.06.021

Rafa, N., Ahmed, S.F., Badruddin, I.A., Mofijur, M., Kamangar, S., 2021. Strategies to produce cost-effective third-generation biofuel from microalgae. *Frontiers in Energy Research*, 9, 749968. https://doi.org/10.3389/FENRG.2021.749968

Raja, K., Shastri, Y., 2025. Sustainability assessment of succinic acid production from cyanobacteria. *Bioresource Technology*, 421, 132108. https://doi.org/10.1016/J.BIORTECH.2025.132108

Richard, T.L., 2010. Challenges in scaling up biofuels infrastructure. *Science (1979)*, 329, 793–796. https://doi.org/10.1126/SCIENCE.1189139

Rueda, E., Senatore, V., Zarra, T., Naddeo, V., García, J., Garfí, M., 2023. Life cycle assessment and economic analysis of bioplastics production from cyanobacteria. *Sustainable Materials and Technologies*, 35, e00579. https://doi.org/10.1016/J.SUSMAT.2023.E00579

Saranya, G., Ramachandra, T. V., 2020. Life cycle assessment of biodiesel from estuarine microalgae. *Energy Conversion and Management: X*, 8, 100065. https://doi.org/10.1016/J.ECMX.2020.100065

Scarlat, N., Dallemand, J.F., Monforti-Ferrario, F., Nita, V., 2015. The role of biomass and bioenergy in a future bioeconomy: Policies and facts. *Environment and Development*, 15, 3–34. https://doi.org/10.1016/j.envdev.2015.03.006

Schmer, M.R., Vogel, K.P., Mitchell, R.B., Perrin, R.K., 2008. Net energy of cellulosic ethanol from switchgrass. *Proceedings of the National Academy of Sciences USA*, 105, 464–469. https://doi.org/10.1073/PNAS.0704767105/SUPPL_FILE/INDEX.HTML

Searchinger, T., Heimlich, R., Houghton, R.A., Dong, F., Elobeid, A., Fabiosa, J., Tokgoz, S., Hayes, D., Yu, T.H., 2008. Use of U.S. croplands for biofuels increases greenhouse gases through emissions from land-use change. *Science (1979)*, 319, 1238–1240. https://doi.org/10.1126/SCIENCE.1151861/SUPPL_FILE/SEARCHINGER.SOM.PDF

Shah, W.U.H., Hao, G., Yan, H., Zhu, N., Yasmeen, R., Dincă, G., 2024. Role of renewable, non-renewable energy consumption and carbon emission in energy efficiency and productivity change: Evidence from G20 economies. *Geoscience Frontiers*, 15, 101631. https://doi.org/10.1016/J.GSF.2023.101631

Shrestha, S., Fonoll, X., Khanal, S.K., Raskin, L., 2017. Biological strategies for enhanced hydrolysis of lignocellulosic biomass during anaerobic digestion: Current status and future perspectives. *Bioresource Technology*, 245, 1245–1257. https://doi.org/10.1016/j.biortech.2017.08.089

SimaPro Craft - PRé Sustainability (n.d.) https://pre-sustainability.com/solutions/tools/simapro-craft/.

Singh, A., Olsen, S.I., 2012. Key issues in life cycle assessment of biofuels. *Green Energy and Technology*, 62, 213–228. https://doi.org/10.1007/978-1-4471-2324-8_11

Singhal, A., Prashant, P., 2020. Biotechnology for biofuels: A sustainable green energy solution. https://doi.org/10.1007/978-981-15-3761-5

Soleymani Angili, T., Grzesik, K., Rödl, A., Kaltschmitt, M., 2021. Life cycle assessment of bioethanol production: A review of feedstock, technology and methodology. *Energies*, 14, 2939. https://doi.org/10.3390/EN14102939

Subramaniam, R., Dufreche, S., Zappi, M., Bajpai, R., 2010. Microbial lipids from renewable resources: Production and characterization. *Journal of Industrial Microbiology & Biotechnology*, 37, 1271–1287. https://doi.org/10.1007/S10295-010-0884-5

Sun, C.H., Fu, Q., Liao, Q., Xia, A., Huang, Y., Zhu, X., Reungsang, A., Chang, H.X., 2019. Life-cycle assessment of biofuel production from microalgae via various bioenergy conversion systems. *Energy*, 171, 1033–1045. https://doi.org/10.1016/J.ENERGY.2019.01.074

Swathe Sriee, A.E., Sharma, Y., Ranjitha, J., Shankar, V., 2024. A comparative assessment of microbial biodiesel and its life cycle analysis. *Folia Microbiologica*, 69, 521–547. https://doi.org/10.1007/S12223-024-01153-4

Tan, W.H., Chong, S., Fang, H.W., Pan, K.L., Mohamad, M., Lim, J.W., Tiong, T.J., Chan, Y.J., Huang, C.M., Yang, T.C.K., 2021. Microbial fuel cell technology—A critical review on scale-up issues. *Processes*, 9, 985. https://doi.org/10.3390/PR9060985

Teillard, F., Maia de Souza, D., Thoma, G., Gerber, P.J., Finn, J.A., 2016. What does life-cycle assessment of agricultural products need for more meaningful inclusion of biodiversity? *Journal of Applied Ecology*, 53, 1422–1429. https://doi.org/10.1111/1365-2664.12683

U.S. Life Cycle Inventory Database | Energy Analysis | NREL. (n.d.) https://www.nrel.gov/analysis/lci.html

Ubando, A.T., Anderson, S., Ng, E., Chen, W.H., Culaba, A.B., Kwon, E.E., 2022. Life cycle assessment of microalgal biorefinery: A state-of-the-art review. *Bioresource Technology*, 360, 127615. https://doi.org/10.1016/J.BIORTECH.2022.127615

Umberto Life Cycle Assessment Software (LCA) (n.d.) https://www.environmental-expert.com/software/umberto-life-cycle-assessment-software-lca-372395

Understanding System Boundaries in Life Cycle Assessment. (n.d.) https://sustained.com/blog/understanding-system-boundaries-in-life-cycle-assessment

USDA Publishes Interim Rule on Technical Guidelines for Climate-Smart Agriculture Crops Used as Biofuel Feedstocks | s (n.d.) https://www.usda.gov/about-usda/news/press-releases/2025/01/15/usda-publishes-interim-rule-technical-guidelines-climate-smart-agriculture-crops-used-biofuel

Vance, C., Sweeney, J., Murphy, F., 2022. Space, time, and sustainability: The status and future of life cycle assessment frameworks for novel biorefinery systems. *Renewable and Sustainable Energy Reviews*, 159, 112259. https://doi.org/10.1016/J.RSER.2022.112259

Vidic, R.D., Brantley, S.L., Vandenbossche, J.M., Yoxtheimer, D., Abad, J.D., 2013. Impact of shale gas development on regional water quality. *Science (1979)*, 340. https://doi.org/10.1126/SCIENCE.1235009/SUPPL_FILE/1235009.MP3

Villacreses-Freire, D., Ketzer, F., Rösch, C., 2022. Advanced metabolic engineering approaches and renewable energy to improve environmental benefits of algal biofuels: LCA of large-scale biobutanol production with Cyanobacteria synechocystis PCC6803. *Bioenergy Research*, 15, 1515–1530. https://doi.org/10.1007/S12155-021-10323-Y

Voloshin, R.A., Rodionova, M. V., Zharmukhamedov, S.K., Nejat Veziroglu, T., Allakhverdiev, S.I., 2016. Review: Biofuel production from plant and algal biomass. *International Journal of Hydrogen Energy*, 41, 17257–17273. https://doi.org/10.1016/j.ijhydene.2016.07.084

Wei, C., Xu, Y., Li, Y., Wei, W., Feng, Y., Li, Z., Xu, L., 2024. Life-cycle assessment of microalgae liquid biofuel production in biofilm cultivation system via conversion technologies of transesterification, hydrothermal liquefaction and pyrolysis. *Journal of Cleaner Production*, 436, 140559. https://doi.org/10.1016/J.JCLEPRO.2024.140559

What are the environmental impacts of biofuel production? (n.d.) https://www.greenfacts.org/en/biofuels/l-3/4-environmental-impacts.htm

Wu, Y., Zhao, F., Liu, S., Wang, L., Qiu, L., Alexandrov, G., Jothiprakash, V., 2018. Bioenergy production and environmental impacts. *Geoscience Letters*, 5, 1–9. https://doi.org/10.1186/S40562-018-0114-Y

Yung, W.K.C., Muthu, S.S., Subramanian, K., 2018. Carbon footprint analysis of printed circuit board. *Environmental carbon footprints: Industrial case studies* (pp. 365–431). https://doi.org/10.1016/B978-0-12-812849-7.00013-1

Zhang, D., An, S., Yao, R., Fu, W., Han, Y., Du, M., Chen, Z., Lei, A., Wang, J., 2022. Life cycle assessment of auto-tropically cultivated economic microalgae for final products such as food, total fatty acids, and bio-oil. *Frontiers in Marine Science*, 9, 990635. https://doi.org/10.3389/FMARS.2022.990635/BIBTEX

12 Integration of Biofuel Production with Other Biorefinery Processes

S. Sree Priya, C. Gayathri, and Rakesh K. Yasarla

12.1 INTRODUCTION

Fossil fuels, such as diesel, petrol, and natural gas, account for approximately 84% of the world's energy needs (Kumar et al., 2023). For decades, humanity has depended on these energy sources to power industries, transportation, and homes (Azni et al., 2023). However, if this level of consumption continues, it could lead to a severe energy crisis in the near future. Fossil fuel usage not only depletes these finite resources but also contributes to the release of harmful pollutants like carbon monoxide, sulfur oxides, and volatile organic compounds. These toxic pollutants negatively impact the human wellness and the nature (Kumar et al., 2023).

One of the most concerning consequences of long-term fossil fuel consumption is the increase in carbon emissions, particularly carbon dioxide (CO_2), which is a leading greenhouse gas. These emissions have played a significant role in accelerating climate change, which has led to rising global temperatures, melting ice caps, and more extreme weather patterns. As fossil fuels become increasingly scarce, their cost is expected to rise, making them less affordable and accessible to poorer populations, particularly those in remote or marginalized areas. This economic strain disproportionately affects communities that are already vulnerable (Azni et al., 2023).

In 2010, global emissions from fossil fuel combustion amounted to 54 billion tons of carbon equivalent. This figure is projected to reach 70 billion tons by 2050 if current trends continue (Hossain et al., 2018). Although fossil fuel-based electricity generation has historically supported economic growth, it remains a significant contributor to global greenhouse gas emissions, accounting for over one-third of total emissions world (Bello et al., 2018). The environmental impacts of these emissions are far-reaching, including air pollution, damage to the ozone layer, and the exacerbation of global warming, which causes heat waves, floods, and other extreme weather events (Azni et al., 2023).

As the negative impacts of fossil fuel use become more evident, there is growing global support for renewable energy sources like wind, solar, and hydroelectric power. These clean energy alternatives are crucial for reducing emissions, combating climate change, and ensuring a sustainable energy future. Shifting from fossil fuels to renewables is essential for both environmental protection and long-term energy

DOI: 10.1201/9781003633709-12

security (Azni et al., 2023; Cherwoo et al., 2023). Reports highlight countries such as China, the United States, and India are the largest carbon emitters (Cherwoo et al., 2023). As a result, there is increasing interest in biofuels, which are derived from organic materials, such as plants and algae. Biofuels emit less carbon than traditional fuels, making them a more environmentally friendly option to replace fossil fuels and reduce global warming (Mat et al., 2020).

The biorefinery concept involves a concept of sustainability and circular bioeconomy by creating a wide spectrum of products from various biomass feedstocks like agricultural residues, forest waste, industrial waste, and marine biomass. The industrial facility for biorefinery concepts can be a new facility creation or establishing with an existing facility with new unit operations for the complete utilization of the feedstock or by introducing new feedstock for recycling through circular and sustainable processes. Figure 12.1 illustrates the integrated circular biorefinery concept where conversion of various sustainable biomass to value-added products is shown. Biorefineries are classified into energy, product, and value-added biochemicals based on the final products and the conversion of the technologies and feedstock of the biomass. Biorefinery is an advancing tool for reducing the dependency on fossil fuel to achieve sustainability and circular bioeconomy, yet there are several challenges, including the operational procedures, feedstock supply, productivity, and financial feasibility in establishing the integrated commercial facility for large-scale production. Several concepts that align with biorefinery are being developed lately, and their implementations are underway, which plays a crucial role in establishing a bioeconomy.

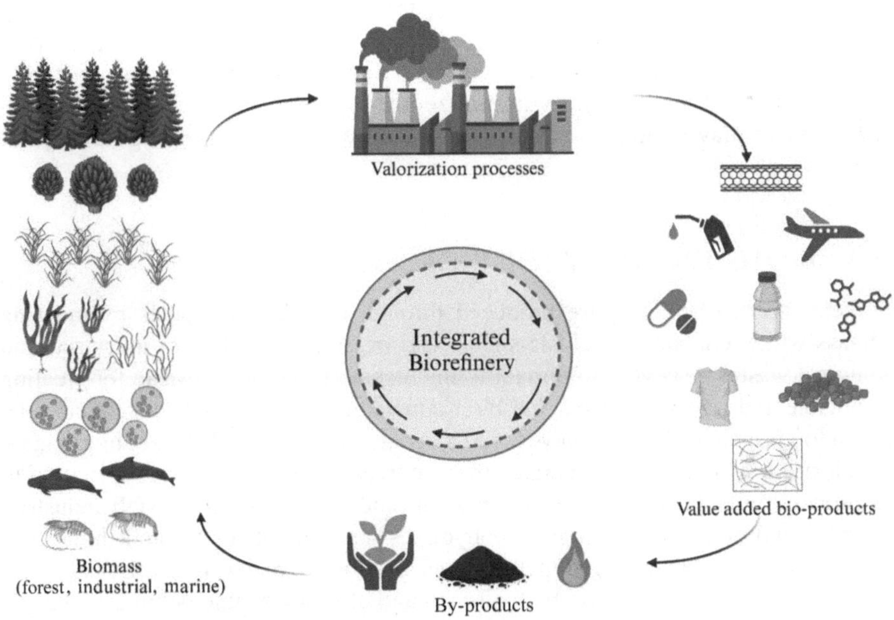

FIGURE 12.1 Integrated circular biorefinery process.

12.2 CIRCULAR BIOECONOMY – BIOMASS AND BIOFUELS

The concept of circular bioeconomy is based on renewable, biological materials and aims at enhancing the core value of the economy by utilizing the resources completely in a loop, maximizing the secondary biomass utilization and resources or new materials produced without depleting the existing natural resources. A sustainable, low carbon bioeconomy with maximum product recovery from biological resources is one of the key principles of a circular bioeconomy. Although in a circular bioeconomy virgin resources are fed with the waste streams in the production cycle, the circular flow is infinite and mimics the natural circle of life. Biomass from agriculture, forest, and industrial waste are the major raw materials. The majority of biomass is processed into feed and food products and smaller portions are converted to bioproducts and biofuels like bioethanol and bioplastics. Biobased products and energy are expected to grow stronger compared to food and feed for the circular bioeconomy. In chemical processing industries biobased chemicals often provide a vital precursor that takes a crucial role in the circular bioeconomy (Korhonen et al., 2018).

Circular Bioeconomy with innovative technologies for the valorization of biomass toward biobased products play a pivotal role in achieving SDGs. Biorefinery integrated with other industries provides a synergy for ensuring sustainable energy (SDG 7) combating climate change (SDG 13), and responsible production and consumption (SDG 12). Biorefineries are often classified based on their conversion route for the product development and biomass utilized (Allen et al., 2018).

According to international agencies like the World Economic Forum (WEF), the International Energy Agency (IEA), and the Organization for Economic Co-operation and Development (OECD), the concept of biorefineries gaining attention from researchers, stakeholders and industries. Biorefineries are perceived as a solution for industries that face hurdles in a global economy that is constantly changing and offers a way to minimize the reliance on conventional fossil fuels and can play a vital role in efficient resource utilization through a circular economy.

12.3 BIOTECHNOLOGY FOR BIOFUELS

Biofuels and biochemicals are produced through fermentative routes by utilizing biomass which contains cellulosic sugars and organic content from industrial and municipal wastes. It is very important to understand the value of waste for creating sustainable and renewable fuels and bioproducts that are essential for reducing the carbon footprint and also to achieve the SDG goals (Smith et al., 2017). In biorefinery, the valorization of the waste generated from sources of the industrial and agricultural, the process unit operations like pretreatment and recalcitration through industrial enzymes, and conversion of the carbohydrates and derivatives through biological processes like fermentation and methanogenesis to make renewable fuels – i.e., bioethanol, biobutanol, biogas and biohydrogen and also degradable polymeric materials like PHA's and PHB's, and composite materials like fiber composite materials for aerospace and fire retardant applications.

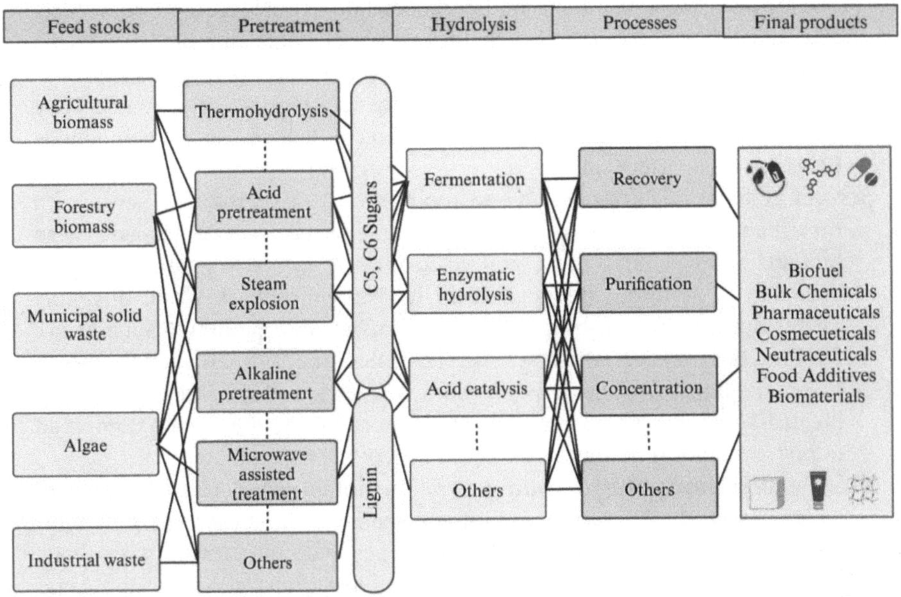

FIGURE 12.2 Biorefinery process integrated with pretreatment technologies for production of biofuels and biochemicals.

With the advancement in the concepts of recycling, reuse and reduce, the business potential for the development of biorefining integrated with separation technologies has been developed to improve the production processes and make it competitive with the existing conventional technologies for waste based biorefining. Biorefineries utilizing microalgae are a fascinating case in this category, as they contribute to innumerable possibilities for the manufacturing of energy and nonenergy goods, such as biodiesel from microalgae. The current production from biobased refinery focuses on a single product, and this approach does not fall under the true biorefinery concept, as the concept suggests, to synthesize multiple products from a production facility. Figure 12.2 describes the biorefinery processes integrated with pretreatment technologies for the production of biofuels and biochemicals.

12.3.1 Classification of Biofuels Production

Biofuels are classified into four generations – first, second, third, and fourth – based on their feedstocks, and are often referred to as advanced biofuels (Lamichhane et al., 2021). While first- and second-generation biofuels are the most commonly used, third- and fourth-generation biofuels have gained attention since 2010 and are currently in the developmental stage (Ale et al., 2019; Chavan et al., 2024).

First-generation biofuels: First-generation biofuels are derived from edible feedstocks, which are divided into starch-based (e.g., potato, barley, corn, wheat) and sugar-based (e.g., sugarcane, sugar beet) categories (Mat et al., 2020).

The most common feedstocks for first-generation biofuels are corn, sugarcane, and wheat. These biofuels include biogas, bio-alcohols, and biodiesel (Saha et al., 2018). First-generation biofuels have been widely commercialized, contributing around 50 billion liters to global biofuel production annually. They are seen as a promising alternative to fossil fuels, as their use can help reduce CO_2 emissions (Mat et al., 2020).

Second-generation biofuels: Due to food security concerns associated with first-generation biofuels, second-generation biofuels were developed. These biofuels, often referred to as "cellulosic ethanol" or "olive greens," are produced from nonfood, lignocellulosic feedstocks such as waste oils, forest residues, industrial by-products, and sustainable biomass (Saha et al., 2018). Common feedstocks for second-generation biofuels include wood, sugarcane bagasse, organic waste, corn stover, rice straw, and fast-growing plants like poplar. These biofuels offer a more sustainable option compared to first-generation biofuels (Ale et al., 2019).

Third-generation biofuels: Third-generation biofuels are derived from algae, which can produce a variety of biofuels, including biodiesel, ethanol, butanol, etc. Algae biofuels offer higher efficiency, with output nearly ten times greater than that of second-generation biofuels (Saha et al., 2018). Algal cultivation is environmentally friendly as it requires minimal land area, produces high oil content, and absorbs atmospheric CO_2, contributing to a balanced ecosystem (Mat et al., 2020).

Fourth-generation biofuels: Fourth-generation biofuels are the outcome of progress in genetic engineering and plant biotechnology, especially in the fields of carbon capture and storage (Chavan et al., 2024). These biofuels are produced from genetically modified microorganisms, such as microalgae, designed to enhance carbon dioxide absorption for photosynthesis and increase biofuel production. This innovative approach aims to create artificial carbon sinks and further improve the efficiency of biofuel production (Mat et al., 2020).

12.4 MICROBIAL PRODUCTION OF BIOFUELS

Bacteria, yeast, fungi, and other microorganisms are essential in the production of biofuels and biochemicals. For example, fungi produce organic acids, yeast generates alcohol and lipids, bacteria contribute to alcohols and biochemicals, and microalgae produce lipids and carotenoids. Advances in molecular biology have made it possible to engineer microbial metabolism more efficiently, improving product yields, substrate use, and redox balance. Technologies like high-throughput sequencing and genome editing have expanded the potential for microbial engineering. The growth of omics fields (such as metabolomics and proteomics) has enhanced our understanding of cellular processes, while metagenomics and microbiomics are helping direct microbial consortia engineering. These data-driven methods are particularly suited for advancing our knowledge and manipulation of biological systems (Decker et al., 2023). Table 12.1 illustrates the various biofuels production from different sources of the waste stream through microbial processes.

TABLE 12.1
Biofuels Production Through Microbial Processes

Biofuel	Feedstocks Used	Microorganisms Involved	References
Bioethanol	sucrose containing feedstocks (sugarcane, sugar beet, and sweet sorghum), starchy compounds (rice, wheat, corn, and barley), and cellulosic biomass (wood, forestry residue, straw, and grasses)	*S. cerevisiae* *Pichia stipites* *Candida shehatae* *Pachysolan tannophilus* *Candida brassicae* *Zymomonas mobilis* *Mucor indicus* *Escherichia coli* *Clostridium thermocellum*	Toor et al. (2020) Chavan et al. (2024)
Biodiesel	vegetable sources (oil seed grains and plant fruits), vegetable oils, petroleum oils, or animal fat	Molds Bacteria *(Streptomyces, Rhodococcus, Nocardia)* Yeasts *(Rhodotorula, Rhizopus, Yarrowia)* and Microalgae act as the lipid sources.	Elgharbawy et al. (2021) Uthandi et al. (2022)
Biogas	energy crops, plant biomass, agricultural residues such as animal manure and crop residues, industrial residues particularly from the food and beverage industries, wastewater treatment and sewage sludge, municipal organic waste, biowaste such as that from paper processing, and other organic waste sources	Methanogens (*Methanosarcina, Methanosaeta Methanococcus*, etc.)	Dölle et al. (2020)
Biobutanol	agricultural residues, energy crops, cellulosic biomass (wood, forestry residue, straw, and grasses), corn, sweet sorghum, molasses	*Escherichia coli* *Corynebacterium glutamicum* *Bacillus subtilis* *Saccharomyces cerevisiae.*	Su et al. (2020)

12.4.1 Integrated Biorefinery for Biofuels and Bioproducts

Integrated biorefineries play a vital role in utilizing and promoting natural resources and these multifunctional facilities integrate various techniques for the extraction and valorization of the biofuels and biochemicals from biomass. Biorefineries encounter several operational challenges due to variations in the feedstock's availability and complexities of the biomass compositions. To overcome these challenges, both technical and economic with continuous innovation and strategic planning are essential for sustainable solutions. One of the challenges is the pretreatment of the biomass to extract the fibers, cellulosic sugars, and lignin. The composition of the cellulose, hemicellulose, and lignin vary from feedstock to feedstock.

Lignocellulosic biomass (LCB) is an ideal feedstock for second-generation bioethanol, offering a strong alternative to first-generation crops. Biofuel and bioproducts from LCB is a promising approach for the production processes, particularly in countries with abundant agricultural and forestry waste (Devi et al., 2022). The production of biofuels from LCB involves two main processes: thermochemical and biochemical. The thermochemical process includes methods like pyrolysis and gasification to produce synthesis gas ($CO + H_2$), which is then used to generate ethanol and co-products with a catalyst. In the biochemical process, microorganisms and enzymes convert lignocellulosic components into sugars, which are then fermented to produce ethanol. The biochemical process involves four steps: pretreatment, hydrolysis, fermentation, and distillation (Devi et al., 2022).

Globally, paper is produced by virgin paper or recycled paper produced from reprocessed pulp or recycled paper; paper recycling minimizes the natural resource consumption and decreases greenhouse gas emissions. Paper waste has an added advantage for the biorefinery industry, as the processes are similar and no additional pretreatments are required, and the biofuels like bioethanol and biobutanol production are possible with not many changes in the existing facilities. Paper rejects and paper sludge is used for the production of the chemical like lysine and bioethanol (Suvachan et al., 2021).

For establishing the consolidated biorefinery with other industries at an industrial scale, critical parameters need to be identified, including raw materials availability and their sustainable production, process design technical and economical evaluations, risk and sensitivity for the feasibility of the plant design, and also integration with other existing industries. The innovation in the development of the products through biorefinery routes should be more eco-friendly, less toxic, sustainable, low-energy consumption, and with new applications.

12.4.2 Pretreatment and Hydrolysis of Biomass

The LCB has a complex and rigid structure, which makes it difficult to access and convert into usable sugars. Hence the pretreatment of LCB is a crucial step in processing for the complete utilization of the feedstock. Various pretreatment methods – physical, chemical, physicochemical, and biological – are employed to make the biomass more accessible for further processing (Devi et al., 2022; Chavan et al., 2024).

Pretreatment methods for biomass include physical, chemical, physicochemical, and biological approaches. Physical methods increase surface area and reduce cellulose crystallinity. Chemical methods use acids or alkalis to break down hemicellulose. Physicochemical methods, like steam explosion, combine physical and chemical treatments to disrupt biomass structure. Biological methods use microorganisms to degrade lignocellulosic material with enzymes. These strategies enhance biomass accessibility, making it easier to convert into sugars for fermentation or other uses (Devi et al., 2022).

The acid pretreatment process hemicellulose is removed, and lignin structure is disrupted, which increases the porosity of the cellulose fibers for the subsequent enzymatic hydrolysis. Acid pretreatment is operated with a wide range of acids, including HCl, H_2SO_4 and also organic acids. The major drawback of the acid pretreatment is during the hydrolysis reaction C5 sugars degrade to furans and furfurals through acid catalyzed reaction. Organic acids and condensation products like lignin also formed during acid pretreatment, which act as inhibitory components during the fermentation process.

The organosolv pretreatment is widely adopted for the biorefinery processes for biofuels production from the LCB. An organic solvent in the pretreatment process disrupts the linkage between the hemicellulose and lignin and enhances the accessibility for the enzymes for the recalcitration of the cellulose pulp. The process is considered an advantageous process due to production of cellulose as residue and hemicellulose derivatives and lignin fractions in aqueous phase. The process is demonstrated with different technical names and was scaled from pilot to industrial scale. Each fraction that is isolated should be able to undergo processes that upgrade them into marketable end products and the development of such processes in an industrial scale remains the key to success for organosolv pretreatments.

During hydrolysis, complex polysaccharides, like cellulose or starch, are broken down into simpler monosaccharides, such as glucose, by the action of acids or enzymes. This process is crucial, as it converts the larger, more complex sugar molecules into fermentable sugars, making them accessible for fermentation. By breaking down these polysaccharides into monosaccharides, the material becomes suitable for further fermentation, where microorganisms can convert these sugars into ethanol or other biofuels (Chavan et al., 2024).

12.4.2.1 Enzymatic Hydrolysis

Enzymatic hydrolysis is a crucial step in biofuel production, especially in second-generation biofuels derived from LCB. It involves the breakdown of complex carbohydrates into fermentable sugars using enzymes. These sugars are then converted into biofuels like ethanol, butanol, and biogas. Enzymatic hydrolysis involves enzymes breaking down polysaccharides (cellulose, hemicellulose, starch) into monosaccharides (glucose, xylose, arabinose). During enzymatic hydrolysis, specific enzymes catalyze the breakdown of polysaccharides into fermentable sugars, and these sugars are later used by microorganisms for fermentation, and sugars are fermented into ethanol, butanol, or other biofuels.

Different types of enzymes target different components of biomass, such as cellulases, which are responsible for cellulose hydrolysis; examples of cellulases are endoglucanases, which break internal bonds of cellulose chains; exoglucanases (cellobiohydrolases), which release cellobiose units from cellulose chains; and β-Glucosidases, which converts cellobiose into glucose for fermentation. Hemicellulose hydrolysis is done by hemicellulases such as xylanases, which are responsible for the breakdown of xylan (major hemicellulose component) into xylose; mannanases, which can degrade mannans into mannose; and arabinofuranosidases, which releases arabinose from hemicellulose. Hydrolysis of starch is performed by amylases, such as α-amylase, which are responsible for hydrolysis of starch into maltose and dextrins and glucoamylase, which can convert maltose to glucose.

Advantages of enzymatic hydrolysis includes its higher efficiency compared to acid hydrolysis; enzymatic hydrolysis is more selective and produces fewer inhibitors; the environmentally friendly nature of enzymes are another reason for promoting enzymes; the use of milder conditions (pH 4.5–5.5, 45°C–55°C) compared to acid/alkaline treatments makes them a best choice for a sustainable future, and the enzymatic process is compatible with fermentation, as they produce clean sugars that microbes can easily ferment into biofuels.

12.4.3 Bioethanol Production in Integrated Biorefinery

Global biofuel production, particularly bioethanol, has risen in recent years, with bioethanol making up about 85% of biofuels used in transportation, cosmetics, pharmaceuticals, and more. It is an ideal alternative to gasoline due to its high burning limit, speed, evaporation temperature, and low greenhouse gas emissions (Kazmi et al., 2025; Igwebuike et al., 2024). Bioethanol is commonly blended with gasoline, with blends like E85 in the United States, E20-E25 in Brazil, and E15 in the Netherlands (Jain and Kumar, 2024). It can also be used in diesel engines through dual fuel systems, fumigation, or biodiesel blends (Kazmi et al., 2025). As fossil fuels contribute to environmental harm, bioethanol offers a cleaner alternative, reducing SO_2 and particulate matter emissions (Jain and Kumar, 2024).

Bioethanol is produced by the fermentative action of microorganisms (yeasts and bacteria) which have the potential to utilize pentose and hexose sugars into ethanol. For the production of bioethanol, mainly separate hydrolysis fermentation (SHF) and simultaneous saccharification and fermentation (SSF) are used. In SHF, saccharification breaks down cellulose into fermentable sugars, followed by fermentation to produce ethanol. SSF combines both steps, leading to higher ethanol yields and lower production costs, but requires optimal conditions, especially temperature. The organism used for ethanol production can be sensitive to lignocellulosic hydrolysate, influenced by fermentation conditions and bacterial strains (Chavan et al., 2024).

Yeast species like *Saccharomyces cerevisiae*, and *Pichia stipitis*, along with bacteria such as *Clostridium thermocellum*, and *Zymomonas mobilis*, are the efficient ethanol producers (Chavan et al., 2024). *S. cerevisiae* has been used for centuries in brewing and winemaking due to its high ethanol yield, productivity, and tolerance to high ethanol concentrations, lowering distillation costs. Today, yeasts are also used to produce fuel ethanol from renewable sources. *S. cerevisiae* is most commonly used in industrial production due to its tolerance to varying pH levels, reducing infection risks, while baker's yeast is favored for its low cost and availability (Mohd Azhar et al., 2017).

Following fermentation, ethanol is often mixed with other impurities, requiring a separation process (Chavan et al., 2024). Distillation is used to efficiently separate a specific substance, such as ethanol, from a miscible liquid mixture like fermentation broth, through a series of selective evaporation and condensation steps based on differences in volatility (Broda et al., 2022). In the distillation process, volatile compounds are heated and evaporated to isolate bioethanol from the remaining components of the mixture (Chavan et al., 2024). The resulting product, which contains about 37% bioethanol, is then further concentrated in a rectification column to around 95% ethanol (Broda et al., 2022). To achieve even higher purity, ethanol is dehydrated, producing a final product with at least 99.5% ethanol by volume, making it suitable for various applications, including fuel and industrial uses (Chavan et al., 2024; Broda et al., 2022).

12.4.4 Microbial Production of Biobutanol

Butanol has almost similar physical and chemical characteristics to gasoline. Butanol is a four-carbon molecule with the presence of hydroxyl group (–OH) attached to one of the carbon atoms, which creates four different butanol isomers of 1-butanol (n-butanol), 2-butanol, isobutanol, and tert- butanol. Selective microorganisms and engineered metabolisms alcohols like n-butanol, iso-butanol, and iso-pentanol can be produced from feedstocks like LCB. Under anaerobic conditions, biological production of acetone, butanol, and ethanol are produced simultaneously and termed ABE (acetone, butanol, and ethanol) fermentation. ABE fermentation has a great industrial production history and is outcompeted by the fossil fuel industry (Birgen et al. 2019; Lu et al., 2022). Butanol is a colorless flammable liquid and can be burned in the presence of oxygen, and the characteristics are very similar to gasoline, making it an alternative and sustainable biofuel to utilize in the automobile industry either by utilizing it directly or by blending with gasoline. Studies also show blending butanol with diesel reduces soot emissions due to the presence of the oxygen molecules. Due to its higher heating value and calorific value of 29.2 MJ dm^3, less ignition issues and melting point at −89.5°C and low volatility, it is advantageous over ethanol (Yusoff et al., 2018).

Clostridia are rod-shaped, strictly anaerobic, Gram positive and forming spore when introduced to nonfavorable conditions with complex life cycle and Clostridium species – i.e., *C. beijerinckii*, *C. acetobutylicum*, *C. beijerinckii*, *C. sporogenes* , *C. butylicum*, *C. acetobutylicum* ATCC 824, *C. acetobutylicum* EA2018, *C. acetobutylicum* DSM 1731, *C. acetobutylicum*, *C. beijerincki*, *C. saccharoperbutylacetonicum*, and *C. saccharobutylicum* were studied extensively with the carbohydrate sources of the lignocellulosic hydrolyzates and crude glycerol for the production of the butanol. The efficient utilization of C5 and C6 sugars was successful for the carbon utilization with the mixed sugar fermentation processes of batch, fed batch and also immobilized cells. For the efficient utilization of the substrate by microbes to improve fermentation and to yield high butanol, the responsible genes for carbon catabolite repression were disrupted and overexpressed with CRISPR Cas9 genome engineering systems (Lu et al., 2022).

Lignocellulosic hydrolyzates during pretreatment releases several inhibitory components also called lignocellulose derived microbial inhibitory compounds (LDMICs), and are classified into weak acids, phenols and furfurals, which are toxic against ABE production. It is observed that clostridia species have different tolerance and stress responses to these inhibitory components. Derivatives from lignin like vanillic acid, ferulic acid, and poly phenolic organic components are more toxic to clostridia species. The combination of the acids, phenolic components and furan aldehydes have a synergistic effect on the butanol fermentation (Lu et al., 2022; Birgen et al., 2019). Butanol production through microbial processes for the efficient yield microbes possess strong physiological and metabolic capabilities during the fermentation process. Some species also change intracellular pH toward the toxic effects of the acids generated during pretreatment and also produced during fermentation processes, and it is observed that higher APT levels divert carbon flux for butanol formation.

Industrial production of the biobutanol through LCB has advantages due to a wide variety of sugar sources from disaccharides to monosaccharides and polysaccharides like starch and cellulose and is an economical and effective strategy for the integrated and circular bioeconomy (Birgen et al., 2019;). In recent times, due to the exceptional compatibility of biobutanol with the combustion engines, along with the bioethanol, an extensive significance was achieved. Effluents from the grain, agro and food processing, paper and oil industries used as an affordable fermentation feedstock. Effluents from agricultural and food processing generate waste like palm oil effluents, crude palm oil, apple pomace, sugar cane, cassava bagasse, soybean hull and corn con bagasse are studied with the clostridia species for the biobutanol production in ABE fermentation. The consolidated bioprocess was developed where single microbe or microbial consortium in a single fermentation process with single carbon source or co-cultivation with mixed sugars were developed for higher yields of the biobutanol production (Palaniswamy et al., 2023;; Riaz et al., 2022). Microalgal lipids can be directed toward biodiesel production via the transesterification reaction, where triacyl glycerides (TAGs) are converted into fatty acid methyl esters (FAMEs, i.e., biodiesel). Meanwhile, microalgal carbohydrates can be directed toward bioethanol or biobutanol production via fermentative processes (Figueroa-Torres et al., 2020).

Clostridium tyrobutyricum Δcat1:adhE2 was used for the production of the biobutanol from the paper mill sludge and corn steep liquor. *Clostridium tyrobutyricum Δcat1:adhE2* strain was developed through genome engineering with the CRISPR Cas system. Paper sludge and corn steep liquor enhanced the butanol, and both feedstock sources are alternative nitrogen sources for fermentation (Cao et al., 2020). With coffee being one of the most important beverages globally, its cultivation and process generates huge amounts of residues from husk, pulp, and spent coffee grounds, which are rich in bioactive components and have potent antioxidant, anticarcinogenic, and antitumor properties. The pretreated spent coffee grounds under optimal conditions resulted in cellulosic material. The cellulosic material was enzymatically hydrolyzed by Cellic Ctec2 and Pectinex Ultra SPl, a cellulolytic and beta-glucosidase enzyme to break the complex sugars into monomers. Further, ABE fermentation with C. beijerinckii for 48 h resulted in achieving biobutanol (López-Linares et al. 2021).

12.4.5 Microbial Production of Organic Acids

12.4.5.1 Lactic Acid

Lactic acid is an important organic acid and is known to hold a wide range of industrial applications. The market for lactic acid is growing enormously due to its wide range of applications. The LCB undergoes processes of fractionation, separation, and hydrolysis, which results in raw materials that are rich in sugars and therefore aids in lactic acid fermentation (Bernado et al., 2016). Microbial production of lactic acid produces two types of enantiomers l- (+)-lactic acid and d- (−)-lactic acid and l-form is the dominant form and has high market potential (Biddy et al., 2016). Lactic acid is used as such in the food industry in the form of an acidulant and can also be used as a starter material for other chemicals such as ethyl lactate, polylactic acid, and potassium lactate (IHS Markit, 2018). The pharmaceutical industry sees lactic acid as an important molecule; approximately 10% of lactic acid produced by industrial processes finds its application in pharmaceuticals, and in biodegradable plastics, about 40% of lactic acid is used.

Therefore, to the pharmaceutical and plastic markets, lactic acid serves as an interesting value chain. Lactic acid production facilities in different sites show that the process is majorly through microbial fermentation. The location of the plant and therefore the geographical site plays a key role in the raw material usage and processing. The raw materials are used for glucose extraction for lactic acid synthesis (Wang et al., 2015). The sugar can vary from cane sugar, corn glucose, or corn starch. The starch content present in the corn is made to undergo enzymatic hydrolysis, and this then results in the production of glucose for lactic acid production through fermentation process. Using different processes (polymerization, esterification, neutralization), lactic acid is further processed.

Lactic acid is produced industrially using microbial fermentation, primarily by lactic acid bacteria (LAB) and some fungal species. The most commonly used organisms include LAB; these bacteria ferment carbohydrates (e.g., glucose, lactose, and starch) to produce lactic acid (Andrea et al., 2017). Homofermentative LAB (produce mainly lactic acid), *Lactobacillus delbrueckii subsp. Bulgaricus*, is used in dairy fermentation and lactic acid production, *Lactobacillus acidophilus* produces high yields of L (+)-lactic acid; *Lactobacillus amylophilus* can ferment starchy substrates; and *Lactococcus lactis* is used in dairy and lactic acid production (Akram et al., 2017). Heterofermentative LAB (produce lactic acid + by-products like ethanol, CO_2) such as *Lactobacillus brevis* is used in fermented foods but is also used for lactic acid production; *Leuconostoc mesenteroides* are capable of fermenting sugars into lactic acid and dextrans; *Weissella spp.* are used in food fermentation but also produce lactic acid. Certain fungal species can also produce lactic acid, especially when fermenting LCB. *Rhizopus oryzae* can produce L(+)-lactic acid and is used in starch-based fermentation. Some strains of *Aspergillus spp.* can be genetically modified to produce lactic acid efficiently (Anil et al., 2019). To improve lactic acid yield and reduce production costs, genetical modification of other microbes are performed. They include *Escherichia coli*, which can be engineered to produce both L(+) and D(-)-lactic acid from various feedstocks; *Saccharomyces cerevisiae* (Yeast), which can be modified to ferment glucose and xylose into lactic acid; and *Corynebacterium glutamicum*, which is engineered for industrial lactic acid production.

12.4.5.2 Citric Acid

Citric acid (CA), 2-hydroxypropane-1,2,3-tricarboxylic acid, is found in plant and animal tissues, such as blood, bone, and muscle. In the case of living organisms, Krebs cycle produces one of the most essential carboxylic acids called citric acid. Citric acid is harmless, and its effect of capturing and sequestering metal ions makes citric acid applicable in food, pharmaceutical, chemical, and even metallurgical industries (Amato et al., 2020; Hu et al., 2019). Market capacity citric acid is growing rapidly, and recent annual global production has reached approximately

2.8 million tons. The reason behind the industry growing rapidly is majorly due to the wide range of applications such as in biopolymer production, environmental protection, and biomedicine (Behera et al., 2021).

Citric acid is primarily produced via microbial fermentation using mainly two different raw materials such as sugar-based feedstocks, LCB and waste stream molasses, sugarcane juice, corn syrup, agricultural residues, forestry waste, fruit peels. The most common microorganism for citric acid production is *Aspergillus niger* (a fungus), but other organisms like *Yarrowia lipolytica* and certain yeast strains are also studied. Citric acid production can be produced by fermentation like submerged (SmF) and solid-state fermentation.

After fermentation, citric acid is separated and purified through filtration to remove biomass, precipitation (using calcium hydroxide) to form calcium citrate, acidification with sulfuric acid to recover citric acid, and crystallization for high-purity citric acid.Top of FormBottom of Form The application of citric acid is seen in various industries such as in food, cosmetics, pharmaceuticals, and in other product formulations such as poly lactic acid (PLA) synthesis.

Citric acid production in industry is done mostly using *Aspergillus niger* due to its high production efficiency (Lende et al., 2021; Lambros et al., 2022). In 1917, Currie initiated the industrial-scale production of citric acid by cultivating the microorganism in media containing sucrose and discovered the process using a filamentous fungi called *Sterigmatocystis nigra* (currently *Aspergillus niger*). These microorganisms are able to attach on substrates and grow quickly utilizing various substrates and are therefore advantageous for citric acid production industrially. The process of growth utilizing the substrates results in citric acid accumulation and the degradation is prevented during Krebs cycle (Sauer et al., 2013; Chen and Nielsen, 2016).

12.4.5.3 Acetic Acid

In an integrated biorefinery, acetic acid is produced as one of the key products, along with other valuable chemicals, biofuels, and materials, leveraging the sustainable use of organic feedstocks. This contrasts with traditional methods that often rely on fossil fuels or petrochemical processes. Acetic acid produced via biotechnological processes using renewable resources generally has a lower carbon footprint compared to traditional petrochemical methods (Chai et al., 2016)

The feedstocks for an acetic acid biorefinery are typically biomass sources like agricultural residues (corn stover, wheat straw), forestry residues, food waste, or even municipal solid waste (MSW). These are rich in sugars and other organic compounds that can be fermented to produce acetic acid. The use of waste feedstocks such as agricultural residues or food waste helps to reduce waste in the environment and can also provide a circular economic solution.

One of the main routes through which acetic acid is produced biologically is through the fermentation of sugars or ethanol by acetic acid bacteria (e.g., *Acetobacter* species). In this process, microorganisms convert sugars or ethanol into acetic acid under aerobic conditions. Some biorefineries also use anaerobic bacteria that can produce acetic acid as a by-product of their metabolism, typically in controlled bioreactors. By integrating various biochemical conversion technologies (like fermentation, hydrolysis, and distillation), acetic acid biorefineries can optimize energy use, making the entire process more sustainable (Gao et al., 2019).

Acetic acid biorefineries often produce other useful co-products, including ethanol, biofuels, organic acids, and biogas. These can be used to meet the energy demands of the plant or be sold to other industries. Acetic acid produced in biorefineries has a wide range of applications, including in the food industry (as a preservative or flavoring agent), in the production of acetic anhydride (used in the synthesis of chemicals), as a solvent in various industrial processes, and in the production of plastics, fibers, and paints.

The challenges faced by the sustainable production of the acetic acid industry include producing acetic acid from biomass or waste sources, which may have higher production costs compared to petrochemical methods. However, technological advances are helping to improve efficiency and reduce costs (Gong et al., 2019). Scaling up lab-based biorefinery processes to industrial levels while maintaining high yield and productivity is a common challenge. As acetic acid is widely produced from fossil-based sources, biorefinery methods need to be competitive in terms of cost and product quality.

Potential benefits of acetic acid biorefineries include reducing reliance on fossil fuels and acetic acid biorefineries contributing to a more sustainable chemical industry. They can make use of renewable resources, including agricultural and food waste, contributing to waste reduction and enhancing resource efficiency. The development of biorefineries could open up new markets and jobs, especially in rural areas where biomass and organic waste are abundant.

Research into more efficient and robust microbial strains for acetic acid production could help lower costs and increase yield. Future biorefineries may integrate multiple processes (e.g., fermentation, hydrolysis, gasification) to maximize resource use and reduce waste. The concept of waste-to-value is at the heart of many biorefinery initiatives, with acetic acid biorefineries playing a key role in creating a more circular and sustainable chemical industry. In short, acetic acid biorefineries hold significant promises for developing more sustainable and resource-efficient production processes while contributing to a circular economy model. However, the technology and economic challenges will need to be addressed for large-scale implementation.

12.4.6 Biopolymers Production Through Microbial Fermentation

Production of sustainable bioproducts like biopolymers through microbial hosts has gained great attention in recent times. Biopolymers produced through microbes are produced through sustainable and green processes at industrial scale. The bioprocessing of the waste stream in the concept of the biorefinery not only address the environmental concerns but also priorities the waste management strategies and utilization of the waste toward the eco-friendly and economically industrial-scale production of the biopolymers with sustainable feedstocks from industries like paper and pulp, beer and winery, forest and agricultural wastes, starch and food processing industries. The growing demand for biodegradable, sustainable and eco-friendly materials biopolymers offered a promising solution to mitigate the impact of fossil fuel derived polymers and plastics. (Almihyawi, et al., 2024; Kaur et al., 2024;). Due to the diverse functional properties biopolymers play a significant role in various industries like food, agriculture, packaging, cosmetics, edible films and medical applications (Almihyawi et al., 2024; Kaur et al., 2024; Sen and Baidurah, 2021).

Biobased materials, specifically biopolymers that are macromolecules – i.e., polylactic acid (PLA), polyhydroxyalkanoates (PHAs), polycaprolactone (PCL), and polyhydroxybutyrate (PHBs) – are formed by the polymerization of the monomers from proteins, nucleic acids and polysaccharides. Microorganisms like *Capriavidus necator, Cupriavidus necator* DSM 545, *Pseudomonas aeruginosa* 42A2 (NCIB 40045), *Alcaligenes eutrophus* (ATCC17699), *Bacillus megaterium* SRKP-3, *Arthrospira*, *Synechocystis*, and *Synechococcus* have been extensively studied for the production of biopolymers like PHA, PHB and PCL. These microbes produce biopolymers by utilizing the wastes of crude glycerol, waste of fried oil, and starch from the dairy and rice bran industries. The microbes while accumulating the biopolymers during fermentation process utilized CO_2 and CO in some cases from syngas and CO_2 from industrial streams (Srisawat et al., 2022; Flórez et al., 2023; Leong et al., 2021; Zikmanis et al., 2021).

Bacterial cellulose (BC) is an exopolysaccharide produced by numerous bacterial species, has an average diameter of 20–100 nm, and is fibrous with different characteristics – i.e., high crystallinity, high tensile strength, high degree of polymerization, and adsorbing capabilities. Several bacterial species such as *Agrobacterium, Aerobacter, Achromobacter, Alcaligenes, Azotobacter, Gluconacetobacter, Komagataeibacter, Rhizobium, Rhodobacter*, and *Sarcina* producing BC through submerged and solid-state fermentation utilize the waste resources from agricultural, food processing industries, and feedstocks like crude glycerol from industrial waste and wine brewers waste (Kaur et al., 2024; Flórez et al., 2023; Tsouko et al., 2024).

12.5 ALGAE BIOFUEL BIOREFINERY

Algae are a distinct group of ecologically important organisms composed of unicellular microscopic forms to large macroscopic forms. They can thrive in freshwater as well as in marine environments. Many advantages of microalgae make them an attractive feedstock; majorly, they can grow photo autotrophically using light energy and inorganic carbon, mixotrophically with sugars, and heterotrophically using organic compounds as energy source (R. Verma et al. 2020). Microalgae became an efficient source for biofuels including biodiesel, biogas, biohydrogen, bioethanol, biochar, and butanol due to a higher rate of productivity and contain a high content of lipids, polysaccharides, and proteins. Some microalgal species can produce higher concentrations of targeted biofuels in specific cultivation conditions. Figure 12.3, a schematic representation of the algal biorefinery, describes the value-added products from algae. *Chlorella protothecoides, Scenedesmus obliques, Nannochloropsis salina, Chlorococum* sp, *Spirulina* sp, etc., are the microalgal species involved in the targeted biofuel production.

12.5.1 Biofuel Production

The third generation of biofuels involves algal biomass, which is a relatively new approach in bioenergetics. Algal biomass can be processed into biofuels or related products using the whole biomass, its extracts, or waste. Technologies for converting

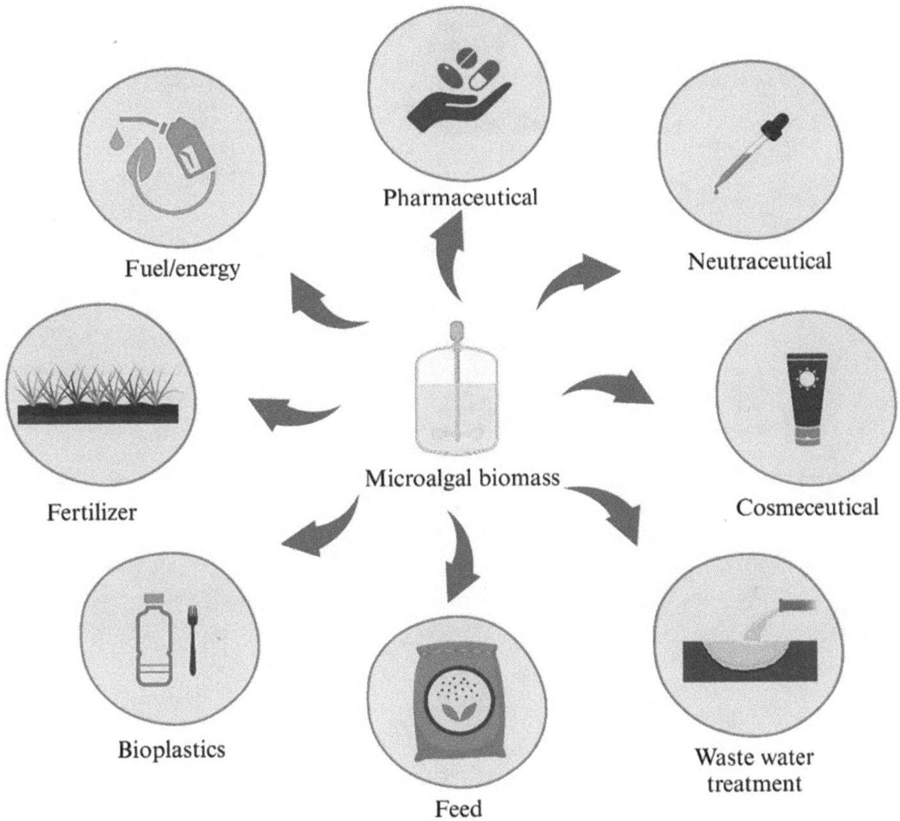

FIGURE 12.3 Algal biorefinery: value-added products from algal biomass.

microalgal biomass into renewable fuels include thermochemical, chemical, and biochemical methods (Voloshin et al., 2016). Figure 12.4 describes the valorization of microalgal biomass to various products, including biofuels, vitamins, lipids, and biomass as a feed.

12.5.2 Thermochemical Conversion Methods

The feedstock is processed under high temperature and pressure to produce compounds with low molecular weight and reduced oxygen content (Voloshin et al., 2016). The thermochemical conversion techniques include combustion, gasification, pyrolysis, and liquefaction. Direct combustion or incineration is used to generate electricity by utilizing the hot gases produced from algal biomass (Peng et al., 2020). The algal biomass is burned in a boiler. This process generates high-pressure steam, which is directed into a steam turbine. As the steam flows over the turbine blades, it causes the turbine and electric generator to spin, thereby producing electricity.

Pyrolysis is a thermal decomposition process occurring with little or no oxygen, producing bio-oil (Ganesan et al., 2020). It is categorized into slow and fast pyrolysis: slow pyrolysis uses slower rates and longer vapor residence times, favoring tar

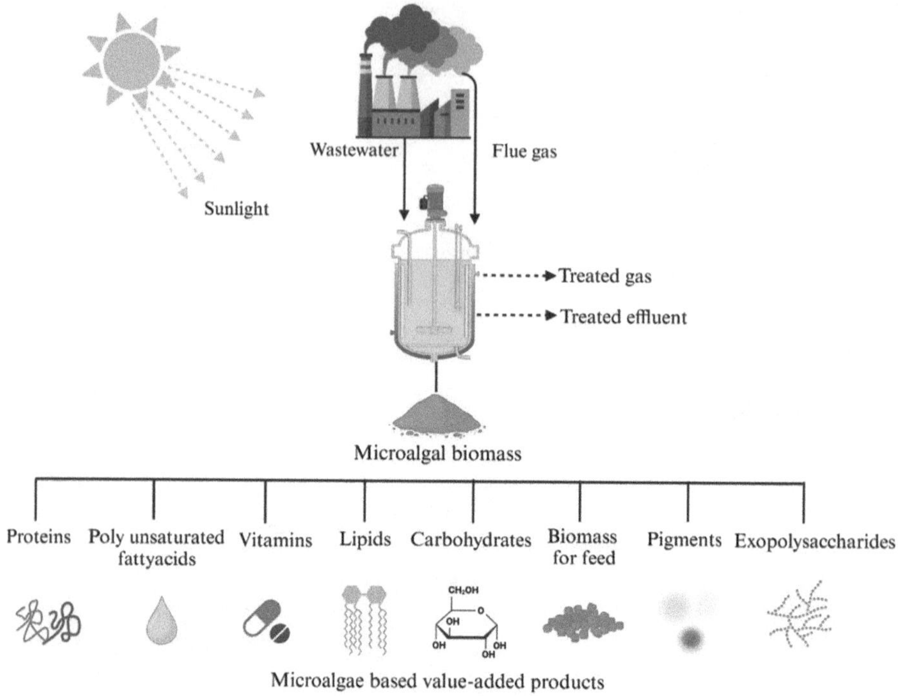

FIGURE 12.4 Microalgal biomass and valorization for value-added products.

and char production, while fast pyrolysis uses higher heating rates and shorter vapor residence times, promoting direct liquid fuel production.

Gasification involves treating biomass at temperatures above 1,000 K, producing a gas mixture that includes carbon dioxide, carbon monoxide, hydrogen gas, methane, nitrogen gas, and other compounds. From this mixture, syngas (a combination of CO and H_2) can be extracted. The Fischer-Tropsch synthesis is a known technology for further converting syngas into Fischer-Tropsch liquids, a mixture of alkanes and alkenes, which can be refined into traditional fuels like jet fuel, clean diesel, and gasoline (Voloshin et al., 2016). Table 12.2 illustrates biofuel production with different species of microalgae. The gasification process is explored in different categories, including steam gasification, plasma gasification, hydrothermal gasification, and supercritical water gasification of biomass (Mishra et al., 2023).

Thermochemical liquefaction is a technique that transforms wet algal biomass into liquid fuel by subjecting it to high temperatures (200°C–500°C) and pressures (above 20 bar) in the presence of a catalyst. This process breaks down the biomass chemically, resulting in the production of bio-oil, a valuable renewable energy source. The high temperature and pressure conditions accelerate the conversion of the complex organic compounds in the algae into simpler liquid hydrocarbons suitable for use as biofuels (Voloshin et al., 2016; Ravichandran et al., 2022).

TABLE 12.2
Biofuels Produced from Microalgae

Microalgae Species	Target Metabolites	Potential Bioproducts	References
Chaetoceros calcitrans	Lipids	Biodiesel	Sivaramakrishnan et al., 2022
Chaetoceros sp. FIKU035	Lipids	Biodiesel	Chaisutyakorn et al. (2018)
Euglena sanguinea	Lipids	Biodiesel	Sivaramakrishnan et al., 2022
Nannochloropsis salina	Lipids	Biodiesel	Sivaramakrishnan et al., 2022
Nannochloropsis oculata	Carbohydrates	Bioethanol	Fetyan et al. (2022)
Tetraselmis suecica	Carbohydrates	Bioethanol	Reyimu and Özçimen (2017)
Dunaliella salina	Lipids	Biodiesel	Sivaramakrishnan et al., 2022

12.5.3 Chemical Conversion Method (transesterification)

Chemical approaches involve initiating specific reactions in the presence of a catalyst to produce a desired product. These methods typically rely on carefully controlled conditions to facilitate the transformation of raw materials into valuable chemicals or fuels. Transesterification is a process that produces fatty acid methyl esters (FAME)/biodiesel by reacting to triglycerides and alcohol in the presence of a catalyst (Peng et al., 2020). Transesterification of microalgal oils for biodiesel production can be achieved using both homogeneous and heterogeneous catalysis. Homogeneous alkaline catalysis is the most common method due to its ability to catalyze the reaction at low temperatures and atmospheric pressure, achieving high conversion yields in a short time (Park et al., 2015).

12.5.4 Biochemical Conversion Methods

Biochemical approaches like hydrolysis and fermentation involve chemical processes within living cells (Voloshin et al., 2016). Algae cells are rich in carbohydrates, with their outer walls made of pectin and alginate, while their nuclei contain cellulose and hemicellulose. Different algae species have significant carbohydrate content, which, when the cells are lysed, can serve as a feedstock for bioethanol. *Chlamydomonas reinhardtii* and *Chlorella vulgaris* are noted for their high carbohydrate content, and the US Renewable Fuels Standard estimates bioethanol production from algae at 36 billion gallons (Ganesan et al., 2020). Some cyanobacteria can also directly produce ethanol through photosynthesis. Additionally, algae can generate alkanes via heterotrophic metabolic pathways, producing hydrocarbons like light crude oil, although in small quantities. This process remains underexplored for biofuel production.

Anaerobic digestion of delipidated algal biomass generates biogas, a mixture of methane and carbon dioxide (Peng et al., 2020). Anaerobic digestion does not require the step of dewatering and drying for moisture removal, and therefore, it is preferred for high-moisture containing biomass like seaweed, and it saves energy. It is a straightforward process from an infrastructure and engineering standpoint. Unlike biodiesel production, which focuses on lipids, anaerobic digestion utilizes all the organic carbon in algae (Milledge et al., 2019).

12.6 INTEGRATION OF WASTEWATER TREATMENT FOR ALGAL BIOFUEL

Microalgae can be utilized for wastewater treatment, and thereby biomass production can also be achieved. This is a low-cost, highly sustainable, and economical process. Also, it reduces the impact of agricultural and industrial waste. Wastewater usually contains organic matter, microorganisms, and inorganic compounds. Microalgae can grow rapidly by utilizing the inorganic nitrogen and phosphorus present in wastewater, and it releases oxygen, which increases the dissolved oxygen in wastewater, facilitating the growth of bacteria. This allows the complete breakdown of organic matter in the wastewater and reduces the biological oxygen demand (BOD) and chemical oxygen demand (COD). In turn, the heterotrophic bacteria present in the wastewater release CO_2 as a result of their metabolism, which is crucial for the microalgal growth. During the biological symbiosis between microalgae and bacteria, the organic and inorganic pollutants are removed by the assimilation of these microorganisms, and it does not cause any secondary pollution. Besides that, the biomass generated through wastewater treatment with microalgae can be utilized for the production of various high-value molecules. Figure 12.5 depicts the process of wastewater treatment integrated with algal biomass production.

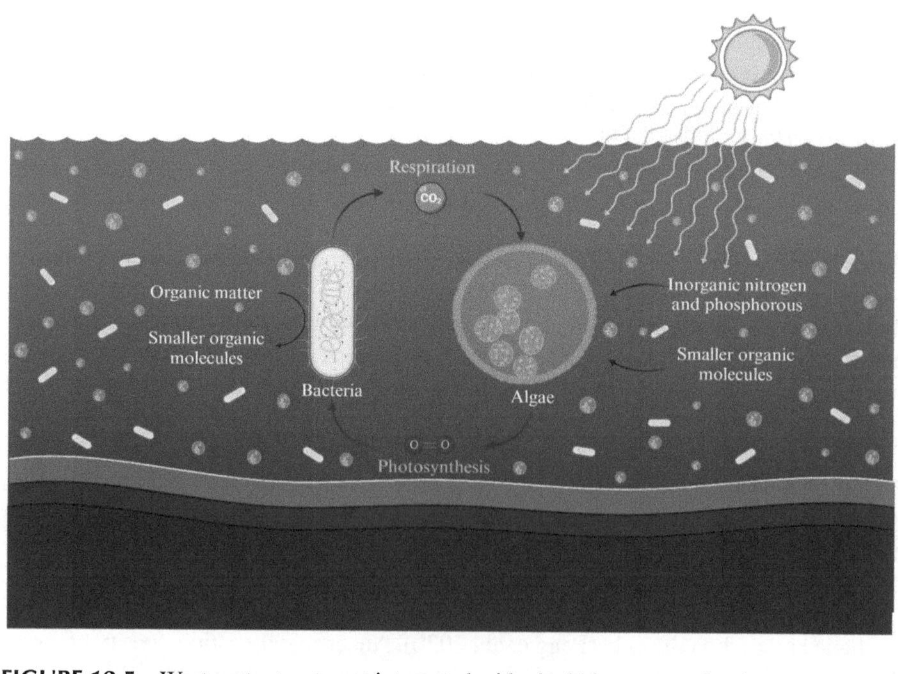

FIGURE 12.5 Wastewater treatment integrated with algal biomass production.

The tertiary and quaternary treatments in domestic wastewater treatment focus on the elimination of inorganic ions (ammonium, nitrate, and phosphate ions), heavy metals, organic compounds, and soluble minerals. Microalgae can combine the tertiary and quaternary sewage treatment and makes it more affordable and economical than the other chemical and biological processes followed. For the microalgal biomass production from wastewater, open raceway ponds are mostly suggested. The closed photo bioreactors and two-stage cultivation system can also be used for biomass production (Rayen et al. 2019).

Wastewater contains high levels of nitrogen (N), phosphorus (P), and organic carbon, which are essential nutrients for algae growth. Using wastewater as a growth medium reduces the need for synthetic fertilizers and helps remove pollutants from the water. Algae can absorb nutrients and contaminants from wastewater through nutrient uptake, wherein algae efficiently remove nitrogen and phosphorus and prevent eutrophication in water bodies, some algal species can sequester heavy metals and other harmful pollutants and thereby remove toxins from wastewater and also help in CO_2 sequestration where algae utilize CO_2 from industrial emissions and reduce greenhouse gases from the atmosphere (Delrue et al., 2016). Algae can be harvested through methods like flocculation, centrifugation, and filtration. The collected biomass undergoes bioconversion into biofuels through processes like lipid extraction and transesterification, which results in the production of biodiesel; hydrothermal liquefaction (HTL) and conversion of algae into bio-crude oil occur, and anaerobic digestion produces biogas (methane) for energy (Dolganyuk et al., 2020).

The process of wastewater treatment through algae generates valuable by-products such as treated water, which can be recycled for industrial/agricultural use and residual biomass, which can be further used as animal feed, fertilizers, or biochar; biogas produced from anaerobic digestion can further enhance energy recovery; a reduction in cost can be seen by using this water for cultivation and sustainable biofuel production, which utilizes nonarable land and prevents competition with food crops (Feng et al., 2016)

12.7 INTEGRATION OF ALGAL BIOREFINERY WITH OTHER INDUSTRY THROUGH CARBON SEQUESTRATION

Algae are emerging as a potential feedstock for biorefineries due to their rapid growth, high biomass productivity, and ability to produce a wide range of valuable compounds. Algae can be utilized for reducing the global impact of carbon dioxide by capturing the gas and utilizing it to produce various types of value-added chemicals. The process of photosynthesis by algae is the crucial process in carbon sequestration from the atmosphere. The process of carbon capturing by algae can be made to occur in large quantities, and this process can result in a huge reduction in the global impact of carbon dioxide, thereby contributing to greenhouse effect mitigation.

Cultivating algae for biofuel production captures carbon dioxide. Using these fuels is carbon neutral, as the emitted carbon dioxide equals what the algae absorbed during growth. Biomass can be used in materials or buried to lock the carbon away. Algae are integrated into industrial processes to capture and utilize carbon dioxide emissions, such as from power plants. Algae hold immense potential in global carbon management, particularly when integrated with renewable energy and sustainable agricultural practices. However, their large-scale deployment requires addressing economic, technical, and environmental challenges.

The microalgal biomass could be used for wide range of value-added products, such as biofuels, pharmaceutical compounds, cosmetics, and nutraceuticals. The method of carbon dioxide capture by microalgae is environmentally sustainable, as it can be utilized for treating wastewater and for the removal of heavy metals. Research is underway into the finding of optimum conditions for the maximum production of compounds via the carbon neutral method of microalgal cultivation. Bioreactor studies of the microalgal system is increasing globally for understanding the individual parameters such as pH, salinity, nutrients, and light source involved in production, and therefore an efficient design for the microalgal bioreactor for carbon capturing is highly required for meeting the global sustainable demand (Verma et al., 2020).

12.7.1 Mechanism of Carbon Dioxide Sequestration

Microalgae undergo photosynthesis for carbon metabolism, and most microalgae mainly depends on the C3 pathway. The photosynthetic reaction that converts solar energy into chemical energy is as follows:

$$6CO_2 + 6H_2O + \text{Sunlight} \longrightarrow C_6H_{12}O_6 + 6O_2$$

Microalgae are capable of utilizing both bicarbonate and carbon dioxide dissolved in the media as carbon source; however in the production, media bicarbonate is the most dominant one to be used as carbon source. The process of photosynthesis results in the accumulation of glucose in the algal cells, which is used as an energy source and is later used for the synthesis of lipids, carbohydrates, and proteins. The excess of glucose accumulation can result in the shift of cell's osmotic balance and thereby glucose conversion to other value-added products can happen (Choi, 2019).

12.8 SUSTAINABLE AVIATION FUEL (SAF)

SAF is a biofuel produced from sustainable feedstocks, including biomass from corn grain, oil seeds, lignocellulosic residues from forestry, agriculture, wood mill waste, MSW streams, and algae, which can power the aircraft with similar properties like conventional jet fuel, which is safe and reliable with a smaller carbon footprint. Advancement in bioeconomy can be achieved by using SAF from renewable and agricultural waste resources, SAF can improve facilities in crop development, aircraft performance and in industrial production sector, thereby aiming for low carbon emissions. Net zero carbon dioxide emissions by airlines by 2050 is the goal set by the 77th International Air Transport Association (IATA) meeting held in 2021; to reach the goal, IATA estimates SAF holds the potential to contribute 65% to the reduction. Air quality is highly impacted by conventional fuel usage by aircraft, and use of SAF by the global aviation industry holds immense power in improving the quality of air by reducing the greenhouse gas emissions and by reducing the air pollutants.

SAFs are liquid hydrocarbon jet fuels that have an advantage over other nonconventional fuels. The shift from conventional fuels to SAF requires very little or no modification at all in the existing fueling infrastructure; therefore, they can be

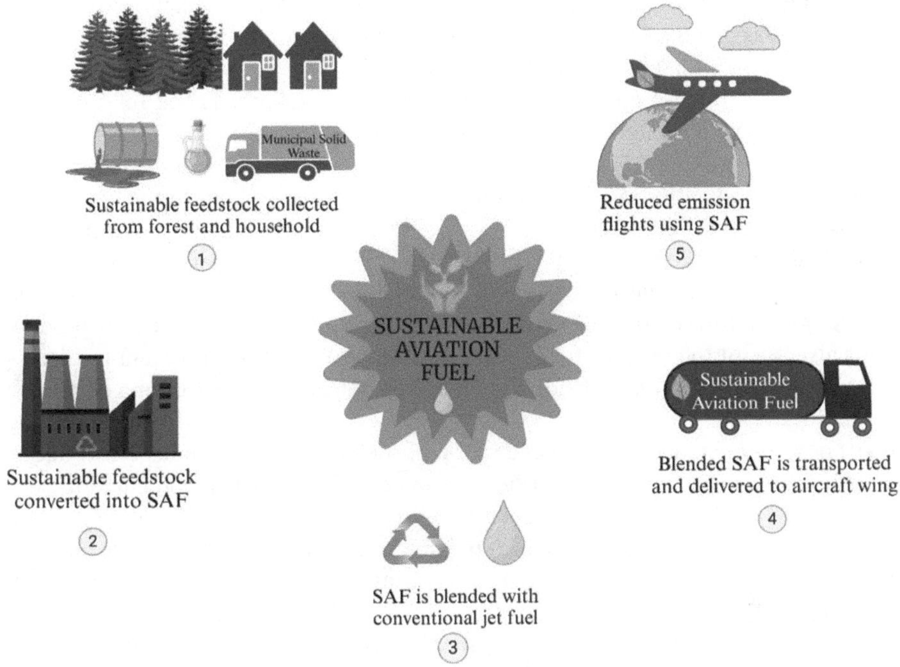

FIGURE 12.6 SAF flowsheet for production and blending.

"dropped in" to current machinery with ease. Meticulous amount of effort has been put in for establishing SAF pathways that are considered safe for use by the aviation industry, yet there remains a major challenge in bringing the cost-effectiveness and making it technically feasible. Figure 12.6 illustrates a flowsheet for the production of SAF for blending with conventional fuel and to reduce carbon emissions.

The SAF that is produced globally in the current scenario is considered to be less in quantity and requires more efforts in this direction for improving the amount of the fuel and also for meeting the global demand to use it in flight operations. The largest SAF customer worldwide is considered to be Lufthansa Group. SAF also accounted for around 0.2% of the Lufthansa Group's total fuel requirements in 2023. The Lufthansa Group wants to increase this share. The company is working on numerous projects worldwide to increase SAF availability, concluding Memorandum of Understandings with SAF manufacturers and examining options for long-term purchase agreements. At the same time, the Lufthansa Group is continuously expanding its offering for more sustainable travel for its customers. However, the aviation industry cannot initiate a self-sustaining market on its own. A targeted policy strategy is needed.

Although SAF can be directly produced from microbes by fermenting sugars, a greener approach would be by using feedstocks that are renewable and sustainable; therefore. the plant biomass that contains fermentable sugars in the form of starch, oils that contain high calorific value, can be used for obtaining SAF. Halophytes are salt-tolerating plants, and they are utilized by the United Arab Emirates by establishing an aviation fuel supply chain. Although the supply chain is still under process and has potential to reach commercial scale, a demonstration of flights using this fuel was performed by Etihad Airways in 2016.

Algae hold immense potential to be used as sustainable feedstock for SAF production. Oil content in algae is higher when compared to other terrestrial plant biomass, with approximately 60% lipids by dry weight. Research toward the cost-effective cultivation of algae and its lipid extraction will aid in the cost-effective production of SAF. According to a 2025 European aviation environmental report, SAF production has only represented 0.53% of global jet fuel usage. This indicates that a significant increase in the area of sustainable production is required for meeting the future goals of net zero carbon emission. The high cost of production is also one of the reasons for less use of SAF; therefore, government initiatives such as tax exemptions and subsidies are required to boost this sector for a better and greener future.

The development and market penetration of SAF requires strenuous effort due to the tighter tolerances and qualification criteria being higher when compared to conventional fuels. Olefinic content, heteroatoms, and oxygenates should not be present in SAF, as these compounds are considered unsafe for aviation fuel performance and can possess safety issues, therefore these metrics are strictly considered, and SAF needs to qualify the hurdles. (Heyne et al., 2021). SAFs approval for usage is monitored via Fast-Track Programs under American Society for Testing and Materials (ASTM) D4054 (DO2 ASTM). The fast-track program currently approves SAF blend with conventional fuel in 50:50 ratio.

12.9 BIODIESEL PRODUCTION IN INTEGRATED BIOREFINERY

Biodiesel is highly sought after for transportation and industrial use due to its eco-friendly nature, nontoxicity, and clean-burning qualities, which reduce harmful emissions like carbon monoxide and sulfur dioxide. It has a higher flash point than petroleum diesel, making it safer to handle and store, and it offers excellent lubricity and combustion efficiency, even when blended with Petro diesel. Biodiesel can be produced cost-effectively from local resources, but its higher production cost – about 1.5 times that of petroleum diesel, primarily due to feedstock oils – limits its widespread commercialization (Gujjala et al., 2019).

Traditional biodiesel is made through transesterification of waste oils, animal fats, and vegetable oils. This eco-friendly fuel can be used in diesel engines without modifications. In the process, glyceride reacts with alcohol (ethanol or methanol) to create fatty acid alkyl esters. Common feedstocks include soybean, palm, and sunflower oils, and sometimes used cooking oils. Depending on the fats, pretreatment steps may be necessary. The transesterification reaction, which involves alcohol, catalysts, and fatty acids, is typically catalyzed by strong acids or bases like potassium methanolate or sodium. The process produces raw biodiesel and glycerol, which are then cleaned in a later step (Hasan et al., 2023).

Microbial lipids offer a promising renewable biofuel source and could help address the issues caused by fossil fuel depletion. Biodiesel is produced by trans esterifying fatty acids with short-chain alcohols to form mono-alkyl esters. To improve this process, it's important to develop microorganisms that produce fatty acids and short-chain alcohols and have more active acyltransferases for short-chain alcohols. Additionally, creating microbes that efficiently use inexpensive substrates is crucial to compete with fossil fuel use (Uthandi et al., 2022).

Oleaginous microorganisms are those that accumulate lipids, making up over 20% of their dry weight under stress conditions like high carbon and low nitrogen. With proper stress, lipid content can exceed 70%. These microbes convert renewable materials into microbial oil, also called single-cell oil, which is used to produce biodiesel via transesterification (Chintagunta et al., 2021). Yeasts and filamentous fungi are typically oleaginous, accumulating triacylglycerols (TAGs), the main lipid components. While bacteria usually store polyhydroxyalkanoates, some actinobacteria, like *Mycobacterium*, *Rhodococcus*, *Streptomyces*, and *Nocardia*, can also produce lipids (Gujjala et al., 2019). Oleaginous yeasts, such as *Rhodotorula*, *Rhizopus*, and *Yarrowia*, can accumulate lipids up to 40% of their biomass, with some reaching 80% under nutrient-limited conditions (Patel et al., 2016; Signori et al., 2016). They are fast-growing, adaptable, and easily cultivated, making them strong candidates for biodiesel production (Uthandi et al., 2022).

Fungi convert external carbon into carbohydrates and then lipids, with some oleaginous fungi storing up to 70% of their biomass as lipids. Lipid production is influenced by carbon and nitrogen sources, C/N ratio, temperature, and pH. Nitrogen deficiency, rather than excess nitrogen, often triggers lipid accumulation (Uthandi et al., 2022). Certain bacteria, like *Rhodococcus*, *Streptomyces*, and *Nocardia*, can accumulate significant lipids. Their TAG composition varies based on bacterial strain and carbon source (Uthandi et al., 2022). These bacteria can use diverse carbon sources such as glucose and glycerol. Under specific conditions, they can accumulate up to 80% lipids, especially during the stationary phase of growth.

Microalgae-based biodiesel is currently a key area of research due to its potential to supply ample oil for global use, offering biodiesel yields much higher per hectare than plant-based feedstocks (Chintagunta et al., 2021; Arunachalam Sivagurulingam et al., 2022). Oleaginous microalgae typically accumulate lipids ranging from 20% to 80%. Common strains like *Chlorella*, *Neochloris*, *Nannochloropsis*, and *Scenedesmus* contain 40%–60% lipids of their total dry weight. Their high productivity, ability to capture industrial CO_2, and non-competition with food crops make microalgae-based oil a promising option. However, a major challenge for scaling up autotrophic microalgae-based biodiesel production is the need for light during cultivation (Uthandi et al., 2022).

After extraction, microbial lipids (e.g, fatty acids) undergo transesterification with methanol to produce biodiesel and glycerol. The process efficiency depends on factors like reactor type, catalyst choice, and feedstock quality. Ideal catalysts should be recyclable, cost-effective, and environmentally friendly. Enzymes, homogenous and heterogeneous catalysts are the commonly used catalysts. Crude biodiesel must be purified to meet standards, removing impurities such as glycerol, water, methanol, catalysts, and fatty acids. Purification methods like distillation, washing, and membrane separation are used, starting with glycerol removal through settling or centrifugation, followed by further purification for high-quality biodiesel (Zhang et al., 2021).

12.10 BIOGAS PRODUCTION IN INTEGRATED BIOREFINERY

Biogas is a colorless, combustible gas produced through the biological breakdown of organic matter in the absence of oxygen (Sawyerr et al., 2019). It is generated through anaerobic digestion, a process that not only produces a renewable energy resource but also helps treat feedstock, resulting in digestate that can serve as an organic fertilizer, offering a sustainable alternative to chemical fertilizers in agriculture (Kabeyi and Olanrewaju, 2022). Biogas mainly consists of methane, carbon dioxide, and 1%–5% other gases, including hydrogen (Sawyerr et al., 2019). Various organic feedstocks can be processed through anaerobic digestion, which can be further enriched to produce biomethane, a natural gas-like substance. The conversion of biomass waste to biogas supports a circular economy by reducing carbon emissions, improving organic waste management, and increasing resource efficiency. The microbial generation of biogas plays a key role in the global carbon cycle, with natural anaerobic biodegradation estimated to release 590-800 million tons of methane into the atmosphere annually (Kabeyi and Olanrewaju, 2022).

The anaerobic digestion process consists of four stages: hydrolysis, acidogenesis, acetogenesis, and methanogenesis, which are carried out by symbiotic bacterial communities. In hydrolysis, bacteria's hydrolytic enzymes break down insoluble polymers, carbohydrates, lipids, and proteins into soluble mono- and oligomers, which are accessible to microorganisms. In acidogenesis, simple sugars, amino acids, and fatty acids are further degraded into acetate, carbon dioxide, hydrogen, volatile fatty acids, and alcohols. Acetogenesis then converts these volatile fatty acids and alcohols into hydrogen and acetic acid. Finally, in methanogenesis, methanogenic microorganisms, primarily from the Archaea domain, convert the mixture of carbon dioxide, hydrogen, formate, methanol, and acetate into methane (Kasinath et al., 2021). Biogas production can utilize several types of biomasses, primarily composed of carbohydrates, proteins, and fats. Suitable substrates for the fermentation process should have high nutritional value and ensure that the resulting digestate is free from pathogens and heavy metals for safe land application. Common substrates include agricultural, municipal, and industrial wastes, as well as energy crops and aquatic biomass, all of which can be used for biogas generation (Atelge et al., 2020).

12.11 CONCLUSION

Sustainable biofuels and biochemicals production with an integrated biorefinery concept is an advancing tool for reducing the dependency on fossil fuels and simultaneously reducing the carbon footprint to achieve the SDGs. The utilization of the waste generated from different sources has potential for the valorization of the value-added products with the existing facilities and unit operation by strategic planning and innovation toward biobased products through chemical or biological processes. The natural resources like agricultural waste, woody forest biomass, and algal biomass, and industrial waste like sustainable biofuels and biochemicals production with an integrated biorefinery concept are advancing tools for reducing the dependency on fossil fuels and simultaneously reducing the carbon footprint to achieve the SDGs. The utilization of the waste generated from different sources has the potential for the valorization of value-added products with the existing facilities and unit operations by strategic planning and innovation toward biobased products through chemical or biological processes. The valorization of the biomass has the potential for the production of specialty chemicals and sustainable products, which are renewable and processed through green ecosystems. Bioeconomy with sustainable circular processes which are integrated with chemical, food processing and waste water treatment facilities can be utilized for the feedstocks for the new raw materials and also for the recycled materials from various sources. The natural resources like agricultural waste, woody forest biomass, algal biomass, and industrial waste like paper and pulp industry waste, sugar and food processing industry waste, and MSW, which are rich in cellulosic and organic substances, have a greater potential for the development of sustainable fuels including bioethanol, biobutanol, aviation fuel, and biogas. Several microbial species, including bacteria, fungi, and algae, have the capability of converting the waste generated in various industries to sustainable products that are alternatives to fossil fuels and products. In the biorefinery processes, the large-scale production of these biofuels and bioproducts are technically feasible and economically viable.

REFERENCES

Akram AA, Abdel-Hamied M, Rasmey MM, Osman NK, Salha GD, 2017. Optimization of lactic acid production by a novel strain Enterococcus faecalis KY072975 isolated from infants' stool in Egypt. *European Journal of Biological Research*, 7(1), 22–30.

Ale, S, Femeena, PV, Mehan, S, Cibin, R, 2019. Environmental impacts of bioenergy crop production and benefits of multifunctional bioenergy systems. In *Bioenergy With Carbon Capture and Storage: Using Natural Resources for Sustainable Development* (pp. 195–217). Academic Press.

Allen C, Metternicht G, Wiedmann T 2018 Initial progress in implementing the Sustainable Development Goals (SDGs): A review of evidence from countries. *Sustainability Science*, *13*, 1453–1467.

Almihyawi RAH, Musazade E, Alhussany N, Zhang S, Chen H, 2024. Production and characterization of bacterial cellulose by Rhizobium sp. isolated from bean root. *Scientific Reports*, *14*(1), 10848.

Amato A, Becci A, Beolchini F, 2020. Citric acid bioproduction: The technological innovation change. *Critical Reviews in Biotechnology 40*, 199–212.

Andrea K, Oliveira JARD, Martins LHDS, Maciel MRW, Maciel FilhoR, 2017. Lactic acid production to purification: A review. *BioResources*, 12, 4364–4383.

Anil K, Thakur A, Panesar PS, 2019. Lactic acid and its separation and purification techniques: A review. *Reviews in Environmental Science and Biotechnology*, 823–853.

Arunachalam Sivagurulingam AP, Sivanandi P, Pandian S, 2022. Isolation, mass cultivation, and biodiesel production potential of marine microalgae identified from Bay of Bengal. *Environmental Science and Pollution Research*, *29*(5), 6646–6655.

Atelge MR, Krisa D, Kumar G, Eskicioglu C, Nguyen DD, Chang SW, Atabani AE, Al-Muhtaseb AH, Unalan S, 2020. Biogas Production from Organic Waste: Recent Progress and Perspectives. *Waste and Biomass Valorization*, *11*(3), 1019–1040.

Azni MA, Md Khalid R, Hasran UA, Kamarudin SK, 2023. Review of the effects of fossil fuels and the need for a hydrogen fuel cell policy in Malaysia. *Sustainability (Switzerland)*, *15*(5), 4033.

Behera BC, Mishra R, Mohapatra S, 2021. Microbial citric acid: Production, properties, application, and future perspectives. *Food Frontiers*, *2*, 62–76.

Bello MO, Solarin SA, Yen YY, 2018. Hydropower and potential for interfuel substitution: The case of electricity sector in Malaysia. *Energy*, *151*, 966–983.

Bernado MP, Coelho LF, Sass DC, Contiro J, 2016. L(+)-Lactic acid production by *Lactobacillus rhamnosus* B103from dairy industry waste. *Brazilian Journal of Microbiology*, 47, 640–646.

Biddy MJ, Scarlata C, Kinchin C 2016. Chemicals from biomass: A market assessment of bioproducts with near-term potential; NREL/TP-5100-65509; National Renewable Energy Lab (NREL).

Birgen C, Dürre P, Preisig HA, Wentzel A, 2019. Butanol production from lignocellulosic biomass: Revisiting fermentation performance indicators with exploratory data analysis. *Biotechnology for Biofuels 12*(1), 1–15.

Broda M, Yelle DJ, Serwańska K, 2022. Bioethanol production from lignocellulosic biomass—Challenges and solutions. *Molecules*, 27(24), 8717.

Cao X, Chen Z, Liang L, Guo L, Jiang Z, Tang F, Yun Y, Wang Y, 2020. Co-valorization of paper mill sludge and corn steep liquor for enhanced n-butanol production with Clostridium tyrobutyricum Δcat1::adhE2. *Bioresource Technology, 296.*

Chai X, He D, Li Y, Niu D, Xie T, Zhao Y, 2016. A kind of method utilizing changing food waste fermentative production acetic acid Thereof. CN103865975 B.

Chaisutyakorn P, Praiboon J, Kaewsuralikhit C, 2018. The effect of temperature on growth and lipid and fatty acid composition on marine microalgae used for biodiesel production. *Journal of Applied Phycology*, *30*(1), 37–45.

Chavan S, Mitra D, Ray A, 2024. Harnessing rice husks: Bioethanol production for sustainable future. *Current Research in Microbial Sciences*, *7*, 100298.

Chen Y, Nielsen J, 2016. Biobased organic acids production by metabolically engineered microorganisms. *Current Opinion in Biotechnology*, *37*, 165–172.

Cherwoo L, Gupta I, Flora G, Verma R, Kapil M, Arya SK, Ravindran B, Khoo KS, Bhatia SK, Chang SW, Ngamcharussrivichai C, Ashokkumar V, 2023. Biofuels an alternative to traditional fossil fuels: A comprehensive review. *Sustainable Energy Technologies and Assessments*, *60*, 103503.

Chintagunta AD, Zuccaro G, Kumar M, Kumar SPJ, Garlapati VK, Postemsky PD, Kumar NSS, Chandel AK, Simal-Gandara J, 2021. Biodiesel production from lignocellulosic biomass using oleaginous microbes: Prospects for integrated biofuel production. *Frontiers in Microbiology*, *12*, 658284.

Choi YY, (2019) Microalgae bioenergy with carbon capture and storage (BECCS): An emerging sustainable bioprocess for reduced CO_2 emission and biofuel production. *Bioresource Technology Reports*, *7*, 100270.

Decker SR, Brunecky R, Yarbrough JM, Subramanian V, 2023. Perspectives on biorefineries in microbial production of fuels and chemicals. *Frontiers in Industrial Microbiology*, *1*, 1202269.

Delrue F, Álvarez-Díaz PD, Fon-Sing S, Fleury G, Sassi J-F, 2016. The environmental biorefinery: Using microalgae to remediate wastewater, a win-win paradigm. *Energies*, *9*(3), 132.

Devi A, Bajar S, Kour H, Kothari R, Pant D, Singh A, 2022. Lignocellulosic biomass valorization for bioethanol production: A circular bioeconomy approach. *Bioenergy Research*, *15*(4), 1820–1841.

Dolganyuk V, Belova D, Babich O, Prosekov A, Ivanova S, Katsero D, Patyukov N, Sukhikh S, 2020. Microalgae: A promising source of valuable bioproducts. *Biomolecules*, *10*(8), 1153.

Dölle K, Hughes T, Kurzmann DE, 2020. From fossil fuels to renewable biogas production from biomass based feedstock – A review of anaerobic digester systems. *Journal of Energy Research and Reviews*, *5*(3), 1–37

Elgharbawy AS, Sadik WA, Sadek OM, Kasaby MA, 2021. A review on biodiesel feedstocks and production technologies. *Journal of the Chilean Chemical Society*, *65*(1), 5098–5109.

Feng P-Z, Zhu L-D, Qin X-X, Li Z-H, 2016. Water footprint of biodiesel production from microalgae cultivated in photobioreactors. *Journal of Environmental Engineering*, *142*(12), 04016067.

Fetyan NAH, El-Sayed AEKB, Ibrahim FM, Attia YA, Sadik MW, 2022. Bioethanol production from defatted biomass of Nannochloropsis oculata microalgae grown under mixotrophic conditions. *Environmental Science and Pollution Research*, *29*(2), 2588–2597.

Figueroa-Torres GM, Wan Mahmood WMA, Pittman JK, Theodoropoulos C, 2020. Microalgal biomass as a biorefinery platform for biobutanol and biodiesel production. *Biochemical Engineering Journal*, *153*, 107396.

Flórez M, Cazón P, Vázquez M, 2023. Selected biopolymers' processing and their applications: A review. *Polymers*, *15*(3), 641.

Ganesan R, Manigandan S, Samuel MS, 2020. A review on prospective production of biofuel from microalgae. *Biotechnology Reports*, *27*, e00509.

Gao B, Pan H, Wang H, Xiao Y, Zhu D, 2019. Thermoresistant acetic acid bacteria and application thereof. CN110195025.

Gong Y, Hu Y, Li D, Liu Z, Qi Y, Shi Y, Bin T, Wang A, Wang C, Wu Q, Xu N, Zhou M, 2019. Immobilized acetic-acid fermentation method for improving acetic-acid producing capability of acetic acid bacteria. CN109182399A.

Gujjala LKS, Kumar SPJ, Talukdar B, Dash A, Kumar S, Sherpa KC, Banerjee R, 2019. Biodiesel from oleaginous microbes: Opportunities and challenges. *Biofuels*, *10*(1), 45–59.

Hasan MR, Anzar N, Sharma P, Malode SJ, Shetti NP, Narang J, Kakarla RR, 2023. Converting biowaste into sustainable bioenergy through various processes. *Bioresource Technology Reports*, *23*, 101542.

Heyne J, Rauch B, Le Clercq P, Colket M, 2021. Sustainable aviation fuel prescreening tools and procedures. *Fuel*, *290*, 120004.

Hossain M, Huda ASN, Mekhilef S, Seyedmahmoudian M, Horan B, Stojcevski A, Ahmed M, 2018. A state-of-the-art review of hydropower in Malaysia as renewable energy: Current status and future prospects. *Energy Strategy Reviews*, *22*, 426–437.

Hu W, Li WJ, Yang HQ, Chen J, 2019. Hong Current strategies and future prospects for enhancing microbial production of citric acid. *Applied Microbiology and Biotechnology*, *103*, 201–209.

Igwebuike CM, Awad S, Andrès Y, 2024. Renewable energy potential: Second-generation biomass as feedstock for bioethanol production. *Molecules*, *29*(7), 1619.

IHS Markit. 2018. *Chemical economics handbook – Lactic acid, its salts and esters*, S&P Global.

Jain S, Kumar S, 2024. A comprehensive review of bioethanol production from diverse feedstocks: Current advancements and economic perspectives. *Energy*, *296*, 131130.

Kabeyi MJB, Olanrewaju OA, 2022. Biogas production and applications in the sustainable energy transition. *Journal of Energy*, *2022*, 1–43.

Kasinath A, Fudala-Ksiazek S, Szopinska M, Bylinski H, Artichowicz W, Remiszewska-Skwarek A, Luczkiewicz A, 2021. Biomass in biogas production: Pretreatment and codigestion. *Renewable and Sustainable Energy Reviews*, *150*, 111509.

Kaur R, Pathak L, Vyas P, 2024. Biobased polymers of plant and microbial origin and their applications – A review. *Biotechnology for Sustainable Materials*, *1*(1), 13.

Kazmi A, Sultana T, Ali A, Nijabat A, Li G, Hou H, 2025. Innovations in bioethanol production: A comprehensive review of feedstock generations and technology advances. *Energy Strategy Reviews*, *57*, 101634.

Korhonen J, Honkasalo A, Seppälä J. 2018. Circular economy: The concept and its limitations. *Ecological Economics*, *143*, 37–46.

Kumar S, Tirlangi S, Kumar A, Imran M, Pillai HPJS, Kumar Koshariya A, Sathish T, Ubaidullah M, Ayub R, Minnam Reddy VR, AlAbdulaal TH, Algarni H, Maiz F, Shkir M, 2023. A review on the contribution of nanotechnology for biofuel production from algal biomass: A bridge to the reduction of carbon footprint.

Lambros M, Tran TH, Fei Q, Nicolaou M, 2022. Citric acid: A multifunctional pharmaceutical excipient. *Pharmaceutics*, *14*, 972.

Lamichhane G, Acharya A, Poudel DK, Aryal B, Gyawali N, Niraula P, Phuyal SR, Budhathoki P, Bk G, Parajuli N, 2021. Recent advances in bioethanol production from lignocellulosic biomass. *International Journal of Green Energy*, *18*(7), 731–744.

Lende SV, Karemore H, Umekar MJ, Lende SV, Karemore H, Umekar MJ, 2021. Review on production of citric acid by fermentation technology. *GSC Biological and Pharmaceutical Sciences*, *17*, 85–93.

Leong HY, Chang CK, Khoo KS, Chew KW, Chia SR, Lim JW, Chang JS, Show PL, 2021. Waste biorefinery towards a sustainable circular bioeconomy: A solution to global issues. *Biotechnology for Biofuels*, *14*(1).

López-Linares JC, García-Cubero MT, Coca M, Lucas S, 2021. A biorefinery approach for the valorization of spent coffee grounds to produce antioxidant compounds and biobutanol. *Biomass and Bioenergy*, *147*.

Lu L, Shen X, Sun X, Yan Y, Wang J, Yuan Q, 2022. CRISPR-based metabolic engineering in non-model microorganisms. In *Current Opinion in Biotechnology* (Vol. *75*). Elsevier Ltd.

Mat Aron NS, Khoo KS, Chew KW, Show PL, Chen WH, Nguyen THP, 2020. Sustainability of the four generations of biofuels – A review. *International Journal of Energy Research*, *44*(12), 9266–9282.

Milledge JJ, Nielsen BV, Maneein S, Harvey PJ, 2019. A brief review of anaerobic digestion of algae for BioEnergy. *Energies*, 12(6).

Mishra K, Singh Siwal S, Kumar Saini A, Thakur VK, 2023. Recent update on gasification and pyrolysis processes of lignocellulosic and algal biomass for hydrogen production. *Fuel*, *332*.

Mohd Azhar SH, Abdulla R, Jambo SA, Marbawi H, Gansau JA, Mohd Faik AA, Rodrigues KF, 2017. Yeasts in sustainable bioethanol production: A review. *Biochemistry and Biophysics Reports*, *10*, 52–61.

Palaniswamy S, Ashoor S, Eskasalam SR, Jang YS, 2023. Harnessing lignocellulosic biomass for butanol production through clostridia for sustainable waste management: Recent advances and perspectives. *Frontiers in Bioengineering and Biotechnology*, 11.

Park JY, Park MS, Lee YC, Yang JW, 2015. Advances in direct transesterification of algal oils from wet biomass. *Bioresource Technology*, *184*, 267–275.

Patel A, Arora N, Sartaj K, Pruthi V, Pruthi PA, 2016. Sustainable biodiesel production from oleaginous yeasts utilizing hydrolysates of various non-edible lignocellulosic biomasses. *Renewable and Sustainable Energy Reviews*, *62*, 836–855.

Peng L, Fu D, Chu H, Wang Z, Qi H, 2020. Biofuel production from microalgae: A review. *Environmental Chemistry Letters*, *18*(2), 285–297.

Ravichandran SR, Venkatachalam CD, Sengottian M, 2022. A review on hydrothermal liquefaction of algal biomass on process parameters, purification and applications. *Fuel*, *313*.

Rayen F, Behnam T, Dominique P, 2019. Optimization of a raceway pond system for wastewater treatment: A review. *Critical Reviews in Biotechnology*, *39*(3), 422–435.

Reyimu Z, Özçimen D, 2017. Batch cultivation of marine microalgae Nannochloropsis oculata and Tetraselmis suecica in treated municipal wastewater toward bioethanol production. *Journal of Cleaner Production*, *150*, 40–46.

Riaz S, Mazhar S, Abidi SH, Syed Q, Abbas N, Saleem Y, Nadeem AA, Maryam M, Essa R, Ashfaq S, 2022. Biobutanol production from sustainable biomass process of anaerobic ABE fermentation for industrial applications. *Archives of Microbiology*, *204*(11).

Saha S, Sharma A, Purkayastha S, Pandey K, Dhingra S, 2018. Bioplastics and biofuel: Is it the way in future development for end users? In *Plastics to energy: Fuel, chemicals, and sustainability implications* (pp. 365–376). William Andrew Publishing.

Sauer M, Mattanovich D, Marx H, 2013. 12-Microbial production of organic acids for use in food. In *Microbial production of food ingredients, enzymes and nutraceuticals* (pp. 288–320). Elsevier.

Sawyerr N, Trois C, Workneh T, Okudoh V, 2019. An overview of biogas production: Fundamentals, applications and future research. *International Journal of Energy Economics and Policy*, *9*(2), 105–116.

Sen KY, Baidurah S, 2021. Renewable biomass feedstocks for production of sustainable biodegradable polymer. *Current Opinion in Green and Sustainable Chemistry*, *27*, 100412.

Signori L, Ami D, Posteri R, Giuzzi A, Mereghetti P, Porro D, Branduardi P, 2016. Assessing an effective feeding strategy to optimize crude glycerol utilization as sustainable carbon source for lipid accumulation in oleaginous yeasts. *Microbial Cell Factories*, *15*(1).

Sivaramakrishnan R, Suresh S, Kanwal S, Ramadoss G, Ramprakash B, Incharoensakdi A, 2022. Microalgal biorefinery concepts' developments for biofuel and bioproducts: Current perspective and bottlenecks. *International Journal of Molecular Sciences*, *23*(5).

Srisawat P, Higuchi-Takeuchi M, Numata K, 2022). Microbial autotrophic biorefineries: Perspectives for biopolymer production. *Polymer Journal*, *54*(10), 1139–1151.

Stafford-Smith M, Griggs D, Gaffney O, Ullah F, Reyers B, Kanie N, Stigson B, Shrivastava P, Leach M, O'Connell D. 2017. Integration: The key to implementing the sustainable development goals. *Sustainability Science*, *12*, 911–919.

Su Y, Zhang W, Zhang A, Shao W, 2020. Biorefinery: The production of isobutanol from biomass feedstocks. *Applied Sciences (Switzerland)*, *10*(22), 1–18.

Suvachan A, Lal R, Madhavan Nampoothiri K, Sindhu R, Bhaskar T, Binod P, Pandey A, Yasarla R 2021. Valorization of paper industry rejects by combined thermo chemical pretreatment and biological conversion to L- Lysine. *Environmental Technology and Innovation Environmental Technology & Innovation*, 24, 101882.

Toor M, Kumar SS, Malyan SK, Bishnoi NR, Mathimani T, Rajendran K, Pugazhendhi A, 2020. An overview on bioethanol production from lignocellulosic feedstocks. *Chemosphere*, *242*.

Tsouko E, Pilafidis S, Kourmentza K, Gomes HI, Sarris G, Koralli P, Papagiannopoulos A, Pispas S, Sarris D, 2024). A sustainable bioprocess to produce bacterial cellulose (BC) using waste streams from wine distilleries and the biodiesel industry: Evaluation of BC for adsorption of phenolic compounds, dyes and metals. *Biotechnology for Biofuels and Bioproducts*, *17*(1).

Uthandi S, Kaliyaperumal A, Srinivasan N, Thangavelu K, Muniraj IK, Zhan X, Gathergood N, Gupta VK, 2022. Microbial biodiesel production from lignocellulosic biomass: New insights and future challenges. *Critical Reviews in Environmental Science and Technology*, *52*(12), 2197–2225.

Verma R, Kumari KVLK, Srivastava A, Kumar A, 2020. Photoautotrophic, mixotrophic, and heterotrophic culture media optimization for enhanced microalgae production. *Journal of Environmental Chemical Engineering*, *8*(5).

Voloshin RA, Rodionova MV, Zharmukhamedov SK, Nejat Veziroglu T, Allakhverdiev SI. 2016. Review: Biofuel production from plant and algal biomass. *International Journal of Hydrogen Energy*, *41*(39).

Wang Y, Tashiro Y, Sonomoto K, 2015. Fermentative production of lactic acid from renewable materials: Recent achievements, prospects, and limits. *Journal of Bioscience and Bioengineering*, *119*.

Yusoff MNAM, Zulkifli NWM, Masjuki HH, Harith MH, Syahir AZ, Khuong LS, Alabdulkarem A, 2018. Comparative assessment of ethanol and isobutanol addition in gasoline on engine performance and exhaust emissions. *Journal of Cleaner Production*, *190*, 483–495.

Zhang L, Loh KC, Kuroki A, Dai Y, Tong YW, 2021. Microbial biodiesel production from industrial organic wastes by oleaginous microorganisms: Current status and prospects. *Journal of Hazardous Materials*, *402*.

Zikmanis P, Kolesovs S, Ruklisha M, Semjonovs P, 2021. Production of bacterial cellulose from glycerol: The current state and perspectives. *Bioresources and Bioprocessing*, 8(1).

Index

Pages in **bold** refer to tables.

M

N

O

P

R

S

T

U

V

W

X

Y

Z

FSC
www.fsc.org
MIX
Papier | Fördert
gute Waldnutzung
FSC® C083411

Kolibri 360 GmbH
Ferdinand-Jühlke-Straße 7
99095 Erfurt, Deutschland
produktsicherheit@kolibri360.de